Figure 1
Presentation of the group G of the Novikov–Boone theorem.

Generators:

q_j s_b r_i x t k

Relations:

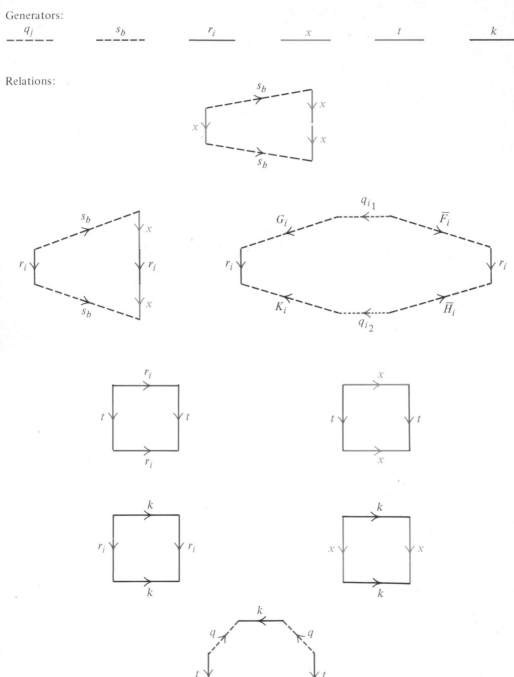

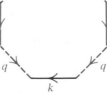

Figure 2
Sufficiency of Boone's lemma

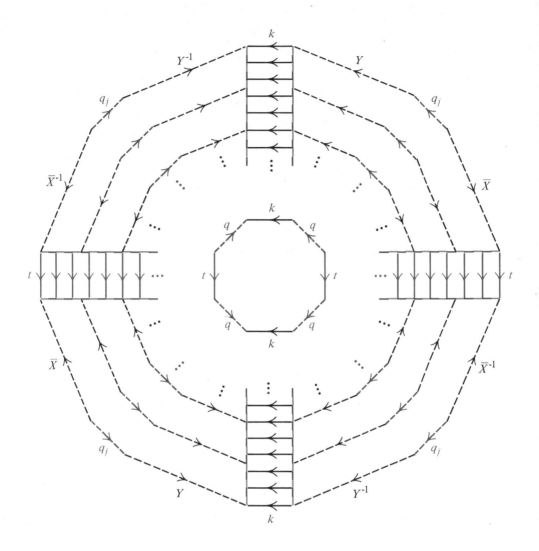

third edition

AN INTRODUCTION TO THE THEORY OF GROUPS

JOSEPH J. ROTMAN

University of Illinois
Urbana

wcb
Wm. C. Brown Publishers
Dubuque, Iowa

לזכרון נצח אבי מורי הנכבד

אליהו בן יוסף יונה הלוי

Contents

Preface

Quand j'ai voulu me restreindre, je suis tombé dans l'obscurité; j'ai préféré passer pour un peu bavard.

H. POINCARÉ, *Analysis situs,*
Journal de l'École Polytechnique, 1895, pp. 1–121.

Although permutation groups had been studied earlier, the theory of groups really began with Galois (1811–1832) who demonstrated that polynomials are best understood by examining certain groups of permutations of their roots. Since that time, groups have arisen in almost every branch of mathematics. Even in this introductory text we shall see connections with number theory, combinatorics, geometry, topology, and logic.

By the end of the nineteenth century, there were two main streams of group theory: topological groups (especially Lie groups) and finite groups. In this century, a third stream has joined the other two: infinite (discrete) groups. It is customary, nowadays, to approach our subject by two paths: "pure" group theory (for want of a better name) and representation theory. This book is an introduction to "pure" (discrete) group theory, both finite and infinite. Of course, the reader who wishes to become an expert must also travel the other path (several appropriate books on representation theory are mentioned in the bibliography).

We assume the reader knows the rudiments of modern algebra, by which we mean that matrices and finite-dimensional vector spaces are friends, while groups, rings, fields, and their homomorphisms are only acquaintances. A familiarity with elementary set theory is also assumed, but some appendices are at the back of the book so that readers may see whether my notation agrees with theirs.

This is the third version of this book. I have reorganized the presentation so that the discussion of finite abelian groups now comes after normal series. There is a new chapter on permutations (multiple transitivity, primitivity) and the Mathieu groups (construction, simplicity, automorphism groups of Steiner systems), repairing a serious omission in the earlier editions. Other additions include outer automorphisms of S_6, fundamental theorem of projective geometry, the transfer and Burnside's normal complement theorem, and the theorems of Boone-Higman and Adian-Rabin. Besides inserting new material into almost every chapter, I have also rewritten many of the older sections: the discussion of Grothendieck groups has been redone; the treatment of amalgams and HNN extensions has been revised (as Lyndon and Schupp have taught me) with pushouts and the Seifert–van Kampen theorem shown to be the basic constructions underlying them. The minor change in the title of the book thus serves to signal the major difference between this edition and its earlier versions.

I am fortunate in having attended lectures on group theory given by I. Kaplansky, S. Mac Lane, and M. Suzuki. Their influence is evident throughout in many elegant ideas and proofs. I am happy to thank once again those who helped me (directly and indirectly) with the first two editions: K. I. Appel, M. Barr, W. W. Boone, J. L. Britton, G. Brown, D. Collins, C. Jockusch, T. McLaughlin, C. F. Miller, III, H. Paley, P. Schupp, F. D. Veldkamp, and C. R. B. Wright. It is a pleasure to thank the following who helped with the present edition: K. I. Appel, W. W. Boone, E. C. Dade, F. Haimo, L. McCulloh, P. M. Neumann, E. Rips, A. Spencer, and J. Walter. I particularly thank F. Hoffman, who read my manuscript, for his valuable comments and suggestions. Finally, I thank Ms. P. Coombs for a splendid job of typing my manuscript.

JJR

To the Reader

Exercises in a text generally have two functions: to reinforce the reader's grasp of the material and to provide puzzles whose solution gives a certain pleasure. Here, exercises also have a third function: to enable the reader to discover important facts, examples, and counterexamples. The serious reader should attempt exercises as they arise (many are not difficult), for subsequent proofs may depend on them. The casual reader should regard the exercises as part of the text proper. Those exercises preceded by * will be cited elsewhere in the book; those preceded by ** will be cited in a proof.

chapter one

Groups and Homomorphisms

Our study of groups begins with the consideration of a rather general setting.

Definition If G is a nonempty set, a **binary operation** on G is a function $\mu: G \times G \to G$.

A binary operation μ assigns to each ordered pair (a, b) of elements of G a third element $\mu(a, b)$ of G. In practice, μ is thought of as a multiplication of elements of G, and instead of $\mu(a, b)$, more suggestive notations used are $ab, a + b, a \circ b$, and $a * b$. In this chapter, we shall use the star notation.

Two remarks must be made here. First, it is quite possible that $a * b$ and $b * a$ are distinct elements of G. Second, the law of substitution – if $a = a'$ and $b = b'$, then $a * b = a' * b'$ – is just the statement that μ is a well defined (i.e., single-valued) function (see Appendix III).

One cannot develop a theory in this rarefied atmosphere; conditions on the binary operation are needed to obtain interesting results. If we are given elements a_1, a_2, a_3 of G (not necessarily distinct), the notation $a_1 * a_2 * a_3$ is ambiguous. Since one can $*$ only two elements of G at a time, there is a choice: form $a_1 * a_2$ first, and then $*$ this new element of G with a_3 to get $(a_1 * a_2) * a_3$; or form $a_1 * (a_2 * a_3)$. In general, these two elements are not the same. For example, if G is the set of all integers (positive, negative, and zero) and if the binary operation is subtraction, i.e.,

1

$a * b = a - b$, then any choice of integers a, b, c with $c \neq 0$ yields an example in which $(a - b) - c \neq a - (b - c)$.

Definition   A binary operation $*$ on G is **associative** if

$$(a * b) * c = a * (b * c)$$

for every three elements a, b, c of G.

Associativity allows one to multiply every ordered triple of elements of G unambiguously; parentheses are unnecessary, and there is no confusion in writing $a * b * c$. If we are confronted by four elements of G or, more generally, by a finite number of elements of G, must we postulate more intricate associativity axioms to avoid parentheses?

Theorem 1.1   *Let $*$ be an associative binary operation on G. Any two ways of multiplying the elements $a_1, a_2, \cdots, a_n$ of G in this order yield the same element of G.*

Proof   We prove the theorem by induction on n, where $n \geq 3$. If $n = 3$, the theorem is true, for we are assuming that $*$ is associative.

Let $n > 3$. How do we multiply n elements? Since a binary operation allows multiplication of only two elements at a time, first choose a pair of adjacent a's and multiply them; then, either multiply this new element by an a adjacent to it or choose two other adjacent a's and multiply them. With each choice, the total number of factors decreases by one, so that eventually only two factors are left. Suppose now that

$$X = (a_1 * \cdots * a_i) * (a_{i+1} * \cdots * a_n)$$

and

$$Y = (a_1 * \cdots * a_j) * (a_{j+1} * \cdots * a_n)$$

are elements of G obtained by two people multiplying the a's together, each having made his or her own choices. The parentheses indicate the final multiplication each has just performed. For notational convenience, we assume that $i \leq j$. Since each of the final two factors in X and in Y contains less than n of the a's, the inductive hypothesis allows us to rearrange parentheses in them. Therefore, we may assume $i < j$ (or we are done) and write

$$X = (a_1 * \cdots * a_i) * ([a_{i+1} * \cdots * a_j] * [a_{j+1} * \cdots * a_n])$$

and

$$Y = ([a_1 * \cdots * a_i] * [a_{i+1} * \cdots * a_j]) * (a_{j+1} * \cdots * a_n).$$

If we denote $a_1 * \cdots * a_i$ by A, $a_{i+1} * \cdots * a_j$ by B, and $a_{j+1} * \cdots * a_n$ by C, then $X = A * (B * C)$ and $Y = (A * B) * C$. Since A, B, and C are unambiguously defined, by induction, and since $*$ is associative, we have $X = Y$, as we wished. ∎

As a result of this theorem, the notation $a_1 * a_2 * \cdots * a_n$ is unambiguous when $*$ is an associative binary operation.

Definition A **semigroup** is an ordered pair $(G, *)$ where G is a set and $*$ is an associative binary operation on G.

Even though a semigroup is an ordered pair, we shall usually say that G is a semigroup and tacitly assume that $*$ is known. The reader must realize, however, that there are many possible operations on a set G making it a semigroup. For example, the set of positive integers is a semigroup under either of the binary operations of ordinary addition or ordinary multiplication.

Definition Let G be a semigroup and let $a \in G$. Define $a^1 = a$; if $n > 1$, then define $a^n = a^{n-1} * a$.

Corollary 1.2 *Let G be a semigroup and let $a \in G$. If m and n are positive integers, $a^m * a^n = a^{m+n}$.*

Proof Both elements arise from a product of $m + n$ factors equal to a; they differ only in the arrangement of parentheses. ∎

The most important semigroups are groups.

Definition A **group** is a semigroup G containing an element e such that:

(i) $e * a = a$ for all $a \in G$;
(ii) for every $a \in G$, there is an element b in G with

$$b * a = e.$$

Lemma 1.3 *If G is a group and $a \in G$, then $a * a = a$ implies $a = e$.*

Proof There is an element $b \in G$ with $b * a = e$. Now

$$b * (a * a) = b * a = e.$$

On the other hand,

$$(b * a) * a = e * a = a.$$

Therefore, $a = e$. ∎

Theorem 1.4 *Let $(G, *)$ be a group. (i) There is a unique element e with $e * a = a$ for all $a \in G$; moreover,*

$$e * a = a = a * e \qquad \text{for every } a \in G.$$

(ii) *For every $a \in G$ there is a unique $b \in G$ with $b * a = e$; moreover,*

$$b * a = e = a * b.$$

Proof We first show that in this particular case, the order in which we multiply makes no difference.

Suppose $b * a = e$. Then $(a * b) * (a * b) = a * (b * a) * b = a * (e * b)$ $= a * b$. By Lemma 1.3, $a * b = e$.

If $a \in G$, then $e * a = a$. Choose b with $b * a = e$. Then $a * e = a * (b * a)$ $= (a * b) * a = e * a = a$, by our calculation above. Therefore $a * e = a$.

We now prove the uniqueness assertions. Suppose $e_0 * a = a$ for all $a \in G$; in particular, $e_0 * e_0 = e_0$. By the lemma, $e_0 = e$.

Finally, suppose $b * a = e$ and $c * a = e$. Then $a * b = e$ and $c = c * e$ $= c * (a * b) = (c * a) * b = e * b = b$. ∎

As a result of the uniqueness assertions of the theorem, we may now give names to e and to b. We call e the **identity** of G, and if $b * a = e$, we call b the **inverse** of a and denote it by a^{-1}.

Corollary 1.5   *If G is a group and $a \in G$, then*

$$(a^{-1})^{-1} = a.$$

Proof $(a^{-1})^{-1}$ is that element $b \in G$ such that $b * a^{-1} = e$. But we have just seen that $a * a^{-1} = e$, so the uniqueness of the inverse implies that $b = a$. ∎

Definition   Let G be a group. For $a \in G$, define the **powers** of a as follows: If n is a positive integer, then define a^n as above; define $a^0 = e$; finally, define a^{-n} as $(a^{-1})^n$.

EXERCISES

1.1.   If G is a group and $a_1, \cdots, a_n \in G$, then

$$(a_1 * \cdots * a_n)^{-1} = a_n^{-1} * \cdots * a_1^{-1}.$$

1.2.   Let G be a group, $a \in G$, and m, n not necessarily positive integers. The following formulas hold:

$$a^m * a^n = a^{m+n};$$
$$(a^m)^n = a^{mn}.$$

In particular,

$$(a^n)^{-1} = a^{-n}.$$

*1.3.   (**Cancellation laws**) In a group G, each of the equations $a * b = a * c$ and $b * a = c * a$ implies $b = c$.

1.4.   A semigroup G is a group if and only if the equations $a * x = b$ and $y * a = b$ have solutions in G for every a and b in G. Moreover, when G is a group these solutions are unique.

1.5.   Let G be a group and let $a \in G$. Define a function $L_a : G \to G$ by $L_a(x) = a * x$ (L_a is called **left translation by a). Prove that L_a is a one-one correspondence.

1.6.   Let **Z** denote the additive group of integers (we shall adhere to this notation from now on). What is the identity of **Z**? If $a \in \mathbf{Z}$, what is its inverse? State Corollary 1.5 for **Z**.

1.7.   For each fixed $n > 0$, prove that $\mathbf{Z}/n\mathbf{Z}$, the set of integers modulo n, is a group if one defines $\bar{a} + \bar{b} = \overline{a + b}$ ($\bar{a}$ is the congruence class of a modulo n, i.e.,

$$\bar{a} = \{b \in \mathbf{Z} : b \equiv a \,(\mathrm{mod}\, n)\} = \{a + kn : k \in \mathbf{Z}\}).$$

Conclude that, for each $n > 0$, there exists a group having exactly n elements.

1.8.   Let G denote the multiplicative group of positive rationals. What is the identity of G? If $a \in G$, what is its inverse?

1.9.   Let n be a positive integer, and let G be the multiplicative group of all nth roots of unity, i.e., all complex numbers of the form $e^{2\pi i k/n}$, where $k \in \mathbf{Z}$. What is the identity of G? If $a \in G$, what is its inverse? How many elements does G have?

1.10.   Let G be the multiplicative group of all 2×2 nonsingular matrices with rational entries. What is the identity of G? If $a \in G$, what is its inverse? Exhibit elements a and b in G such that $a * b \neq b * a$.

*1.11.   Prove that the following four matrices form a multiplicative group:

$$\begin{bmatrix} 1 & 0 \\ 0 & 1 \end{bmatrix}, \begin{bmatrix} -1 & 0 \\ 0 & 1 \end{bmatrix}, \begin{bmatrix} 1 & 0 \\ 0 & -1 \end{bmatrix}, \begin{bmatrix} -1 & 0 \\ 0 & -1 \end{bmatrix}.$$

*1.12.   Show that the six functions $\{x, 1/x, 1 - x, 1/(1 - x), x/(x - 1), (x - 1)/x\}$ form a group with composition as binary operation (we agree that $1/0 = \infty$, $1/\infty = 0$, $1 - \infty = \infty = \infty - 1$, and $\infty/\infty = 1$).

*1.13.   Let X be a nonempty set and let G be the set of all one-one correspondences of X onto itself. If f and g are in G, define $f * g$ to be the composite of f and g:

$$(f * g)(x) = f(g(x))$$

for every $x \in X$. Prove that G is a group. (If $X = \{1, 2, \cdots, n\}$, then G is denoted by S_n.)

When does one "know" a group G? One answer (but not the only answer, as we shall see) is that all the elements of G are known and all possible products can be computed. This answer can be made to look more sophisticated in the special case when G is finite. Let $a_1, a_2, \cdots, a_n$ be a list with no repetitions of all the elements in G. A **multiplication table** of G is the $n \times n$ matrix whose ij entry is $a_i * a_j$:

G	a_1	a_2	$\cdots$	a_n
a_1	$a_1 * a_1$	$a_1 * a_2$	$\cdots$	$a_1 * a_n$
a_2	$a_2 * a_1$	$a_2 * a_2$	$\cdots$	$a_2 * a_n$
$\vdots$	$\vdots$	$\vdots$		$\vdots$
a_n	$a_n * a_1$	$a_n * a_2$	$\cdots$	$a_n * a_n$

Therefore, we "know" a finite group if we can write a multiplication table for it. (We say "a" multiplication table rather than "the" multiplication table, since the matrix depends on the particular listing of the elements of G.) One also speaks of a multiplication table of an infinite group, but in this case, of course, the matrix is infinite.

Let $[a_i * a_j]$ be the multiplication table of a group G having elements $a_1, \cdots, a_n$. The cancellation laws (Exercise 1.3) imply no element a_i occurs twice in any row or in any column. An $n \times n$ matrix of this type (the entries lie in a set X with n elements, and each row and column is a permutation of X) is called a **Latin square**. Not every Latin square comes from a group, however. If $G = \{a_1, \cdots, a_n\}$ is a group and $e = a_i$, say, then the ith row of the multiplication table is $(a_i * a_1, a_i * a_2, \cdots, a_i * a_n) = (a_1, \cdots, a_n)$; similarly, the ith column is also $(a_1, \cdots, a_n)$. It follows that the Latin square

$$\begin{bmatrix} a & b & c \\ c & a & b \\ b & c & a \end{bmatrix}$$

is not the multiplication table of any group. When writing a multiplication table, one usually lets $a_1 = e$ so that the first row and the first column display the group elements in order. Given a table for a group, it is easy to find the inverse of any element; it is tedious, however, to check the associative law from a table.

Let us now consider two almost trivial examples of groups. Let G be the multiplicative group with elements $\{1, -1\}$; let $H = \mathbf{Z}/2\mathbf{Z}$, the additive group of integers modulo 2. Compare multiplication tables of these two groups:

G	1	-1
1	1	-1
-1	-1	1

H	$\bar{0}$	$\bar{1}$
$\bar{0}$	$\bar{0}$	$\bar{1}$
$\bar{1}$	$\bar{1}$	$\bar{0}$

It is quite clear that G and H are distinct groups; it is equally clear that there is no significant difference between them. We formalize this idea.

Definition Let $(G, *)$ and $(H, \circ)$ be groups. A **homomorphism** $f: G \to H$ is a function such that

$$f(a * b) = f(a) \circ f(b) \qquad \text{for all } a, b \in G.$$

An **isomorphism** is a homomorphism that is also a one-one correspondence.† Two groups G and H are **isomorphic**, denoted by $G \cong H$, if there is an isomorphism $f: G \to H$.

The groups G and H whose multiplication tables are given above are isomorphic: define $f: G \to H$ by $f(1) = \bar{0}$ and $f(-1) = \bar{1}$. It is important to realize that

---

† Many authors use "isomorphism" to mean "homomorphism and one-one"; here, isomorphisms are always onto as well.

a given group has many multiplication tables, one for each way of listing the elements of the group. In particular, a list $x_1, x_2, \cdots$ of the elements of a group G determines a multiplication table for G. If $f: G \to H$ is a one-one correspondence, then $f(x_1), f(x_2), \ldots$ determines a multiplication table for H. To say that f is an isomorphism is to say that, upon superimposing the first multiplication table (for G) upon the second table (for H), the tables "match". In this sense, one abuses language and says that isomorphic groups have the same multiplication table. Moreover, one regards isomorphic groups as "essentially" the same, differing only in notation.

EXERCISES

*1.14. Write a multiplication table for S_3 and show S_3 is isomorphic to the group of Exercise 1.12 (see Exercise 1.13).

1.15. Isomorphic groups have the same number of elements. Prove that the converse is false by showing that $\mathbf{Z}/4\mathbf{Z}$ is not isomorphic to the group of four matrices in Exercise 1.11.

1.16. For each positive integer n, prove there are only finitely many distinct groups having exactly n elements (we regard isomorphic groups as being the same).

Two basic problems occurring in mathematics are (1) the classification of all systems of a given kind, e.g., all groups, all vector spaces, all topological spaces; and (2) the classification of all the transformations of one system into another. By a classification of systems, one usually means a scheme that distinguishes essentially different systems, or to say it another way, a scheme that tells when two systems are essentially the same. A classification of transformations is more subtle, and we needn't discuss it further now. As an illustration, consider the collection of all finite-dimensional vector spaces over a field. In this case, the first problem is answered by the theorem that two such spaces are isomorphic if and only if they have the same dimension. Even the second problem has been answered. The transformations of vector spaces are, of course, the linear transformations; these give rise to similarity classes of matrices, which are classified by the canonical forms. The same problems arise in group theory: (1) When are two groups isomorphic? (2) How may one describe the homomorphisms from one group to another? In contrast to our illustration, both problems are exceedingly difficult (if not impossible) and are only partially solved.

EXERCISES

1.17. If $f: G \to H$ and $g: H \to K$ are homomorphisms, then so is $g \circ f: G \to K$.

1.18. If $\mathscr{A}$ is a class of groups, then the relation of isomorphism is an equivalence relation on $\mathscr{A}$.

1.19. Let $f: G \to H$ be a homomorphism, and let $a \in G$. For every $n \in \mathbf{Z}$, $f(a^n) = f(a)^n$. Note the special cases $n = 0$ and $n = -1$.

1.20. Let G be a group and let X be a set having the same number of elements as G. If $f: G \to X$ is a one-one correspondence, there is a unique binary operation that can be defined on X so that X is a group and f is an isomorphism.

1.21. Let G be the multiplicative group of positive reals and let H be the additive group of all reals. Show that log: $G \to H$ is an isomorphism. (HINT: Exhibit a function inverse to log.)

*1.22. Let G be the additive group of reals and let **T** be the **circle group**, i.e., the multiplicative group of all complex numbers of absolute value 1. For a fixed real number y, prove that $f: G \to \mathbf{T}$ defined by $f(x) = e^{iyx}$ is a homomorphism. Which real numbers x are such that $f(x) = 1$? (It can be proved that these f are the only *continuous* homomorphisms from G to **T**.)

1.23. If G is the multiplicative group of all $n \times n$ nonsingular matrices with entries in a field F, and if $F^{\#}$ is the multiplicative group of nonzero elements of F, then determinant is a homomorphism from G to $F^{\#}$.

*1.24. Let a be a fixed element of a group G. Prove that $\gamma_a: G \to G$ defined by

$$\gamma_a(x) = a * x * a^{-1}$$

is an isomorphism (γ_a is called **conjugation** by a). One often writes $a * x * a^{-1} = x^a$.

1.25. If $\gamma_a: G \to G$ is conjugation by a, prove that $\gamma_a \gamma_b = \gamma_{ab}$, where $a, b \in G$. (This is the reason conjugation is defined by $a * x * a^{-1}$ instead of by $a^{-1} * x * a$.)

1.26. Let G be the multiplicative group of all nth roots of unity; prove that $G \cong \mathbf{Z}/n\mathbf{Z}$.

1.27. Describe all the homomorphisms from $\mathbf{Z}/12\mathbf{Z}$ to itself. Which of these are isomorphisms?

*1.28. Let G be the additive group of $\mathbf{Z}[x]$, all polynomials in x having coefficients in $\mathbf{Z}$, and let H be the multiplicative group of positive rationals. Prove that $G \cong H$. (HINT: Use the fundamental theorem of arithmetic to construct an isomorphism.)

Having solved Exercise 1.28, the reader may wish to reconsider the answer to the question: When does one "know" a group? Surely the reader could compute multiplication tables for the two groups G and H there, but it was not obvious at the outset that these groups are essentially the same. As an alternative answer to the question of "knowing", we suggest that one knows a group G when one can determine, given any other group H, whether G and H are isomorphic.

Definition   Two elements a and b in a semigroup G **commute** if $a * b = b * a$. A semigroup is **abelian†** (or commutative) if every two elements in G commute.

It is easy to see that a group G is abelian if and only if every one of its multiplication tables is a symmetric matrix.

EXERCISES

1.29.   Let $a_1, a_2, \cdots, a_n$ be elements of an abelian semigroup G. If $b_1, b_2, \cdots, b_n$ is a rearrangement of the a_i, then

$$a_1 * a_2 * \cdots * a_n = b_1 * b_2 * \cdots * b_n.$$

*1.30.   Let a and b lie in a semigroup G. If a and b commute, then $(a * b)^n = a^n * b^n$ for every integer $n > 0$. If, further, G is a group, the equation holds for all $n \in \mathbf{Z}$.

**1.31.   Let G be a group in which the square of every element is the identity. Prove that G is abelian.

1.32.   A group G is abelian if and only if the function $f: G \to G$ defined by $f(x) = x^{-1}$ is a homomorphism.

Definition   A semigroup G is a **cancellation semigroup** if, for every $a, b, c \in G$, each of the equations $a * b = a * c$ and $b * a = c * a$ implies $b = c$.

EXERCISES

1.33.   Prove that every finite cancellation semigroup is a group.

1.34.   Prove that every abelian cancellation semigroup G can be imbedded in a group. (HINT: As in the construction of the rationals from the integers, consider all ordered pairs $(a, b) \in G \times G$ under the equivalence relation

$$(a, b) \equiv (a', b') \qquad \text{if } a * b' = a' * b.)$$

[Mal'cev (1937) has exhibited a cancellation semigroup that cannot be imbedded in a group (see P. M. Cohn, *Universal Algebra*, p. 269).]

1.35.   (**Kaplansky**)   An element x in a ring R has a *left quasi-inverse* y in R if $x + y - yx = 0$. Prove that a ring R in which every element but one has a left quasi-inverse is a division ring. (HINT: Let R' be the set R with the exceptional element deleted; consider the binary operation on R' defined by $x \circ y = x + y - yx$.)

---

† After N. H. Abel who proved (1826) that (in modern language) a polynomial whose Galois group is commutative is solvable by radicals. This result was superseded by Galois's magnificent theorem (1830) that a polynomial is solvable by radicals if and only if its Galois group is a solvable group.

*1.36. Let G be a finite abelian group containing no elements $a \neq e$ with $a^2 = e$. Evaluate $a_1 * \cdots * a_n$, where $a_1, \cdots, a_n$ is a list with no repetitions of all the elements of G.

1.37. Prove Wilson's theorem: If p is a prime, then

$$(p - 1)! \equiv -1 \pmod{p}.$$

(HINT: The nonzero elements of $\mathbf{Z}/p\mathbf{Z}$ form a group under multiplication.)

*1.38. Let G be a finite group and $f: G \to G$ an isomorphism. If f has no nontrivial fixed points (i.e., $f(x) = x$ implies $x = e$) and if $f \circ f$ is the identity function on G, then $f(x) = x^{-1}$ for all x and G is abelian. [HINT: Prove that every element in G has the form $x^{-1} * f(x)$.]

chapter two

The Isomorphism Theorems

SUBGROUPS

From now on, we shall usually write ab instead of $a * b$, and 1 will denote the identity element instead of e.

Definition A nonempty subset S of a group G is a **subgroup** of G if $s \in S$ implies $s^{-1} \in S$ and $s, t \in S$ implies $st \in S$.

The next theorem shows that a subgroup of G is a group. However, a subgroup is not merely a group contained in another group. For example, if G is the additive group of reals and S is the multiplicative group of positive rationals, then S is a group whose elements lie in G, but S is not a subgroup of G.

Theorem 2.1 *A subgroup S of G is a group, and the inclusion $i: S \rightarrow G$ is a homomorphism. Conversely, if $(S, \circ)$ and $(G, *)$ are groups with $S \subset G$ and if the inclusion $i: S \rightarrow G$ is a homomorphism, then S is a subgroup of G.*

Proof Since S is nonempty, it contains an element s; since S is a subgroup, it also contains s^{-1} and $1 = ss^{-1}$. As the associative law holds for all products of three elements in G, it surely holds for products of three elements in S. It follows that S is a group (with binary operation the restriction of that on G).

The inclusion $i: S \rightarrow G$ is defined by $i(s) = s$ for all $s \in S$. Plainly $i(st) = st$ $= i(s)i(t)$.

That the inclusion $i: S \rightarrow G$ is a homomorphism says $i(s \circ t) = i(s) * i(t)$; simply stated, $s \circ t = s * t$ (the products agree). We conclude that $s, t \in S$ implies $s * t \in S$. By Exercise 1.19, $i(s^0) = i(1) = 1$, so S and G have the same identity; moreover, $i(s^{-1}) = i(s)^{-1}$; the inverse of $s \in S$ is the same in S as in G. Thus $s \in S$ implies $s^{-1} \in S$ and S is a subgroup of G. ∎

Theorem 2.2 *If S is a subset of a group G, then S is a subgroup of G if and only if S is nonempty, and whenever $a, b \in S$, then $ab^{-1} \in S$.*

Proof Since S is nonempty, it contains an element a. Hence, $aa^{-1} = 1 \in S$. If $b \in S$, then $1b^{-1} = b^{-1} \in S$, and if $a, b \in S$, then $a(b^{-1})^{-1} = ab \in S$. ∎

Example 1 If G is a group, G itself and $\{1\}$ are subgroups of G; any other subgroup of G is called **proper**.

Example 2 If $a \in G$, then the set of all powers of a, denoted by $\langle a \rangle$, is a subgroup, called the **cyclic subgroup generated by** a; a is called a **generator** of $\langle a \rangle$. A cyclic subgroup may have several different generators, e.g., $\langle a \rangle = \langle a^{-1} \rangle$.

Definition A group G is **cyclic** if $G = \langle a \rangle$ for some $a \in G$.

Example 3 Let G and H be groups and let $f: G \rightarrow H$ be a homomorphism. Define

$$\mathbf{kernel}\, f = \{x \in G : f(x) = 1\}.$$

We show that kernel f is a subgroup of G. Kernel f is nonempty because it contains 1; if $f(x) = 1$, then $f(x^{-1}) = f(x)^{-1} = 1$, so that $x \in$ kernel f implies $x^{-1} \in$ kernel f; if $f(x) = 1 = f(y)$, then $f(xy) = f(x)f(y) = 1 \cdot 1 = 1$, so that $x, y \in$ kernel f implies $xy \in$ kernel f.

Example 4 If $f: G \rightarrow H$ is a homomorphism, define

$$\mathbf{image}\, f = \{y \in H : y = f(x) \quad \text{for some } x \in G\}.$$

The reader may check that image f is a subgroup of H.

Notation We will usually write **ker** f and **im** f instead of kernel f and image f.

We now consider several more ways of manufacturing subgroups.

Theorem 2.3 *The intersection of any family of subgroups of G is again a subgroup of G.*

Proof Let $\{S_i : i \in I\}$ be a family of subgroups of G and let $S = \cap S_i$. Since $1 \in S_i$ for every i, $1 \in S$, and so S is nonempty. Suppose $a, b \in S$. These elements got into S by being in every S_i. Since each S_i is a subgroup, $ab^{-1} \in S_i$ for every i, and so $ab^{-1} \in S$. ∎

Theorem 2.4 *If X is a subset of a group G, there is a smallest subgroup H of G that contains X, i.e., if S is any other subgroup containing X, then $S \supset H$.*

Proof Subgroups of G containing X do exist, e.g., G itself. Let H be the intersection of all the subgroups of G that contain X. By the preceding theorem, H is a subgroup and it clearly contains X. The fact that H is the smallest such subgroup follows from the observation that the intersection is contained in each of the subgroups being intersected. ∎

Definition If X is a subset of a group G, the smallest subgroup of G containing X is denoted by $\langle X \rangle$ and is called the **subgroup of G generated by** X. We say that X **generates** $\langle X \rangle$.

If X is a finite set, then $X = \{a_1, a_2, \cdots, a_n\}$. For notational convenience, we write $\langle X \rangle = \langle a_1, a_2, \cdots, a_n \rangle$ instead of $\langle \{a_1, a_2, \cdots, a_n\} \rangle$.

Definition If X is a subset of a group G, a **word** on X is either 1 or an element w of G of the form

$$w = x_1^{e_1} \cdots x_n^{e_n},$$

where $x_i \in X$ and $e_i = \pm 1$.

Theorem 2.5 *Let X be a subset of a group G. If $X = \varnothing$, then $\langle X \rangle = \{1\}$; if $X \neq \varnothing$, then $\langle X \rangle$ is the set of all words on X.*

Proof If $X = \varnothing$, then $X \subset \{1\}$ and so $\langle X \rangle = \{1\}$. Assume $X \neq \varnothing$ and let W denote the set of all words on X. Plainly $X \subset W$ (so $W \neq \varnothing$); moreover, the inverse of a word is a word and the product of two words is a word. Therefore W is a subgroup of G containing X and $\langle X \rangle \subset W$. The reverse inclusion follows from the observation that any subgroup of G that contains X must contain every word on X. ∎

EXERCISES

2.1. Let G be a group and let $a \in G$. If $X = \{a\}$, then $\langle X \rangle = \langle a \rangle$, the cyclic subgroup generated by a.

2.2. The set-theoretic union of two subgroups is a subgroup if and only if one is contained in the other. Is this true if we replace "two subgroups" by "three subgroups"?

2.3. If H and K are subgroups of G, denote $\langle H \cup K \rangle$ by $H \vee K$. Prove that $H \vee K$ is the smallest subgroup of G containing H and K.

2.4. Let S be a subgroup of G with $S \neq G$. If $G - S$ is the complement of S in G, then $\langle G - S \rangle = G$.

2.5. The multiplicative group of positive rationals is generated by all rationals of the form $1/p$, where p is prime.

Definition   Let S and T be nonempty subsets of a group G.

$$ST = \{st : s \in S \text{ and } t \in T\}.$$

In particular, if we write t for $\{t\}$, then

$$St = \{st : s \in S\}.$$

Definition   Let S be a subgroup of G. A **right coset** of S in G is a subset St (a **left coset** is tS).† We say that t is a **representative** of St (and also of tS).

Suppose we consider the plane $\mathbf{R}^2$ as an additive abelian group. A one-dimensional subspace S is a subgroup; geometrically S is a line through the origin. A coset $t + S \neq S$ is a line in $\mathbf{R}^2$ parallel to S.

EXERCISES

**2.6.   The multiplication of nonempty subsets of G defined above is associative.

**2.7.   Let S be a subgroup of G. Prove that $St = S$ if and only if $t \in S$.

**2.8.   If S is a subgroup of G, then $SS = S$. If S is a finite nonempty subset of G and $SS = S$, then S is a subgroup of G. Is this true if S is infinite?

**2.9.   Let H be a subgroup of G having exactly two right cosets. Show that $g^2 \in H$ for every $g \in G$. (See Exercise 2.44.)

**2.10.   There is a one-one correspondence between S and St; conclude that any two right cosets of S have the same number of elements.

2.11.   (Product formula**)   If S, T are subgroups of G, then

$$|ST|\,|S \cap T| = |S|\,|T|.$$

(The subset ST may not be a subgroup.)

**2.12.   Let $\{S_i : i \in I\}$ be a family of subgroups of a group G, let $\{x_i S_i : i \in I\}$ be a family of left cosets, and let $T = \bigcap_{i \in I} S_i$. Prove that if $\bigcap_{i \in I} x_i S_i \neq \varnothing$ and contains an element y, then $\bigcap x_i S_i = yT$.

A right coset St has many representatives, namely, all elements of the form st, where $s \in S$. The next lemma gives a criterion for determining whether two right cosets of S are the same when a representative of each is known.

Lemma 2.6   *Let S be a subgroup of G. Then $Sa = Sb$ if and only if $ab^{-1} \in S$ ($aS = bS$ if and only if $a^{-1}b \in S$).*

Proof   If $Sa = Sb$, then $a \in Sb$ and $a = sb$ for some $s \in S$. Therefore, $ab^{-1} = s$ and $ab^{-1} \in S$. Conversely, if $ab^{-1} = s \in S$, then $a = sb$, so that $Sa = Ssb = Sb$, by Exercise 2.7. ∎

† Our right cosets are called left cosets by some authors.

Theorem 2.7 *If S is a subgroup of G, then any two right cosets of S in G are either identical or disjoint.*

Proof We show that if there is an element x in $Sa \cap Sb$, then $Sa = Sb$. Such an element has the form $x = s_1 a = s_2 b$, where s_1 and s_2 lie in S. Hence $ab^{-1} = s_1^{-1} s_2 \in S$, so that Lemma 2.6 gives $Sa = Sb$. ∎

Theorem 2.7 may be paraphrased to say that a subgroup S induces a partition of the group G (into right cosets). This being true, there must be an equivalence relation on G lurking somewhere in the background. Indeed, such a relation on G is defined by $a \sim b$ if $ab^{-1} \in S$; it is an equivalence relation on G whose equivalence classes are the right cosets of S.

In Exercise 1.14, we asked the reader to give a multiplication table for the group $G = S_3$. The six elements of S_3 may be denoted by

$$\{1, a, b, b^2, ab, ab^2\},$$

where $a^2 = 1 = b^3$ and $ba = ab^2$. If $H = \langle a \rangle = \{1, a\}$, then the right cosets of H in S_3 are

$$H = \{1, a\}, \quad Hb = \{b, ab\}, \quad Hb^2 = \{b^2, ab^2\}.$$

Note that the left cosets of H in S_3 are

$$H = \{1, a\}, \quad bH = \{b, ba\} = \{b, ab^2\}, \quad b^2 H = \{b^2, ab\}.$$

Thus, right cosets and left cosets need not coincide (for example, $Hb \neq bH$ in this case; indeed bH is not a right coset of H at all).

Theorem 2.8 *If S is a subgroup of G, the number of right cosets of S in G equals the number of left cosets of S in G.*

Proof Let $\mathscr{R}$ denote the set of all right cosets of S in G, and let $\mathscr{L}$ denote the set of all left cosets. We exhibit a one-one correspondence $f : \mathscr{R} \to \mathscr{L}$. If $Sa \in \mathscr{R}$, your first guess is to define $f(Sa) = aS$, but this does not work. Your second guess, $f(Sa) = a^{-1} S$, is correct. It must be verified that f is well defined, i.e., if $Sa = Sb$, then $a^{-1} S = b^{-1} S$ (this is why your first guess is wrong). We also leave to the reader the verification that f is a one-one correspondence. ∎

Definition If S is a subgroup of G, the **index** of S in G, denoted by $[G:S]$, is the number of right cosets of S in G.

Theorem 2.8 tells us that there is no need to define a "right index" and a "left index", for the number of right cosets is the same as the number of left cosets.

[It is a remarkable theorem (P. Hall, 1935) in combinatorics that one can always choose a **common system of representatives** for the right and left cosets of a subgroup: If H is a subgroup of a finite group G with $[G:H] = n$, then there exist $t_1, \cdots, t_n$ in G so that $\{t_1 H, \cdots, t_n H\}$ is the set of all left cosets of H in G and $\{Ht_1, \cdots, Ht_n\}$ is the set of all right cosets of H in G.]

Definition   The **order** of G, denoted by $|G|$, is the number of elements in G.

Theorem 2.9   (Lagrange)†   *If S is a subgroup of a finite group G, then $[G:S] = |G|/|S|$ (so that $|S|$ divides $|G|$).*

Proof   The right cosets of S partition G into $[G:S]$ parts, each of which has precisely $|S|$ elements, by Exercise 2.10. Therefore $|G| = [G:S]|S|$.   ∎

Corollary 2.10   *If $|G| = p$, where p is prime, then G is cyclic.*

Proof   Since p is prime, Lagrange's theorem says that G can have no proper subgroups. Choose $a \in G, a \neq 1$. Then $\langle a \rangle$ is a subgroup of G and $\langle a \rangle \neq \{1\}$, so that $\langle a \rangle = G$.   ∎

Definition   If $a \in G$, the **order** of a is $|\langle a \rangle|$. (Thus, the order of a is either a positive integer or infinity.)

Corollary 2.11   *If G is a finite group and $a \in G$, then the order of a divides $|G|$.*

The next theorem gives an important characterization of the order of an element.

Theorem 2.12   *Let $a \in G$ have finite order m. Then m is the least positive integer such that $a^m = 1$.*

Proof   Since a has finite order, not all the powers of a can be distinct elements of G. There is thus an integer m such that the entries in the list $1, a, a^2, \cdots, a^{m-1}$ are distinct elements of G, while a^m is equal to some element on this list. If $a^m = a^i$ for $0 < i \leq m-1$, then $a^{m-i} = 1$, contradicting the list's having no repetitions. Therefore we must have $i = 0$, i.e., $a^m = 1$. Plainly m is the least positive integer with $a^m = 1$. The theorem is proved if $\{1, a, a^2, \cdots, a^{m-1}\}$ is a subgroup of G (for it has order m and must be $\langle a \rangle$). This is easy: if $0 \leq i \leq m-1$, then $(a^i)^{-1} = a^{m-i}$; if $0 \leq j \leq m-1$, then $a^i a^j = a^{i+j}$ or a^{i+j-m}, depending on whether $i+j \leq m-1$ or $i+j \geq m$.   ∎

EXERCISES

**2.13.   Assume $a \in G$ has finite order n and that $n = mk$ (where m and k are positive integers). Show that a^m has order k.

*2.14.   If $|G| = 4$, then G is either cyclic or isomorphic to the group of Exercise 1.11. Conclude that a group of order 4 must be abelian.

**2.15.   Let G be a finite group with subgroups H and K. If $K \subset H \subset G$, then $[G:K] = [G:H][H:K]$.

† The idea of a group arose in the early 1800s. In 1770, Lagrange wrote (what amounts to the statement) that the order of a subgroup of the symmetric group S_n divides $n!$ (his proof is not quite correct). This special case of Theorem 2.9, as well as Corollary 2.11, was proved by Abbati in 1803. Theorem 2.9 itself, however, was probably first proved by Galois about 1830.

2.16. If $a^n = 1$, then the order of a divides n.

2.17. Let $a \in G$ have finite order and let $f: G \to H$ be a homomorphism. Prove that the order of $f(a)$ divides the order of a.

2.18. If G is a group of order $2n$, then the number of elements in G of order 2 is odd. Conclude that G must contain an element of order 2.

2.19. If two elements a and b commute, and if $a^m = 1 = b^n$, then $(ab)^k = 1$, where $k = \text{lcm } \{m, n\}$. (The order of ab may be less than k; for example, let $b = a^{-1}$.) Conclude that if a and b have finite order and if a and b commute, then their product ab also has finite order.

*2.20. This exercise shows that a product of two elements of finite order may have infinite order; of course this cannot happen in a finite group.

Let G be the multiplicative group of all 2×2 nonsingular matrices with rational entries. Let

$$a = \begin{bmatrix} 0 & -1 \\ 1 & 0 \end{bmatrix} \quad \text{and} \quad b = \begin{bmatrix} 0 & 1 \\ -1 & -1 \end{bmatrix}.$$

Show that $a^4 = 1$ and $b^3 = 1$, but that ab has infinite order.

2.21. If H and K are (not necessarily distinct) subgroups of G, then a **double coset** of H, K is a subset of G of the form HtK, where $t \in G$. Prove that the family of all double cosets of H, K partitions G. (HINT: Define an equivalence relation on G by $t' \sim t$ if $t' = htk$ for some $h \in H$ and $k \in K$.)

2.22. Suppose H, K are subgroups of a finite group G, and G is the disjoint union

$$G = \bigcup_{i=1}^{n} Ht_iK.$$

Prove that $[G:K] = \sum_{i=1}^{n} [H : H \cap t_i K t_i^{-1}]$. (Note that Lagrange's theorem is the special case of this when $K = \{1\}$.)

2.23. Let H be a subgroup of G and assume $x \in G - H$ has order 2 and $xHx = H$. Prove that H has index 2 in $\langle H, x \rangle$. (HINT: Show that $\langle H, x \rangle = H \cup Hx$.)

2.24. Let $G = \langle a \rangle$ have order n. Prove that a^k is a generator of G if and only if $(k, n) = 1$. Conclude that G has exactly $\varphi(n)$ generators, where $\varphi(n)$ is the Euler φ-function defined below.

Definition The **Euler φ-function** is defined as follows: $\varphi(1) = 1$; if $h > 1$, then $\varphi(h)$ is the number of integers k such that $1 \le k < h$ and $(k, h) = 1$.

EXERCISES

2.25. Let $(r, s) = 1$. If $G = \langle x \rangle$ has order rs, then $x = yz$, where y has order r, z has order s, and y and z commute; the factors y and z are unique. Conclude that if $(r, s) = 1$, then

$$\varphi(rs) = \varphi(r)\varphi(s).$$

2.26. If $h = p^n$, where p is prime, then

$$\varphi(h) = p^n\left(1 - \frac{1}{p}\right).$$

If the distinct prime divisors of h are $p_1, \cdots, p_m$, then

$$\varphi(h) = h\left(1 - \frac{1}{p_1}\right) \cdots \left(1 - \frac{1}{p_m}\right).$$

2.27. Let $U(\mathbf{Z}/n\mathbf{Z}) = \{\bar{a} \in \mathbf{Z}/n\mathbf{Z}: (a, n) = 1\}$. Prove that $U(\mathbf{Z}/n\mathbf{Z})$ is a multiplicative group of order $\varphi(n)$.

2.28. Prove that every subgroup of a cyclic group is cyclic. (HINT: Use the division algorithm.)

*2.29. Prove that two cyclic groups are isomorphic if and only if they have the same order.

Notation   The cyclic group of order n is denoted by $\mathbf{Z}(n)$.

EXERCISES

2.30.   (**Fermat**)   Let p be a prime; for any integer n, show that $n^p \equiv n \pmod{p}$. (HINT: If $\mathbf{Z}/p\mathbf{Z}$ denotes the integers modulo p, then the nonzero elements of $\mathbf{Z}/p\mathbf{Z}$ form a multiplicative group of order $p - 1$.) Generalize to obtain **Euler's theorem**: If $(m, n) = 1$, then $n^{\varphi(m)} \equiv 1 \pmod{m}$.

*2.31.   (**H. B. Mann**)   Let G be a finite group, and let S and T be (not necessarily distinct) nonempty subsets. Prove that either $G = ST$ or $|G| \geq |S| + |T|$.

2.32.   Using Exercise 2.31, prove that every element in a finite field F is a sum of two squares.

NORMAL SUBGROUPS AND QUOTIENT GROUPS

There is one kind of subgroup that is especially interesting, for it is intimately related to homomorphisms.

Definition   A subgroup S of G is a **normal subgroup** of G, denoted by $S \lhd G$, if $aSa^{-1} = S$ for every $a \in G$.

The first properties of normal subgroups are contained in the following exercises.

EXERCISES

2.33.   A subgroup S of G is normal in G if and only if $aSa^{-1} \subset S$ for every $a \in G$.

2.34.   If G is abelian, then every subgroup of G is normal. The converse is false, as the group of quaternions shows (see Exercise 4.32).

2.35. Show that a subgroup H of a group G is normal if and only if whenever $xy \in H$ (where $x, y \in G$), then $yx \in H$.

2.36. Any subgroup S in G of index 2 is normal.

2.37. A subgroup S of a group G is normal if and only if $aS = Sa$ for all $a \in G$.

**2.38. If $S \lhd G$, $a \in G$, and $s \in S$, then there exists an element $s' \in S$ with $as = s'a$ (this is a partial commutativity).

Definition Let $x \in G$. A **conjugate** of x is an element of the form axa^{-1}, where $a \in G$. One often denotes axa^{-1} by x^a.

EXERCISES

*2.39. A subgroup S of G is normal if and only if, whenever $x \in S$, all conjugates of x also lie in S. Conclude that $S \lhd G$ if and only if $\gamma(S) \subset S$ for every conjugation γ (see Exercise 1.24).

2.40. Let G be the multiplicative group of all $n \times n$ nonsingular matrices over a field F. Prove that the set of all matrices of determinant 1 is a normal subgroup of G.

2.41. Any intersection of normal subgroups of G is itself a normal subgroup of G. Conclude that if X is any subset of G, there is a smallest normal subgroup of G containing X; this subgroup is called the **normal subgroup of G generated by** X (or the **normal closure** of X).

2.42. If X is empty, the normal subgroup generated by X is $\{1\}$. If X is nonempty, the normal subgroup generated by X is the set of all words on the conjugates of elements in X.

**2.43. If H and K are normal subgroups of G, then so is $H \vee K$.

Lemma 2.13 *Let G be a cyclic group of order n. For each divisor d of G, there exists a unique subgroup of G of order d.*

Proof If a is a generator of G, then $\langle a^{n/d} \rangle$ is a subgroup of order d. Assume $\langle b \rangle$ is a subgroup of order d (necessarily cyclic, by Exercise 2.28). Now $b^d = 1$ and $b = a^m$ for some m; hence $a^{md} = 1$, $md = nk$ for some k, and $b = a^m = (a^{n/d})^k$. Therefore $\langle b \rangle \subset \langle a^{n/d} \rangle$, and the inclusion is equality because both subgroups have order d. ∎

Theorem 2.14 *If n is a positive integer, then*

$$n = \sum_{d \mid n} \varphi(d),$$

where the sum is over all divisors d of n $(1 \leq d \leq n)$.

Proof If C is a cyclic subgroup of a group G, let $g(C)$ denote the set of its generators. It is clear that G is the disjoint union

$$G = \cup\, g(C),$$

where C varies over all the cyclic subgroups of G. When G is cyclic of order n, we have just seen that there is a unique cyclic subgroup C_d of order d for every divisor d of n. Therefore, $n = |G| = \sum_{d\,|\,n} |g(C_d)|$. In Exercise 2.24, however, we saw that $|g(C_d)| = \varphi(d)$. ∎

We can now characterize finite cyclic groups.

Theorem 2.15 *A group G of order n is cyclic if and only if, for each divisor d of n, there is at most one cyclic subgroup of G of order d.*

Proof If G is cyclic, the result is Lemma 2.13. For the converse, recall from the previous proof that $G = \cup\, g(C)$, where C ranges over all the cyclic subgroups of G, whence $n = |G| = \Sigma|g(C)|$. By hypothesis, for each divisor d of n, there is at most one such C of order d (and $|g(C)| = \varphi(d)$). Hence $\Sigma|g(C)| \leq \sum_{d\,|\,n} \varphi(d)$ $= n$, by Theorem 2.14. We conclude that G must have exactly one cyclic subgroup of order d for every divisor d of n. In particular, G has a cyclic subgroup of order $d = n$ and G is cyclic. ∎

Observe that the hypothesis of Theorem 2.15 is satisfied if, for every divisor d of n, there are at most d solutions in G of the equation $x^d = 1$ (two cyclic subgroups of order d would contain more than d solutions). Frobenius (1895) proved a much deeper result: If d divides the order of an arbitrary finite group G, then the number of elements in G with $x^d = 1$ is a multiple of d.

Theorem 2.16 *If F is a finite field and $F^{\#}$ is the multiplicative group of its nonzero elements, then $F^{\#}$ is cyclic.*

Proof $F^{\#}$ is a group of order n, where $n = |F| - 1$. If d is a divisor of n and $a \in F^{\#}$, then $a^d = 1$ if and only if a is a root of the polynomial $x^d - 1$. Since a polynomial of degree d over a field has at most d roots, our observation above shows the hypothesis of Theorem 2.15 is satisfied, and so $F^{\#}$ is cyclic. ∎

REMARK The proof of Theorem 2.16 does not prove Wedderburn's theorem (1905), stating that every finite division ring is a field, for a polynomial of degree d with coefficients in a division ring may have more than d roots: if

$$\mathbf{H} = \{a + bi + cj + dk\colon a,\, b,\, c,\, d \in \mathbf{R}\}$$

is the ring of *real quaternions* (in which $i^2 = j^2 = k^2 = -1$, $ij = k$, $jk = i$, $ki = j$, and $ji = -k$, $kj = -i$, $ik = -j$), then $\mathbf{H}$ is a division ring containing 8 roots of $x^4 - 1$, namely, ± 1, $\pm i$, $\pm j$, $\pm k$. These last 8 elements form a multiplicative group (the quaternions) that we will discuss in Chapter 4.

The construction of a **quotient group** in the following theorem is of fundamental importance.

Theorem 2.17 *If $S \lhd G$, then the cosets of S in G form a group, denoted by G/S, of order $[G:S]$.*

Proof In order to define a group, we must present a set and a binary operation. The set is the collection of all cosets of S in G; note that since $S \lhd G$, we need not bother with the adjectives "right" and "left". As multiplication, we propose the multiplication of nonempty subsets defined on page 14. Recall that we proved then (Exercise 2.6) that this multiplication is associative. Now $(Sa)(Sb) = Sa(a^{-1}Sa)b$ (because S is normal) $= S(aa^{-1})Sab = SSab = Sab$ (because S is a subgroup). Thus, we have proved that $SaSb = Sab$, so that a product of two cosets of S is itself a coset of S. We leave to the reader the proof that the identity is S and that the inverse of Sa is Sa^{-1}. We have constructed a group (which is denoted by G/S). Finally, the definition of index says that $|G/S| = [G:S]$. ∎

Example 5 Let m be a fixed positive integer. If k is an integer, recall that the congruence class of k modulo m is

$$\bar{k} = \{k + im : i = 0, \pm 1, \pm 2, \cdots\}.$$

The group of integers modulo m, denoted by $\mathbf{Z}/m\mathbf{Z}$, is the set of all these congruence classes under the binary operation

$$\bar{k} + \bar{l} = \overline{k + l}.$$

If $\mathbf{Z}$ is the additive group of integers, then the cyclic subgroup generated by m is $m\mathbf{Z} = \{im : i = 0, \pm 1, \pm 2, \cdots\}$. Thus $\bar{k} = k + m\mathbf{Z}$ is a coset of $m\mathbf{Z}$ in $\mathbf{Z}$. Now $m\mathbf{Z}$ is a normal subgroup of $\mathbf{Z}$, since $\mathbf{Z}$ is abelian, so that the quotient group $\mathbf{Z}/m\mathbf{Z}$ is defined. Addition in $\mathbf{Z}/m\mathbf{Z}$ is given by

$$(k + m\mathbf{Z}) + (l + m\mathbf{Z}) = k + l + m\mathbf{Z}.$$

Thus $\mathbf{Z}/m\mathbf{Z}$ is the quotient group of $\mathbf{Z}$ by its (normal) subgroup $m\mathbf{Z}$. Because of this example, an arbitrary quotient group G/H is often called G modulo H.

Example 6 Let G be the multiplicative group of all $n \times n$ nonsingular matrices over a field F, and let H be the normal subgroup of all matrices of determinant 1. We shall show that $G/H \cong F^{\#}$, the multiplicative group of all nonzero elements in F.

If $a \in G$, then $\det(a) \in F^{\#}$, since a is nonsingular. Define a function $d: G/H \to F^{\#}$ by

$$d(Ha) = \det(a).$$

This function is well defined, for if $Hb = Ha$, then $b = ha$ for some $h \in H$; therefore, $\det(h) = 1$ and

$$\det(b) = \det(ha) = \det(h)\det(a) = \det(a);$$

d is a homomorphism, for

$$d(HaHb) = d(Hab) = \det(ab) = \det(a)\det(b) = d(Ha)d(Hb);$$

d is one-one, for if $d(Ha) = d(Hb)$, then $\det(a) = \det(b)$. Therefore, $\det(ab^{-1}) = 1$ and $ab^{-1} \in H$, so that $Ha = Hb$. Finally, d is onto, for if $x \in F^{\#}$, then the diagonal matrix a that has x in the upper left corner and 1 elsewhere on the diagonal has determinant x and $d(Ha) = x$. Thus, d is an isomorphism.

THE ISOMORPHISM THEOREMS

There are three theorems that describe the relationship among quotient groups, normal subgroups, and homomorphisms. The reader should be warned that the numbering of these theorems is not canonical, so that one person's first isomorphism theorem may be another's second. A second remark is that a testimony to the elementary character of these theorems is that analogs of them are true for almost every type of algebraic system, e.g., groups, rings, and vector spaces.

Theorem 2.18   (First Isomorphism Theorem)†   *Let $f: G \to H$ be a homomorphism with kernel K. Then K is a normal subgroup of G and $G/K \cong$ image f.*

Proof   We have already seen that K is a subgroup of G. To see that K is a normal subgroup, we must show $aKa^{-1} \subset K$ for every $a \in G$ (Exercise 2.33). If $x = aka^{-1} \in aKa^{-1}$, then $f(x) = f(a)f(k)f(a)^{-1} = 1$ and $x \in K$, as desired.

The remainder of the proof is patterned after the example of determinants given above. Define $F: G/K \to H$ by

$$F(Ka) = f(a).$$

Now F is well defined, for if $Ka = Kb$, then $ab^{-1} \in K$, $f(ab^{-1}) = 1$, and so $f(a) = f(b)$. F is a homomorphism, for

$$F(KaKb) = F(Kab) = f(ab)$$
$$= f(a)f(b) = F(Ka)F(Kb).$$

F is one-one, for if $F(Ka) = F(Kb)$, then $f(a) = f(b)$; therefore, $f(ab^{-1}) = 1$, $ab^{-1} \in K$, and so $Ka = Kb$. (That F is one-one is the converse of the statement that F is well defined.) Clearly, image $F =$ image f. Therefore, F is the desired isomorphism.   ∎

The first isomorphism theorem thus says that there is no significant difference between a quotient group and a homomorphic image. Given a homomorphism, one should always try to identify its kernel and image; the first isomorphism theorem will then provide an isomorphism.

Here is an illustration of the use of the first isomorphism theorem: we give a

---

† The isomorphism theorems 2.18, 2.21, and 2.22 are due to E. Noether (1882–1935).

proof of Exercise 2.29 that two cyclic groups of the same order are isomorphic. If $G = \langle a \rangle$ is cyclic of order n, then $f: \mathbf{Z} \to G$ defined by $f(i) = a^i$ is a homomorphism with image G and kernel $n\mathbf{Z}$. It follows that $G \cong \mathbf{Z}/n\mathbf{Z}$. Similarly, if H is cyclic of order n, then $H \cong \mathbf{Z}/n\mathbf{Z}$ and so $G \cong H$.

The next theorem explains our interest in normal subgroups.

Theorem 2.19   *A subgroup H of a group G is normal if and only if H is the kernel of some homomorphism.*

Proof   We have just seen that kernels are normal subgroups. Suppose that $H \lhd G$. Define $\pi: G \to G/H$ by $\pi(a) = Ha$. Clearly π is onto and the formula $HaHb = Hab$ shows π is a homomorphism. The following are equivalent for $a \in G$: $a \in \ker \pi$; $\pi(a) = H$; $Ha = H$; $a \in H$. Therefore $\ker \pi = H$.   ∎

We point out that different homomorphisms can have the same kernel. For example, let $f: \mathbf{Z} \to \mathbf{Z}$ be the identity map and let $g: \mathbf{Z} \to \mathbf{Z}$ be defined by $g(n) = -n$. The kernel of each of these maps is $\{0\}$, but clearly $f \neq g$.

Definition   If $H \lhd G$, the homomorphism $\pi: G \to G/H$ defined by $\pi(a) = Ha$ is called the **natural map**.

EXERCISES

*2.44.   Let $H \lhd G$ have index n. If $y \in G$, prove that $y^n \in H$. Give an example to show this may be false when H is not normal. (See Exercise 2.9.)

**2.45.   Let $f: G \to H$ be a homomorphism with kernel K. For each $h \in \operatorname{im} f$, choose $l(h) \in G$ with $f(l(h)) = h$. Show the inverse of the isomorphism F in Theorem 2.18 is $h \mapsto Kl(h)$.

2.46.   Let $H \lhd G$ and let $\pi: G \to G/H$ be the natural map. Assume X is a subset of G with $\pi(X)$ generating G/H. Prove that $G = \langle H \cup X \rangle$.

2.47.   Let $f: G \to H$ be a homomorphism with kernel K. Prove that there exists a homomorphism $g: G/K \to H$ with $f = g \circ \pi$, where π is the natural map.

2.48.   Let $N \lhd G$ and let $f: G \to H$ be a homomorphism whose kernel contains N. Then f induces a homomorphism $f_{\#}: G/N \to H$, namely $f_{\#}(Na) = f(a)$.

2.49.   Let $f: G \to H$ be a homomorphism; f is one-one if and only if kernel $f = \{1\}$.

**2.50.   Let $f: G \to H$ be a homomorphism and let S be a subgroup of H. Then $f^{-1}(S) = \{x \in G : f(x) \in S\}$ is a subgroup of G and contains $\ker f$.

Definition If $a, b \in G$, the **commutator** of a and b, denoted by $[a, b]$, is the element $aba^{-1}b^{-1}$. The **commutator subgroup** of G, denoted by G', is the subgroup of G generated by all the commutators in G.

EXERCISES

2.51. Prove that G' is a normal subgroup of G. (HINT: Use Exercise 2.39.)

**2.52 If H is any normal subgroup of G, then G/H is abelian if and only if $G' \subset H$.

2.53. Let G be a finite group of odd order, and let x be the product of all the elements of G in some order. Prove that $x \in G'$. (HINT: Exercise 1.36.)

2.54. **(Yff)** For any group G, show its commutator subgroup G' is the set

$$G' = \{a_1 a_2 \cdots a_n a_1^{-1} a_2^{-1} \cdots a_n^{-1} : a_i \in G \text{ and } n \geq 2\}.$$

[HINT (**Weichsel**): $(aba^{-1}b^{-1})(cdc^{-1}d^{-1}) = a(ba^{-1})b^{-1}c(dc^{-1})d^{-1}a^{-1}(ab^{-1})bc^{-1}(cd^{-1})d$.]

2.55. The fact that the set of commutators in a group need not be a subgroup is an old result; the following simple example is due to Cassidy (1979).

  (i) For a field k, let $k[x, y]$ denote the ring of all polynomials in two variables, and let $k[x]$ and $k[y]$ be the subrings of all polynomials in x and in y, respectively. Let G be the set of all matrices of the form

$$A = \begin{bmatrix} 1 & f(x) & h(x, y) \\ 0 & 1 & g(y) \\ 0 & 0 & 1 \end{bmatrix}$$

  where $f(x) \in k[x]$, $g(y) \in k[y]$, and $h(x, y) \in k[x, y]$. Prove that G is a (multiplicative) group and that its commutator subgroup G' consists of all such matrices with $f = 0 = g$. (HINT: Denote A by (f, g, h) so that $(f, g, h)(f', g', h') = (f + f', \ g + g', \ h + h' + fg')$. If $h = h(x, y) = \Sigma a_{ij} x^i y^j$, then $(0, 0, h) = \Pi_{i,j} [(a_{ij} x^i, 0, 0), (0, y^j, 0)]$.)

  (ii) If $(0, 0, h)$ is a commutator, then there are polynomials $f(x)$, $f'(x) \in k[x]$ and $g(y), g'(y) \in k[y]$ with $h(x, y) = f(x)g'(y) - f'(x)g(y)$.

  (iii) Show that $h(x, y) = x^2 + xy + y^2$ does not possess a decomposition as in part (ii) and conclude that $(0, 0, h) \in G'$ is not a commutator. [HINT: If $f(x) = \Sigma b_i x^i$ and $f'(x) = \Sigma c_i x^i$, then there are equations

$$b_0 g'(y) - c_0 g(y) = y^2$$
$$b_1 g'(y) - c_1 g(y) = y$$
$$b_2 g'(y) - c_2 g(y) = 1.$$

Considering $k[x, y]$ as a vector space over k, one obtains the contradiction that the independent set $\{1, y, y^2\}$ is in the subspace spanned by $\{g', g\}$.] (One can modify this construction to obtain a finite example G if one knows some ring theory: replace $k[x, y]$ by its quotient ring in which all degree 3 terms x^3, y^3, x^2y, xy^2 are 0. If $k = \mathbf{Z}/2\mathbf{Z}$, the corresponding group G has order 2^{12}.)

Theorem 2.20 *If S and T are subgroups of G and if one of them is normal, then $ST = S \vee T = TS$.*

Proof Recall that ST is just the set of products of the form st, where $s \in S$ and $t \in T$; hence, ST and TS are always subsets of $S \vee T$ that contain S and T. Thus, we need show only that ST and TS are themselves subgroups in order to show that $S \vee T$ is a subset of ST (and of TS). Let us suppose $T \lhd G$. If $s_1 t_1$ and $s_2 t_2 \in ST$, then

$$(s_1 t_1)(s_2 t_2)^{-1} = s_1(t_1 t_2^{-1} s_2^{-1}) = s_1(s_2^{-1} t_3)$$

for some $t_3 \in T$ (this is the partial commutativity from the normality of T [Exercise 2.38]), and $s_1(s_2^{-1} t_3) = (s_1 s_2^{-1}) t_3 \in ST$. A similar proof shows that TS is a subgroup. ∎

EXERCISES

**2.56. Let S and T be subgroups of G. Then ST is a subgroup of G if and only if $ST = TS$.

2.57. (**Modular Law**) Let $A, B,$ and C be normal subgroups of G with $A \subset B$. If $A \cap C = B \cap C$ and $AC = BC$, then $A = B$.

*2.58. (**Dedekind Law**) Let $H, K,$ and L be normal subgroups of G with $H \subset L$. Then

$$HK \cap L = H(K \cap L).$$

Suppose $T \subset H \subset G$ are subgroups with $T \lhd G$. Then $T \lhd H$ and the quotient group H/T is defined: it is the subgroup of G/T consisting of all cosets Th, where $h \in H$. In particular, if $T \lhd G$ and S is any subgroup of G, then $T \subset TS \subset G$, and TS/T is the subgroup of G/T consisting of all cosets Tts, where $ts \in TS$. Since $Tts = Ts$, it follows that TS/T consists precisely of all the cosets of T having a representative in S.

It follows at once from Theorem 2.20 and the product formula (Exercise 2.11) that if S and T are subgroups of G, one of which is normal, then $|TS|/|T| = |S|/|S \cap T|$. This suggests the following theorem.

Theorem 2.21 (**Second Isomorphism Theorem**) *Let S and T be subgroups of G with T normal. Then $S \cap T$ is normal in S and $S/(S \cap T) \cong TS/T$.*

REMARK   The following diagram is a mnemonic for this theorem.

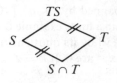

Proof Let $\pi: G \to G/T$ be the natural map and let π_0 be the restriction of π to S. Since π_0 is a homomorphism whose kernel is $S \cap T$, Theorem 2.18 tells us that $S \cap T \lhd S$ and that $S/(S \cap T) \cong$ image π_0. By our remarks above, image π_0 is just the collection of all cosets of T having representatives in S; these are precisely the cosets in TS/T. ∎

Theorem 2.22 (Third Isomorphism Theorem) *Let $K \subset H \subset G$, where both H and K are normal subgroups of G. Then H/K is a normal subgroup of G/K and*

$$(G/K)/(H/K) \cong G/H.$$

Proof Again we let the first isomorphism theorem do the dirty work. Define $f: G/K \to G/H$ by $f(Ka) = Ha$. The reader can check that f is a well defined homomorphism whose kernel is H/K and whose image is G/H. ∎

THE CORRESPONDENCE THEOREM

The main theorem of this section could justifiably be called the fourth isomorphism theorem.

Let G and H be sets, and let $f: G \to H$ be a function. The reader is aware that f induces a "forward motion" and a "backward motion" between the subsets of G and the subsets of H. The forward motion assigns to every subset S of G its image $f(S)$ in H; the backward motion assigns to every subset L of H its inverse image $f^{-1}(L) = \{x \in G: f(x) \in L\}$ in G. Now, if f is onto, these motions define a one-one correspondence between all the subsets of H and some of the subsets of G. The following theorem is the group-theoretic translation of this observation.

Theorem 2.23 (Correspondence Theorem) *Let $K \lhd G$ and let $\pi: G \to G/K$ be the natural map; π defines a one-one correspondence between the set of those subgroups of G containing K and the set of all subgroups of G/K.*

If the subgroup of G/K corresponding to $S \subset G$ is denoted by S^*, then

(i) $S^* = S/K = \pi(S)$;
(ii) $T \subset S$ if and only if $T^* \subset S^*$, and then $[S:T] = [S^*:T^*]$;
(iii) $T \lhd S$ if and only if $T^* \lhd S^*$, and then $S/T \cong S^*/T^*$.

REMARK   A mnemonic diagram for this theorem is

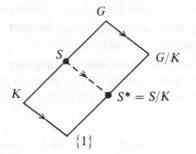

Proof We first show that the correspondence is one-one. Suppose that S and T are subgroups of G containing K and that $S/K = T/K$; we must prove that $S = T$. If $s \in S$, then $Ks = Kt$ for some $t \in T$; hence $s = kt$ for some $k \in K \subset T$, and so $S \subset T$. A symmetric argument proves the reverse inclusion.

We now prove that the correspondence is onto. If A is a subgroup of G/K, we must find a subgroup S of G containing K for which $S/K = A$. By Exercise 2.50, if we define $S = \pi^{-1}(A)$, then S is a subgroup of G that contains K; furthermore, $S/K = \pi(S) = \pi(\pi^{-1}(A)) = A$, since π is onto.

It is obvious that this correspondence preserves inclusions. To prove $[S:T] = [S^*:T^*]$, we must show that the set of cosets Ts (where $s \in S$) is in one-one correspondence with the set of cosets T^*s^* (where $s^* \in S^*$). Such a correspondence is given by $Ts \leftrightarrow T^*\pi(s)$; the verification is left to the reader.†

If $T \lhd S$, then we conclude from the third isomorphism theorem that $T/K \lhd S/K$ and $(S/K)/(T/K) \cong S/T$, i.e., $T^* \lhd S^*$ and $S^*/T^* \cong S/T$. Suppose, conversely, that $T^* \lhd S^*$. Let $\eta: S^* \to S^*/T^*$ be the natural map and let $\pi_0 = \pi|S$. The reader may verify that T is the kernel of $\eta\pi_0: S \to S^*/T^*$, which implies that $T \lhd S$. ∎

EXERCISES

2.59. Let H be a subgroup of G that contains G', the commutator subgroup of G. Prove that $H \lhd G$.

2.60. (Zassenhaus**) Let G be a finite group such that, for some fixed integer $n > 1$, $(xy)^n = x^n y^n$, for all $x,\ y \in G$. Let $G_n = \{z \in G: z^n = 1\}$ and $G^n = \{x^n : x \in G\}$. Show that both G_n and G^n are normal subgroups of G and that $|G^n| = [G:G_n]$. (NOTE: The wise reader lets homomorphisms do most of the work.)

2.61. Let G be a finite group with a normal subgroup H such that $(|H|, [G:H]) = 1$. Show that H is the unique subgroup of G having order $|H|$. (HINT: If K is another such subgroup, what happens to K in G/H?)

---

† If all groups are finite, we may prove that $[S:T] = [S^*:T^*]$ as follows: $[S^*:T^*] = |S^*|/|T^*| = |S/K|/|T/K| = (|S|/|K|)/(|T|/|K|) = |S|/|T| = [S:T]$.

2.62. If $H \lhd G$, need G contain a subgroup isomorphic to G/H?

2.63. Prove that the circle group **T** is isomorphic to $\mathbf{R}/\mathbf{Z}$, where **R** is the additive group of real numbers. (HINT: See Exercise 1.22.)

2.64. Prove that H is a **maximal normal subgroup** of G (i.e., H is a proper normal subgroup of G and there is no normal subgroup strictly between G and H) if and only if G/H has no proper normal subgroups.

Definition   A group G is **simple** if $G \neq \{1\}$ and G contains no proper normal subgroups.

If $H \lhd G$, then G/H is simple if and only if H is a maximal normal subgroup of G.

EXERCISES

2.65. Prove that an abelian group is simple if and only if it is finite and of prime order.

2.66. (**Schur**) Let $f: G \to H$ be a nontrivial homomorphism, i.e., f does not send every element into 1. If G is simple, then f is one-one.

DIRECT PRODUCTS

Definition   If H and K are groups, the (**external**) **direct product** of H and K, denoted by $H \times K$, is the set of all ordered pairs (h, k), where $h \in H$ and $k \in K$, with the binary operation

$$(h, k)(h', k') = (hh', kk').$$

It is easy to check that $H \times K$ is a group containing isomorphic copies of H and K, namely, $H \times \{1\}$ and $\{1\} \times K$.

EXERCISES

2.67. The operation of direct product is commutative and associative: for any groups H, K, and L,

$$H \times K \cong K \times H$$

and

$$(H \times K) \times L \cong H \times (K \times L).$$

Conclude that the notations $H_1 \times \cdots \times H_m$ and $\Pi_{i=1}^m H_i$ are unambiguous.

2.68. $H \times \{1\}$ and $\{1\} \times K$ are normal subgroups of $H \times K$; these two subgroups generate $H \times K$, their intersection is $\{(1, 1)\}$, and the elements $(h, 1)$ and $(1, k)$ commute for all $h \in H$ and $k \in K$.

2.69. Let $\mu: G \times G \to G$ be the binary operation of a group G. If $G \times G$ is regarded as a group (the direct product), prove that μ is a homomorphism if and only if G is abelian.

2.70. Let $G = H \times K$ and let N be a normal subgroup of $H \times \{1\}$. Then N is a normal subgroup of G.

2.71. Prove that a direct product of groups is abelian if and only if each factor is abelian.

*2.72. Let G be an abelian group and let $f: H \to G$ and $g: K \to G$ be homomorphisms. Prove that there exists a unique homomorphism $F: H \times K \to G$ that extends both f and g, i.e., $F(h, 1) = f(h)$ and $F(1, k) = g(k)$. Show that this may be false if G is not abelian.

2.73. Prove that $\mathbf{Z}(6) \cong \mathbf{Z}(2) \times \mathbf{Z}(3)$.

**2.74. If $(m, n) = 1$, then $\mathbf{Z}(mn) \cong \mathbf{Z}(m) \times \mathbf{Z}(n)$.

2.75. If p is prime, prove that $\mathbf{Z}(p^2) \not\cong \mathbf{Z}(p) \times \mathbf{Z}(p)$.

We now adopt another point of view. It is easy to multiply two polynomials together; it is harder to factor a given polynomial. We have just seen how to "multiply" two groups together; can we "factor" a given group?

Theorem 2.24 *Let G be a group with normal subgroups H and K; if $H \cap K = \{1\}$ and $HK = G$, then $G \cong H \times K$.*

Proof Let $a \in G$. Since $HK = G$, $a = hk$ for some $h \in H$ and $k \in K$, and we claim that h and k are uniquely determined by a. If $a = h_1 k_1$, then $hk = h_1 k_1$ and $h^{-1} h_1 = k k_1^{-1}$. This element is simultaneously in H and K, i.e., in $H \cap K = \{1\}$; hence, $h = h_1$ and $k = k_1$.

Define $f: G \to H \times K$ by $f(a) = (h, k)$, where $a = hk$. Is f a homomorphism? If $a = hk$ and $a' = h'k'$, then $aa' = hkh'k'$, which is not in the proper form for evaluating f. Were it true that $kh' = h'k$, however, then we would be done. Indeed, we prove that for any $h \in H$ and $k \in K$, $hk = kh$. Consider the commutator $hkh^{-1}k^{-1}$. Now $(hkh^{-1})k^{-1} \in K$, since K is normal, and $h(kh^{-1}k^{-1}) \in H$, since H is normal. Therefore, this commutator is in $H \cap K = \{1\}$, and so $hk = kh$. We let the reader prove that f is a one-one correspondence. ∎

We pause to give an example which shows that all the hypotheses in Theorem 2.24 are necessary. Let $G = S_3$, H the (normal) subgroup of order 3, and K any subgroup of order 2. It is easy to check that $H \cap K = \{1\}$ and $HK = G$. Note that H is normal, but that K is not. Were $S_3 \cong H \times K$, then $S_3 \cong \mathbf{Z}(3) \times \mathbf{Z}(2)$, which is abelian, a contradiction.

Theorem 2.25 *Let* $G = H \times K$, *and let* $H_1 \lhd H$ *and* $K_1 \lhd K$. *Then* $H_1 \times K_1 \lhd G$ *and*

$$G/(H_1 \times K_1) \cong (H/H_1) \times (K/K_1).$$

Proof Let $\pi: H \to H/H_1$ and $\rho: K \to K/K_1$ be the natural maps. Define $F: G \to (H/H_1) \times (K/K_1)$ by $F(h, k) = (\pi h, \rho k)$. The kernel of F is $H_1 \times K_1$ and the image of F is $(H/H_1) \times (K/K_1)$. ∎

Corollary 2.26 *If* $G = H \times K$, *then* $G/(H \times \{1\}) \cong K$.

The elements of an external direct product are ordered pairs, a rather restrictive condition. We say that a group G is the (**internal**) **direct product** of H and K if H and K are normal subgroups of G with $H \cap K = \{1\}$ and $HK = G$. The emphasis here is that the factors themselves, not merely isomorphic copies of them, lie in G. (If $G = H \times K$ is an external direct product, then it is also the internal direct product of $H \times \{1\}$ and $\{1\} \times K$, but it is not the internal direct product of H and K. For example, $\mathbf{Z}/6\mathbf{Z} = \langle \overline{2} \rangle \times \langle \overline{3} \rangle$; thus $\mathbf{Z}/6\mathbf{Z} \cong \mathbf{Z}/3\mathbf{Z} \times \mathbf{Z}/2\mathbf{Z}$, but $\mathbf{Z}/6\mathbf{Z} \neq \mathbf{Z}/3\mathbf{Z} \times \mathbf{Z}/2\mathbf{Z}$.) The two versions of direct product, of course, yield isomorphic groups. In the future, we shall not distinguish between external and internal, and we shall use "direct product" without an adjective. In almost all cases, however, our point of view is internal. For example, we shall write Corollary 2.26 as $(H \times K)/H \cong K$.

EXERCISES

*2.76. Show that a noncyclic group of order 4 is isomorphic to $\langle a \rangle \times \langle b \rangle$, where $a^2 = 1$ and $b^2 = 1$ (see Exercise 2.14).

2.77. Show it is possible that $G = H \times K = H \times L$ (internal direct products) with $K \neq L$ (of course, $K \cong L$, by Corollary 2.26). (HINT: Take G as in Exercise 2.76 and show that $G = \langle a \rangle \times \langle ab \rangle$.)

2.78. If p is a prime and G is an abelian group of order p^2, prove that either G is cyclic or $G \cong \mathbf{Z}(p) \times \mathbf{Z}(p)$. (We shall see in Chapter 4 that a group of order p^2 must be abelian.)

*2.79. Denote by 0 the trivial homomorphism that sends every element into the identity. Prove that $G \cong H \times K$ if and only if there exist homomorphisms

$$H \underset{q}{\overset{i}{\rightleftarrows}} G \underset{j}{\overset{p}{\rightleftarrows}} K$$

with $qi = 1_H$ (the identity function on H), $pj = 1_K$, $pi = 0$, $qj = 0$, and

$$i(q(x))j(p(x)) = x \qquad \text{for all } x \in G.$$

2.80. Let G be a group with normal subgroups H and K. Prove that $G = H \times K$ (internal direct product) if and only if each $a \in G$ has a unique expression $a = hk$, where $h \in H$ and $k \in K$.

**2.81. If G is a group with normal subgroups $H_1, \cdots, H_m$, define $G = \prod_{i=1}^{m} H_i$ (internal) if $G = \langle \bigcup_{i=1}^{m} H_i \rangle$ and, for all j, $H_j \cap \langle \bigcup_{i \neq j} H_i \rangle = \{1\}$. Prove that $G \cong H_1 \times \cdots \times H_m$.

*2.82. Let G be a group with normal subgroups $H_1, \cdots, H_m$. Prove that $G = H_1 \times \cdots \times H_m$ (internal direct product) if and only if each $a \in G$ has a unique expression $a = h_1 \cdots h_m$, where $h_i \in H_i$ for $i = 1, \cdots, m$.

2.83. Let $N \lhd G = H \times K$. Prove that either N is abelian or N intersects one of the factors H or K nontrivially.

2.84. Give an example of an abelian group $H \times K$ that contains a nontrivial subgroup N such that $N \cap H = \{1\}$ and $N \cap K = \{1\}$. Conclude that if $N \subset H \times K$, then $N \neq (N \cap H) \times (N \cap K)$ is possible.

2.85. Let G be a group having a simple subgroup H of index 2. Prove that either H is the unique proper normal subgroup of G or G contains a normal subgroup K of order 2 with $G = H \times K$. (HINT: Second isomorphism theorem.)

chapter three

Permutation
Groups

PERMUTATIONS

The reader knows the formula giving the roots of a quadratic polynomial and is aware of similar formulas giving the roots of polynomials of degree 3 and degree 4. Many mathematicians put their faith in a faulty induction and tried to find a formula that would give the roots of an arbitrary polynomial of degree 5. In the early 1800s, Ruffini and Abel, independently, proved, by studying permutations of the roots of quintic polynomials, that no such formula exists, and this result led Galois to his discovery of the intimate relationship between polynomials and certain groups of permutations of their roots. Influenced by the beauty of the work of Abel and Galois, most nineteenth-century mathematicians considered only those groups whose elements are permutations (we shall see presently that this is no restriction at all). We now proceed to develop this point of view.

Definition If X is a nonempty set, a **permutation** of X is a function $\alpha: X \to X$ that is a one-one correspondence.

Definition If X is a nonempty set, the **symmetric group** on X, denoted by S_X, is the group whose elements are the permutations of X and whose binary operation is composition of functions.

Of particular interest is the special case when X is finite. If $X = \{1, 2, \cdots, n\}$, we write S_n instead of S_X, and we call S_n the symmetric group of degree n, or the symmetric group on n letters. Note that $|S_n| = n!$.

EXERCISES

3.1.   Write a multiplication table for S_4.

**3.2.   Let X and Y be two sets and let $f: X \to Y$ be a one-one correspondence. Prove that $\alpha \mapsto f \circ \alpha \circ f^{-1}$ defines an isomorphism between S_X and S_Y.

3.3.   Suppose that $X \subset Y$, where X is nonempty. Prove that S_X can be imbedded in S_Y, i.e., S_X is isomorphic to a subgroup of S_Y.

Let X denote the set $\{1, 2, \cdots, n\}$. One way of denoting a permutation α of X is by displaying its values:

$$\alpha = \begin{pmatrix} 1 & 2 & \cdots & n \\ \alpha 1 & \alpha 2 & \cdots & \alpha n \end{pmatrix}.$$

Thus, $\alpha = \begin{pmatrix} 1 & 2 & 3 \\ 3 & 2 & 1 \end{pmatrix}$ and $\beta = \begin{pmatrix} 1 & 2 & 3 \\ 2 & 3 & 1 \end{pmatrix}$ are permutations of $\{1, 2, 3\}$. Their product is $\alpha\beta = \begin{pmatrix} 1 & 2 & 3 \\ 2 & 1 & 3 \end{pmatrix}$ (we compute this product by first applying β and then α).† Note that $\beta\alpha = \begin{pmatrix} 1 & 2 & 3 \\ 1 & 3 & 2 \end{pmatrix}$, so that $\alpha\beta \neq \beta\alpha$. It follows that S_3, and hence any larger symmetric group, is not abelian.

CYCLES

In this section, we consider some factorizations of permutations in S_n as products of simpler permutations called *cycles*.

Definition   Let $x \in X$ and $\alpha \in S_X$; we say α **fixes** x if $\alpha(x) = x$ and α **moves** x if $\alpha(x) \neq x$.

Definition   Let $i_1, i_2, \cdots, i_r$ be distinct integers between 1 and n. If $\alpha \in S_n$ fixes the other integers and

$$\alpha(i_1) = i_2, \quad \alpha(i_2) = i_3, \quad \cdots, \quad \alpha(i_{r-1}) = i_r, \quad \alpha(i_r) = i_1,$$

then α is an **r-cycle**. One also says α is a cycle of **length** r.

---

† Some authors compute the product $\alpha\beta$ in the reverse order. These are the same authors who put functions on the right, i.e., who write $(x)f$ instead of $f(x)$.

Every 1-cycle is the identity on X because it fixes every element of X.

Another way to denote the r-cycle α besides the cumbersome two-rowed notation introduced earlier is $\alpha = (i_1 i_2 \cdots i_r)$ [if necessary for clarity, we may insert commas and write, e.g., $\alpha = (123) = (1, 2, 3)$]. Thus

$$\begin{pmatrix} 1 & 2 & 3 & 4 \\ 2 & 3 & 4 & 1 \end{pmatrix} = (1\ 2\ 3\ 4),$$

$$\begin{pmatrix} 1 & 2 & 3 & 4 & 5 \\ 5 & 1 & 4 & 2 & 3 \end{pmatrix} = (1\ 5\ 3\ 4\ 2),$$

and

$$\begin{pmatrix} 1 & 2 & 3 & 4 & 5 \\ 2 & 3 & 1 & 4 & 5 \end{pmatrix} = (1\ 2\ 3)(4)(5) = (1\ 2\ 3)$$

(one usually suppresses all 1-cycles).

The cycle notation is very convenient for multiplication. For example, let us compute $\gamma = \alpha\beta$, where $\alpha = (12)$ and $\beta = (13425)$. Since multiplication is composition, $\gamma(1) = \alpha(\beta(1))$; thus, $\beta(1) = 3$ and $\alpha(3) = 3$ implies $\gamma(1) = 3$. What is $\gamma(3)$? $\beta(3) = 4$ and $\alpha(4) = 4$, so $\gamma(3) = 4$. Continuing this way

$$(12)(13425) = (134)(25).$$

The cycles on the right are disjoint as defined below.

Definition   Two permutations α and β in S_X are **disjoint** if every x moved by one is fixed by the other. In symbols, if $\alpha(x) \neq x$, then $\beta(x) = x$ and if $\beta(y) \neq y$, then $\alpha(y) = y$ [it is possible that $\alpha(z) = z = \beta(z)$ for some $z \in X$]. A set of permutations is **disjoint** if each pair of them is disjoint.

EXERCISES

3.4.   Prove that $(1\ 2\ 3 \cdots r) = (2\ 3 \cdots r\ 1) = (3\ 4 \cdots r\ 1\ 2) = \cdots = (r\ 1\ 2 \cdots r-1)$. Conclude that there are exactly r such notations for this cycle.

3.5.   The order of an r-cycle is r.

3.6.   Exhibit two 2-cycles whose product is a 3-cycle. This example shows that if α and β do not commute, nothing intelligent can be said about the order of $\alpha\beta$ in terms of the orders of the factors.

**3.7.   Let α and β be r-cycles in S_X. If there is an $x_0 \in X$ such that (i) both α and β move x_0 and (ii) $\alpha^t(x_0) = \beta^t(x_0)$ for all integers t, then $\alpha = \beta$. (HINT: If α is an r-cycle and $0 \leq j < r$, then $\alpha^j(i_1) = i_{1+j}$.)

3.8.   Let $\alpha = (i_1\ i_2 \cdots i_r)$ and $\beta = (j_1 j_2 \cdots j_s)$. Prove that α and β are disjoint if and only if $\{i_1, i_2, \cdots, i_r\} \cap \{j_1, j_2, \cdots, j_s\} = \varnothing$.

**3.9.   If α and β are disjoint, then $\alpha\beta = \beta\alpha$.

*3.10.   A permutation $\alpha \in S_n$ is **regular** if it is 1 or if it has no fixed points and is the product of disjoint cycles of the same length. Prove that α is regular if and only if α is a power of an n-cycle.

*3.11. Let $\alpha = \beta_1 \beta_2 \cdots \beta_m$, where the β_i are disjoint r_i-cycles. Prove that the order of α is lcm $\{r_1, r_2, \cdots, r_m\}$. Conclude that each r_i divides the order of α. Conclude further that if p is prime, then every power of a p-cycle is a p-cycle, or 1.

*3.12. If α is an n-cycle, then α^k is a product of (n, k) disjoint cycles, each of length $n/(n, k)$. (HINT: Use Exercises 3.10 and 3.11.)

Consider the permutation α defined by

$$\alpha = \begin{pmatrix} 1 & 2 & 3 & 4 & 5 & 6 & 7 & 8 & 9 \\ 6 & 4 & 7 & 2 & 5 & 1 & 8 & 9 & 3 \end{pmatrix};$$

we write α as a product of disjoint cycles. Since $\alpha(1) = 6$ and $\alpha(6) = 1$, α involves the 2-cycle (16). The first integer not yet considered is 2. Because $\alpha(2) = 4$ and $\alpha(4) = 2$, α also involves the 2-cycle (24). Now because $\alpha(3) = 7$, $\alpha(7) = 8$, $\alpha(8) = 9$, and $\alpha(9) = 3$, we have a 4-cycle (3789). Finally, $\alpha(5) = 5$. The permutation α is the product

$$\alpha = (16)(24)(3789)(5) = (16)(24)(3789)$$

(remember that two functions f and g on a set X, e.g., α and the above product of cycles, are equal if and only if $f(x) = g(x)$ for every $x \in X$). The next two theorems assert that such factorizations exist for every permutation in S_n, and they are essentially unique.

We shall use the following convention: if $Y \subset X$ and $\alpha \in S_Y$, then α may be regarded as a permutation of X that fixes every $x \notin Y$.

Theorem 3.1 *Every permutation α in S_n is either a cycle or a product of disjoint cycles.*

Proof We do an induction on k, where $k \geq 0$ is the number of letters moved by α. If $k = 0$, then α is the 1-cycle 1. Assume $k \geq 1$ and that α moves i_1, say. If $i_2 = \alpha(i_1)$, $i_3 = \alpha(i_2)$, $\cdots$, there is at least $r > 1$ with $\alpha(i_r) = i_1$ [note α one-one implies $\alpha(i_r) \neq \alpha(i_j)$, $1 < j < r$]. Define $\alpha' = \alpha \,|\, \{i_1, \cdots, i_r\}$. If $r = k$, then $\alpha = \alpha'$ is a cycle. Otherwise, define $\alpha'' = \alpha \,|\, Y$, where Y consists of the remaining $k - r$ integers. By induction, α'' is a cycle or a product of disjoint cycles. Finally, observe that α' and α'' are disjoint and that $\alpha = \alpha'\alpha''$ (just evaluate each side on any $i \in \{1, 2, \cdots, n\}$). ∎

The factorization of Theorem 3.1 of α into disjoint cycles will contain one 1-cycle for each i fixed by α. When any such factors are omitted, the factorization is essentially unique.

Theorem 3.2 *Let $\alpha \in S_n$ and suppose $\alpha = \beta_1 \beta_2 \cdots \beta_t$ is a factorization into disjoint cycles each of length ≥ 2. Then this factorization is unique except for the order in which the cycles are written.*

Proof   Assume $\alpha = \gamma_1 \gamma_2 \cdots \gamma_s$ is a second factorization of α into disjoint cycles of length ≥ 2. Let β_1 move i_1. Since $\beta_1(i_1) = \alpha(i_1)$, some γ_k moves i_1; since disjoint cycles commute, we may assume γ_1 moves i_1. For each j,

$$\beta_1^j(i_1) = \alpha^j(i_1) = \gamma_1^j(i_1).$$

By Exercise 3.7, $\beta_1 = \gamma_1$. We may now cancel these cycles, and the proof is completed by an induction on $\max\{t, s\}$. ∎

EXERCISES

3.13.   Show that $4x^2 - 3x^7$ is a permutation of $\mathbf{Z}/11\mathbf{Z}$, write it as a product of disjoint cycles, and compute its order in S_{11}.

**3.14.   If p is a prime, prove that the only elements in S_n of order p are p-cycles or products of disjoint p-cycles.

3.15.   How many elements of order 2 are in S_6?

3.16.   If α commutes with β and β commutes with γ, show that α may not commute with γ.

**3.17.   Show that

$$(ab)(ac_1 \cdots c_k bd_1 \cdots d_m) = (ac_1 \cdots c_k)(bd_1 \cdots d_m),$$

where $a, b, c_1, \cdots, c_k, d_1, \cdots, d_m$ are all distinct and k or m may be zero (i.e., there may be no c's or no d's).

Definition   A 2-cycle is also called a **transposition.**

Of all the permutations, surely the transposition, which merely interchanges two points, is the simplest.

Theorem 3.3   *Every $\alpha \in S_n$ is a product of transpositions.*

Proof   By Theorem 3.1 it suffices to factor any r-cycle into a product of transpositions. This is done in the following way:

$$(1 \ 2 \ \cdots \ r) = (1 \ r)(1 \ r-1)\cdots(1 \ 2). \quad ∎$$

Any permutation can thus be realized as a sequence of interchanges. This factorization, however, is not as nice as the factorization into disjoint cycles. First of all, the transpositions occurring need not commute; e.g., $(123) = (13)(12) \neq (12)(13)$. Second, the factors are not uniquely determined; e.g., $(123) = (13)(12) = (23)(13)$ $= (13)(42)(12)(14) = (13)(42)(12)(14)(23)(23)$. Is there any uniqueness in such a factorization? We shall prove that the number of factors is always even or always odd.

Definition A permutation α in S_n is **even** if α has a factorization as a product of an even number of transpositions; otherwise α is **odd**. We say the **parity** of α is even or odd.

It is not obvious whether any odd permutations exist, for it is conceivable that a permutation having a factorization into an odd number of transpositions might have a second factorization into an even number of transpositions.

If $\alpha = \beta_1 \beta_2 \cdots \beta_t$ is a factorization in S_n of α into disjoint cycles and $\Sigma = \Sigma$ length (β_i), then the number $\Sigma - t$ depends only on α (not on the factorization) even if some of the β_i are distinct 1-cycles.

Definition The **sign** function sgn: $S_n \to \{\pm 1\}$ is defined by

$$\text{sgn}\,(\alpha) = (-1)^{\Sigma - t}.$$

Note that sgn $(1) = 1$ and sgn $(\tau) = -1$ for every transposition τ.

Lemma 3.4 *If $\alpha \in S_n$ and τ is a transposition, then*

$$\text{sgn}\,(\tau\alpha) = -\,\text{sgn}\,(\alpha).$$

Proof Write $\alpha = \beta_1 \beta_2 \cdots \beta_t$ as a product of disjoint cycles (as in Theorem 3.1) in which $\Sigma = \Sigma$ length $(\beta_i) = n$; thus, every integer between 1 and n occurs in exactly one of the β_i (which may be a 1-cycle). Let $\tau = (ab)$. If τ and α are disjoint, then we may apply the definition of the sign function:

$$\text{sgn}\,(\tau\alpha) = (-1)^{(\Sigma + 2) - (t + 1)} = (-1)^{\Sigma - t + 1} = -\,\text{sgn}\,(\alpha).$$

Assume, therefore, that τ and α are not disjoint. If a and b occur in the same β_i, then $\beta_i = (ac_1 \cdots c_k b d_1 \cdots d_m)$. By Exercise 3.17,

$$\tau\beta_i = (bd_1 \cdots d_m)(ac_1 \cdots c_k).$$

Since τ is disjoint from the other cycles, it commutes with them, and sgn $(\tau\alpha)$ $= (-1)^{n - (t + 1)} = -\,\text{sgn}\,(\alpha)$. The other possibility is that a and b occur in different cycles: $\beta_i = (ac_1 \cdots c_k)$ and $\beta_j = (bd_1 \cdots d_m)$. But now τ commutes with the other cycles and

$$\tau\beta_i\beta_j = (ac_1 \cdots c_k b d_1 \cdots d_m)$$

(just multiply both sides of the permutation equation above by τ); hence sgn $(\tau\alpha) = (-1)^{n - (t - 1)} = -\,\text{sgn}\,(\alpha)$. ∎

Theorem 3.5 sgn: $S_n \to \{\pm 1\}$ *is a homomorphism.*

Proof Assume $\alpha \in S_n$ is given and $\alpha = \tau_1 \cdots \tau_m$ is a factorization into transpositions with m minimal. We prove by induction on m that sgn $(\alpha\beta)$

$= \text{sgn}(\alpha)\,\text{sgn}(\beta)$ for every $\beta \in S_n$. If $m = 1$, this is precisely Lemma 3.4. If $m > 1$, note that $\tau_2 \cdots \tau_m$ is minimal (if $\tau_2 \cdots \tau_m = \tau_1' \cdots \tau_q'$ with $q < m-1$ and each τ_j' a transposition, then $\tau_1\tau_2 \cdots \tau_m = \tau_1\tau_1' \cdots \tau_q'$ violates the minimality of m). Therefore

$$\begin{aligned}
\text{sgn}(\alpha\beta) &= \text{sgn}(\tau_1\tau_2 \cdots \tau_m\beta) = -\text{sgn}(\tau_2 \cdots \tau_m\beta) \quad \text{(Lemma 3.4)}\\
&= -\text{sgn}(\tau_2 \cdots \tau_m)\,\text{sgn}(\beta) \quad \text{(induction)}\\
&= \text{sgn}(\alpha)\,\text{sgn}(\beta) \quad \text{(Lemma 3.4)}. \quad \blacksquare
\end{aligned}$$

Corollary 3.6 *A permutation α in S_n is even if $\text{sgn}(\alpha) = 1$ and odd if $\text{sgn}(\alpha) = -1$. The number of factors occurring in any factorization of α into transpositions is always even or always odd.*

Proof Assume $\alpha = \tau_1\tau_2 \cdots \tau_q$ is a factorization into transpositions. Since $\text{sgn}(\tau) = -1$ for every transposition τ and sgn is a homomorphism, $\text{sgn}(\alpha) = (-1)^q$. Thus, $\text{sgn}(\alpha) = 1$ implies q is even and hence α is even. If $\text{sgn}(\alpha) = -1$, then α cannot be a product of an even number k of transpositions lest $\text{sgn}(\alpha) = (-1)^k = +1$; hence α is odd. The same argument proves the second statement. $\blacksquare$

Definition The **alternating group** of degree n, denoted by A_n, is the subgroup of S_n consisting of all even permutations.

Theorem 3.7 *For $n \geq 2$, A_n is a normal subgroup of S_n of order $n!/2$.*

Proof sgn is a homomorphism with image $\{\pm 1\}$ and kernel A_n. The first isomorphism theorem says that A_n is a normal subgroup of S_n and that $S_n/A_n \cong \{\pm 1\}$, whence $|A_n| = n!/2$. $\blacksquare$

EXERCISES

*3.18. If $z_1, \cdots, z_n$ are distinct complex numbers, define

$$d = \prod_{i<j}(z_j - z_i)$$

[when $f(x)$ is a polynomial of degree n with roots $z_1, \cdots, z_n$, d^2 is called the **discriminant** of $f(x)$.] If σ is a permutation of $\{z_1, \cdots, z_n\}$, prove that $\prod_{i<j}(\sigma z_j - \sigma z_i) = \pm d$; moreover, the product is equal to $+d$ if and only if σ is even. [HINT: Let **2** denote the multiplicative group $\{1, -1\}$; define $\varphi: S_n \to \mathbf{2}$ by $\varphi(\sigma) = \pm 1$ according as the product above is $\pm d$. Show that φ is a homomorphism onto **2** whose kernel must be A_n.]

3.19. An $n \times n$ **permutation matrix is a matrix obtained from the $n \times n$ identity matrix E by permuting its columns. If P_n is the set of all $n \times n$ permutation matrices, prove that P_n is a multiplicative group and that $\theta: S_n \to P_n$ is an isomorphism, where $\theta(\alpha)$ is the matrix obtained from E by permuting its columns according to α. Prove that α is even (or odd) if and only if $\det \theta(\alpha)$ is 1 (or -1).

3.20. An r-cycle is even if and only if r is odd.

**3.21. If $n > 2$, then A_n is generated by the 3-cycles. [HINT: $(ij)(jk) = (ijk)$ and $(ij)(km) = (ij)(jk)(jk)(km)$.]

3.22. If a subgroup G of S_n contains an odd permutation, then $|G|$ is even and exactly half the elements of G are odd permutations.

3.23. Imbed S_n as a subgroup of A_{n+2}, but show that for $n \geq 2$, S_n cannot be imbedded in A_{n+1}.

**3.24. Prove that S_n can be generated by (12), (13), $\cdots$, $(1n)$.

3.25. Prove that S_n can be generated by (12), $\cdots$, $(i, i+1)$, $\cdots$, $(n-1, n)$.

*3.26. Prove that S_n can be generated by two elements (12) and $(12 \cdots n)$.

CONJUGATES

In this section we study conjugates and conjugacy classes, first for arbitrary groups and then for the special case of symmetric groups.

We begin by defining another equivalence relation.

Definition   If G is a group, the relation "x is a conjugate of y in G" is an equivalence relation on G; the equivalence classes are called **conjugacy classes**.

As an example, if G is the multiplicative group of all $n \times n$ nonsingular matrices over a field, then two matrices lie in the same conjugacy class if and only if they are similar.

Now x and y lie in the same conjugacy class if there is an element $a \in G$ with $y = axa^{-1} = x^a$. There is thus an isomorphism $\gamma: G \to G$ (namely, conjugation by a) with $y = \gamma(x)$. It follows that all elements in the same conjugacy class have the same order. An interesting consequence is that, for any two elements a and b in G, ab and ba have the same order.

An element $x \in G$ is the sole resident of its conjugacy class if $x = axa^{-1}$ for all $a \in G$, i.e., x commutes with every element in G. In an abelian group, therefore, conjugacy classes are not of much interest.

Definition   The **center** of G, denoted by $Z(G)$, is the set of all $x \in G$ that commute with every element in G.

EXERCISES

3.27. $Z(G)$ is a normal, abelian subgroup of G.

**3.28. If G is a nonabelian group, then $G/Z(G)$ is not cyclic.

**3.29. A subgroup S of G is a normal subgroup of G if and only if S is a (set theoretic) union of conjugacy classes of G.

*3.30.   Let G be a group containing an element of finite order $n > 1$ and exactly
two conjugacy classes. Prove that G is a cyclic group of order 2.

It is very useful to count the number of elements in a conjugacy class. To this
end, we introduce the following subgroup.

Definition   If $x \in G$, the **centralizer** of x in G, denoted by $C_G(x)$, is the set of all $a \in G$
that commute with x.

It is immediate that $C_G(x)$ is a subgroup of G. When the meaning is clear from
the context, we shall abbreviate $C_G(x)$ by $C(x)$.

Theorem 3.8   *The number of conjugates of x in G is $[G : C_G(x)]$, and hence this
number is a divisor of $|G|$ when G is finite.*

Proof   If a and b are elements of G, then the following statements are
equivalent:

$axa^{-1} = bxb^{-1}$;
$a^{-1}b$ commutes with x;
$a^{-1}b \in C(x)$;
a and b lie in the same left coset of $C(x)$, i.e., $aC(x) = bC(x)$.

The function ψ, defined by $\psi(axa^{-1}) = aC(x)$, is thus a well defined one-one
correspondence between the set of distinct conjugates of x and the left cosets of
$C(x)$.   ∎

We return to symmetric groups, and we ask when two permutations are
conjugate in S_n.

Lemma 3.9   *If α and β are in S_n, then $\alpha\beta\alpha^{-1}$ is the permutation which has
the same cycle structure as β and that is obtained by applying α to the symbols in β.*

Proof   Let $\beta = \gamma_1\gamma_2 \cdots (\cdots ij \cdots) \cdots \gamma_t$ be a factorization of β into
disjoint cycles. To prove that two permutations are equal, we must show that
they have the same effect on each symbol. We have pictured β so that $\beta(i) = j$.
Suppose $\alpha(i) = k$ and $\alpha(j) = m$. The instructions in the lemma say that we
should send k into m. On the other hand, $\alpha\beta\alpha^{-1}$ sends $k \mapsto i \mapsto j \mapsto m$, so that
both permutations have the same effect on k. Since $k = \alpha(i)$ is a typical integer
between 1 and n (α is onto), the two permutations are equal.   ∎

Example   If $\beta = (13)(247)$ and $\alpha = (256)(143)$, then $\alpha\beta\alpha^{-1} = (\alpha1\ \alpha3)(\alpha2\ \alpha4\ \alpha7)$
$= (41)(537)$.

Theorem 3.10 α *and* β *in* S_n *are conjugate in* S_n *if and only if they have the
same cycle structure.*

Proof Our lemma shows that conjugate permutations have the same cycle structure.

Conversely, let α and β have the same cycle structure. Define $\gamma \in S_n$ as follows: place α over β so that the cycles of the same length correspond; let γ: top $\to$ bottom. By the lemma, $\gamma\alpha\gamma^{-1} = \beta$, so that α and β are conjugate. ∎

For example, if

$$\alpha = (231)(45)(6)$$

and

$$\beta = (462)(31)(5),$$

then $\gamma\alpha\gamma^{-1} = \beta$, where

$$\gamma = \begin{pmatrix} 1 & 2 & 3 & 4 & 5 & 6 \\ 2 & 4 & 6 & 3 & 1 & 5 \end{pmatrix} = (1\ 2\ 4\ 3\ 6\ 5).$$

If $1 < k \leq n$, then there are $(1/k)[n(n-1) \cdots (n-k+1)]$ distinct k-cycles in S_n. This formula may be used to compute the number of permutations α in S_n of a given cycle structure if one is careful about the case when several factors of α have the same length. For example, the number of permutations in S_4 of the form $(ab)(cd)$ is

$$\frac{1}{2}\left(\frac{4 \times 3}{2} \times \frac{2 \times 1}{2}\right),$$

the factor $\frac{1}{2}$ occurring so that we do not count $(ab)(cd) = (cd)(ab)$ twice.

Let us now examine S_4 in some detail, using the accompanying table.

| | S_4 | | |
Cycle Structure	Number	Order	Parity
(1)	1	1	Even
(12)	$6 = (4 \times 3)/2$	2	Odd
(123)	$8 = (4 \times 3 \times 2)/3$	3	Even
(1234)	$6 = 4!/4$	4	Odd
(12)(34)	$3 = \frac{1}{2}\left(\dfrac{4 \times 3}{2} \times \dfrac{2 \times 1}{2}\right)$	2	Even
	$24 = 4!$		

Thus, the 12 elements of A_4 are eight 3-cycles, three products of disjoint transpositions, and the identity. These elements are: (123), (132), (234), (243), (341), (314), (412), (421), (14)(23), (12)(34), (13)(24), and the identity 1. Observe that the identity and the three elements of order 2 in A_4 comprise an abelian subgroup of S_4 of order 4; this subgroup is called the **4-group** and is denoted by **V**.

We now show that the converse of Lagrange's theorem is false.

Theorem 3.11 A_4 *is a group of order* 12 *having no subgroup of order* 6.

Proof† If H is such a subgroup, then $[A_4:H] = 2$ and $\alpha^2 \in H$ for every $\alpha \in A_4$, by Exercise 2.9. In particular, for every 3-cycle α, we have $\alpha = \alpha^4 = (\alpha^2)^2 \in H$, and this gives eight elements in H, a contradiction. ∎

It is now easy to give an example showing that the equality in Theorem 2.20 may be false when neither subgroup is normal. Let S and T be cyclic subgroups of A_4 of orders 2 and 3, respectively. Then $|ST| = |S||T|/|S \cap T| = 6$, but $|S \vee T| = 12$ since A_4 has no subgroup of order 6.

EXERCISES

**3.31. Prove that A_4 is the only subgroup of S_4 having order 12.

**3.32. Let G be a finite group with subgroup H of index 2. If $x \in H$ has m conjugates in G, then x has either m or $m/2$ conjugates in H. [HINT: Use the product formula (Exercise 2.11) or the second isomorphism theorem and the observation $C_H(x) = C_G(x) \cap H$.]

3.33. If $x \in G$ has k conjugates and a power of x, say, x^n, has m conjugates, then m is a divisor of k. [HINT: $C_G(x) \subset C_G(x^n)$.]

**3.34. For every x and $a \in G$, show that $C_G(axa^{-1}) = aC_G(x)a^{-1}$.

THE SIMPLICITY OF A_n

We shall prove that A_n is simple when $n \geq 5$. Since the 4-group **V** contains all the permutations in S_4 of a given cycle structure, **V** is a normal subgroup of S_4, *a fortiori*, **V** is a normal subgroup of A_4. Therefore, A_4 is not simple. Let us now examine S_5 and A_5.

Cycle Structure	S_5 Number	Order	Parity
(1)	1	1	Even
(12)	$10 = (5 \times 4)/2$	2	Odd
(123)	$20 = (5 \times 4 \times 3)/3$	3	Even
(1234)	$30 = (5 \times 4 \times 3 \times 2)/4$	4	Odd
(12345)	$24 = 5!/5$	5	Even
(12)(34)	$15 = \dfrac{1}{2}\left(\dfrac{5 \times 4}{2} \times \dfrac{3 \times 2}{2}\right)$	2	Even
(123)(45)	$20 = \dfrac{5 \times 4 \times 3}{3} \times \dfrac{2 \times 1}{2}$	6	Odd

$$120 = 5!$$

† This proof is due to T.-L. Sheu.

Cycle Structure	A_5 Number	Order	Parity
(1)	1	1	Even
(123)	20	3	Even
(12345)	24	5	Even
(12)(34)	15	2	Even
	60		

Lemma 3.12 A_5 *is simple.*

Proof

(i) *All 3-cycles are conjugate in A_5.* (We know this is true in S_5, but now we are only allowed to conjugate by even permutations.)

In S_5 a 3-cycle α has 20 conjugates (for our table shows that there are 20 3-cycles). Hence, $C_S(\alpha)$ has index 20 and order 6. We can exhibit those 6 elements that commute with α: if, for example, $\alpha = (1\,2\,3)$, then the elements of $C_S(\alpha)$ are 1, (123), (132), (45), (123)(45), (132)(45). Now only the first three of these are even, so that $C_A(\alpha)$ has order 3 and hence index 20 in A_5. Therefore, α has 20 conjugates in A_5, so that all 3-cycles are conjugate in A_5.

(ii) *All products of disjoint transpositions are conjugate in A_5.*

If, for example, $\alpha = (12)(34)$, then our table shows that α has 15 conjugates in S_5. By Exercise 3.32, α has either 15 or 15/2 conjugates in A_5, and the latter is clearly impossible.

(iii) *There are two conjugacy classes of 5-cycles in A_5, each of which has 12 elements.*

In S_5, $\alpha = (12345)$ has 24 conjugates, so that $C_S(\alpha)$ has 5 elements, and these must be the powers of α. Therefore, $|C_A(\alpha)| = 5$, and so the number of conjugates in A_5 is 60/5 = 12.

We have now surveyed all the conjugacy classes occurring in A_5. If H is a normal subgroup of $A_5 \neq \{1\}$, then H is a union of conjugacy classes of A_5. The order of H is thus a sum of certain of the numbers 1, 20, 15, 12, 12. Since H contains 1, it is easily checked that no such sum is a proper divisor of 60. Therefore, $|H| = 60$ and $H = A_5$; it follows that A_5 is simple. ∎

Lemma 3.13   *Let $H \lhd A_n$, where $n \geq 5$. If H contains a 3-cycle, then $H = A_n$.*

Proof   We shall first show that (123) and $(i\,j\,k)$ are conjugate in A_n (and thus that all 3-cycles are conjugate in A_n). If these cycles are not disjoint, at most 5 symbols are moved by them. Let $A_* \subset A_n$ be the alternating group on these symbols (if the cycles move less than 5 symbols, just supplement them to get 5 symbols). As in part (i) of the proof of Lemma 3.12, (123) and (ijk) are conjugate in A_*; *a fortiori*, they are conjugate in A_n. If the cycles are disjoint, then we have just seen that (123) is conjugate to $(i\,2\,3)$ and $(i\,2\,3)$ is conjugate to (ijk); therefore, (123) is conjugate to (ijk).

Since H is normal in A_n, it must contain all 3-cycles once it contains one of them. But A_n is generated by the 3-cycles; therefore, $H = A_n$. ∎

For later use, we give a complete description of S_6, even though we need only a bit of it for the next lemma.

	Cycle Structure	S_6 Number	Order	Parity
C_1	(1)	1	1	Even
C_2	(12)	15	2	Odd
C_3	(123)	40	3	Even
C_4	(1234)	90	4	Odd
C_5	(12345)	144	5	Even
C_6	(123456)	120	6	Odd
C_7	(12)(34)	45	2	Even
C_8	(12)(345)	120	6	Odd
C_9	(12)(3456)	90	4	Even
C_{10}	(12)(34)(56)	15	2	Odd
C_{11}	(123)(456)	40	3	Even
		720 = 6!		

Cycle Structure	A_6 Number	Order	Parity
(1)	1	1	Even
(123)	40	3	Even
(12345)	144	5	Even
(12)(34)	45	2	Even
(12)(3456)	90	4	Even
(123)(456)	40	3	Even
	360		

Lemma 3.14 A_6 *is a simple group.*

Proof Let $H \neq \{1\}$ be a normal subgroup of A_6 and let $\alpha \in H$ be distinct from 1. Suppose $\alpha(i) = i$ for some i. If $F(i) = \{\beta \in A_6 : \beta(i) = i\}$, then $F(i) \cong A_5$ and $\alpha \in H \cap F(i)$. Since A_5 is simple and $H \cap F(i) \lhd F(i)$ (second isomorphism theorem), $H \cap F(i) = F(i)$, i.e., $F(i) \subset H$. Therefore H contains a 3-cycle, $H = A_6$ by Lemma 3.13, and we are done.

We may now assume no α in H (other than 1) fixes any i, $1 \leq i \leq 6$. A glance at the table of A_6 shows that α has cycle structure $\alpha = (12)(3456)$ or $\alpha = (123)(456)$. In the first case, $\alpha^2 \neq 1$ lies in H and fixes 1 and 2, a contradiction. In the second case, take $\beta = (234)$; then $\alpha(\beta\alpha\beta^{-1}) \in H$, it is not 1, and it fixes 1, another contradiction. ∎

Theorem 3.15 A_n is a simple group for all $n \geq 5$.

Proof Let $H \neq \{1\}$ be a normal subgroup of A_n, where $n \geq 5$. If $\beta \in H$ and $\beta \neq 1$, there is a symbol i moved by β:$\beta\{(i) = j \neq i$. If α is any 3-cycle that fixes i and moves j, then α and β do not commute: $\beta\alpha(i) = \beta(i) = j$ and $\alpha\beta(i) = \alpha(j) \neq j$. Therefore the commutator $\alpha\beta\alpha^{-1}\beta^{-1} \neq 1$; moreover, $(\alpha\beta\alpha^{-1})\beta^{-1}$ lies in the normal subgroup H and is a product $\alpha(\beta\alpha^{-1}\beta^{-1})$ of two 3-cycles. Thus, $\alpha(\beta\alpha^{-1}\beta^{-1})$ moves at most 6 symbols, say, $i_1, \cdots, i_6$. If $F = \{\beta \in A_n: \beta$ fixes the complement of $\{i_1, \cdots, i_6\}\}$, then $F \cong A_6$ and $\alpha\beta\alpha^{-1}\beta^{-1} \in H \cap F$ F. Since A_6 is simple, $H \cap F = F$, whence $F \subseteq H$. Therefore H contains a 3-cycle, $H = A_n$ by Lemma 3.13, and we are done. ∎

EXERCISES

*3.35.   If $n \neq 4$, A_n is the only proper normal subgroup of S_n.

*3.36.   Consider the chain of subgroups:

$$S_4 \supset A_4 \supset \mathbf{V} \supset W \supset \{1\},$$

where $W = \langle (12)(34) \rangle$. We have already seen that $\mathbf{V}$ (the 4-group) is normal in A_4. Show that W is normal in $\mathbf{V}$, but that W is not normal in A_4 Conclude that normality need not be transitive.

3.37.   A group G is called **centerless if $Z(G) = \{1\}$. Prove that S_n is centerless for $n \geq 3$ and that A_n is centerless for $n \geq 4$.

3.38.   The following is an example of an infinite simple group. Let X be the set of positive integers and let S be the group of all permutations of X; let F be the subgroup of S consisting of all α that move only finitely many elements of X.

The subgroup of F generated by all 3-cycles, denoted by A_∞, is the **infinite alternating group**. Prove that A_∞ is an infinite simple group. (HINT: Use the proof of Theorem 3.15.)

SOME REPRESENTATION THEOREMS

A very useful technique in studying a group is to represent it in terms of something familiar and concrete. If, for example, the elements of a group happen to be permutations or matrices, we may be able to obtain results by using this extra information. In this section, we give some elementary theorems on representations, i.e., on homomorphisms into familiar groups.

The following theorem shows that the study of groups of permutations is no less general than the study of arbitrary groups.

Theorem 3.16 **(Cayley, 1878)** *Every group G is isomorphic to a subgroup of S_G. In particular, every finite group of order n is isomorphic to a subgroup of S_n.*

Proof For each $a \in G$, define $L_a : G \to G$ by $L_a(x) = ax$ (L_a is left translation by a). In Exercise 1.5, we saw that L_a is a one-one correspondence, so that $L_a \in S_G$. The function λ sending a into L_a is thus a function from G to S_G.

We claim that λ is one-one and a homomorphism. If $a \neq b$ are elements of G, then $L_a(1) = a \neq b = L_b(1)$, so that $L_a \neq L_b$ and $\lambda(a) \neq \lambda(b)$; therefore, λ is one-one. Finally, consider $\lambda(ab) = L_{ab}$ and $\lambda(a)\lambda(b) = L_a L_b$. To show that these permutations are the same, we must show that they assign the same value to each $x \in G$. But $L_{ab}(x) = (ab)x$, and $L_a L_b(x) = L_a(bx) = a(bx)$, and these are the same by associativity. ∎

EXERCISES

3.39. Prove that A_∞ contains an isomorphic copy of every finite group.

3.40. If G is a finite group, each L_a in the above proof is a regular permutation. (HINT: The orbits of L_a are the right cosets of $\langle a \rangle$ in G.) For this reason, the homomorphism λ is called the **(left) regular representation** of G.

3.41. Why doesn't the proof of Cayley's theorem show that every cancellation semigroup can be imbedded in a group?

3.42. For $a \in G$, define **right translation** $R_a : G \to G$ by $x \mapsto xa^{-1}$, and define $\rho : G \to S_G$ by $\rho(a) = R_a$ (ρ is called the **(right) regular representation** of G). Show that ρ is a one-one homomorphism.

3.43. Prove that if λ and ρ are the left and right regular representations of G, then $\lambda(a)$ commutes with $\rho(b)$ for all $a, b \in G$.

*3.44. Prove that every finite group can be imbedded in a group that can be generated by two elements.

*3.45. Let G be a group of order $2^m k$, where k is odd. If G contains an element of order 2^m, then the set of all elements in G having odd order is a (normal) subgroup of G. (HINT: Consider G as permutations via Cayley's theorem and show that it contains an odd permutation.) Conclude that a finite nonabelian simple group of even order must have order divisible by 4.

Corollary 3.17 *Let G be a finite group of order n and let F be a field. Then G can be imbedded in $GL(n, F)$, the multiplicative group of all nonsingular $n \times n$ matrices with entries in F.*

Proof By Cayley's theorem, there is an imbedding of G into S_n. By Exercise 3.19, there is an isomorphism $S_n \cong P_n$, the multiplicative group of all $n \times n$

permutation matrices. Since the entries of a permutation matrix are only 0 and 1, and since it is only the multiplicative properties of 0 and 1 that are needed in establishing the isomorphism of Exercise 3.19, we may regard 0 and 1 as lying in the field F. ∎

Theorem 3.18 *Let B be a subgroup of index n in a group G. There is a homomorphism $\rho: G \to S_n$ whose kernel is contained in B. Indeed, $\ker \rho = \bigcap_{x \in G} xBx^{-1}$.*

Proof Let X be the family of all left cosets of B in G. If $a \in G$, define a function $\rho_a: X \to X$ by $gB \mapsto agB$, for all g. It is easy to check that each ρ_a is a permutation and that $\rho: G \to S_X \cong S_n$ defined by $a \mapsto \rho_a$ is a homomorphism.

If $a \in \ker \rho$, then ρ_a is the identity permutation, i.e., $agB = gB$ for all $g \in G$; thus $g^{-1}ag \in B$ and $a \in gBg^{-1}$ for all $g \in G$, and so $\ker \rho \subset \bigcap_{g \in G} gBg^{-1}$. The reverse inclusion is easy to prove and is left to the reader. ∎

It follows immediately that $\bigcap_{x \in G} xBx^{-1}$ is a normal subgroup of G, but this also follows easily from the definitions.

Definition The homomorphism of Theorem 3.18 is called the **representation of G on the cosets of B**.

Corollary 3.19 *If an infinite group G contains a proper subgroup B of finite index, then G contains a proper normal subgroup of finite index.*

Proof Let B have index n. By Theorem 3.18, there is a homomorphism $\rho: G \to S_n$ whose kernel K is contained in B. Therefore, $K \lhd G$, and since G/K is isomorphic to a subgroup of the finite group S_n, K has finite index in G. ∎

Corollary 3.20 *A simple group G containing a proper subgroup B of index n can be imbedded in S_n.*

Proof By Theorem 3.18, there is a homomorphism $\rho: G \to S_n$ whose kernel K is contained in B. Since G is simple, $K = \{1\}$ or $K = G$; since B is proper, $K \subset B \neq G$. Therefore, $K = \{1\}$ and ρ is one-one. ∎

Corollary 3.20 provides a substantial improvement over Cayley's theorem, at least for simple groups. For example, if $G \cong A_5$, then Cayley's theorem asserts that G can be imbedded in S_{60}. On the other hand, G contains a subgroup B of order 12 (isomorphic to A_4) and hence of index 5; therefore, Corollary 3.20 asserts that G can be imbedded in S_5.

EXERCISES

3.46. (**Poincaré**) If H and K are subgroups of G having finite index, then $H \cap K$ has finite index in G. (HINT: $[G:H \cap K] \leq [G:H][G:K]$.)

3.47. Let G be a finite group containing a subgroup H with $[G:H] = p$, where p is the smallest prime divisor of $|G|$. Prove that H is normal in G.

3.48. Prove there exists a finite group having no subgroups of prime index. (HINT: Try A_6.)

3.49. (i) Let H be a proper subgroup of G. Prove there is a group K and distinct homomorphisms $f, g : G \to K$ with $f|H = g|H$. (HINT: Let X be the family of all left cosets of H in G together with an additional element denoted by ∞, and let $K = S_X$. Define $f : G \to S_X$ as follows: If $t \in G$, then $f_t(\infty) = \infty$ and $f_t(aH) = taH$. Define $g : G \to S_X$ by $g = \beta \circ f$, where $\beta : S_X \to S_X$ is conjugation by the transposition that interchanges H and ∞.)

(ii) A homomorphism $h : A \to G$ is onto if and only if for every group K and every pair of homomorphisms $f, g : G \to K$, $f \circ h = g \circ h$ implies $f = g$.

One may conjugate subgroups as well as elements.

Definition If H is a subgroup of G and $g \in G$, then the subgroup $gHg^{-1} = \{ghg^{-1} : h \in H\}$ is called the **conjugate** of H by g; one often denotes this subgroup by H^g.

EXERCISES

3.50. A subgroup H of G is normal if and only if it has only one conjugate.

3.51. If H and K are conjugate subgroups of G, then they are isomorphic. Give an example to show the converse may be false.

3.52. If λ and ρ are the left and right regular representations of S_3, then $\lambda(S_3)$ and $\rho(S_3)$ are conjugate subgroups of S_6.

Definition If H is a subgroup of G, its **normalizer** is the subgroup

$$N_G(H) = \{g \in G : gHg^{-1} = H\}.$$

We may write $N(H)$ in place of $N_G(H)$ if no confusion can result.

It is evident that H is a normal subgroup of $N_G(H)$ and the normalizer is the largest subgroup of G in which H is normal.

Definition An $n \times n$ matrix $M = [m_{ij}]$ over a field K is **monomial** if there is $\beta \in S_n$ and (not necessarily distinct) nonzero elements $x_1, \cdots, x_n \in K$ such that

$$m_{ij} = \begin{cases} x_i & \text{if } j = \beta(i) \\ 0 & \text{otherwise.} \end{cases}$$

Monomial matrices thus have only one nonzero entry in any row or column. Of course, a monomial matrix in which each $x_i = 1$ is a permutation matrix.

EXERCISES

3.53. If $G = GL(n, K)$ and T is the subgroup of G of all (nonsingular) diagonal matrices, prove that $N_G(T)$ consists of all monomial matrices over K and that $N_G(T)/T \cong S_n$.

3.54. Let G be a finite group with proper subgroup H. Prove that G is not the set-theoretic union of all the conjugates of H. Give an example in which H is not normal and this union is a subgroup.

**3.55. (i) Assume $H \subset K \subset G$. Show that $N_K(H) = N_G(H) \cap K$.
(ii) Prove that $N_G(xHx^{-1}) = xN_G(H)x^{-1}$.

3.56. If H and K are subgroups of G, then $N_G(H \cap K) \supset N_G(H) \cap N_G(K)$; give an example in which the inclusion is proper.

Here is another permutation representation.

Theorem 3.21 *Let H be a subgroup of G and let X be the set of all the conjugates of H in G. There is a homomorphism $\varphi: G \to S_X$ whose kernel is contained in $N_G(H)$. Indeed, $\ker \varphi = \bigcap_{a \in G} aN_G(H)a^{-1}$.*

Proof If $a \in G$, define $\varphi_a : X \to X$ by $\varphi_a(gHg^{-1}) = agHg^{-1}a^{-1}$. If $b \in G$, then

$$\varphi_a\varphi_b(gHg^{-1}) = \varphi_a(bgHg^{-1}b^{-1}) = abgHg^{-1}b^{-1}a^{-1} = \varphi_{ab}(gHg^{-1}).$$

We conclude that φ_a has inverse $\varphi_{a^{-1}}$, so that $\varphi_a \in S_X$. Furthermore, the equation above gives $\varphi_a\varphi_b = \varphi_{ab}$, so that $\varphi: G \to S_X$ defined by $\varphi(a) = \varphi_a$ is a homomorphism. We leave the computation of $\ker \varphi$ to the reader. ∎

Definition The homomorphism φ of Theorem 3.21 is called the **representation of G on the conjugates of H.**

EXERCISE

3.57. Let G be an infinite group containing an element $x \neq 1$ having only finitely many conjugates. Prove that G is not simple.

COUNTING ORBITS

We have encountered several situations in which elements of a group may be regarded as permutations of a set.

Definition A group G **acts** on a set X if there is a homomorphism $\varphi: G \to S_X$. If $a \in G$, write φ_a for $\varphi(a) \in S_X$ (so that $\varphi_a : X \to X$) and, if $x \in X$, write ax instead of $\varphi_a(x)$.

[The early mathematicians who studied group-theoretic problems, e.g., Lagrange, were concerned with the following question. Assume $f(x_1, \cdots, x_n)$ is a function of n variables; what happens when the variables are permuted? If $f(x_{\sigma 1}, \cdots, x_{\sigma n}) = f(x_1, \cdots, x_n)$ for all $\sigma \in S_n$, then f is called a **symmetric function**. Such functions arise, for example, in the study of polynomials; if $g(x) = x^n + a_{n-1} x^{n-1} + \cdots + a_0$ has the factorization $g(x) = \prod_{i=1}^{n} (x - \alpha_i)$, then expanding this product shows each of the coefficients of $g(x)$ is a symmetric function of the roots $\alpha_1, \cdots, \alpha_n$. Other functions of the roots are not quite symmetric. For example, the function d of Exercise 3.18 (whose square is the discriminant) satisfies $d(x_{\sigma 1}, \cdots, x_{\sigma n}) = \pm \, d(x_1, \cdots, x_n)$ for all $\sigma \in S_n$. Indeed, it is **alternating**, i.e., its value is unchanged for $\sigma \in A_n$ while its sign changes otherwise. This suggests a slight change in viewpoint. Given f, find all $\sigma \in S_n$ for which $f(x_{\sigma 1}, \cdots, x_{\sigma n}) = f(x_1, \cdots, x_n)$. It is easy to see that the set of all such permutations, for a given function f, is a subgroup of S_n; when f is symmetric, the subgroup is S_n; when f is alternating, the subgroup is A_n. This same type of problem concerns modern mathematicians. Suppose X is a set on which a group G acts. Then the set of all $f: X \to X$ such that $f(\sigma x) = f(x)$ for all $x \in X$ and all $\sigma \in G$ is often valuable in analyzing X. Indeed, this is the idea underlying Klein's *Erlangen Program* (1872) which classifies different forms of geometry.]

Example 1   If G is a subgroup of S_X, then the inclusion homomorphism $i: G \to S_X$ shows that G acts on X.

Example 2   Cayley's theorem shows that G acts on itself by left translation; the left regular representation $\lambda: G \to S_G$ is a homomorphism. Similarly, the right regular representation $\rho: G \to S_G$ shows another way in which G acts on itself.

Example 3   If H is a subgroup of G, then H acts on G by left translation: just restrict the left regular representation $\lambda: G \to S_G$ to H. Thus, if $h \in G$, then $\varphi_h: g \mapsto hg$ for every $g \in G$.

Example 4   Theorem 3.18 shows that if H is a subgroup of G and $G /\!/ H$ is the set of all left cosets of H in G, then G acts on $G /\!/ H$. Note that this is the left regular representation when $H = \{1\}$.

Example 5   Theorem 3.21 shows that if H is a subgroup of G and X is the set of all conjugates of H, then G acts on X.

Example 6   If X is the set of all subgroups of G, then it is easy to see (as in Theorem 3.21) that G acts on X by conjugation: if S is a subgroup of G and $a \in G$, define $\varphi_a(S) = aSa^{-1}$.

Example 7   The reader may verify (as in Theorem 3.21) that every group G acts on itself by conjugation: if $a \in G$, then $\varphi_a(x) = axa^{-1}$ for every $x \in G$.

There are two fundamental ideas in this situation.

Definition   If G acts on X and $x \in X$, the **orbit** of x is $\{gx: g \in G\}$.

It is easy to see that the orbits form a partition of X. In fact, one may define an equivalence relation on X by $x_1 \sim x_2$ if $x_2 = gx_1$ for some $g \in G$, and the equivalence classes of this relation are the orbits.

Definition   If G acts on X and $x \in X$, then the **stabilizer** of x, denoted by G_x, is the subgroup

$$G_x = \{g \in G: gx = x\}.$$

Let us see some examples of orbits and stabilizers.

Example 8   Let α be a permutation of X and let $\langle \alpha \rangle \subset S_X$. The orbit of $x \in X$ is $\{\alpha^i x: i \in \mathbf{Z}\}$ and thus consists of the elements involved in one of the cycles in the cycle decomposition of α.

Example 9   In the left regular representation of G, there is only one orbit (if $x, y \in G$, there is $g \in G$ with $y = gx$); also, the stabilizer of any $x \in G$ is $\{1\}$.

Example 10   If a subgroup H acts on G by left translation, its orbits are the right cosets of H.

Example 11   If G acts on its subgroups by conjugation, then the orbit of a subgroup H is the set of its conjugates and the stabilizer of H is its normalizer $N_G(H)$.

Example 12   If G acts on itself by conjugation, then the orbit of $x \in X$ is its conjugacy class and the stabilizer of x is its centralizer $C_G(x)$.

Theorem 3.22   *Let G act on a set X. Then the number of elements in the orbit of $x \in X$ is $[G: G_x]$.*

Proof   Observe first that the following statements are equivalent: $gx = hx$; $g^{-1}hx = x$; $g^{-1}h \in G_x$; $gG_x = hG_x$. If O_x denotes the orbit $\{gx: g \in G\}$ and $G /\!/ G_x$ denotes the family of left cosets of G_x in G, it follows that $\beta: O_x \to G /\!/ G_x$ given by $\beta(gx) = gG_x$ is a well defined function that is one-one. Since β is plainly onto, it is a one-one correspondence; this proves the desired equality.   ∎

Corollary 3.23   *If a finite group G acts on a finite set X, then the number of elements in any orbit is a divisor of $|G|$.*   ∎

Corollary 3.24 (= Theorem 3.8)   *If G is a finite group and $x \in G$, then the number of conjugates of x is $[G: C_G(x)]$.*

Proof We saw in Example 12 that when G acts on itself by conjugation, the orbit of $x \in G$ is its conjugacy class and the stabilizer of x is $C_G(x)$. ∎

Corollary 3.25 *If H is a subgroup of a finite group G, then the number of conjugates of H in G is $[G: N_G(H)]$.*

Proof We saw in Example 11 that when G acts on its subgroups by conjugation, the orbit of H is the set of its conjugates and the stabilizer of H is $N_G(H)$. ∎

Example 13 Let Σ_n be the regular solid having n (congruent) faces, each of which has k edges, and let $G(\Sigma_n)$ be the group of all rotations of Σ_n. Regarding $G(\Sigma_n)$ as acting on the n faces of Σ_n, we see that congruence of the faces implies there is only one orbit. Moreover, the stabilizer of a face β can consist only of the k rotations of β about its center. We conclude from Theorem 3.22 that $G(\Sigma_n)$ has order nk.

It is a classical result that there are only five regular solids: Σ_4, the tetrahedron having 4 triangular faces; Σ_6, the cube having 6 square faces; Σ_8, the octahedron having 8 triangular faces; Σ_{12}, the dodecahedron having 12 pentagonal faces; Σ_{20}, the icosahedron having 20 triangular faces. The rotation groups of these solids thus have orders 12, 24, 24, 60, and 60, respectively.

EXERCISE

*3.58. (**Hamilton–von Dyck**)† (i) Prove that $G(\Sigma_4) \cong A_4$.
(ii) Prove that $G(\Sigma_6) \cong S_4 \cong G(\Sigma_8)$.
(iii) Prove that $G(\Sigma_{12}) \cong A_5 \cong G(\Sigma_{20})$.
[HINT: A_4 is generated by elements s and t such that $s^2 = t^3 = (st)^3 = 1$; S_4 is generated by elements s, t with $s^2 = t^3 = (st)^4 = 1$; A_5 is generated by elements s, t with $s^2 = t^3 = (st)^5 = 1$.] Because of the isomorphisms above, A_4 is also called the **tetrahedral group**, S_4 is called the **octahedral group**, and A_5 is called the **icosahedral group**.

Theorem 3.26 (Cauchy-Frobenius)†† *Let a group G act on a finite set X. If N is the number of orbits of G, then*

$$N = \frac{1}{|G|} \sum_{t \in G} F(t),$$

where $F(t)$ is the number of x in X that are fixed by t.

Proof In the sum $\Sigma_{t \in G} F(t)$, each $x \in X$ is counted $|G_x|$ times. If x and y lie in the same orbit, then Theorem 3.22 implies that $|G_x| = |G_y|$, for $[G:G_x] = [G:G_y]$. Therefore the $[G:G_x]$ elements constituting the orbit of x are, in the above sum, collectively counted $[G:G_x]|G_x| = |G|$ times. Each orbit thus contributes $|G|$ to the sum, so that $\Sigma_{t \in G} F(t) = N|G|$. ∎

---

† Hamilton (1856) found the description of A_5; von Dyck (1882) found the descriptions of A_4 and S_4.
†† Frobenius proved this in 1887, but the idea of the proof may be found in work of Cauchy of 1845.

Corollary 3.27 *Let X be a finite set with $|X| > 1$ and let G be a finite group acting transitively on X,* i.e., *there is only one orbit. Then there exists $t \in G$ having no fixed points.*

REMARK Compare Exercise 3.62.

Proof Since G acts transitively on X, the Cauchy-Frobenius theorem gives

$$1 = \frac{1}{|G|} \sum_{t \in G} F(t).$$

Now $F(1) = |X| > 1$; it follows that if $F(t) > 0$ for every $t \in G$, then the right-hand side is too large. ∎

EXERCISES

3.59. Let G be a group of permutations on a set X and let $x, y \in X$. If $t(x) = y$ for some $t \in G$, then G_x and G_y are isomorphic.

3.60. Use Theorem 3.26 to prove Lagrange's theorem.

*3.61. If G is a finite group and c is the number of conjugacy classes in G, then

$$c = \frac{1}{|G|} \sum_{x \in G} |C_G(x)|.$$

*3.62. Let G be a group of order p^n, where p is prime, and assume G acts on a set X with $|X|$ not divisible by p. Prove there exists $x \in G$ with $gx = x$ for all $g \in G$.

3.63. If V is a vector space, let $GL(V)$ be the group of all nonsingular linear transformations from V to itself. Assume V is a finite-dimensional vector space over $\mathbf{Z}/p\mathbf{Z}$ and that G is a subgroup of $GL(V)$ having order p^n. Prove that there is a nonzero vector $v \in V$ with $gv = v$ for all $g \in G$.

Theorem 3.26 was used by G. Pólya (1937) to solve some interesting combinatorial problems in chemistry; we illustrate his technique here. Given q distinct colors, how many striped flags are there having n stripes (of equal width)? Clearly, the two flags below are the same.

If, then, X is the set of all ordered sets of n colored boxes, then the cyclic group $G = \langle t \rangle$ operates on X, where t is the permutation

$$t = \begin{pmatrix} 1 & 2 & \cdots & n \\ n & n-1 & \cdots & 1 \end{pmatrix},$$

and a flag is just an orbit of G. In order to apply Theorem 3.26, we need compute only $F(1)$ and $F(t)$. There are q^n elements in X and each is fixed by the identity; hence $F(1) = q^n$. To compute $F(t)$, we first observe that, since t has order 2, Exercise 3.14 says that t is a product of disjoint transpositions. In fact, t is a product of k disjoint transpositions, where $k = [(n+1)/2]$, the greatest integer in $(n+1)/2$. If $n = 2k$, then $t = (1,n)(2,n-1)\cdots(k,k+1)$; if $n = 2k+1$, then $t = (1,n)(2,n-1)\cdots(k,k+2)$. An ordered set of $n = 2k$ colored boxes is thus fixed by t if and only if $c_1 = c_n, c_2 = c_{n-1}, \cdots, c_k = c_{k+1}$; there is a similar statement if $n = 2k+1$. We conclude that

$$F(t) = q^k = q^{[(n+1)/2]}.$$

The number of flags is thus

$$\tfrac{1}{2}(q^n + q^{[(n+1)/2]}).$$

The next two exercises are counting problems of this type; see also Exercise 4.23.

EXERCISES

3.64. If there are q colors available, prove that there are

$$\tfrac{1}{4}(q^{n^2} + 2q^{[(n^2+3)/4]} + q^{[(n^2+1)/2]})$$

distinct $n \times n$ colored chessboards. (HINT: The group is a cyclic group of order 4 consisting of rotations of $0°, 90°, 180°$, and $270°$. The generator of the group is a permutation of n^2 ordered colored boxes and it is a product of disjoint 4-cycles.)

$n = 4$

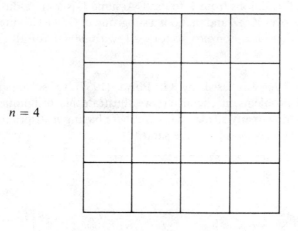

*3.65. If there are q colors available, prove that there are

$$\frac{1}{n}\sum_{d|n} \varphi\left(\frac{n}{d}\right) q^d$$

colored roulette wheels having n compartments, each a circular sector. In this formula, φ is the Euler φ-function, and the summation ranges over all divisors d of n. [HINT: The group is a cyclic group of order n operating by rotating by multiples of $(360/n)°$. Using Exercise 3.12 for $\alpha = (1\ 2 \cdots n)$, show that there are

$$\frac{1}{n} \sum_{0 \le k < n} q^{(n,\ k)}$$

roulette wheels. The desired formula arises from this one by collecting terms having the same exponent.]

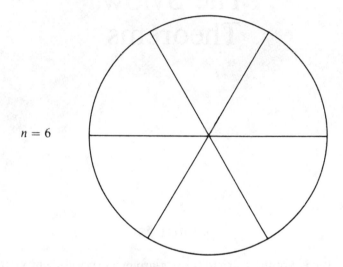

$n = 6$

chapter four

The Sylow Theorems

p-GROUPS

Definition If p is a prime, a group G is a **p-group** if every element $x \in G$ has order some power of p.

We begin with a characterization of finite p-groups (Corollary 4.3).

Lemma 4.1 *If G is a finite abelian group whose order is divisible by a prime p, then G contains an element of order p.*

Proof Let $x \in G$ be distinct from 1. If the order of x is pm, then Exercise 2.13 shows x^m is an element in G of order p.

Suppose that x has order t, where $(p, t) = 1$. Since G is abelian, $\langle x \rangle$ is a normal subgroup and $G/\langle x \rangle$ is an abelian group of order $|G|/t$. Since $|G|/t$ is divisible by p and is less than $|G|$, an induction on $|G|$ provides an element $\bar{y} \in G/\langle x \rangle$ of order p. Hence, if $y \in G$ goes into $\bar{y}$ under the natural map, the order of y is a multiple of p (Exercise 2.17). We have returned to the first case. ∎

Theorem 4.2 (Cauchy, 1845) *If G is a finite group whose order is divisible by a prime p, then G contains an element of order p.*

Proof If $x \in G$, the number of conjugates of x is $[G:C(x)]$, where $C(x)$ is the centralizer of x in G. Now if $x \notin Z(G)$, then $|C(x)| < |G|$, so that if p divides $|C(x)|$, we are done, by induction. Therefore, we may assume that p does not divide $|C(x)|$ for all $x \notin Z(G)$; better, since $|G| = [G:C(x)]|C(x)|$, we may assume that p does divide $[G:C(x)]$ for all $x \notin Z(G)$ (it is here we use the fact that p is prime).

We partition G into its conjugacy classes and count:

$$|G| = |Z(G)| + \sum [G:C(x)],$$

where the summation ranges over a complete set of nonconjugate x not in $Z(G)$. Now p divides both $|G|$ and $\sum [G:C(x)]$, so that p divides $|Z(G)|$. Since $Z(G)$ is an abelian group, it contains an element of order p, by Lemma 4.1. ∎

Definition If G is a finite group, its **class equation** is

$$|G| = |Z(G)| + \sum_i [G:C_G(x_i)],$$

where one x_i is chosen from each conjugacy class having more than one element.

Corollary 4.3 *A finite group G is a p-group if and only if $|G|$ is a power of p.*

Proof Suppose there is a prime $q \neq p$ dividing $|G|$. By Cauchy's theorem, there is an element $a \in G$ of order q, contradicting the fact that G is a p-group.

The converse follows immediately from Lagrange's theorem (Theorem 2.9). ∎

Theorem 4.4 *If G is a finite p-group with more than one element, then $Z(G)$, the center of G, has more than one element.*

Proof Partitioning G into its conjugacy classes gives the class equation:

$$|G| = |Z(G)| + \sum [G:C(x_i)],$$

where one x_i is chosen from each conjugacy class not in $Z(G)$. Since $x_i \notin Z(G)$, we see $C(x_i)$ is a proper subgroup of G, and so, by Corollary 4.3, $[G:C(x_i)]$ is a power of p. Thus p divides each $[G:C(x_i)]$, and therefore p divides $|Z(G)|$. ∎

Corollary 4.5 *If p is a prime, then every group G of order p^2 is abelian.*

Proof Suppose G is not abelian, i.e., $G \neq Z(G)$. Then $Z(G)$ has order p (for this order cannot be 1). Therefore, $G/Z(G)$ has order p and so is cyclic, contradicting Exercise 3.28. ∎

EXERCISES

4.1. Let $H \lhd G$. If both H and G/H are p-groups, then G is a p-group.

4.2. Let $|G| = p^n$, where p is prime. If $0 \leq k \leq n$, prove that G contains a normal subgroup of order p^k.

4.3.   Let G be a finite p-group and let $\{1\} \neq H \lhd G$. Prove that $H \cap Z(G) \neq \{1\}$.

4.4.   If G is a finite p-group, then every normal subgroup of order p is contained in $Z(G)$.

4.5.   Let G be a finite p-group and let H be a proper subgroup of order p^k. Then H can be imbedded in a subgroup of order p^{k+1}. Conclude that every maximal subgroup of G has index p.

Since each term in the class equation of a finite group G is a divisor of $|G|$, multiplying each side by $|G|^{-1}$ gives an equation of the form $1 = \Sigma_j (1/i_j)$ with each i_j a positive integer; moreover, $|G|$ is the maximum value of i_j occurring in this expression.

Lemma 4.6   (Landau, 1903)   *Given $n > 0$ and $q \in \mathbf{Q}$, there are only finitely many n-tuples $(i_1, \cdots, i_n)$ of natural numbers such that $q = \sum_{j=1}^{n} (1/i_j)$.*

Proof   We do an induction on n; the case $n = 1$ is obviously true. There being only $n!$ permutations of n objects, it suffices to prove there are only finitely many n-tuples satisfying the equation $q = \Sigma(1/i_j)$ with $i_1 \leq i_2 \leq \cdots \leq i_n$. Given such an n-tuple, we have $q \leq n/i_1$, for

$$q = \frac{1}{i_1} + \cdots + \frac{1}{i_n} \leq \frac{1}{i_1} + \cdots + \frac{1}{i_1} = \frac{n}{i_1};$$

therefore $i_1 \leq n/q$. But for each natural number $k \leq n/q$, induction shows there are only finitely many $(n-1)$-tuples $(i_2, \cdots, i_n)$ of natural numbers with $\sum_{j=2}^{n} (1/i_j) = q - (1/k)$. This completes the proof, for there are only finitely many such k.   ∎

Although we are interested in the case $q = 1$, we need the stronger result of the lemma to establish the inductive step of the proof.

Theorem 4.7   *For each $n > 0$, there are only finitely many finite groups having exactly n conjugacy classes.*

Proof   Assume G is a finite group with exactly n conjugacy classes. Now $|Z(G)| = m$, say, and the class equation is

$$|G| = |Z(G)| + \sum_{j=m+1}^{n} [G: C_G(x_j)].$$

If we define $i_j = |G|$ for $1 \leq j \leq m$ and $i_j = |G|/[G: C_G(x_j)] = |C_G(x_j)|$ for $m + 1 \leq j \leq n$, then $1 = \Sigma(1/i_j)$. By the lemma, there are only finitely many such n-tuples $(i_1, \cdots, i_n)$, and so there is a maximum value for any i_j, say M. It follows that a finite group with exactly n conjugacy classes has order at most M. This completes the proof, for there are only finitely many groups of any given order.   ∎

THE SYLOW THEOREMS

The major results in this section are basic for understanding the structure of a finite group. We shall prove the existence of largest possible p-subgroups of a finite group G. Now several such subgroups may exist, and we can count the number of them (within a congruence). On the other hand, these subgroups are unique in the sense that any two are isomorphic via a conjugation of G.

Definition Let p be a prime. A subgroup P of G is a **Sylow p-subgroup** of G if it is a maximal p-subgroup of G.

Thus, if P is a Sylow p-subgroup of G and if Q is a p-subgroup of G such that $Q \supset P$, then $Q = P$. Observe that every p-subgroup of G is contained in a Sylow p-subgroup of G. This statement is obvious when G is finite; if G is infinite, this statement may be proved by Zorn's lemma.

EXERCISE

**4.6. Every conjugate of a Sylow p-subgroup of G is itself a Sylow p-subgroup of G. Conclude that if, for some fixed prime p, G has only one Sylow p-subgroup P, then P is normal in G.

Lemma 4.8 *Let P be a Sylow p-subgroup of G.*

(i) *$N_G(P)/P$ has no elements $\neq 1$ whose order is a power of p.*
(ii) *If $a \in G$ has order a power of p, then $aPa^{-1} = P$ implies $a \in P$.*

Proof

(i) Suppose $\bar{x} \in N(P)/P$ has order a power of p. If S^* is the subgroup of $N(P)/P$ generated by $\bar{x}$, then S^* is a p-group. By the correspondence theorem (Theorem 2.23), there is a subgroup S of G containing P such that $S/P \cong S^*$. By Exercise 4.1, S is a p-group containing P, so that the maximality of P implies $S = P$. Therefore, $S^* = \{1\}$ and $\bar{x} = 1$.

(ii) Clearly, $a \in N(P)$. If $\pi: N(P) \to N(P)/P$ is the natural map, then $\pi(a)$ has order a power of p (since a has). By part (i), $\pi(a) = 1$, i.e., $a \in \ker \pi = P$. ∎

We are ready to prove the main theorems of this chapter.

Theorem 4.9 **(Sylow, 1872)** *Let G be a finite group with Sylow p-subgroup P. All Sylow p-subgroups of G are conjugate to P, and the number of these subgroups is a divisor of $|G|$ that is $\equiv 1 \pmod{p}$.*

Proof The basic idea is the realization (Exercise 4.6) that conjugation by any element of G sends a Sylow p-subgroup into a Sylow p-subgroup.

Let X be the set of all the conjugates of P; say,

$$X = \{P_1, P_2, \cdots, P_r\}$$

(where our notation sets $P = P_1$). In Theorem 3.21, we saw that G acts on X by conjugation; there is a homomorphism $\alpha\colon G \to S_X$ sending $a \mapsto \alpha_a$, where $\alpha_a(P_i) = aP_ia^{-1}$. If we restrict α to P, then P acts on X, and Corollary 3.23 shows the size of each orbit is a power of p. Now what does it mean to say that one of these sizes is 1? There would be an i with $\alpha_a(P_i) = P_i$ for all α_a, i.e., $aP_ia^{-1} = P_i$ for all $a \in P$. By Lemma 4.8 (ii), for all $a \in P = P_1$, $a \in P_i$, so that $P \subset P_i$. Since P is a Sylow p-subgroup, we must have $P = P_i$. Conclusion: Each orbit has size an "honest" power of p save $\{P_1\}$, which has size 1. Therefore, $r \equiv 1 \pmod{p}$.

Suppose now that Q is a Sylow p-subgroup of G which is not a conjugate of P, i.e., Q is not a P_i for any i. Restrict the map α to a homomorphism $Q \to S_X$. Again Corollary 3.23 tells us that the size of each orbit of Q is a power of p, and again we ask if any of these have size 1. The same argument as before shows that if there were such an orbit $\{P_i\}$, then $Q = P_i$, and this is contrary to the choice of Q. But now every orbit has size an honest power of p, so that p divides r, i.e., $r \equiv 0 \pmod{p}$. This contradicts our previous congruence, so that no such Q can exist. Therefore, every Sylow p-subgroup of G is conjugate to P.

Finally, the number of conjugates of P is the index of its normalizer, and so it is a divisor of $|G|$. ∎

It follows that the converse of Exercise 4.6 is true: If a group G has a normal Sylow p-subgroup P, then P is the unique Sylow p-subgroup of G.

Theorem 4.10   **(Sylow, 1872)**   *Let G be a finite group of order $p^k m$, where $(p, m) = 1$. Every Sylow p-subgroup P of G has order p^k.*

Proof We first prove that $[G\colon P]$ is prime to p. Since $[G\colon P] = [G\colon N(P)][N(P)\colon P]$, it suffices to prove that each of these factors is prime to p. Now $[G\colon N(P)]$ is the number of conjugates of P, which we have just seen is $\equiv 1 \pmod{p}$. Also, $[N(P)\colon P] = |N(P)/P|$, and $N(P)/P$ has no elements of order p by Lemma 4.8(i); by Cauchy's theorem, $|N(P)/P|$ is prime to p. It follows that $[G\colon P]$ is prime to p.

By Lagrange's theorem, $|P| = p^n$ for $n \leq k$, so that $|G|/|P| = mp^{k-n}$. But $|G|/|P| = [G\colon P]$ is prime to p, so that $k = n$. ∎

Corollary 4.11   *Let G be a finite group and let p be a prime. If p^n divides $|G|$, then G contains a subgroup of order p^n.*

Proof If P is a Sylow p-subgroup of G, then p^n divides $|P|$, by the preceding theorem. By Exercise 4.2, P (hence G) contains a subgroup of order p^n. ∎

We have now seen how much of the converse of Lagrange's theorem can be salvaged. If m divides $|G|$ and m is a power of a prime, then G contains a subgroup of order m; if m has two distinct prime factors, however, we have exhibited a group G, namely A_4, such that m divides $|G|$ and such that G contains no subgroup of order m.

The next theorem will be useful.

Theorem 4.12 (**Frattini argument**) *Let G be a finite group with normal subgroup K. If P is a Sylow p-subgroup of K (for some prime p), then*

$$G = KN_G(P).$$

Proof If $g \in G$, then $gPg^{-1} \subset gKg^{-1} = K$, for $K \lhd G$. It follows that gPg^{-1} is a Sylow p-subgroup of K, so there exists $k \in K$ with $kPk^{-1} = gPg^{-1}$. Therefore $P = (k^{-1}g)P(k^{-1}g)^{-1}$, i.e., $k^{-1}g \in N_G(P)$ and $g \in KN_G(P)$. ∎

EXERCISES

4.7.   Let G be a finite group and, for each prime p, choose a Sylow p-subgroup of G. Prove that G is generated by these subgroups.

**4.8.   Let G be a finite group with a Sylow p-subgroup P. Prove that any subgroup of G that contains $N_G(P)$ (inclusion not necessarily proper) is equal to its own normalizer.

**4.9.   Let G be a finite group, and suppose that, for every prime p, every Sylow p-subgroup of G is normal. Prove that G is the direct product of its Sylow subgroups.

4.10.   Let G be a finite group with normal subgroup H, and let P be a Sylow p-subgroup of G. Prove that $H \cap P$ is a Sylow p-subgroup of H, and that HP/H is a Sylow p-subgroup of G/H. (HINT: Compare orders.)

*4.11.   Prove that a Sylow 2-subgroup of A_5 has exactly 5 conjugates.

*4.12   Let $GL(3, p)$ denote the multiplicative group of all 3 x 3 nonsingular matrices with entries in $\mathbf{Z}/p\mathbf{Z}$. Show that the set of all matrices of the form

$$\begin{bmatrix} 1 & a & b \\ 0 & 1 & c \\ 0 & 0 & 1 \end{bmatrix},$$

where $a, b, c \in \mathbf{Z}/p\mathbf{Z}$, is a Sylow p-subgroup of $GL(3, p)$. (HINT: Count.)

4.13.   Show there is a group G with Sylow p-subgroups A, B, and C (for some prime p) such that $A \cap B = \{1\}$ and $A \cap C \neq \{1\}$. (HINT: Take $G = S_3 \times S_3$.)

4.14.   Prove that every subgroup of S_4 having order 8 contains the 4-group **V**.

SOME APPLICATIONS OF THE SYLOW THEOREMS

We shall now illustrate the power of the Sylow theorems by classifying the groups of small order.

Definition   The **dihedral group** D_n, $n \geq 2$, is a group of order 2n generated by two elements s and t that satisfy the relations

$$s^n = 1, \quad t^2 = 1, \quad \text{and} \quad tst = s^{-1}.$$

EXERCISES

4.15.   Let A be a regular polygon with vertices $v_1, v_2, \cdots, v_n$, and let G be the set of all rigid motions of A that send vertices into vertices. In particular, let S be a clockwise rotation sending each vertex into the adjacent one, and let T be a reflection of A about the line joining v_1 with the center of A. As usual, multiply in G by first performing one motion and then the other. Prove that $G \cong D_n$.

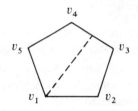

*4.16.   Prove that a Sylow 2-subgroup of S_5 is isomorphic to D_4 and that a Sylow 2-subgroup of S_6 is isomorphic to $D_4 \times \mathbf{Z}(2)$.

4.17.   Let $\omega = e^{2\pi i/n}$ be a primitive nth root of unity. Prove that the matrices

$$A = \begin{bmatrix} \omega & 0 \\ 0 & \omega^{-1} \end{bmatrix} \quad \text{and} \quad B = \begin{bmatrix} 0 & 1 \\ 1 & 0 \end{bmatrix}$$

generate a subgroup of $GL(2, \mathbf{C})$ isomorphic to D_n.

4.18.   Show that the set of all matrices

$$\begin{bmatrix} \pm 1 & a \\ 0 & 1 \end{bmatrix}, \quad a \in \mathbf{Z}/n\mathbf{Z},$$

is a multiplicative group isomorphic to D_n.

4.19.   What is the center of D_n? (HINT: First show that every element in D_n has a factorization $s^i t^j$.) Show the center is nontrivial if and only if n is even.

4.20.   Let G be a finite group and let $a, b \in G$ be distinct elements of order 2. Prove that $\langle a, b \rangle \cong D_n$ for some n.

*4.21. D_4 is not generated by its elements of largest order (compare Exercise 6.3).

4.22.   Let G be a finite group with normal subgroups H and K. If $G/H \cong G/K$, is $H \cong K$?

*4.23.   How many bracelets are there having n beads each of which may be painted any one of q colors? (HINT: See Exercise 3.65; the group D_n is acting.)

Theorem 4.13   *Let p be an odd prime. Every group G of order 2p is either cyclic or dihedral.*

Proof By Cauchy's theorem, G contains an element s of order p and an element t of order 2. If $H = \langle s \rangle$, then $H \lhd G$, since it has index 2. Therefore, $tst = s^i$ for some i. Now $s = t^2st^2 = t(tst)t = ts^it = s^{i^2}$. Hence, $i^2 \equiv 1 \pmod{p}$, so that, since p is prime, $i \equiv \pm 1 \pmod{p}$. Thus, $tst = s$ or $tst = s^{-1}$. In the first case, s and t commute, G is abelian, and so $G \cong \mathbf{Z}(p) \times \mathbf{Z}(2) \cong \mathbf{Z}(2p)$. In the second case, we have $G \cong D_p$. ∎

Theorem 4.14 *Let $|G| = pq$, where $p > q$ are primes. Then either G is cyclic or G is generated by two elements a and b satisfying the following relations:*

$$b^p = 1; \quad a^q = 1; \quad a^{-1}ba = b^r,$$

where $r \not\equiv 1 \pmod{p}$ but $r^q \equiv 1 \pmod{p}$. The second possibility can occur only if q divides $p - 1$.

Proof G contains an element b of order p; let $S = \langle b \rangle$. Since S is a Sylow p-subgroup of G, the number of its conjugates is $1 + up$ for some $u \geq 0$. But $1 + up = [G: N(S)]$ which divides $|G| = pq$. Since $(1 + up, p) = 1, 1 + up$ divides q. Since $q < p$, $u = 0$ and $S \lhd G$.

Now G contains an element a of order q; let $T = \langle a \rangle$. T is a Sylow q-subgroup of G, so that $[G: N(T)] = 1 + kq$ for some $k \geq 0$. As above, $1 + kq$ divides p, so that either $k = 0$ or q divides $p - 1$. If $k = 0$, $T \lhd G$ so that $G \cong S \times T$, by Exercise 4.9. Hence, $G \cong \mathbf{Z}(p) \times \mathbf{Z}(q) \cong \mathbf{Z}(pq)$.

We now assume that T is not normal (and so q divides $p - 1$). Since S is normal, $a^{-1}ba = b^r$; furthermore, we may assume $r \not\equiv 1 \pmod{p}$ lest we return to the abelian case. By induction on j, the reader may prove that $a^{-j}ba^j = b^{r^j}$. In particular, if $j = q$, we have $b = b^{r^q}$ so that $r^q \equiv 1 \pmod{p}$. ∎

Corollary 4.15 *If $|G| = pq$, where $p > q$ are primes, then G has a normal subgroup of order p. Furthermore, if q does not divide $p - 1$, then G is cyclic.*

Definition† The **quaternions**, denoted by Q, is a group of order 8 having two generators x and y that satisfy the relations

$$x^2 = y^2; \quad xyx = y.$$

EXERCISES

4.24. Prove that A_5 contains no subgroups of order 15 or order 30.

4.25. If G is a group of order 8 generated by two elements a and b satisfying the relations: $a^4 = 1$; $b^2 = a^2$; $b^{-1}ab = a^{-1}$, then $G \cong Q$.

---

† In this definition, as in the definition of D_n, we describe a group by generators and relations. At this point, it is fair to ask whether there *is* any group fitting this description (this question will be answered when we deal with extensions and with free groups). Therefore, it is necessary here to exhibit particular groups (motions, matrices, etc.) that do satisfy the conditions.

*4.26. Let G be the multiplicative group of all 2×2 nonsingular complex matrices, and let H be the subgroup of G generated by

$$A = \begin{bmatrix} 0 & i \\ i & 0 \end{bmatrix} \quad \text{and} \quad B = \begin{bmatrix} 0 & 1 \\ -1 & 0 \end{bmatrix}.$$

Prove that $H \cong Q$.

4.27. Consider the set $\{\pm 1, \pm i, \pm j, \pm k\}$ with the multiplication rules: $i^2 = j^2 = k^2 = -1; ij = k; jk = i; ki = j; ji = -k; kj = -i; ik = -j;$ and the usual rules for multiplying by ± 1. Prove that we have described a group isomorphic to Q. (See the remark after Theorem 2.16.)

4.28. What is the center of Q?

4.29. Q contains exactly one subgroup of order 2.

*4.30. $Q/Z(Q)$ is abelian; moreover, Q contains no subgroup isomorphic to $Q/Z(Q)$.

4.31. Prove that Q is not isomorphic to D_4.

*4.32. Every subgroup of Q is normal.

4.33. **(Dedekind, 1897)** Let $G = Q \times A \times B$, where Q is the quaternions, A is a (necessarily abelian) group in which every element $\neq 1$ has order 2, and B is an abelian group in which every element has odd order. Prove that every subgroup of G is normal. (Nonabelian groups in which every subgroup is normal are called **hamiltonian**, after W. R. Hamilton, who discovered the quaternions; all hamiltonian groups are of the form described above.)

4.34. Show that a Sylow 2-subgroup of $SL(2, 5)$, the group of all 2×2 matrices over $\mathbf{Z}/5\mathbf{Z}$ having determinant 1, is isomorphic to Q.

Theorem 4.16 Q and D_4 are the only nonabelian groups of order 8.

Proof If G is a nonabelian group of order 8, then G has no element of order 8 and not every element of G has order 2 (Exercise 1.31). Thus, G contains an element a of order 4. Suppose $b \in G$ and $b \notin \langle a \rangle$. Since $\langle a \rangle \lhd G$ (it has index 2) and $G/\langle a \rangle \cong \mathbf{Z}(2)$, we must have $b^2 \in \langle a \rangle$. If $b^2 = a$ or $b^2 = a^3$, then b has order 8, a contradiction. Hence,

$$b^2 = a^2 \quad \text{or} \quad b^2 = 1.$$

Furthermore, since $\langle a \rangle$ is normal, $b^{-1}ab \in \langle a \rangle$. Thus,

$$b^{-1}ab = a \quad \text{or} \quad b^{-1}ab = a^3$$

(because a^2 has order 2). The first case is ruled out, for then a and b commute and G is abelian. The following possibilities remain:

(i) $a^4 = 1, b^2 = a^2$, and $b^{-1}ab = a^3$.
(ii) $a^4 = 1, b^2 = 1$, and $b^{-1}ab = a^3$.

Since $a^3 = a^{-1}$, (i) describes the quaternions Q and (ii) describes the dihedral group D_4. ∎

Theorem 4.17 *Every group G of order 12 that is not isomorphic to A_4 contains an element of order 6.*

Proof If B is a Sylow 3-subgroup of G, then $B = \langle b \rangle$, where b has order 3. Since B has index 4, Theorem 3.18 gives a homomorphism $\pi: G \to S_4$ whose kernel K is a subgroup of B. Now B has no proper subgroups, so that $K = \{1\}$ or $K = B$. If $K = \{1\}$, then π is one-one and G is isomorphic to a subgroup of S_4 of order 12, which must be A_4, by Exercise 3.31. Since $G \not\cong A_4$, $K = B$, and so B is normal in G. It follows from Exercise 4.6 that B is the unique Sylow 3-subgroup of G. Therefore, there are only two elements in G of order 3, namely, b and b^2.

Let $C(b)$ be the centralizer of b in G. Now $[G: C(b)] = 1$ or 2, for a conjugate of b must have order 3. Therefore $|C(b)| = 12$ or 6; in either case, $C(b)$ contains an element a of order 2. Since a commutes with b, the element ab has order 6. ∎

Corollary 4.18 *If G has order 12 and $G \not\cong A_4$, then G has a normal Sylow 3-subgroup.*

EXERCISES

4.35. For every divisor d of 24, show there is a subgroup of S_4 having order d. Moreover, if $d \neq 4$, then any two subgroups of order d are isomorphic (of course, a subgroup of order 4 is either cyclic or isomorphic to the 4-group). Show that a Sylow 2-subgroup of S_4 is isomorphic to D_4.

4.36. Exhibit all the proper subgroups of S_4; there are allegedly 28 of them.

4.37. Prove that $D_3 \cong S_3$.

4.38. Prove that $D_6 \cong S_3 \times \mathbf{Z}(2)$.

*4.39. Prove that there are only two nonabelian groups of order 12 that contain a subgroup of order 6. One of these is D_6; the other, denoted by T, has generators a and b satisfying the relations $a^6 = 1$ and $b^2 = a^3 = (ab)^2$. (We shall examine the group T more carefully in Chapter 7.)

*4.40. If G is a nonabelian group of order p^3, where p is a prime, then $Z(G) = G'$, the commutator subgroup of G.

*4.41. Let p be an odd prime. Prove that there are at most two nonabelian groups of order p^3; one has generators a and b satisfying the relations $a^{p2} = 1$, $b^p = 1$, and $bab^{-1} = a^{1+p}$; the other has generators a, b, and c satisfying the relations $c = aba^{-1}b^{-1}$, $ca = ac$, $cb = bc$, and $a^p = b^p = c^p = 1$.

*4.42. Let G be a Sylow p-subgroup of $GL(3, p)$ (see Exercise 4.12). If p is an odd prime, then $x^p = 1$ for all $x \in G$. [HINT: If $A \in G$ and E is the identity matrix, expand $(A - E)^p$ by the binomial theorem.]

4.43. What is the order of a Sylow p-subgroup of $GL(n, p)$?

*4.44. Give an example of two nonisomorphic groups G and H such that, for each integer $d > 0$, the number of elements in G of order d is the same as the number of elements in H of order d (compare Exercise 6.17).

*4.45.   Let G be a finite group and let P be a Sylow p-subgroup of G. If H is a normal subgroup of G containing P, then $P \lhd H$ implies $P \lhd G$.

*4.46.   If G is a nonabelian group of square-free order (more generally, if every Sylow subgroup of G is cyclic), it is known that G is not simple. (We know this in the special case when $|G|$ is even, by Exercise 3.45; the general result is Theorem 7.33). Use this fact to prove that if $|G| = p_1 p_2 \cdots p_t$, where $p_1 < p_2 < \cdots < p_t$ are primes, then G contains a normal Sylow p_t-subgroup. (Compare Corollary 4.15.)

*4.47.   If p and q are primes, prove that a group of order $p^2 q$ cannot be simple.

Although there may exist many nonisomorphic groups of the same order (there are over 2000 groups of order 128), one can often deduce properties of a group from its order alone (as we have seen). Here are two more illustrations.

Example 1   There is no simple group of order 30.

Such a group would have 6 Sylow 5-subgroups (there are r such subgroups, where $r \equiv 1 \bmod 5$, r divides 30, and $r \neq 1$). Aside from the identity, this accounts for 24 elements. Similarly, there are 10 Sylow 3-subgroups, accounting for 20 elements, and we have exceeded 30.

Example 2   There is no simple group of order 36.

If G is such a group and P is a Sylow 3-subgroup, then $[G: P] = 4$. Representing G on the cosets of P gives a homomorphism $\varphi: G \to S_4$ which must be one-one because G is simple. This is impossible, for $36 > 24$.

EXERCISES

*4.48.   Prove that there are no nonabelian simple groups of order less than 60. (Do not use the unproven assertion in Exercise 4.46.)

*4.49.   Prove that any simple group G of order 60 is isomorphic to A_5. (HINT: Use Exercise 4.11 to see that G has a subgroup of index 5.) In Chapter 8, we shall exhibit two nonisomorphic finite simple groups having the same order.

Definition   A **generalized quaternion group** (or *dicyclic group*) Q_n, $n \geq 3$, is a group of order 2^n having generators x and y and relations:

$$x^{2^{n-2}} = y^2; \quad xyx = y$$

EXERCISES

4.50. If G is a group of order 2^n generated by two elements a and b satisfying the relations: $a^{2^{n-2}} = b^2 = (ab)^2$, then $G \cong Q_n$.

*4.51. Let G be a group of order 2^n, which is generated by elements a and b. If

$$a^{2^{n-1}} = 1, \quad bab^{-1} = a^{-1}, \quad \text{and} \quad b^2 = a^{2^{n-2}},$$

then $G \cong Q_n$.

*4.52. Prove that Q_n has a unique subgroup of order 2 which is $Z(Q_n)$.

4.53. Prove that $Q_n/Z(Q_n) \cong D_{2^{n-2}}$.

4.54. Let G be the multiplicative group of all 2×2 nonsingular complex matrices, and let H be the subgroup of G generated by

$$A = \begin{bmatrix} 0 & \omega \\ \omega & 0 \end{bmatrix} \quad \text{and} \quad B = \begin{bmatrix} 0 & 1 \\ -1 & 0 \end{bmatrix},$$

where $\omega = \exp(2\pi i/2^{n-1})$ is a 2^{n-1} th root of unity (here $n \geq 3$; also, $\exp(x)$ is another notation for e^x). Prove that $H \cong Q_n$ (compare Exercise 4.26).

TABLE OF GROUPS OF SMALL ORDER

Order	Number of Distinct Groups	Groups
4	2	cyclic and 4-group
6	2	$\mathbf{Z}(6) \cong \mathbf{Z}(2) \times \mathbf{Z}(3)$, S_3
8	5	3 are abelian, D_4, Q
9	2	abelian
10	2	cyclic and dihedral
12	5	2 are abelian, $\mathbf{Z}(2) \times S_3$, A_4, T
14	2	cyclic and dihedral
15	1	cyclic (Corollary 4.15)

There are 14 nonisomorphic groups of order 16, so we end our list here. No one knows a formula giving $G(n)$, the number of groups of order n, but here is a table of values for $16 \leq n \leq 32$:

n	16	17	18	19	20	21	22	23	24	25	26
$G(n)$	14	1	5	1	5	2	2	1	15	2	2

n	27	28	29	30	31	32
$G(n)$	5	4	1	4	1	51

chapter five

Normal
Series

SOME GALOIS THEORY

We begin this chapter with a brief history of the study of polynomials. It was well known to mathematicians of the Middle Ages (even to the Babylonians!) that the roots of a quadratic $x^2 + bx + c$ are given by the formula

$$x = \tfrac{1}{2}[-b \pm (b^2 - 4c)^{1/2}].$$

Here is the derivation of the **cubic formula** (due to del Ferro about 1515; this formula was also discovered, about the same time, by Tartaglia).† A cubic $x^3 + ax^2 + bx + c$ can be transformed into a cubic of the form

$$x^3 + qx + r$$

by replacing x with $x - a/3$. Once we have a root α of the second polynomial, then $\alpha + a/3$ is a root of the original one.

Suppose $\alpha^3 + q\alpha + r = 0$; write $\alpha = \beta + \gamma$, where β and γ are to be determined. Now $\alpha^3 = (\beta + \gamma)^3 = \beta^3 + \gamma^3 + 3(\beta^2\gamma + \beta\gamma^2) = \beta^3 + \gamma^3 + 3\alpha\beta\gamma$. Therefore

(1) $$\beta^3 + \gamma^3 + (3\beta\gamma + q)\alpha + r = 0.$$

---

† This formula is often called *Cardan's formula*, for it first appeared in print in a book by Cardan.

Two constraints should determine β and γ, and we have only imposed one: $\beta + \gamma = \alpha$. Let us further assume

$$\beta\gamma = -q/3,$$

thus making the middle term of (1) vanish. Equation (1) now reads

$$\beta^3 + \gamma^3 = -r.$$

We also know

$$\beta^3\gamma^3 = -q^3/27,$$

and we proceed to find β^3 and γ^3. Substituting,

$$\beta^3 - q^3/27\beta^3 = -r,$$

so the quadratic formula yields

$$\beta^3 = \tfrac{1}{2}[-r \pm (r^2 + 4q^3/27)^{1/2}].$$

Since $\beta^3 + \gamma^3 = -r$, we also have

$$\gamma^3 = \tfrac{1}{2}[-r \mp (r^2 + 4q^3/27)^{1/2}].$$

Therefore,

$$\alpha = \beta + \gamma = \left[\frac{-r}{2} + \left(\frac{r^2}{4} + \frac{q^3}{27}\right)^{1/2}\right]^{1/3} + \left[-\frac{r}{2} - \left(\frac{r^2}{4} + \frac{q^3}{27}\right)^{1/2}\right]^{1/3},$$

where the cube roots are chosen so the product $\beta\gamma = -q/3$.

About 1545, Ferrari discovered the **quartic formula**; we present the derivation of this formula due to Descartes (1637). First of all, replacing x with $x - a/4$ converts a polynomial $x^4 + ax^3 + bx^2 + cx + d$ to one of the form

$$x^4 + qx^2 + rx + s.$$

Factor this into quadratics:

$$x^4 + qx^2 + rx + s = (x^2 + kx + l)(x^2 - kx + m)$$

(the coefficient of x in the second factor must be $-k$ because there is no cubic term). If k, l, m can be found, the problem is reduced to the quadratic case. Expanding the right-hand side and equating coefficients of like powers of x, we obtain

$$l + m - k^2 = q;$$
$$k(m - l) = r;$$
$$lm = s.$$

The first two equations yield

(2) $$2m = k^2 + q + r/k$$

and

(3) $$2l = k^2 + q - r/k.$$

Substituting into the third equation $lm = s$ gives

$$(k^3 + qk + r)(k^3 + qk - r) = 4sk^2,$$

and this becomes

(4) $$k^6 + 2qk^4 + (q^2 - 4s)k^2 - r^2 = 0,$$

a cubic in k^2. We may now solve for k^2 using the cubic formula. Equations (2) and (3) yield m and l, so we have found the quadratic factors we sought.

Ruffini (1799)† and Abel (1827) proved the nonexistence of a similar formula for the roots of the general quintic polynomial, thus ending nearly three centuries of searching for a generalization of the work of del Ferro, Tartaglia, and Ferrari. Several years later, Galois (1811–1832) (essentially) invented group theory and showed that properties of a certain group of permutations associated with each polynomial explain why a formula for the roots of this polynomial may or may not exist.

Let us discuss some Galois theory, for it is the cradle of group theory. For a field F, denote the ring of all polynomials having coefficients in F by $F[x]$; if $f(x) \in F[x]$, then $f(x) = a_0 + a_1 x + \cdots + a_n x^n$, where $a_i \in F$ for all i. Informally, there is a formula for the roots of $f(x)$ if they can be obtained from $a_0, a_1, \cdots, a_n$ in a finite number of steps by the field operations (addition, subtraction, multiplication, division) and by extraction of roots. This informal idea will now be made precise. We shall restrict ourselves to subfields of the complex numbers $\mathbf{C}$ even though we shall need some of these ideas for more general fields in Chapter 8.

Definition   Let F be a subfield of $\mathbf{C}$ and let $\alpha \in \mathbf{C}$. The field obtained by **adjoining** α to F, denoted by $F(\alpha)$, is the set of all quotients $f(\alpha)/g(\alpha)$, where $f(\alpha)$ and $g(\alpha)$ are polynomials in α with coefficients in F, and $g(\alpha) \neq 0$.

It is easy to check that $F(\alpha)$ is a subfield of $\mathbf{C}$ containing F and α; indeed, it is the smallest such subfield. Note that $F(\alpha) = F$ if and only if $\alpha \in F$.

The process of adjoining a number to F may be iterated: If $\alpha_1, \alpha_2, \cdots, \alpha_n$ are in $\mathbf{C}$, define $F(\alpha_1, \alpha_2, \cdots, \alpha_n) = K(\alpha_n)$, where $K = F(\alpha_1, \alpha_2, \cdots, \alpha_{n-1})$.

EXERCISES

5.1.   Every subfield of $\mathbf{C}$ contains the rational numbers, $\mathbf{Q}$.

5.2.   Let F be a subfield of $\mathbf{C}$ and let $\{\alpha_1, \alpha_2, \cdots, \alpha_n\} \subset \mathbf{C}$. Prove that $F(\alpha_1, \cdots, \alpha_n) = F(\alpha_{\pi(1)}, \cdots, \alpha_{\pi(n)})$, where π is a permutation of $\{1, 2, \cdots, n\}$. (HINT: Prove that $F(\alpha_1, \alpha_2, \cdots, \alpha_n)$ is the intersection of all the subfields of $\mathbf{C}$ containing F and $\{\alpha_1, \cdots, \alpha_n\}$.)

If F is any field, a polynomial $f(x) \in F[x]$ of degree n has at most n roots in F. If $F = \mathbf{C}$, the fundamental theorem of algebra says that there are complex numbers

---

† Actually, Ruffini's work, while correct in outline, contained gaps and was not accepted by his contemporaries.

$\alpha_1, \alpha_2, \cdots, \alpha_n$ (not necessarily distinct) such that

$$f(x) = (x - \alpha_1)(x - \alpha_2) \cdots (x - \alpha_n)$$

(we are assuming, with no loss of generality, that $f(x)$ is **monic**, i.e., the coefficient of x^n is 1).

Definition If E and E' are fields, a function $\sigma: E \to E'$ is a **field map** if, for all $\alpha, \beta \in E$:

$$\sigma(1) = 1;$$
$$\sigma(\alpha + \beta) = \sigma(\alpha) + \sigma(\beta);$$
$$\sigma(\alpha\beta) = \sigma(\alpha)\sigma(\beta).$$

If σ is a one-one correspondence, then σ is an **isomorphism**; if $E = E'$ an isomorphism σ is called an **automorphism** of E.

EXERCISES

5.3. Show that every field map $\sigma: E \to E'$ is one-one.

**5.4. Assume F is a subfield of $\mathbf{C}$ and $\{\alpha_1, \cdots, \alpha_n\} \subset \mathbf{C}$.
If $\sigma_i: F(\alpha_1, \cdots, \alpha_n) \to \mathbf{C}$, $i = 1, 2$, are field maps with $\sigma_1|F = \sigma_2|F$ and $\sigma_1(\alpha_j) = \sigma_2(\alpha_j)$ for all j, then $\sigma_1 = \sigma_2$.

Definition Let F be a subfield of $\mathbf{C}$ and let $f(x) \in F[x]$. The **root field** (or *splitting field*) of $f(x)$ **over** F is the field $E = F(\alpha_1, \cdots, \alpha_n)$, the subfield obtained from F by adjoining all the roots $\alpha_1, \cdots, \alpha_n$ of $f(x)$.

Note that the root field of $f(x)$ over F does depend on F. For example, if $f(x) = x^2 + 1$, then the root field of $f(x)$ over $\mathbf{Q}$ is $\mathbf{Q}(i) = \{a + bi: a, b \in \mathbf{Q}\}$, but the root field over $\mathbf{R}$ is $\mathbf{R}(i) = \mathbf{C}$. We can now give the precise definition we need.

Definition If $f(x) \in F[x]$ has root field E over F, then $f(x)$ is **solvable by radicals** if there is a chain of subfields of $\mathbf{C}$,

$$F = K_0 \subset K_1 \subset \cdots \subset K_t,$$

in which each K_{i+1} is obtained from K_i by adjoining a root of an element in K_i and with $E \subset K_t$.

When we say there is a formula for the roots of $f(x)$, we really mean that $f(x)$ is solvable by radicals. Let us illustrate this by considering the quadratic, cubic, and quartic formulas.

If $f(x) = x^2 + bx + c$, set $F = \mathbf{Q}(b, c)$, where $\mathbf{Q}$ is the rationals; observe that $b^2 - 4c \in F$. Let $\alpha = (b^2 - 4c)^{1/2}$, and set $K_1 = F(\alpha)$. The quadratic formula says that K_1 is the root field of $f(x)$ over F.

If $f(x) = x^3 + qx + r$, set $F = \mathbf{Q}(q,r)$; observe that $r^2/4 + q^3/27 \in F$. Let

$$\alpha = \left(\frac{r^2}{4} + \frac{q^3}{27}\right)^{1/2},$$

and set $K_1 = F(\alpha)$. Observe further that $-r/2 + \alpha \in K_1$. Let

$$\beta = \left(-\frac{r}{2} + \alpha\right)^{1/3},$$

and set $K_2 = K_1(\beta)$. Finally, set $K_3 = K_2(\omega)$, where ω is a complex cube root of unity. Note that $\gamma = -q/3\beta$ is a cube root of $-(r/2) - \alpha$, so the cubic formula shows K_3 contains the root field of $f(x)$ over F (we need the cube root of unity to allow us to choose any cube root of β we may need). Note that K_3 may be strictly larger than the root field of $f(x)$.

If $f(x) = x^4 + qx^2 + rx + s$, set $F = \mathbf{Q}(q,r,s)$. Equation (4) above shows there is a cubic polynomial in $F[x]$ having k^2 as a root. As in the paragraph above, there is a chain of subfields $F = K_0 \subset K_1 \subset K_2 \subset K_3$ with $k^2 \in K_3$, $K_1 = F(\alpha)$, and $K_2 = K_1(\beta)$. Define $K_4 = K_3(k)$, $K_5 = K_4(\sqrt{\gamma})$ where $\gamma = k^2 - 4l$, and $K_6 = K_5(\sqrt{\delta})$ where $\delta = k^2 - 4m$. Equations (2) and (3) show that $l, m \in K_4$. Our discussion of the quartic formula shows that K_6 contains the root field of $f(x)$ over F.

Definition   If F is a subfield of a field K, then an automorphism σ of K **fixes** F if $\sigma(a) = a$ for every $a \in F$.

Lemma 5.1   *Let $f(x) = a_0 + a_1 x + \cdots + a_n x^n \in F[x]$ and let K be the root field of $f(x)$ over F. If $\sigma \colon K \to K$ is an automorphism fixing F and if α is a root of $f(x)$, then $\sigma(\alpha)$ is also a root of $f(x)$.*

Proof   Applying σ to the equation $f(\alpha) = 0$ gives

$$\sigma(a_0) + \sigma(a_1)\sigma(\alpha) + \cdots + \sigma(a_n)\sigma(\alpha)^n = 0.$$

Since σ fixes F, we have $\sigma(a_i) = a_i$ for all i and $f(\sigma(\alpha)) = 0$. ∎

Lemma 5.2   *Let F be a subfield of K. The set of all automorphisms of K that fix F forms a group (in which the binary operation is composition of functions).*

Proof   An easy verification. ∎

Definition   If $f(x) \in F[x]$ and K is the root field of $f(x)$ over F, then the **Galois group** of $f(x)$ over F, denoted by $\mathrm{Gal}(K/F)$, is the group of all automorphisms of K that fix F.

Theorem 5.3   *Let $X = \{\alpha_1, \cdots, \alpha_m\}$ be the set of all the distinct roots of a polynomial $f(x) \in F[x]$ and let $K = F(\alpha_1, \cdots, \alpha_m)$ be the root field of $f(x)$ over F. Then the function $\sigma \mapsto \sigma|X$ is an imbedding $\mathrm{Gal}(K/F) \to S_X \cong S_m$.*

Proof   If $\sigma \in \mathrm{Gal}(K/F)$, then Lemma 5.1 shows $\sigma|X \colon X \to X$. Indeed, $\sigma|X \in S_X$ because σ is one-one and X is finite. It is easy to show this restriction map is a homomorphism; it is one-one by Exercise 5.4. ∎

Not every permutation of the roots of a polynomial $f(x)$ need correspond to some $\sigma \in \text{Gal}(K/F)$. For example, let $f(x) = (x^2 - 2)(x^2 - 3) \in \mathbf{Q}[x]$. Every $\sigma \in \text{Gal}(\mathbf{Q}(\sqrt{2}, \sqrt{3})/\mathbf{Q})$ must permute $\{\sqrt{2}, -\sqrt{2}, \sqrt{3}, -\sqrt{3}\}$, but it is plain there is no such σ with $\sigma(\sqrt{2}) = \sqrt{3}$.

Let K be a field with subfield F. Clearly K satisfies the axioms for a vector space over F if we take as scalar multiplication the given multiplication between elements of F and K. In particular, one may ask about the dimension of K over F, denoted by $[K:F]$.

EXERCISES

**5.5. Let $p(x) \in F[x]$ be an irreducible polynomial of degree n. If $\alpha \in \mathbf{C}$ is a root of $p(x)$, prove that $\{1, \alpha, \alpha^2, \cdots, \alpha^{n-1}\}$ is a basis of $F(\alpha)$ viewed as a vector space over F. Conclude that $[F(\alpha):F] = n$.

**5.6. Let $F \subset K \subset R$ be fields, where $[K:F]$ and $[R:K]$ are finite. Then

$$[R:F] = [R:K][K:F].$$

(HINT: If $\{\alpha_1, \cdots, \alpha_r\}$ is a basis of K over F and $\{\beta_1, \cdots, \beta_s\}$ is a basis of R over K, then the set of rs elements of the form $\alpha_i \beta_j$ is a basis of R over F.)

**5.7. Let $F \subset K \subset E$ be fields, with K the root field of $f(x) \in F[x]$ and E the root field of $g(x) \in F[x]$. If $\sigma \in \text{Gal}(E/F)$, then $\sigma|K \in \text{Gal}(K/F)$. (HINT: Lemma 5.1.)

Lemma 5.4 *Let $\lambda: F \to F'$ be an isomorphism of fields, and let $p(x) \in F[x]$ be irreducible with root $\alpha \in \mathbf{C}$. Then there is an irreducible polynomial $q(x) \in F'[x]$ with root $\beta \in \mathbf{C}$ and an isomorphism*

$$\lambda^*: F(\alpha) \to F'(\beta)$$

with $\lambda^*|F = \lambda$ and $\lambda^*(\alpha) = \beta$.

Proof The map $\varphi: F[x] \to F'[x]$ defined by

$$b_0 + b_1 x + \cdots + b_m x^m \mapsto \lambda(b_0) + \lambda(b_1)x + \cdots + \lambda(b_m)x^m$$

is an isomorphism of rings. Define $q(x) = \varphi(p(x))$, which is irreducible since $p(x)$ is; let $\beta \in \mathbf{C}$ be any root of $q(x)$.

Since $p(x)$ is irreducible, Exercise 5.5 says $\{1, \alpha, \cdots, \alpha^{n-1}\}$ is a basis of $F(\alpha)$ over F, where $n = \text{degree } p(x)$. Define $\lambda^*: F(\alpha) \to F'(\beta)$ by

$$\lambda^*(b_0 + b_1 \alpha + \cdots + b_{n-1} \alpha^{n-1}) = \lambda(b_0) + \lambda(b_1)\beta + \cdots + \lambda(b_{n-1})\beta^{n-1}.$$

It is easy to check that λ^* is a field map with $\lambda^*|F = \lambda$ and $\lambda^*(\alpha) = \beta$. To see that λ^* is an isomorphism, construct its inverse in the same manner, beginning with $\lambda^{-1}: F' \to F$. ∎

Lemma 5.5 *Let $f(x) \in K[x]$, where K is a subfield of F and of F'; let R be the root field of $f(x)$ over F and let R' be the root field of $f(x)$ over F'. If $\lambda\colon F \to F'$ is an isomorphism fixing K, then there is an isomorphism $\lambda^*\colon R \to R'$ with $\lambda^*|F = \lambda$.*

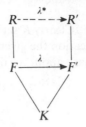

Proof We perform an induction on $d = [R\colon F]$. If $d = 1$, then $R = F$ and every root $\alpha_1, \cdots, \alpha_n$ of $f(x)$ lies in F. Since λ fixes K, each $\lambda(\alpha_i)$ is also a root of $f(x)$; therefore λ permutes the α_i and every root of $f(x)$ lies in F'. Thus $R' = F'$ and we may take $\lambda^* = \lambda$. If $d > 1$, then $R \neq F$ and there is some root α of $f(x)$ not lying in F (of course $\alpha \in R$, by definition of root field). Now α is a root of some irreducible factor $p(x)$ of $f(x)$; since $\alpha \notin F$, degree $p(x) = k > 1$. By Lemma 5.4, there is $\beta \in \mathbf{C}$ and an isomorphism $\lambda_1\colon F(\alpha) \to F'(\beta)$ that extends λ. In particular, λ_1 fixes K, so that β is a root of $f(x)$ and $F'(\beta) \subset R'$. By Exercise 5.6, $[R\colon F(\alpha)] = d/k < d$. Furthermore, R arises from $F(\alpha)$ by adjoining all the roots of $f(x)$, so that R is the root field of $f(x)$ over $F(\alpha)$; similarly, R' is the root field of $f(x)$ over $F'(\beta)$. With all the inductive hypotheses being verified, λ_1, a fortiori λ, can be extended to an isomorphism $\lambda^*\colon R \to R'$. ∎

Lemma 5.6 *Let $F \subset K \subset R$, where R and K are each root fields over F. Then $\mathrm{Gal}\,(R/K) \lhd \mathrm{Gal}\,(R/F)$ and*

$$\mathrm{Gal}\,(R/F)/\mathrm{Gal}\,(R/K) \cong \mathrm{Gal}\,(K/F).$$

Proof Define $\Phi\colon \mathrm{Gal}\,(R/F) \to \mathrm{Gal}\,(K/F)$ by

$$\Phi(\lambda) = \lambda|K.$$

By Exercise 5.7, Φ is a well defined function that is easily seen to be a homomorphism. The kernel of Φ is the set of all automorphisms of R that fix K, i.e., $\mathrm{Gal}\,(R/K)$. We claim that Φ is onto. Suppose $\lambda \in \mathrm{Gal}\,(K/F)$, i.e., λ is an automorphism of K that fixes F. Since R is a root field of F, Lemma 5.5 says λ can be extended to an automorphism λ^* of R. Therefore, $\lambda^* \in \mathrm{Gal}\,(R/F)$ and

$$\Phi(\lambda^*) = \lambda^*|K = \lambda.$$

The first isomorphism theorem completes the proof. ∎

EXERCISES

**5.8.   Assume F contains the kth roots of unity, and let $R = F(\alpha)$, where α is a root of $x^k - a$ for some $a \in F$. Prove that there exist intermediate fields

$$F = K_0 \subset K_1 \subset \cdots \subset K_t = R,$$

where $K_{i+1} = K_i(\beta_i)$, β_i is a root of $x^{p(i)} - b_i$ for some $b_i \in K_i$, and $p(i)$ is prime.

5.9. If p is prime, if F contains the pth roots of unity, and if $a \in F$, then either $x^p - a$ has a root in F or it is irreducible over F.

**5.10. Assume p is prime and that F contains all pth roots of unity. If α is a root of $x^p - a$ for some $a \in F$, prove that $\mathrm{Gal}(F(\alpha)/F)$ is a cyclic group of order p or 1.

We summarize this investigation in the next theorem.

Theorem 5.7 **(Galois, 1830)**   *Assume F is a subfield of $\mathbf{C}$ and R is the root field of $f(x) \in F[x]$. If $f(x)$ is solvable by radicals, then there exist subgroups G_i of $G = \mathrm{Gal}(R/F)$ such that:*

(i) $G = G_0 \supset G_1 \supset \cdots \supset G_n = \{1\}$;
(ii) $G_{i+1} \lhd G_i$ *for all* i;
(iii) G_i/G_{i+1} *is cyclic of prime order.*

Proof   Define $G_i = \mathrm{Gal}(R/K_i)$, where the K_i are as in Exercise 5.8. It is only necessary to remind the reader of previous work: (i) follows immediately from the definition of Galois group; (ii) follows from Lemma 5.6; (iii) follows from Exercises 5.8 and 5.10 and Lemma 5.6. (We remark that, in its use of Exercises 5.8 and 5.10, this proof needs the added hypothesis that F contains appropriate roots of unity; it may be shown that this hypothesis is not necessary.) ∎

The converse of this theorem is true also; both are due to Galois (1830) and are a remarkable achievement. We remind the reader of a remark made in Chapter 1. In 1826, Abel proved that a polynomial whose Galois group is commutative is solvable by radicals; this is why abelian groups are so called.

Definition   A **normal series** of G is a chain of subgroups

(*) $G = G_0 \supset G_1 \supset \cdots \supset G_n = \{1\}$

in which $G_{i+1} \lhd G_i$ for all i. The **factor groups** of (*) are the groups G_i/G_{i+1} for $i = 0$, $1, \cdots, n - 1$; the **length** of the series (*) is the number of strict inclusions.

Alternatively, the length of the series is the number of factor groups with more than one element. Note that the factor groups are the only quotient groups we can always form from this series, for we have seen (Exercise 3.36) that normality need not be transitive.

Definition   A finite group G is **solvable**† if it has a normal series whose factor groups are cyclic of prime order.

---

† Some authors call such groups "soluble".

In this terminology, Theorem 5.7 and its converse say that a polynomial $f(x)$ is solvable by radicals if and only if its Galois group $\text{Gal}(R/F)$ is a solvable group. Now Ruffini and Abel proved that there exists a polynomial $f(x)$ of degree 5 whose Galois group† is S_5. Since S_5 is not a solvable group (as we shall soon see), this polynomial is not solvable by radicals. Thus, the classical problem of the determination of the roots of a polynomial led inevitably to groups and to normal series, and there its solution (or nonsolution) lies. Expositions of Galois theory can be found in the books of Artin, Birkhoff and Mac Lane, Jacobson, Kaplansky, Lang, and van der Waerden listed in the bibliography.

The work of Abel and Galois has not only enriched the study of polynomials and fields, it has also contributed a new point of view to the study of groups. Let us give a brief review of what we have learned so far. Our first results arose from examining properties of a single subgroup via Lagrange's theorem. The second, deeper set of results arose from examining properties of a family of subgroups via the Sylow theorems (Theorems 4.9 and 4.10); this family of subgroups consists of the conjugates of a single subgroup, and so each member has the same order. Normal series will give results by allowing us to examine a family of subgroups of distinct orders, thus providing an opening wedge for an inductive proof.

THE JORDAN-HÖLDER THEOREM

Definition   A normal series

$$G = H_0 \supset H_1 \supset H_2 \supset \cdots \supset H_m = \{1\}$$

is a **refinement** of a normal series

$$G = G_0 \supset G_1 \supset G_2 \supset \cdots \supset G_n = \{1\}$$

if the list $G_0, G_1, \cdots$ is a sublist of $H_0, H_1, \cdots$

A refinement is thus a normal series that contains each of the terms of the original series; the length of a refinement is thus at least as great as that of the original series (a term may be repeated).

Definition   A normal series

$$G = G_0 \supset G_1 \supset \cdots \supset G_n = \{1\}$$

is a **composition series** if either each G_{i+1} is a maximal normal subgroup of G_i or $G_{i+1} = G_i$.

EXERCISES

5.11.   A normal series is a composition series if and only if each of its factor groups is simple or $\{1\}$.

---

† Galois outlined Galois theory in 1830, when he was only 19; he was killed in a duel two years later.

5.12. A composition series is a normal series of maximal length.

5.13. Every finite group has a composition series.

5.14. An abelian group has a composition series if and only if it is finite.

5.15. If G is a finite group having a normal series with factor groups $H_1, \cdots, H_m$, then $|G| = \Pi |H_i|$.

5.16. Give an example of an infinite group with a composition series. (A necessary and sufficient condition that a group have a composition series is given in Exercise 6.59.)

Consider the group $G = \mathbf{Z}(30)$ with generator x; we write two composition series for G (normality is automatic since G is abelian):

(*) $G \supset \langle x^5 \rangle \supset \langle x^{10} \rangle \supset \{1\}$;

(**) $G \supset \langle x^2 \rangle \supset \langle x^6 \rangle \supset \{1\}$.

The factor groups of (*) are $G/\langle x^5 \rangle$, $\langle x^5 \rangle / \langle x^{10} \rangle$, and $\langle x^{10} \rangle$, i.e., $\mathbf{Z}(5)$, $\mathbf{Z}(2)$, and $\mathbf{Z}(3)$. The factor groups of (**) are $G/\langle x^2 \rangle$, $\langle x^2 \rangle / \langle x^6 \rangle$, and $\langle x^6 \rangle$, i.e., $\mathbf{Z}(2)$, $\mathbf{Z}(3)$, and $\mathbf{Z}(5)$. In this case, these two composition series of G have the same length, and the factor groups can be "paired isomorphically" after rearranging them. We give a name to this phenomenon.

Definition   Two normal series of G are **equivalent** if there is a one-one correspondence between the factor groups of each such that corresponding factor groups are isomorphic.

Observe that equivalent normal series have the same length.

The two composition series for $\mathbf{Z}(30)$ exhibited above are equivalent; the amazing fact is that this is true for every (possibly infinite) group that has a composition series!

The next rather technical result, a generalization of the second isomorphism theorem, is used in proving equivalence of composition series.

Lemma 5.8 (Zassenhaus lemma, 1935)   *Let $A \triangleleft A^*$ and $B \triangleleft B^*$ be subgroups of a group G. Then*

$$A(A^* \cap B) \triangleleft A(A^* \cap B^*),$$
$$B(B^* \cap A) \triangleleft B(B^* \cap A^*),$$

and there is an isomorphism

$$\frac{A(A^* \cap B^*)}{A(A^* \cap B)} \cong \frac{B(B^* \cap A^*)}{B(B^* \cap A)}.$$

REMARK   Note that the hypothesis and conclusion are not changed by interchanging the symbols A and B.

Proof Let D be the subset $(A^* \cap B)(A \cap B^*)$ of $A^* \cap B^*$. Now $B \lhd B^*$ implies $B(B^* \cap A^*)$ is a subgroup of B^* (Theorem 2.20) and $B \lhd B(B^* \cap A^*) \subset B^*$. The second isomorphism theorem with $T = B$ and $S = B^* \cap A^*$ shows $T \cap S = (B \cap A^*) \lhd (B^* \cap A^*)$. Similarly, interchanging the symbols A and B, one sees $(A \cap B^*) \lhd (A^* \cap B^*)$. It follows from Theorem 2.20 and Exercise 2.43 that $D = (A^* \cap B)(A \cap B^*)$ is a normal subgroup of $A^* \cap B^*$.

If $x \in B(B^* \cap A^*)$, then $x = bc$ for $b \in B$ and $c \in B^* \cap A^*$. Define $f: B(B^* \cap A^*) \to (A^* \cap B^*)/D$ by $f(x) = f(bc) = cD$. To see f is well defined, suppose $x = bc = b_1 c_1$, where $b_1 \in B$ and $c_1 \in B^* \cap A^*$; then $c_1 c^{-1} = b_1^{-1} b \in (B^* \cap A^*) \cap B = B \cap A^* \subset D$. One may check quickly that f is a homomorphism with image $(A^* \cap B^*)/D$ and kernel $B(B^* \cap A)$. The first isomorphism theorem gives $B(B^* \cap A) \lhd B(B^* \cap A^*)$ and

$$\frac{B(B^* \cap A^*)}{B(B^* \cap A)} \cong \frac{A^* \cap B^*}{D}.$$

Interchanging the symbols A and B gives $A(A^* \cap B) \lhd A(A^* \cap B^*)$ and the corresponding quotient group isomorphic to $(A^* \cap B^*)/D$. Therefore, the two quotient groups in the statement of the lemma are isomorphic. ∎

Theorem 5.9   (Schreier, 1926)   *Any two normal series of an arbitrary group G have refinements that are equivalent.*

Proof   Let

(*) $G = G_0 \supset G_1 \supset \cdots \supset G_n = \{1\}$

and

(**) $G = H_0 \supset H_1 \supset \cdots \supset H_m = \{1\}$

be normal series. In the first series, between each G_i and G_{i+1} insert a "copy" of the second series. More precisely, define $G_{ij} = G_{i+1}(G_i \cap H_j)$, where $0 \leq j \leq m$. Thus

$$G_{ij} = G_{i+1}(G_i \cap H_j) \supset G_{i+1}(G_i \cap H_{j+1}) = G_{i,j+1}.$$

Note that $H_0 = G$ implies $G_{i0} = G_i$ and $H_m = \{1\}$ implies $G_{im} = G_{i+1}$. Moreover, setting $A = G_{i+1}$, $A^* = G_i$, $B = H_{j+1}$, and $B^* = H_j$ in the Zassenhaus lemma shows $G_{i,j+1} \lhd G_{ij}$. It follows that the G_{ij} form a normal series (with mn terms) that is a refinement of (*). Similarly, if we define $H_{ij} = H_{j+1}(H_j \cap G_i)$, then $H_{ij} \supset H_{i+1,j}$ and the subgroups H_{ij} form a refinement of (**) with mn terms. Finally, the function that pairs the factor group $G_{ij}/G_{i,j+1}$ of the first refinement with the factor group $H_{ij}/H_{i+1,j}$ of the second refinement is a one-one correspondence, and the Zassenhaus lemma asserts corresponding factor groups are isomorphic. Therefore, these two refinements of (*) and (**), respectively, are equivalent. ∎

Theorem 5.10 (**Jordan-Hölder**)† *Any two composition series of a group G are equivalent.*

Proof Since composition series are normal series, any two composition series of G have equivalent refinements. But a composition series, being a normal series of maximal length, admits no nontrivial refinement (i.e., a refinement merely repeats some of the terms). Thus, two composition series are already equivalent. ∎

Definition The factor groups of a composition series of G are called the **composition factors** of G.

The Jordan-Hölder theorem may be regarded as a kind of unique factorization theorem. In a sense to be made explicit in Chapter 7, a group G is a "product" of its composition factors. Let us support this viewpoint by using the Jordan-Hölder theorem to prove the fundamental theorem of arithmetic (Appendix VI): The factorization of an integer $n > 1$ into primes is unique. Let $n = p_1 p_2 \cdots p_m$, where the p_i are (not necessarily distinct) primes. If $G = \mathbf{Z}(n)$ has generator x, then

$$G = \langle x \rangle \supset \langle x^{n/p_1} \rangle \supset \langle x^{n/p_1 p_2} \rangle \supset \cdots \supset \langle x^{p_m} \rangle \supset \{1\}$$

is a normal series. Since the factor groups $\mathbf{Z}(p_1), \mathbf{Z}(p_2), \cdots, \mathbf{Z}(p_m)$ have prime order, this is a composition series. The Jordan-Holder theorem says that these factor groups, hence their orders, depend only on $G = \mathbf{Z}(n)$ and not on the choice of composition series.

The proof given above is valid, *mutatis mutandis*, for vector spaces over a field F. In particular, if V is finite-dimensional (i.e., V is spanned by a finite number of vectors), then a composition series for V has the form

$$V = V_0 \supset V_1 \supset \cdots \supset V_n = \{0\},$$

where the quotient spaces V_i/V_{i+1} are either $\{0\}$ or isomorphic to F (for F itself is the only simple vector space). The Jordan-Hölder theorem asserts that the length of such a series depends only on V. However, if $\{x_1, \cdots, x_m\}$ is a basis of V, there is a composition series of length m defined for $i \geq 0$ by $V_i = \langle x_{i+1}, x_{i+2}, \cdots, x_m \rangle$, the subspace spanned by the displayed vectors. It follows that m is the length of this composition series and, thus, any two bases of V have the same size: dimension is well defined. Generalizations of the Jordan-Hölder theorem to more general situations, e.g., operator groups or lattices, can be found in Zassenhaus.

EXERCISES

**5.17. Let G and H be groups having composition series. If there are normal series of G and H having the same set of factor groups, then G and H have the same composition factors.

† Jordan (1868) proved that the orders of the factor groups of a composition series of G depend only on G; Hölder (1889) proved that the factor groups depend only on G.

5.18. The normal series of S_4 displayed in Exercise 3.36 is a composition series.

5.19. S_n is solvable for $n \le 4$.

__*5.20.__ If $G = H_1 \times \cdots \times H_m = K_1 \times \cdots \times K_n$, where each H_i and K_j is simple, then $m = n$ and there is a permutation α of $\{1, \cdots, n\}$ with $H_i \cong K_{\alpha(i)}$ for all i. (HINT: Construct two composition series of G.)

Recall that a finite group G is solvable if it has a normal series with factor groups of prime order. Clearly such a series is a composition series. To test whether a particular group G is solvable, choose a composition series and check its factor groups. It is conceivable, without Jordan-Hölder, that one might have to look at all composition series of a group G before concluding G is not solvable.

Theorem 5.11 S_n *is not solvable for* $n \ge 5$.

Proof A normal series for S_n is

$$S_n \supset A_n \supset \{1\}.$$

Since the factor groups are $\mathbf{Z}(2)$ and A_n and since A_n is simple for $n \ge 5$, $S_n \supset A_n \supset \{1\}$ is a composition series. By the Jordan-Hölder theorem, S_n is not solvable. ∎

We confess that we really don't need Jordan-Hölder to prove Theorem 5.11, for it is easy to prove that S_n has a unique composition series for $n \ge 5$ (see Exercise 3.35).

SOLVABLE GROUPS

Solvable groups have been defined in connection with Galois theory. Since these groups form a large class of groups of purely group-theoretic interest as well, we give another definition of solvability that is more convenient to work with and that is equivalent to our earlier definition when G is finite.

Definition A group G is **solvable** if it has a normal series with abelian factor groups. Such a normal series is called a **solvable series** of G.

EXERCISES

__**5.21.__ Any refinement of a solvable series is a solvable series.

5.22. A finite group is solvable (definition just given) if and only if it has a composition series with cyclic factor groups of prime order (earlier definition).

5.23. A solvable group with a composition series is finite.

5.24. Let $H \lhd G$, where G has a composition series. Prove that G has a composition series one of whose terms is H.

Let us see how to manufacture solvable groups.

Theorem 5.12 *Any subgroup H of a solvable group G is solvable.*

Proof Let $G = G_0 \supset G_1 \supset \cdots \supset G_n = \{1\}$ be a solvable series for G; we claim that

$$H = H_0 \supset (H \cap G_1) \supset \cdots \supset (H \cap G_n) = \{1\}$$

is a solvable series for H. By the second isomorphism theorem,

$$H \cap G_{i+1} = (H \cap G_i) \cap G_{i+1} \lhd H \cap G_i,$$

so we do have a normal series. Moreover, we have the diagram

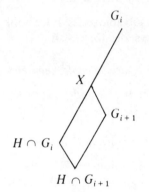

where $X = G_{i+1}(H \cap G_i)$. Therefore,

$$\frac{H \cap G_i}{H \cap G_{i+1}} \cong \frac{X}{G_{i+1}} \subset \frac{G_i}{G_{i+1}}.$$

Since G_i/G_{i+1} is abelian, so is its subgroup X/G_{i+1}, as desired. ∎

Theorem 5.13 *If G is solvable and $H \lhd G$, then G/H is solvable.*

Proof Let $G = G_0 \supset G_1 \supset \cdots \supset G_n = \{1\}$ be a solvable series. Since $H \lhd G$, the subsets HG_i are subgroups of G; since $G_{i+1} \lhd G_i$, we see that $HG_{i+1} \lhd HG_i$. Therefore

$$G = HG_0 \supset HG_1 \supset \cdots \supset HG_n = H \supset \{1\}$$

is a normal series. Consider the chain of subgroups

(*) $$G/H = HG_0/H \supset HG_1/H \supset \cdots \supset H/H = \{1\}.$$

Since H and HG_{i+1} are normal subgroups of HG_i, the third isomorphism theorem asserts $HG_{i+1}/H \lhd HG_i/H$, whence (*) is a normal series; moreover $(HG_i/H)/(HG_{i+1}/H) \cong HG_i/HG_{i+1} = (HG_{i+1})G_i/HG_{i+1} \cong G_i/(G_i \cap HG_{i+1})$ (second isomorphism theorem). Finally, the third isomorphism theorem shows the last group is a quotient group of the abelian group G_i/G_{i+1}, hence is itself abelian. Therefore (*) is a solvable series and G/H is solvable. ∎

Theorem 5.14 *Let $H \lhd G$. If both H and G/H are solvable, then G is solvable.*

Proof Let

$$G/H \supset K_1^* \supset K_2^* \supset \cdots \supset \{1\}$$

be a solvable series. By the correspondence theorem, we can construct the beginning of a solvable series from G to H, i.e., there are subgroups K_i with $K_{i+1} \lhd K_i$, K_i/K_{i+1} abelian, and

$$G = G_0 \supset K_1 \supset K_2 \supset \cdots \supset H.$$

Since H is solvable, it has a solvable series. If we hook these two series together at H, we obtain a solvable series for G. ∎

Corollary 5.15 *If H and K are solvable, then $H \times K$ is solvable.*

Proof If $G = H \times K$, then $H \lhd G$ and $G/H \cong K$, so that G is solvable, by Theorem 5.14. ∎

Corollary 5.16 *Every finite p-group G is solvable.*

Proof We perform an induction on $|G|$. By Theorem 4.4, $|Z(G)| \neq 1$. Therefore, $G/Z(G)$ is a p-group of order $< |G|$, and so is solvable, by induction. Because every abelian group is solvable, $Z(G)$ is solvable. By Theorem 5.14, G is solvable. ∎

EXERCISES

5.25. If S and T are solvable subgroups of G with $S \lhd G$, then ST is a solvable subgroup of G.

5.26. (i) Any group of order p^2q is solvable, where p and q are primes (compare Exercise 4.47).
(ii) If p and q are primes with $p < q$, then any group G of order pq^n is solvable. (HINT: Use Sylow's theorem.)

5.27. It is known that no conjugacy class in a finite simple group has cardinal a prime power > 1. Use this fact (proved by Burnside) to prove **Burnside's theorem** (1904): Every group of order $p^\alpha q^\beta$ is solvable.

5.28. The dihedral groups D_n are solvable.

5.29. Prove that S_4 has no normal series

$$S_4 = G_0 \supset G_1 \supset \cdots \supset G_m = \{1\}$$

with each factor group G_i/G_{i+1} cyclic and with each G_{i+1} normal in G (not merely in G_i). (A group G having a normal series as described above is called **supersolvable**. This exercise thus shows a solvable group need not be supersolvable.)

*5.30. The following two statements are equivalent: (1) Every group of odd order is solvable; (2) every finite nonabelian simple group has even order. (In a deep paper in 1963, W. Feit and J. Thompson proved the truth of these statements.)

Another approach to solvable groups is with commutator subgroups; that we are dealing with abelian quotient groups suggests this approach at once.

Definition Define the **higher commutator subgroups** of G inductively:

$$G^{(0)} = G; \quad G^{(i+1)} = G^{(i)\prime},$$

i.e., $G^{(i+1)}$ is the commutator subgroup of $G^{(i)}$. The series of higher commutator subgroups is called the **derived series** of G.

Let us begin by proving the higher commutator subgroups of G are normal subgroups of G. To see this, it is convenient to introduce characteristic subgroups.

Definition A subgroup H of G is **characteristic** in G if $\quad(H) \subseteq H$ for every isomorphism $\quad :G \to G$. It is plain that every characteristic subgroup is normal, for conjugations are automorphisms.

EXERCISES

5.31. Let $G \neq \{1\}$ be a finite group. If G is solvable, then G contains a normal abelian subgroup $H \neq \{1\}$; if G is not solvable, then G contains a normal subgroup $H \neq \{1\}$ such that $H = H'$.

**5.32. If $K \subset H \lhd G$ and K is characteristic in H, then $K \lhd G$. (Compare Exercise 4.45.)

**5.33. Being characteristic is transitive: if K is characteristic in H and H is characteristic in G, then K is characteristic in G.

5.34. For any group G, the center $Z(G)$ is characteristic.

**5.35. If G is an abelian p-group, then $G[p]$, the set of all elements in G of order $\leq p$, is characteristic in G.

**5.36. Every normal Sylow p-subgroup of a finite group G is characteristic.

**5.37. If $K \lhd G$ and $(|K|, [G:K]) = 1$, then K is a characteristic subgroup in G.

Lemma 5.17 *If $G = G_0 \supset G_1 \supset \cdots \supset G_n = \{1\}$ is a solvable series, then $G_i \supset G^{(i)}$ for all i.*

Proof We prove the lemma by induction on i. If $i = 0$, $G_i = G^{(i)} = G$. Suppose now that $G_i \supset G^{(i)}$; then $G_i' \supset G^{(i)\prime} = G^{(i+1)}$. Since G_i/G_{i+1} is abelian, $G_{i+1} \supset G_i'$, by Exercise 2.52. Therefore, $G_{i+1} \supset G^{(i+1)}$, as desired. ∎

Theorem 5.18 *A group G is solvable if and only if $G^{(n)} = \{1\}$ for some integer n.*

Proof Let $G = G_0 \supset G_1 \supset \cdots \supset G_n = \{1\}$ be a solvable series. By the lemma, $G_n \supset G^{(n)}$, and so $G^{(n)} = \{1\}$.

If $G^{(n)} = \{1\}$ for some n, then the series

$$G = G^{(0)} \supset G' \supset \cdots \supset G^{(n)} = \{1\}$$

is a solvable series for G. ∎

The reader should supply alternative proofs of Theorems 5.12 and 5.13 using this new characterization of solvability.

Theorem 5.19 *For every group G, the subgroups $G^{(i)}$ are characteristic, hence normal, subgroups in G.*

Proof We prove the theorem by induction on i: if $i = 1$, then $G^{(1)} = G' = \langle xyx^{-1}y^{-1}: x, y \in G \rangle$. For every isomorphism $\varphi: G \to G$, we see $\varphi(xyx^{-1}y^{-1})$ is a commutator, $\varphi(G') \subset G'$, and G' is characteristic in G. Since $G^{(i+1)} = G^{(i)\prime}$ is characteristic in $G^{(i)}$ and $G^{(i)}$ is characteristic in G, by induction, the result follows from Exercise 5.33. ∎

What are the groups G whose only characteristic subgroups are the trivial ones $\{1\}$ and G?

Theorem 5.20 *A finite group $G \neq \{1\}$ whose only characteristic subgroups are $\{1\}$ and G is either simple or a direct product of isomorphic simple groups.*

Proof Choose a normal subgroup H of G with $H \neq \{1\}$ and whose order is minimal among such subgroups. Write $H = H_1$ and consider all subgroups of the form $H_1 \times H_2 \times \cdots \times H_n$ ($n \geq 1$), where $H_i \lhd G$ and $H_i \cong H$ for each i; choose one such direct product of largest order and call it M. Note that $M \lhd G$, for it is generated by normal subgroups. We show $M = G$ by showing it is characteristic in G; we show M is characteristic by showing $\varphi(H_i) \subset M$ for every i and every isomorphism $\varphi: G \to G$. Now $\varphi(H_i) \lhd G$ and $\varphi(H_i) \cong H$. If $\varphi(H_i)$ is not contained in M, then $\varphi(H_i) \cap M \subsetneqq \varphi(H_i)$ and $|\varphi(H_i) \cap M| < |\varphi(H_i)| = |H|$. But $\varphi(H_i) \cap M \lhd G$, so minimality of $|H|$ implies $\varphi(H_i) \cap M = \{1\}$. The subgroup $\langle M, \varphi(H_i) \rangle$ is a subgroup of the same form as M but of larger order, a contradiction. It follows that $\varphi(H_i) \subset M$, all i, and $M = G$. Finally, $H = H_1$ must be simple, for $N \lhd H_1$ implies $N \lhd H_1 \times \cdots \times H_n = G$, contradicting the minimality of $|H|$. ∎

Exercise 5.20 shows that G determines the subgroup H to isomorphism.

Definition A normal subgroup $H \neq \{1\}$ of G is a **minimal normal subgroup** of G if it contains no normal subgroups of G other than itself and $\{1\}$.

Every nontrivial finite group contains minimal normal subgroups. In particular, a minimal normal subgroup of a finite simple group G is G itself.

Corollary 5.21   *A minimal normal subgroup H of a finite group G is either simple or a direct product of isomorphic simple groups.*

Proof   If N is a proper characteristic subgroup of H, then Exercise 5.32 shows $N \lhd G$, contradicting the choice of H.   ∎

Corollary 5.22   *If H is a minimal normal subgroup of a finite solvable group G, there is a prime p with H either cyclic of order p or a direct product of cyclic groups of order p.*

Proof   A simple subgroup of a solvable group must be cyclic of prime order.   ∎

REMARK   There is a simpler proof of Corollary 5.22 avoiding Theorem 5.20. Use Exercise 5.32 with $K = H'$ to see H is abelian, then use this exercise with K any Sylow subgroup of H to see H is a p-group for some prime p, and finally, use this exercise with $K = \{x \in H : x^p = 1\}$ to see H is as described.

A THEOREM OF P. HALL

The main result of this section is a generalization of the Sylow theorems that holds for (and, in fact, characterizes) finite solvable groups.

Theorem 5.23   **(P. Hall, 1928)**   *Let G be a solvable group of order ab, where $(a, b) = 1$. Then G contains at least one subgroup of order a, and any two such are conjugate.*

Proof   The proof proceeds by induction on $|G|$, the theorem holding when $|G| = 1$.

Case (i): G contains a normal subgroup H of order $a_1 b_1$, where a_1 divides a, b_1 divides b, and $b_1 < b$.

G/H is a solvable group of order $(a/a_1) \cdot (b/b_1)$ and so contains a subgroup $\hat{A}/H$ of order a/a_1. The subgroup $\hat{A}$ of G has order $(a/a_1)|H| = ab_1 < ab$. Since $\hat{A}$ is also solvable, it contains a subgroup of order a, as desired.

Suppose A and A_1 are subgroups of G of order a. Let us compute $k = |AH|$. By Lagrange's theorem, $k | ab = |G|$ while $a_1 b_1 | k$ and $a | k$. By the second isomorphism theorem (actually, by the product formula), $k | aa_1 b_1 = |A| |H|$. Since $(a, b) = 1$, it follows that $k = |AH| = ab_1$. In a similar manner, $|A_1 H| = ab_1$ also. Thus AH/H and $A_1 H/H$ are subgroups of G/H of order a/a_1; by induction, these subgroups are conjugate (say, by $\bar{x} \in G/H$). If $x \in G$ goes into $\bar{x}$ under the natural map, it is quickly checked that $xAHx^{-1} = A_1 H$. Therefore, xAx^{-1} and A_1 are subgroups of $A_1 H$ of order a and so are conjugate, by induction. This completes case (i).

If there is some proper normal subgroup of G whose order is not divisible by b, the theorem is proved. We may therefore assume that b divides $|H|$ for every proper normal subgroup H. If H is a minimal normal subgroup, however, $|H| = p^m$ for some prime p, by Corollary 5.22. We conclude that $b = p^m$. Thus, H is a Sylow p-subgroup of G; that H is normal implies that H is the unique such subgroup. The problem has now been reduced to the following case.

Case (ii): G has a unique minimal normal subgroup H (and H is an abelian group of order p^m and index prime to p).

We finish the proof now, but note that this case follows immediately from the Schur-Zassenhaus lemma, which will be proved in Chapter 7.

Observe that since G is finite, every normal subgroup of G contains a minimal normal subgroup of G; in the present case, every proper normal subgroup of G must contain H.

Let K/H be a minimal normal subgroup of G/H. By Corollary 5.22, $|K/H| = q^n$, so that $|K| = p^m q^n$. Let S be a Sylow q-subgroup of K and let $N^* = N_G(S)$ be the normalizer of S in G. We shall show that $|N^*| = a$.

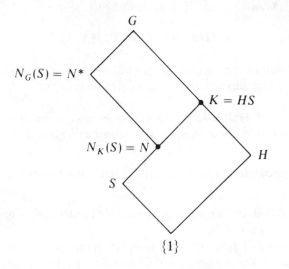

Let $N = K \cap N^* = N_K(S)$. The Frattini argument (Theorem 4.12) shows that $K = HN$.

Since $K \lhd G$, $N^* = N_G(S)$, and S is a Sylow subgroup of K, the Frattini argument also gives $G = KN^*$. There is an isomorphism

$$G/K = KN^*/K \cong N^*/K \cap N^* = N^*/N;$$

hence $|G||N|/|K| = |N^*|$. But $|K| = |HN| = |H||N|/|H \cap N|$, so that

$$|N^*| = (|G|/|H|)|H \cap N| = a|H \cap N|.$$

Thus $|N^*| = a$ if we can show $H \cap N = \{1\}$. We do this in two stages: (1) $H \cap N \subset Z(K)$; (2) $Z(K) = \{1\}$.

Let $x \in H \cap N$. If $k \in K$, then $k = hs$, where $h \in H$ and $s \in S$. Since $x \in H$, x commutes with h, for H is abelian. Hence, we need show only that x commutes with s. Now $(x^{-1}s^{-1}x)s \in S$, since $x \in N = N_K(S)$; $x^{-1}(s^{-1}xs) \in H$, since H is normal. Therefore, $x^{-1}s^{-1}xs \in S \cap H = \{1\}$, and x commutes with s, as desired.

Finally, $Z(K)$ is a characteristic subgroup of K and $K \lhd G$, so that $Z(K) \lhd G$. If $Z(K) \neq \{1\}$, then $Z(K)$ contains a minimal normal subgroup of G; thus $H \subset Z(K)$. This, together with $K = HS$, tells us that $S \lhd K$. By Exercise 5.32, $S \lhd G$ and so S, too, contains H, a contradiction. We conclude $|N^*| = a$.

Suppose A is another subgroup of G of order a. Since $|AK|$ is divisible by a and by $|K| = p^m q^n$, $|AK| = |G| = ab$ and $AK = G$.

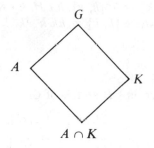

Therefore,

$$G/K = AK/K \cong A/A \cap K$$

implies $|A \cap K| = q^n$. By the Sylow theorem, $A \cap K$ is conjugate to S. Since conjugate subgroups have conjugate normalizers (Exercise 3.55), $N^* = N_G(S)$ is conjugate to $N_G(A \cap K)$. Therefore, $a = |N^*| = |N_G(A \cap K)|$. Since $A \cap K \lhd A$, we have $A \subset N_G(A \cap K)$. It follows that $A = N_G(A \cap K)$, for both have order a. Hence N^* is conjugate to A. ∎

Let us compare Hall's theorem to the Sylow theorems. Suppose $|G| = p^m a$, where $(p^m, a) = 1$. The Sylow theorems say that G has at least one subgroup of order p^m and that any two such are conjugate; Hall's theorem says that if G is solvable, G contains at least one subgroup of order a, and any two such are conjugate. Subgroups of (not necessarily solvable) groups whose order and index are relatively prime are often called **Hall subgroups**.

Definition Let G be a group of order $p^m a$, where $(p^m, a) = 1$. A subgroup of G of order a is a **p-complement** of G.

Hall's theorem tells us that every finite solvable group has a p-complement for every p. The converse of this statement is also true: If G is a finite group containing a p-complement for every p, then G is solvable. Notice that this converse gives an immediate proof of Burnside's theorem: If $|G| = p^m q^n$, then G is solvable. In this case, the p- and q-complements are just Sylow q- and p-subgroups. Unfortunately, the proof of the converse (also due to P. Hall, 1937) makes use of Burnside's theorem which we cannot prove here.

CENTRAL SERIES AND NILPOTENT GROUPS

The Sylow theorems show that knowledge of p-groups gives information about arbitrary finite groups. Moreover, p-groups have a rich supply of normal subgroups, and hence admit many homomorphisms; this suggests that normal series might be a powerful tool in their study.

Definition   Let H and K be subgroups of G. Then

$$[H, K] = <[h, k]: h \, \epsilon \, H, k \, \epsilon \, K>,$$
$$\text{where } [h, k] = hk \, h^{-1}k^{-1}.$$

EXERCISES

5.38.   For every two subgroups H and K of G,

$$[H, K] = [K, H].$$

5.39. $G' = [G, G]$ and, for all i, $G^{(i+1)} = [G^{(i)}, G^{(i)}]$.

5.40.   A subgroup $K \subset N_G(H)$ if and only if $[H, K] \subset H$. One says that K **normalizes H in this case.

5.41. $[H, K] = \{1\}$ if and only if each $k \in K$ commutes with every element of H. One says that K **centralizes** H in this case.

Definition   If H is a subgroup of G, the **centralizer** of H in G is

$$C_G(H) = \{a \in G : [a, h] = 1 \text{ for all } h \in H\}.$$

Of course, $a \in C_G(H)$ if and only if $ah = ha$ for all $h \in H$.

EXERCISES

**5.42.   Let $K \triangleleft G$ and $K \subset H \subset G$. Then $[H, G] \subset K$ if and only if $H/K \subset Z(G/K)$.

**5.43.   Let $f: G \to H$ be a homomorphism onto. If $A \subset Z(G)$, then $f(A) \subset Z(H)$.

Definition   Define a chain of subgroups $\gamma_i(G)$ inductively:

$$\gamma_1(G) = G;$$
$$\gamma_{i+1}(G) = [\gamma_i(G), G].$$

It is easy to see that $\gamma_{i+1}(G) \subset \gamma_i(G)$. Since $[\gamma_i(G), G] = \gamma_{i+1}(G)$, Exercise 5.40 says that $\gamma_{i+1}(G)$ is a normal subgroup of G for all i. Thus, if $x \in G$ and $y \in \gamma_i(G)$, then x and y commute modulo $\gamma_{i+1}(G)$.

Definition   The **descending central series** of G is the normal series

$$G = \gamma_1(G) \supset \gamma_2(G) \supset \cdots.$$

Definition   Define a chain of subgroups inductively: $Z^0(G) = \{1\}$; $Z^{i+1}(G)$ is the subgroup of G corresponding to the center of $G/Z^i(G)$: $Z^i(G)$ is the ith **higher center** of G.

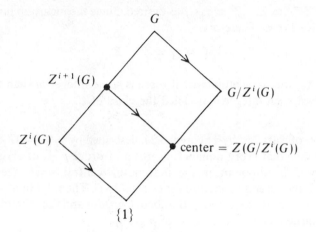

By the correspondence theorem, each $Z^i(G) \subset Z^{i+1}(G)$, and $Z^i(G)$ is a normal subgroup of G.

Definition   The **ascending central series** of G is

$$\{1\} = Z^0(G) \subset Z^1(G) \subset \cdots.$$

When no confusion can occur, we shall abbreviate $Z^i(G)$ by Z^i and $\gamma_i(G)$ by γ_i.

Theorem 5.24   *For any group G, $Z^m(G) = G$ if and only if $\gamma_{m+1}(G) = \{1\}$. Moreover,*

$$\gamma_{i+1}(G) \subset Z^{m-i}(G) \qquad \text{for all } i.$$

Proof Assuming $Z^m = G$, we shall prove the inclusion holds by an induction on i. Both terms equal G when $i = 0$, so the induction begins. If $\gamma_{i+1} \subset Z^{m-i}$, then

$$\gamma_{i+2} = [\gamma_{i+1}, G] \subset [Z^{m-i}, G] \subset Z^{m-i-1},$$

the last inclusion following from Exercise 5.42. Since the inclusion holds for all i, it holds for $i = m$. Therefore,

$$\gamma_{m+1} \subset Z^0 = \{1\}.$$

Assuming that $\gamma_{m+1} = \{1\}$, we shall prove by induction on j that $\gamma_{m+1-j} \subset Z^j$ (this is the same inclusion as in the statement of the theorem, for the sum of the indices is $m+1$). Both terms equal $\{1\}$ when $j = 0$, so the induction begins. If $\gamma_{m+1-j} \subset Z^j$, then there is a homomorphism

$$G/\gamma_{m+1-j} \to G/Z^j$$

that is onto. Now $\gamma_{m-j}/\gamma_{m+1-j} \subset Z(G/\gamma_{m+1-j})$, since $[\gamma_{m-j}, G] \subset \gamma_{m+1-j}$ and Exercise 5.42 holds. By Exercise 5.43,

$$\gamma_{m-j}Z^j/Z^j \subset Z(G/Z^j) = Z^{j+1}/Z^j.$$

Therefore, $Z^{j+1} \supset \gamma_{m-j}Z^j \supset \gamma_{m-j}$, as desired. Since the inclusion holds for all j, it holds for $j = m$. Therefore,

$$G = \gamma_1 \subset Z^m. \quad \blacksquare$$

Definition   A group G is **nilpotent**† if there is some integer m such that $\gamma_{m+1}(G) = \{1\}$. The least such integer m is called the **class** of G.

A group G of class 2 is, by Theorem 5.24, described by $\gamma_2(G) = G' \subset Z(G)$. Thus Exercise 4.40 says that every nonabelian group of order p^3 is of class 2.

A group G is nilpotent if the descending central series reaches $\{1\}$ or, equivalently, if the ascending central series reaches G. The nilpotent groups form a class of groups lying strictly between the abelian groups and the solvable groups, and should be regarded as generalizations of p-groups.

It is easy to prove by induction that $G^{(i)} \subset \gamma_i(G)$ for all i; it follows that if $\gamma_m(G) = \{1\}$ for some m, then $G^{(m)} = \{1\}$, i.e., if G is nilpotent, then G is solvable. The group S_3 is a solvable group; we claim that it is not nilpotent. Every nilpotent group G has a nontrivial center: if m is the first integer for which $\gamma_{m+1}(G) = \{1\}$, then $\{1\} \neq \gamma_m(G) \subset Z^1(G) = Z(G)$. Therefore, S_3 is not nilpotent, for it is centerless.

Theorem 5.25   *Every subgroup of a nilpotent group G is nilpotent.*

Proof   It is easily proved by induction that if $H \subset G$, then $\gamma_i(H) \subset \gamma_i(G)$ for all i. Therefore, $\gamma_m(G) = \{1\}$ implies $\gamma_m(H) = \{1\}$. $\quad \blacksquare$

Theorem 5.26   *If G is nilpotent and $H \lhd G$, then G/H is nilpotent.*

Proof   If $f: G \to L$ is a homomorphism onto, then it is easily proved by induction that $\gamma_i(L) \subset f(\gamma_i(G))$ for all i. The theorem follows if we take f to be the natural map of G onto G/H. $\quad \blacksquare$

We have just proved the analogs for nilpotent groups of Theorems 5.12 and 5.13. Is the analog of Theorem 5.14 true? If $H \lhd G$ and both H and G/H are nilpotent, is G nilpotent? Again, S_3 is a counterexample; both A_3 and S_3/A_3 are abelian and hence nilpotent, but we have already seen that S_3 is not nilpotent. The analog of Corollary 5.15, however, is true.

---

† One may define the analog of the descending central series for a Lie algebra. Engel's theorem asserts that a Lie algebra is nilpotent (i.e., the descending central series reaches $\{0\}$) if and only if its regular representation consists of nilpotent matrices. This is why nilpotent groups are so called.

Theorem 5.27 *A direct product G of a finite number of nilpotent groups is nilpotent.*

Proof An induction on the number of direct factors allows us to assume that $G = H \times K$. Another induction proves that

$$\gamma_i(H \times K) \subset \gamma_i(H) \times \gamma_i(K) \qquad \text{for all } i.$$

Let $M = \max\{m, n\}$, where $\gamma_m(H) = \{1\} = \gamma_n(K)$. Then $\gamma_M(H \times K) = \{1\}$, so that $H \times K$ is nilpotent. ∎

Lemma 5.28 *Every finite p-group G is nilpotent.*

Proof We know, by Theorem 4.4, that G and all its nontrivial quotients have nontrivial centers. Therefore, if $Z^i \neq G$ for some i, then $Z^i \subsetneqq Z^{i+1}$. Since G is finite, we cannot have this inequality for all i. It follows that $Z^i = G$ for some i, i.e., G is nilpotent. ∎

This lemma is false without the finiteness assumption, for there exist infinite p-groups that are centerless.

Lemma 5.29 *Let G be nilpotent and let H be a proper subgroup. Then $H \neq N_G(H)$.*

Proof There exists an i such that $H \supset \gamma_{i+1}(G)$ but $H \not\supset \gamma_i(G)$ (this is true for any chain of subgroups starting at G and ending at $\{1\}$). Now

$$[\gamma_i, H] \subset [\gamma_i, G] = \gamma_{i+1} \subset H,$$

so that $\gamma_i \subset N_G(H)$, by Exercise 5.40. There is thus some element in $N_G(H)$ not in H. ∎

Theorem 5.30 *If G is nilpotent, then every subgroup H of prime index is normal.*

Proof Since $p = [G:H] = [G:N_G(H)][N_G(H):H]$, one of the factors is 1. By Lemma 5.29, $N_G(H) \neq H$. Therefore $G = N_G(H)$ and $H \lhd G$. ∎

The following theorem is a satisfying characterization of finite nilpotent groups that indicates they are not too far from p-groups.

Theorem 5.31 *A finite group G is nilpotent if and only if it is the direct product of its Sylow subgroups.*

Proof If G is the direct product of its Sylow subgroups, then Theorem 5.27 and Lemma 5.28 show G is nilpotent.

To prove the converse, take a Sylow subgroup P of G and let $N = N_G(P)$. In Exercise 4.8, we saw that N is its own normalizer. By Lemma 5.29, N cannot be a proper subgroup of G; thus $N_G(P) = G$ and $P \lhd G$. Since all Sylow subgroups of G are normal, it follows from Exercise 4.9 that G is the direct product of its Sylow subgroups. ∎

EXERCISES

5.44. If G is nilpotent of class 2 and $a \in G$, then the function $[a, \quad]: G \to G$, defined by $b \mapsto [a, b]$, is a homomorphism.

5.45. The group of triangular matrices in Exercise 4.12 is nilpotent of class 2.

*5.46. G is nilpotent if and only if there is a normal series

$$G = G_0 \supset G_1 \supset \cdots \supset G_n = \{1\}$$

in which each $G_i \lhd G$ and such that $G_i/G_{i+1} \subset Z(G/G_{i+1})$ for all i. (Such a series is called a **central series**.)

*5.47. If $H \subset Z(G)$ and G/H is nilpotent, then G is nilpotent.

5.48. The dihedral group D_n is nilpotent if and only if n is a power of 2.

5.49. Let G be a finite nilpotent group of order n, and suppose m divides n. Prove that G contains a subgroup of order m.

5.50. If H, K are normal nilpotent subgroups of a finite group G, then HK is also a normal nilpotent subgroup of G. Conclude that every finite group G contains a unique maximal normal nilpotent subgroup (this subgroup is called the **Fitting subgroup** of G).

**5.51. A finite abelian p-group has a unique subgroup of order p if and only if it is cyclic. (HINT: Use Theorem 2.15.)

5.52. For a field F, let G be the subgroup of $GL(4, F)$ consisting of all matrices of the form

$$\begin{bmatrix} 1 & * & * & * \\ 0 & 1 & * & * \\ 0 & 0 & 1 & * \\ 0 & 0 & 0 & 1 \end{bmatrix}$$

where the $*$'s denote entries in F. Show that G is nilpotent of class 3 and compute G', G'', and $Z(G)$. Generalize to $n \times n$ matrices.

There are many **commutator identities** that are quite useful even though they are quite simple, e.g., $[x, y]^{-1} = [y, x]$. We will need two slightly more complicated identities. Recall that $x^a = axa^{-1}$ and $[x, y] = xyx^{-1}y^{-1}$.

EXERCISES

5.53. If $x, y, z \in G$, prove that

$$[x, yz] = [x, y][x, z]^y$$

and

$$[xy, z] = [y, z]^x[x, z].$$

5.54. **(Jacobi identity)** If $x, y, z \in G$, denote $[x, [y, z]]$ by $[x, y, z]$. Prove that

$$[x, y^{-1}, z]^y [y, z^{-1}, x]^z [z, x^{-1}, y]^x = 1.$$

5.55. **(Three subgroups lemma)** If H, K, and L are subgroups of G, let $[H, K, L]$ denote $\langle [h, k, l] : h \in H, k \in K, l \in L \rangle$. Show that $[H, K, L] = \{1\}$ and $[K, L, H] = \{1\}$ implies $[L, H, K] = \{1\}$. Conclude that if G is a group with $G = G'$, then $G/Z(G)$ is centerless. [HINT: Take $H = Z^2(G)$ and $K = L = G$, and use Exercise 5.42.]

Lemma 5.32　*Let $x, y \in G$ and suppose $[x, y]$ commutes with both x and y. Then*

(i) $[x, y]^n = [x^n, y] = [x, y^n]$ *for all $n \in \mathbf{Z}$;*
(ii) $(xy)^n = [y, x]^{n(n-1)/2} x^n y^n$ *for all $n \geq 0$.*

Proof　To prove (i) for $n \geq 0$, note that

$$
\begin{aligned}
[x, y]^n [x, y] &= x[x, y]^n y x^{-1} y^{-1}, &&\text{by hypothesis}\\
&= x[x^n, y] y x^{-1} y^{-1}, &&\text{by induction}\\
&= x(x^n y x^{-n} y^{-1}) y x^{-1} y^{-1}\\
&= [x^{n+1}, y].
\end{aligned}
$$

By hypothesis, $x[x, y] = [x, y]x$, and this yields $xyx^{-1}y^{-1} = yx^{-1}y^{-1}x$, i.e., $[x, y] = [y, x^{-1}]$. Therefore

$$[x, y]^{-1} = [y, x^{-1}]^{-1} = [x^{-1}, y]$$

(because $[a, b]^{-1} = [b, a]$ always). That $[x, y]^n = [x^n, y]$ for all $n \in \mathbf{Z}$ follows easily.

The second equality in (i) is proved as follows:

$$[x, y^n] = [y^n, x]^{-1} = [y, x]^{-n} = [x, y]^n.$$

To prove the second identity for $n \geq 0$, observe that, by induction,

$$
\begin{aligned}
(xy)^n (xy) &= [y, x]^{n(n-1)/2} x^n y^n x y\\
&= [y, x]^{n(n-1)/2} x^{n+1} [x^{-1}, y^n] y^{n+1}\\
&= [y, x]^{n(n-1)/2} x^{n+1} [y, x]^n y^{n+1}\\
&= [y, x]^{n(n-1)/2} [y, x]^n x^{n+1} y^{n+1}\\
&= [y, x]^{(n+1)n/2} x^{n+1} y^{n+1}. \quad\blacksquare
\end{aligned}
$$

Theorem 5.33　*Let G be a p-group with a unique subgroup of order p and more than one cyclic subgroup of index p. Then $G \cong Q$, the quaternions.*

Proof　If H is a subgroup of G having index p, then $H \triangleleft G$, by Theorem 5.30. Thus, if $x \in G$, then $xH \in G/H$, a group of order p. Therefore, $x^p \in H$.

Let $A = \langle a \rangle$ and $B = \langle b \rangle$ be distinct cyclic subgroups of index p, and let $D = A \cap B$; $D \triangleleft G$, being the intersection of normal subgroups. We collect facts about D. First of all, our initial remarks show the subset

$$G^p = \{x^p: x \in G\} \subset D.$$

Now A and B being distinct maximal normal subgroups implies $G = AB$. The product formula gives

$$[G:D] = p^2.$$

It follows that G/D is abelian, so that

$$G' \subset D,$$

by Exercise 2.52. Further, $G = AB$ means that each $x \in G$ is a product of powers of a and b. Thus, any element $d \in D$, being simultaneously a power of a and a power of b, must commute with each $x \in G$. Therefore

$$D \subset Z(G).$$

Now $G' \subset D \subset Z(G)$, so the hypothesis of Lemma 5.32 holds. Hence, for every $x, y \in G$

$$[y, x]^p = [y^p, x] = 1,$$

because $y^p \in D \subset Z(G)$, and so

$$(xy)^p = [y, x]^{p(p-1)/2} x^p y^p.$$

If p is odd, we have $(xy)^p = x^p y^p$, i.e., $x \mapsto x^p$ is a homomorphism. Setting $G_p = \{x \in G: x^p = 1\}$; Exercise 2.60 says G_p and G^p (defined above) are subgroups and $[G:G_p] = |G^p|$. Thus

$$|G_p| = [G:G^p] = [G:D][D:G^p] \geq p^2.$$

But G_p, containing an elementary abelian subgroup with at least p^2 elements, has more than one subgroup of order p, a contradiction. Therefore, $p = 2$.

When $p = 2$, we have $D = \langle a^2 \rangle = G^2 \subset Z(G)$, $[G:D] = 4$, and, since $[y, x]^2 = 1$ for all $x, y \in G$,

$$(xy)^4 = [y, x]^6 x^4 y^4 = x^4 y^4.$$

Hence

$$|G_4| = [G:G^4] = [G:D][D:G^4] = 8$$

(for $D = \langle a^2 \rangle = G^2$ and $\langle a^4 \rangle = G^4$). If G_4 had only one cyclic subgroup of order 4, it would contain more than one element of order 2. There are thus distinct cyclic subgroups $\langle u \rangle$ and $\langle v \rangle$ of order 4 in G_4. If $a^4 \neq 1$, we may assume $\langle u \rangle \subset \langle a^2 \rangle \subset Z(G)$ and $\langle u \rangle \langle v \rangle$ is an abelian subgroup of G. But $\langle u \rangle \langle v \rangle$ contains two distinct elements of order 2: either $u^2 \neq v^2$ or $u^2 \neq uv^{-1}$. This contradiction shows $a^4 = 1$. It follows that $|D| = 2$ and $|G| = 8$. A survey of the groups of order 8 shows that only the quaternions Q has a unique element of order 2 (and Q does have distinct cyclic subgroups of order 4, hence index 2). Therefore, $G \cong Q$. ∎

Before stating the next theorem, we do a little exercise in congruences.

Lemma 5.34 *Let H_m be the multiplicative group*

$$H_m = \{\bar{a} \in \mathbf{Z}/2^m \mathbf{Z}: a \text{ is odd}\}.$$

If $m \geq 3$, then

$$H_m = \langle -\bar{1}, \bar{5} \rangle \cong \mathbf{Z}(2) \times \mathbf{Z}(2^{m-2}).$$

Proof First note that, by Exercise 2.26, $|H_m| = \varphi(2^m) = 2^{m-1}$. Induction shows that

$$5^{2^{m-3}} = (1+4)^{2^{m-3}} \equiv 1 + 2^{m-1} \pmod{2^m}.$$

Since H_m is a 2-group, $\bar{5}$ has order 2^s, where $s \geq m-2$ [for $1 + 2^{m-1} \not\equiv 1$ $\pmod{2^m}$].

Now $-\bar{1}$ has order 2. We claim $\langle -\bar{1} \rangle \cap \langle \bar{5} \rangle = \{1\}$. Otherwise $\bar{5}^v = -\bar{1}$ for some v, i.e., $5^v \equiv -1 \pmod{2^m}$. Since $m \geq 3$, this congruence implies $5^v \equiv -1 \pmod 4$. But it is easy to see $5^v \equiv 1 \pmod 4$, a contradiction. It follows that these two cyclic subgroups generate their direct product, which is a subgroup of order $2 \cdot 2^s \geq 2 \cdot 2^{m-2} = 2^{m-1} = \varphi(2^m)$. This subgroup is thus all of H_m. ∎

Corollary 5.35 *If G is a group containing elements α and β such that α has order 2^m (where $m \geq 3$), $\beta^2 \in C_G(\alpha)$, and $\beta^{-1} \alpha \beta = \alpha^t$, then*

$$t = \pm 1 \quad or \quad t = \pm 1 + 2^{m-1}.$$

Proof Since β^2 commutes with α, we have

$$\alpha = \beta^{-2} \alpha \beta^2 = \beta^{-1}(\alpha^t)\beta = \alpha^{t^2}$$

so that $t^2 \equiv 1 \pmod{2^m}$. Thus, $\bar{t}$ is an element of order 2 in H_m. If $m \geq 3$, Lemma 5.34 says that there are four such elements:

$$t = \pm 1 \quad \text{and} \quad t = \pm 1 + 2^{m-1}. \quad ∎$$

We remark that if $t = 1$ in Corollary 5.35, then α and β commute.

Theorem 5.36 *A finite p-group G having a unique subgroup of order p is either cyclic or generalized quaternion.*

Proof The proof is by induction on n, where $|G| = p^n$; of course the theorem is true when $n = 0$.

Assume first that p is odd. If $n > 0$, then G has a subgroup H of index p, by Exercise 4.2, which must be cyclic, by induction. There can be no other subgroup of index p lest it, too, be cyclic and G be the quaternions. Therefore, H is the unique maximal subgroup of G and so it contains every proper subgroup of G. Suppose G is not cyclic. Then, for every $x \in G$, $\langle x \rangle$ is a proper subgroup. Hence $\langle x \rangle \subset H$ and $G \subset H$, a contradiction.

Assume that $p = 2$. We wish to show that G contains a cyclic subgroup of index 2, but first let us show it contains a normal cyclic subgroup of index 4. If $|G| = 16$, then G contains a normal subgroup N of index 4 (Exercise 4.2), and thus of order 4; N is cyclic, for the 4-group has 3 elements of order 2. If $|G|$

$= 2^n > 16$, then a subgroup H of G having index 2 is either cyclic or generalized quaternion, by induction. In either case, H contains a unique (Theorem 5.33) cyclic subgroup $\langle a \rangle$ of order 2^{n-2}. Therefore $\langle a \rangle$ is a characteristic subgroup of H and hence is a normal cyclic subgroup of G of index 4.

Clearly we may suppose $|G| = 2^n$ where $n \geq 4$. To satisfy our claim that G contains a cyclic subgroup of index 2, let us consider the group $G/\langle a \rangle$ of order 4. If $G/\langle a \rangle \cong \langle \bar{b} \rangle$ and $b \in G$ with $b \mapsto \bar{b}$, then $G = \langle a, b \rangle$ and $\langle a, b^2 \rangle$ is a subgroup of G of index 2. If $\langle a, b^2 \rangle$ is abelian, it is cyclic (Exercise 5.51) and our claim holds; if $\langle a, b^2 \rangle$ is not abelian, it is generalized quaternion (by induction). In particular, we may assume $b^{-2}ab^2 = a^{-1}$. Now $\langle a \rangle \triangleleft G$ gives $b^{-1}ab = a^i$ for some i. Therefore

$$a^{-1} = b^{-2}ab^2 = b^{-1}(b^{-1}ab)b = b^{-1}a^i b = a^{i^2},$$

so that $i^2 \equiv -1 \pmod{2^{n-2}}$. However no such congruence holds (if $n - 2 = 2$, merely compute modulo 4; if $n - 2 \geq 3$, use Lemma 5.34). We may thus assume $G/\langle a \rangle$ is the 4-group, so there exist elements $b, c \in G$ with $b, c, b^{-1}c \notin \langle a \rangle$ and with $\langle a, b \rangle$, $\langle a, c \rangle$, and $\langle a, b^{-1}c \rangle$ each of index 2 in G. If either $\langle a, b \rangle$ or $\langle a, c \rangle$ is cyclic, the claim is true; otherwise, both are generalized quaternion, by induction. In particular, there are equations $b^{-1}ab = a^{-1} = c^{-1}ac$. It follows that $b^{-1}c \in C_G(a)$ and $\langle a, b^{-1}c \rangle$ is abelian, hence cyclic.

We have shown that G must contain a cyclic subgroup $\langle \alpha \rangle$ of index 2. Choose $\beta \in G$, $\beta \notin \langle \alpha \rangle$. Since $[G : \langle \alpha \rangle] = 2$, we have $\beta^2 \in \langle \alpha \rangle$. Changing generator of $\langle \alpha \rangle$ if necessary (Exercise 2.23), we have

$$\beta^2 = \alpha^{2^r}.$$

Note that $r \leq n - 2$ (if $r > n - 2$, then β is a second element of order 2). Furthermore, we may suppose α and β do not commute, otherwise G is abelian, and hence cyclic. Corollary 5.35 gives

$$\beta^{-1}\alpha\beta = \alpha^t,$$

where either $t = -1$, $-1 + 2^{n-2}$, or $1 + 2^{n-2}$ (since α has order 2^{n-1}). We eliminate the last two possible values of t.

Suppose $t = 1 + 2^{n-2}$. For any integer m,

$$(\alpha^m\beta)^2 = \alpha^m\beta^2(\beta^{-1}\alpha^m\beta) = \alpha^{2s},$$

where $s = 2^{r-1} + m(1 + 2^{n-3})$. Since $1 + 2^{n-3}$ is odd ($n \geq 4$), we may solve the congruence

$$2^{r-1} + m(1 + 2^{n-3}) \equiv 0 \bmod 2^{n-2}.$$

For this choice of m, we have $(\alpha^m\beta)^2 = \alpha^{2s} = \alpha^{2^{n-1}} = 1$. Thus, $\alpha^m\beta \notin \langle \alpha \rangle$ is a new element of order 2.

Suppose $t = -1 + 2^{n-2}$. As in the proof of Corollary 5.35,

$$2^r \equiv 2^r(-1 + 2^{n-2}) \pmod{2^{n-1}}$$

so that

$$2^r \equiv 0 \pmod{2^{n-2}}$$

and $r = n - 2$. But now

$$(\alpha\beta)^2 = \alpha\beta^2(\beta^{-1}\alpha\beta) = \alpha^{1+t+2^r}$$
$$= \alpha^{2^{n-2}}\alpha^{2^{n-2}} = 1,$$

so a second element of order 2 has been exhibited.

Therefore, $G = \langle \alpha, \beta \rangle$, where

$$\alpha^{2^{n-1}} = 1, \quad \beta^{-1}\alpha\beta = \alpha^{-1}, \quad \beta^2 = \alpha^{2^r}.$$

We need only show $r = n - 2$ to complete the proof. This follows from Lemma 5.34: since $t = -1$, we have $2^r \equiv -2^r \pmod{2^{n-1}}$ so that $2^{r+1} \equiv 0 \pmod{2^{n-1}}$, and $r = n - 2$. ∎

EXERCISES

5.56. Every subgroup of Q_n is either cyclic or generalized quaternion.

*5.57. (**Wielandt, 1939**) A finite group G is nilpotent if and only if every maximal subgroup is normal. (HINT: Use Exercise 4.8.) Conclude that in this case every maximal subgroup has prime index.

*5.58. Let $\Phi(G)$ be the intersection of all the maximal subgroups of G [$\Phi(G)$ is called the **Frattini subgroup** of G]. Prove that $\Phi(G) \lhd G$ [indeed, $\Phi(G)$ is a characteristic subgroup of G].

5.59. If G is finite, prove that $\Phi(G)$ is nilpotent.

**5.60. An element $x \in G$ is called a 'nongenerator' if whenever a subset Y satisfies $G = \langle Y, x \rangle$, then $G = \langle Y \rangle$. Prove that $\Phi(G)$ is the set of all nongenerators of G.

**5.61. If G is a finite p-group, then $\Phi(G) = G'G^p$, where G' is the commutator subgroup of G and $G^p = \langle x^p : x \in G \rangle$. Conclude that $G/\Phi(G)$ is a vector space over $\mathbf{Z}/p\mathbf{Z}$.

Theorem 5.37 (**Burnside basis theorem, 1912**) *If G is a finite p-group, then any two minimal generating sets have the same number of elements. Moreover, every $x \notin \Phi(G)$ belongs to some minimal generating set.*

Proof If $\{x_1, \cdots, x_n\}$ is a minimal generating set of G (no proper subset generates G), we must show n depends only on G. Now the cosets $\bar{x}_1, \cdots \bar{x}_n$ span the vector space $G/\Phi(G)$. If $\{\bar{x}_1, \cdots, \bar{x}_n\}$ is dependent, then one of them, say $\bar{x}_1$, lies in $\langle \bar{x}_2, \cdots, \bar{x}_n \rangle$. There is thus $y \in \langle x_2, \cdots, x_n \rangle$ with $x_1 y^{-1} \in \Phi(G)$. Since $\{x_1 y^{-1}, x_2, \cdots, x_n\}$ is a generating set of G, so is $\{x_2, \cdots, x_n\}$, by Exercise 5.60, and this contradicts minimality. Therefore $n = \dim G/\Phi(G)$.

If $x \notin \Phi(G)$, then $\bar{x} \neq 0$ in $G/\Phi(G)$ and so $\bar{x}$ is part of a basis, say $\{\bar{x}, \bar{x}_2, \cdots, \bar{x}_n\}$. If x_i lies in the coset $\bar{x}_i$ for $i \geq 2$, then $G = \langle x, x_2, \cdots, x_n, \Phi(G) \rangle$, and Exercise 5.60 shows $G = \langle x, x_2, \cdots, x_n \rangle$. Now $\{x, x_2, \cdots, x_n\}$ is a minimal generating set, for the cosets of a proper subset do not generate $G/\Phi(G)$. ∎

The hypothesis that G is a p-group is necessary. If $G = \langle x \rangle \cong \mathbf{Z}(6)$, then two minimal generating sets of G are $\{x\}$ and $\{x^2, x^3\}$.

The reader familiar with ring theory will note the analogy of the Frattini subgroup with the Jacobson radical and the similarity of the Burnside basis theorem with the notion of the rank of a noetherian local ring.

chapter six

Finite
Direct Products

The main result of this chapter is a complete description of finite abelian groups as direct products of cyclic p-groups. The essential uniqueness of this factorization is then extended to nonabelian groups that are direct products of "indecomposable" groups.

THE BASIS THEOREM

For the next three sections, we shall deal exclusively with abelian groups. As is the usual custom, we now shift from multiplicative notation to additive notation. The following dictionary should prove useful:

ab	$a+b$
1	0
a^{-1}	$-a$
a^n	na
ab^{-1}	$a-b$
HK	$H+K$
aH	$a+H$
direct product	direct sum
$H \times K$	$H \oplus K$
$\prod\limits_{i=1}^{m} H_i$	$\sum\limits_{i=1}^{m} H_i$

If $G = H \times K$, H is called a **direct factor** of G; in additive notation, H is a **direct summand** of G.

Two remarks that hold for abelian groups greatly simplify our study.

1. If $a, b \in G$ and $n \in \mathbf{Z}$, then $n(a + b) = na + nb$ (Exercise 1.30).
2. If X is a nonempty subset of G, then $\langle X \rangle$ is the set of all (finite) linear combinations of elements of X with coefficients in $\mathbf{Z}$ (Theorem 2.5).

Definition   Let p be a prime. A group G is *p-primary* if every element in G has order a power of p.

If working wholly in the context of abelian groups, use the term *p-primary*; otherwise, the usage of *p-group* is preferred.

Theorem 6.1   (Primary Decomposition)   *Every finite abelian group G is a direct sum of p-primary groups.*

Proof   There is an integer n (namely, $n = |G|$) with $nx = 0$ for all $x \in G$. For each prime divisor p of n, define G_p to be the set of all elements in G of order some power of p. Now $0 \in G_p$, and, since G is abelian, G_p is a subgroup of G. We claim that $G = \Sigma G_p$, where the indices range over all primes p dividing $|G|$; we use the criterion of Exercise 2.81.

(i) Let $x \in G$, $x \neq 0$, and let the order of x be n. By the fundamental theorem of arithmetic, $n = p_1^{e_1} p_2^{e_2} \cdots p_k^{e_k}$, where the p_i are distinct primes and the exponents $e_i \geq 1$. Set $n_i = n/p_i^{e_i}$, and observe that the gcd $(n_1, n_2, \cdots, n_k)$ $= 1$ (for what prime could be a common divisor of the n_i?).† Therefore, there exist integers m_i such that $\Sigma m_i n_i = 1$; hence, $\Sigma (m_i n_i x) = x$. Note that $p_i^{e_i}(m_i n_i x) = m_i n x = 0$, so that $m_i n_i x \in G_{p_i}$. We conclude that the collection of G_p generates G.

(ii) Suppose $x \in G_p \cap \langle \bigcup_{q \neq p} G_q \rangle$. On the one hand, $p^e x = 0$ for some e; on the other hand, $x = \Sigma x_q$, where $q^{e_q} x_q = 0$ for exponents e_q. If we set $t = \Pi q^{e_q}$, then $tx = 0$. Clearly, $(p^e, t) = 1$, so that there are integers a and b with $ap^e + bt = 1$. Therefore, $x = ap^e x + btx = 0$.   ∎

Definition   The subgroups G_p of G are called the **primary components**†† of G.

We plan to show that every finite abelian group is a direct sum of cyclic groups. Theorem 6.1 allows us to consider, without loss of generality, the special case of finite p-primary abelian groups.

---

† A common divisor of a set of integers $\{n_1, \cdots, n_k\}$ is a nonzero integer c that divides each n_i; the greatest common divisor (gcd), denoted by $(n_1, \cdots, n_k)$, is a positive common divisor that is divisible by every common divisor. It may be shown, in a manner analogous to the special case of a set of two integers, that the gcd exists and is a linear combination of $n_1, \cdots, n_k$ (see Appendix VI).
†† Of course the p-primary component of G is its Sylow p-subgroup. Moreover, that abelian groups are nilpotent shows the primary decomposition follows immediately from Theorem 5.31.

EXERCISE

**6.1. Let G be a p-primary abelian group and let $y_1, \cdots, y_t$ be nonzero elements of G for which

$$\langle y_1, \cdots, y_t \rangle = \langle y_1 \rangle \oplus \cdots \oplus \langle y_t \rangle.$$

(a) If $z_1, \cdots, z_t$ are elements of G with $pz_i = y_i$, for all i, then

$$\langle z_1, \cdots, z_t \rangle = \langle z_1 \rangle \oplus \cdots \oplus \langle z_t \rangle.$$

(b) If $k_1, \cdots, k_t$ are integers with $k_i y_i \neq 0$, then

$$\langle k_1 y_1, \cdots, k_t y_t \rangle = \langle k_1 y_1 \rangle \oplus \cdots \oplus \langle k_t y_t \rangle.$$

Definition If G is an abelian group and m is a positive integer, then

$$mG = \{mx: x \in G\}.$$

Lemma 6.2 *An abelian group G with $pG = \{0\}$ is a vector space over $\mathbf{Z}/p\mathbf{Z}$, and it is a direct sum of cyclic groups of order p when G is finite.*

Proof Let $\bar{k}$ denote the congruence class of the integer k in $\mathbf{Z}/p\mathbf{Z}$. Define a scalar multiplication on G by

$$\bar{k}x = kx, \qquad \text{where } x \in G.$$

This operation is well defined, for if $k \equiv k' \pmod{p}$, then $k - k' = mp$ for some integer m, so that

$$(k - k')x = mpx = 0;$$

hence,

$$kx = k'x.$$

It is easily checked that G is a vector space over $\mathbf{Z}/p\mathbf{Z}$, and as such has a basis $\{x_1, x_2, \cdots, x_t\}$ when G is finite. We let the reader prove, using Theorem 2.5, that G is the direct sum of the $\langle x_i \rangle$. ∎

Theorem 6.3 (Basis Theorem)† *Every finite abelian group G is a direct sum of primary cyclic groups.*

Proof By Theorem 6.1, we may assume that G is p-primary. We perform an induction on m, where m is an integer such that $p^m G = \{0\}$. If $m = 1$, the theorem is just Lemma 6.2.

Suppose that $p^{m+1} G = \{0\}$. If $H = pG$, then $p^m H = \{0\}$, so that induction gives

$$H = pG = \Sigma \langle y_i \rangle.$$

---

† The basis theorem was first proved by E. Schering (1868) and Kronecker (1870).

Since $y_i \in pG$, there are elements $z_i \in G$ with $pz_i = y_i$. If L is the subgroup of G generated by the z_i, then Exercise 6.1(a) shows that

$$L = \Sigma \langle z_i \rangle.$$

[We interrupt the proof to understand the next step. Suppose the theorem were true, so that $G = \Sigma C_k$, where each C_k is cyclic. In considering pG, we are neglecting all C_k of order p (if any), for they were destroyed by multiplication by p. Now the construction of L has recaptured the C_k of order bigger than p, but nothing has yet been done to revive those C_k of order exactly p. Let us resume the proof.]

We claim that L is a direct summand of G, and so we produce a subgroup M of G such that $L \oplus M = G$.

If $G[p] = \{x \in G : px = 0\}$, then $p(G[p]) = \{0\}$, so that $G[p]$ is a vector space over $\mathbf{Z}/p\mathbf{Z}$, by Lemma 6.2. If k_i is the order of y_i (of course, k_i is a power of p), then $k_i z_i$ has order p, and so

$$k_i z_i \in G[p].$$

By Exercise 6.1(b), the set of $k_i z_i$ is an independent subset of the vector space $G[p]$. Therefore, we can extend this set to a basis of $G[p]$, i.e., there are elements $\{x_1, x_2, \cdots, x_s\}$ such that

$$\{\text{the } k_i z_i, x_1, x_2, \cdots, x_s\}$$

is a basis of $G[p]$. Let $M = \langle x_1, x_2, \cdots, x_s \rangle$. Observe that, as in the proof of Lemma 6.2,

$$M = \Sigma \langle x_i \rangle.$$

(Of course, M is composed of the resurrected summands of order p.)

We now show that $G = L \oplus M$, which will complete the proof.

(i) $L \cap M = \{0\}$. If $x \in L \cap M$, then $x = \Sigma b_i z_i = \Sigma a_j x_j$. Now $px = 0$ (since $x \in M$), so that $\Sigma pb_i z_i = 0$, which implies

$$0 = pb_i z_i = b_i y_i, \qquad \text{for all } i.$$

Hence, $b_i = b_i' k_i$ for some integer b_i', by Exercise 2.16. Therefore,

$$0 = \Sigma b_i' k_i z_i - \Sigma a_j x_j,$$

so that independence implies each term is 0; thus, $x = 0$.

(ii) $G = L + M$. Let $x \in G$. Now

$$px = \Sigma c_i y_i = \Sigma pc_i z_i,$$

so that

$$p(x - \Sigma c_i z_i) = 0,$$

and

$$x - \Sigma c_i z_i \in G[p].$$

Therefore,

$$x - \Sigma c_i z_i = \Sigma a_j x_j + \Sigma b_i k_i z_i,$$

i.e.,

$$x = \Sigma (c_i + b_i k_i) z_i + \Sigma a_j x_j \in L + M. \quad \blacksquare$$

Corollary 6.4 *Every finite abelian group G can be decomposed into a direct sum of cyclic groups*

$$G = \mathbf{Z}(m_1) \oplus \mathbf{Z}(m_2) \oplus \cdots \oplus \mathbf{Z}(m_s),$$

where $m_i | m_{i+1}$ for all $i = 1, \cdots, s - 1$.

REMARK   Some authors say G is an abelian group of **type** $(m_1, m_2, \cdots, m_s)$.

Proof G has a primary decomposition

$$G = G_{p_1} \oplus \cdots \oplus G_{p_t},$$

where $p_1, \cdots, p_t$ are distinct primes. By the basis theorem, we may assume each primary component is a direct sum of cyclic groups. For each i, let $\mathbf{Z}(p_i^{e_i})$ be a cyclic summand in the given decomposition of G_{p_i} having largest order. Note that if $\mathbf{Z}(p_i^{f_i})$ is any other cyclic summand in the decomposition, $f_i \leq e_i$, so that $p_i^{f_i} | p_i^{e_i}$. Since direct sum is commutative and associative, we have

$$G = H \oplus (\mathbf{Z}(p_1^{e_1}) \oplus \cdots \oplus \mathbf{Z}(p_t^{e_t})),$$

where H is the direct sum of the remaining cyclic summands. By Exercise 2.74,

$$G = H \oplus \mathbf{Z}(m),$$

where $m = \Pi p_i^{e_i}$. Repeating this process on H, we may write $H = H_1 \oplus \mathbf{Z}(n)$. Moreover, $n | m$ because we chose primary summands of largest order. Clearly this process ends in a finite number of steps. $\blacksquare$

Call a decomposition of G as in Corollary 6.4 a **canonical decomposition**.

EXERCISES

6.2.   Let G be a finite abelian group of order n. If m divides n, show that G contains a subgroup of order m. (Compare with Theorem 3.11.)

*6.3.   A finite p-primary abelian group is generated by its elements of largest order (compare Exercise 4.21).

**6.4.   If $G = \sum_{i=1}^{n} H_i$, then $mG = \sum_{i=1}^{n} mH_i$.

6.5.   If $G = \sum_{i=1}^{n} H_i$, then $G[p] = \sum_{i=1}^{n} (H_i[p])$.

6.6.   Let G be a finite p-primary abelian group and let $x \in G$ have order maximal among the elements of G. Prove that $\langle x \rangle$ is a direct summand of G. Use this result to give a new proof of the basis theorem.

6.7. Let $G = \mathbf{Z}(m_1) \oplus \cdots \oplus \mathbf{Z}(m_s)$ be a canonical decomposition. Show that $|G| = \Pi m_i$ and that m_s is the least positive integer n for which $nG = \{0\}$. (The least $n > 0$ with $nG = \{0\}$ is called the **exponent of G; this term is also used for nonabelian groups.)

*6.8. Let H be a finite abelian group with $pH = \{0\}$ for some prime p (H is called an **elementary** abelian group [see Lemma 6.2]). Prove that any two decompositions of H into a direct sum of cyclic groups have the same number of summands. Denote this number by $d(H)$. (The astute reader will note that $d(H)$ is the dimension of H considered as a vector space over $\mathbf{Z}/p\mathbf{Z}$.)

**6.9. Let G be a direct sum of b copies of cyclic groups of order p^k. If $n < k$, then $d(p^nG/p^{n+1}G) = b$.

**6.10. If H and K are elementary p-primary abelian groups, then $d(H \oplus K) = d(H) + d(K)$.

THE FUNDAMENTAL THEOREM OF FINITE ABELIAN GROUPS

We now have quite a bit of information about finite abelian groups, but we still have not answered the basic question: If G and H are finite abelian groups, when are they isomorphic? Since both G and H are direct sums of cyclic groups, your first guess is that $G \cong H$ if they have the same number of summands of each kind. There are two things wrong with this guess. First of all, since, e.g., $\mathbf{Z}(6) \cong \mathbf{Z}(3) \oplus \mathbf{Z}(2)$, we had better require that G and H have the same number of *primary* summands of each kind. Our second objection is much more serious. How can we count summands at all? To do so would require a unique factorization theorem analogous to the fundamental theorem of arithmetic, where the analog of a prime number is a primary cyclic group. Such an analog does exist; it is called the fundamental theorem of finite abelian groups, and it is this theorem we now discuss.

We have already seen, in Exercise 6.8, that the number of cyclic summands occurring in a decomposition of an elementary abelian group depends only on G and not on the particular decomposition; this number is denoted by $d(G)$. For any nonnegative integer n and any finite p-primary abelian group G, the quotient group $p^nG/p^{n+1}G$ is elementary, so that $d(p^nG/p^{n+1}G)$ is defined.

Lemma 6.5 *Let $G = \Sigma C_i$ be a decomposition of a finite p-primary abelian group into a direct sum of cyclic groups. Then $d(p^nG/p^{n+1}G)$ is the number of cyclic summands C_i having order $\geq p^{n+1}$.*

Proof Let B_k denote the direct sum of all those C_i in the decomposition having order exactly p^k, if any, and let b_k be the number of C_i in B_k (of course, B_k may be zero for some k). Thus

$$G = B_1 \oplus \cdots \oplus B_t.$$

By Exercise 6.4,

$$p^nG = p^nB_{n+1} \oplus p^nB_{n+2} \oplus \cdots \oplus p^nB_t$$

and

$$p^{n+1}G = p^{n+1}B_{n+2} \oplus \cdots \oplus p^{n+1}B_t.$$

By Theorem 2.25,

$$p^n G/p^{n+1}G \cong p^n B_{n+1} \oplus (p^n B_{n+2}/p^{n+1}B_{n+2}) \oplus \cdots \oplus p^n B_t/p^{n+1}B_t.$$

Hence $d(p^n G/p^{n+1}G) = b_{n+1} + b_{n+2} + \cdots + b_t$, by Exercises 6.9 and 6.10. ∎

Definition   If G is a finite p-primary abelian group and $n \geq 0$, define

$$U(n, G) = d(p^n G/p^{n+1}G) - d(p^{n+1}G/p^{n+2}G).$$

It follows from Exercise 6.8 that, for each n, the integer $U(n, G)$ depends only on G and not on any particular choice of decomposition of G into a direct sum of cyclic groups.

Theorem 6.6   *Let G be a finite p-primary abelian group. Any two decompositions of G into direct sums of cyclic groups have the same number of summands of each order. In fact, the number of cyclic summands of order p^{n+1} is $U(n, G)$.*

Proof   Choose a decomposition of G. It follows from Lemma 6.5 that $U(n, G)$ is the number of cyclic summands of order exactly p^{n+1}. The first statement now follows, because $U(n, G)$ does not depend on the choice of decomposition. ∎

Corollary 6.7   *Let G and H be finite p-primary abelian groups. Then $G \cong H$ if and only if $U(n, G) = U(n, H)$ for all $n \geq 0$.*

Proof   It is easy to check that an isomorphism between G and H induces isomorphisms $p^n G/p^{n+1}G \cong p^n H/p^{n+1}H$ for all n. Hence $U(n, G) = U(n, H)$ for all n.

Conversely, $G \cong H$ because they have direct sum decompositions into cyclic groups of the same type, by Theorem 6.6. ∎

We have to delete only the adjective "p-primary" in Theorem 6.6 and Corollary 6.7 to finish our discussion. The hard work has already been done, and the following three theorems (whose easy proofs may be supplied by the reader) complete the picture.

Lemma 6.8   *Let G and H be finite abelian groups, and let $f: G \rightarrow H$ be a homomorphism. For each p,*

$$f(G_p) \subset H_p.$$

Theorem 6.9   *Let G and H be finite abelian groups; $G \cong H$ if and only if $G_p \cong H_p$ for all primes p.*

Theorem 6.10   (Fundamental Theorem of Finite Abelian Groups)† *Let G be a finite abelian group. Any two decompositions of G into direct sums of primary cyclic groups have the same number of summands of each order.*

EXERCISES

6.11. Let G and H be finite p-primary abelian groups. Then, for all $n \geq 0$,

$$U(n, G \oplus H) = U(n, G) + U(n, H).$$

**6.12. (i) Let G and H be finite abelian groups. If $G \oplus G \cong H \oplus H$, prove that $G \cong H$.

(ii) Suppose A, B, and C are finite abelian groups. If $A \oplus B \cong A \oplus C$, then $B \cong C$.

Corollary 6.11   *Let G be a finite abelian group. If*

$$G = \mathbf{Z}(m_1) \oplus \cdots \oplus \mathbf{Z}(m_s)$$

and

$$G = \mathbf{Z}(n_1) \oplus \cdots \oplus \mathbf{Z}(n_t)$$

are canonical decompositions, then $s = t$ and $m_i = n_i$ for all i.

Proof   By Exercise 6.7, $m_s = n_t$, for both equal the exponent of G. The proof is completed by induction on max $\{s, t\}$, for the complementary summands are isomorphic by Exercise 6.12(ii). ∎

EXERCISES

6.13. If n is a positive integer, a **partition** of n is a sequence of positive integers $i_1 \leq i_2 \leq \cdots \leq i_s$ with $n = i_1 + i_2 + \cdots + i_s$. Prove that the number of nonisomorphic abelian groups of order p^n, where p is prime, is the number of partitions of n.

6.14. Prove that the number of conjugacy classes in the symmetric group S_n is the number of partitions of n. (See Exercise 3.61)

6.15. How many nonisomorphic abelian groups are there of order 720?

*6.16. Let H be a subgroup of a finite abelian group G. Prove that G contains a subgroup isomorphic to G/H (compare Exercises 4.30 and 11.14).

*6.17. Let G and H be finite abelian groups such that, for each k, both G and H have the same number of elements of order k. Then $G \cong H$ (compare Exercise 4.44).

† This theorem was proved in 1878 by Frobenius and Stickelberger.

MODULES AND MATRICES†

We digress from our study of groups to apply Theorems 6.3 and 6.10 to linear algebra; we shall prove the existence and uniqueness of the rational canonical form of a square matrix over an arbitrary field F. At this stage, this project is merely one of translation, so that we need only introduce a new vocabulary. Our exposition is complete, but since we are assuming that the reader is comfortable with linear algebra, our pace is not leisurely.

Definition   Let R be a commutative ring with 1. An **ideal** I in R is a nonempty subset of R such that:

(i) $a, b \in I$ imply $a - b \in I$;
(ii) $a \in I$ and $r \in R$ imply that $ra \in I$.

An important example of an ideal is the set of all multiples of a fixed $r_0 \in R$ by elements of R; this ideal is denoted by (r_0) and is called the **principal ideal generated** by r_0. Thus,

$$(r_0) = \{x \in R: x = rr_0 \quad \text{for some } r \in R\}.$$

The principal ideal (r_0) is sometimes denoted by Rr_0.

Definition   A **principal ideal domain** is a domain in which every ideal is a principal ideal (a **domain** is a commutative ring with 1 in which $ab = 0$ implies $a = 0$ or $b = 0$).

Example 1   In $\mathbf{Z}$, condition (ii) in the definition of ideal follows from (i); since every subgroup of $\mathbf{Z}$ is cyclic, every ideal in $\mathbf{Z}$ is principal.

Example 2   If F is any field, then the only ideals in F are $\{0\}$ and F itself; since $\{0\} = (0)$ and $F = (1)$, F is a principal ideal domain.

Example 3   Let F be a field and let $F[x]$ be the ring of polynomials in x with coefficients in F. Using the division algorithm for polynomials, the reader may prove that any nonzero ideal I in $F[x]$ consists precisely of all the multiples (by polynomials) of the **monic** polynomial (i.e., leading coefficient is 1) in I of least degree.

Many properties of the ring of integers can be generalized to any principal ideal domain R. An element $u \in R$ is a **unit** if there is an element $v \in R$ with $uv = 1$. In $\mathbf{Z}$, the only units are ± 1; in F, every nonzero element is a unit; in $F[x]$, the nonzero constants are the units. A nonzero element $p \in R$ is **irreducible** if p is not a unit, and in every factorization $p = ab$ in R, either a or b is a unit. In $\mathbf{Z}$, the irreducibles are the primes (positive and negative); in F, there are no irreducibles; in $F[x]$, the irreducibles are the irreducible polynomials.

† This section will not be used until Chapter 8; however, the reader may be amused to see that much of the linear algebra learned in a first course can now be covered in a few pages.

A **common divisor** of $a_1, \cdots, a_n \in R$ is an element $c \in R$ that divides each a_i (i.e., $cb_i = a_i$ for some $b_i \in R$); a **greatest common divisor**, gcd, of $a_1, \cdots, a_n$ is a common divisor that is divisible by every common divisor.

The following two theorems are proved in Appendix VI.

Theorem A   *Every finite set of elements* $a_1, \cdots, a_n$ *in a principal ideal domain* R *has a gcd; moreover, a gcd is a linear combination of* $a_1, \cdots, a_n$ *with coefficients in* R.

Theorem B   *If* d *and* d' *are gcd's of* $a_1, \cdots, a_n$ *in a principal ideal domain* R, *then* d *and* d' *are* **associates**, *that is,* $d' = ud$ *for some unit* $u \in R$.

Because of Theorem B, we may choose gcd's in $\mathbf{Z}$ to be positive and gcd's in $F[x]$ to be monic polynomials; such choices make gcd's unique.

The fundamental theorem of arithmetic may be generalized to principal ideal domains (a proof is in Appendix VI):

Theorem F   *If* R *is a principal ideal domain, then*

  (i) *every nonzero* $a \in R$ *that is not a unit is a product of irreducible elements;*
  (ii) *this factorization is unique in the following sense: if* $p_1 \cdots p_m = q_1 \cdots q_n$, *where the* p *and* q *are irreducible, then* $m = n$ *and there is a one-one correspondence between the factors with corresponding factors associate.*

We shall need a generalization of $\mathbf{Z}/p\mathbf{Z}$.

Theorem 6.12   **(Construction of a Quotient Ring)**   *Let* R *be a commutative ring with* 1 *and let* I *be an ideal in* R. *There exists a ring* S *and a ring homomorphism* π *of* R *onto* S *whose kernel is* I.

Proof   Under addition, R is an abelian group and I is a subgroup. Therefore, R/I is an additive abelian group, and the natural map $\pi: R \to R/I$ is a group homomorphism of R onto R/I whose kernel is I.[†]

For $S = R/I$ to be a ring, define

$$(r + I)(r' + I) = rr' + I.$$

This is a well defined multiplication, for if $r + I = s + I$ and $r' + I = s' + I$, then $rr' - ss' = r(r' - s') + (r - s)s' \in I$; hence, $rr' + I = ss' + I$. The reader should verify that R/I is a ring with 1 under the given operations, and that π is a ring homomorphism. ∎

Definition   The ring S just constructed is denoted by R/I and is called the **quotient ring** of R modulo I.

[†] The zero element in R/I is thus $I = 0 + I$.

EXERCISES

6.18. If $R = \mathbf{Z}$ and $I = (m)$, then the ring R/I is the ring $\mathbf{Z}/m\mathbf{Z}$.

6.19. The first isomorphism theorem (Theorem 2.18) holds for commutative rings: If $f: R \to S$ is a ring homomorphism with kernel I, then I is an ideal and $R/I \cong$ image f.

**6.20. The correspondence theorem (Theorem 2.23) holds for commutative rings if one replaces "normal subgroup" by "ideal".

6.21. If R is a commutative ring with 1, then R is a field if and only if R contains no proper ideals (an ideal I in R is **proper if $I \neq R$).

Definition An ideal I in R is **prime** if $I \neq R$ and $rr' \in I$ implies either r or r' lies in I.

EXERCISE

**6.22. If R is a principal ideal domain, then the nonzero prime ideals are the ideals (p), where p is irreducible.

Theorem 6.13 *Let R be a commutative ring with 1. An ideal I in R is a prime ideal if and only if R/I is a domain.*

Proof If I is a prime ideal, we must show that R/I contains no zero divisors. Suppose $(r + I)(r' + I) = 0$, i.e., $rr' + I = I$. Then $rr' \in I$; since I is a prime ideal, one of these factors, say r, lies in I. Hence, $r + I = 0$.

Suppose R/I is a domain. If $rr' \in I$, then $(r + I)(r' + I) = 0$ in R/I, so that one of the factors is 0, i.e., either r or r' lies in I. Thus I is a prime ideal. ∎

Definition Let R be a commutative ring with 1. An ideal I in R is a **maximal ideal** if $I \neq R$ and there is no larger proper ideal of R that contains I.

Theorem 6.14 *Let R be a commutative ring with 1. An ideal I in R is a maximal ideal if and only if R/I is a field.*

Proof If I is a maximal ideal, then the correspondence theorem for rings (Exercise 6.20) implies that R/I has no proper ideals. By Exercise 6.21, R/I is a field. To prove the converse, just reverse this argument. ∎

Corollary 6.15 *Every maximal ideal in R is a prime ideal.*

In general, the converse of this corollary is false. For example, if $R = \mathbf{Z}[x]$, one may verify that (x) is a prime ideal that is contained in the proper ideal consisting of all polynomials in $\mathbf{Z}[x]$ having even constant term. However, the converse of the corollary is true when R is a principal ideal domain.

Theorem 6.16 *If R is a principal ideal domain, every nonzero prime ideal I is a maximal ideal.*

Proof Let J be an ideal with $I \subsetneq J$. Since R is a principal ideal domain, there are elements a and b in R with $I = (a)$ and $J = (b)$. Now $a \in J = (b)$, so there is an $r \in R$ with $a = rb$. Since I is a prime ideal, either $r \in I$ or $b \in I$. Were b in I, then $J \subset I$, a contradiction. Therefore, $r \in I$, so that $r = sa$ for some $s \in R$. Hence, $a = rb = sab$, and $1 = sb$. The ideal (b) thus contains 1, and so $J = (b) = R$. We conclude that I is a maximal ideal. ∎

Corollary 6.17 *If R is a principal ideal domain and $p \in R$ is irreducible, then $R/(p)$ is a field.*

Proof (p) is a prime ideal, by Exercise 6.22, which is maximal, by Theorem 6.16. ∎

Definition An abelian group V is an **R-module** (where R is a ring) if a "scalar multiplication" is defined, i.e., there is a function $R \times V \to V$ (whose values we write in multiplicative notation) that satisfies:

(i) $(rs)\alpha = r(s\alpha)$;
(ii) $(r+s)\alpha = r\alpha + s\alpha$;
(iii) $r(\alpha + \beta) = r\alpha + r\beta$;
(iv) $1\alpha = \alpha$

for every $\alpha, \beta \in V$ and $r, s, 1 \in R$.

Thus, an R-module is just like a vector space except that the scalars are allowed to be in a ring R instead of a field.

Example 4 If $R = \mathbf{Z}$, an R-module is an abelian group, for axioms (i) to (iv) are always true for scalars in $\mathbf{Z}$. As a consequence, abelian groups are often called $\mathbf{Z}$-modules.

Example 5 If R is a field F, an R-module is a vector space over F.

Example 6 Let V be a vector space over F, and let $T: V \to V$ be a linear transformation. We make V into an $F[x]$-module, denoted by V^T, by defining

$$(a_0 + a_1 x + a_2 x^2 + \cdots + a_n x^n)\alpha = a_0\alpha + a_1 T\alpha + a_2 T^2\alpha + \cdots + a_n T^n\alpha$$

(T^n is the composite of T with itself n times). The reader should check that we have defined a scalar multiplication.

Just as a principal ideal domain R is a generalization of $\mathbf{Z}$, so are R-modules generalizations of abelian groups. Almost any theorem that can be proved for abelian groups has a true analog for R-modules; moreover, the proofs of the theorems for R-modules are exact translations of the proofs for abelian groups. Here is the dictionary from the language of abelian groups to that of modules.

A **submodule** of an R-module V is a subgroup W of V that is closed under scalar multiplication: if $\alpha \in W$ and $r \in R$, then $r\alpha \in W$.

Let W be a submodule of V. Remember that modules are just abelian groups with added structure. If we forget the added structure for a moment, then W is a subgroup of the abelian group V, and so V/W is an abelian group. Define the **quotient module** V/W to be the abelian group V/W with scalar multiplication

$$r(v + W) = rv + W.$$

EXERCISES

6.23. R itself is an R-module. Prove that the submodules of R are its ideals.

6.24. Let V be a vector space over F, let $T: V \to V$ be a linear transformation, and let V^T be the $F[x]$-module of Example 6. Prove that $W \subset V$ is a submodule if and only if W is a subspace of V with $T(W) \subset W$ (W is called an **invariant subspace).

6.25. If V is an R-module and $\alpha \in V$, then

$$\langle \alpha \rangle = \{r\alpha : r \in R\}$$

is a submodule of V.

6.26. The intersection of any family of submodules of V is again a submodule of V.

6.27. If X is a subset of an R-module V, then there is a smallest submodule W of V containing X, denoted by $\langle X \rangle$. Further, if X is nonempty,

$$\langle X \rangle = \{\Sigma r_i x_i : r_i \in R \text{ and } x_i \in X\}.$$

An R-module V is **finitely generated** in case it contains a finite number of elements $\alpha_1, \alpha_2, \cdots, \alpha_k$ such that every element in V is a linear combination of these α with coefficients in R. In particular, an R-module V is **cyclic** if it can be generated by one of its elements.

If V and W are R-modules, their **direct sum** $V \oplus W$ is the usual direct sum made into an R-module by $r(v, w) = (rv, rw)$. There is also an internal version: If V is an R-module with submodules W_1 and W_2, then V is the **(internal) direct sum** if $W_1 \cap W_2 = \{0\}$ and $W_1 + W_2 = V$.

EXERCISES

6.28. If V is an R-module with submodules $W_1, \cdots, W_n$, define $V = \Sigma W_i$ (internal) if $V = W_1 + \cdots + W_n$ and, for all j, $W_j \cap \langle \bigcup_{i \neq j} W_i \rangle = \{0\}$. Prove $V \cong W_1 \oplus \cdots \oplus W_n$.

**6.29. An abelian group G is finite if and only if G is finitely generated and every element in G has finite order.

Exercise 6.29 tells us how to translate "finite abelian group" once we can translate "order" of an element. Before doing this, we note that Exercise 6.29 is false if we do not assume that the group G is abelian. Burnside (1911) proved that if a finitely generated group G comprised of elements of finite order can be imbedded in $GL(n, \mathbf{C})$, where $\mathbf{C}$ is the complex numbers, then G is finite. In 1964, Golod and Šafarevič exhibited an infinite finitely generated p-group; in 1968, Adian and Novikov exhibited an infinite finitely generated group of finite exponent (so this latter group has a uniform bound on the orders of its elements).[†]

Let $\alpha \in V$, where V is an R-module. The **order ideal** of α is $\{r \in R: r\alpha = 0\}$. It is quickly verified that the order ideal is an ideal of R. If R is a principal ideal domain, this order ideal consists of all the multiples of a fixed element in R. In $\mathbf{Z}$, we choose this fixed element to be positive (and we get the usual definition of order); in $F[x]$, we choose this fixed element to be the monic polynomial of smallest degree in the ideal. Thus, if $\alpha \in V^{T}$, its order ideal is generated by the monic polynomial $m(x)$ of least degree for which $m(T)\alpha = 0$.

We say that $\alpha \in V$ has **finite order** if its order ideal is nonzero. Call an R-module V **finite** if V is finitely generated and every element of V has finite order. Note that if V is finite, there is a nonzero element $r \in R$ with $rv = 0$ for every $v \in V$: if $V = \langle \alpha_1, \cdots, \alpha_n \rangle$ and the order ideal of α_i is (r_i), set $r = \Pi r_i$. (The module analog of the order of V will be seen below.) Finally, an R-module is p-**primary** if the order ideal of each element is generated by some power of an irreducible element p.

Theorem 6.18 *Let R be a principal ideal domain and let V be a finitely generated R-module in which every element has finite order. Then*

$$V = \langle \alpha_1 \rangle \oplus \cdots \oplus \langle \alpha_s \rangle.$$

Moreover, the cyclic summands may be chosen to satisfy either of the following conditions. If the order ideal of α_i is (r_i),

(i) *each r_i is a power of an irreducible element in R;*
or

(ii) *$r_i | r_{i+1}$ for $i = 1, \cdots, s-1$.*

Proof Our proofs of the corresponding group theorems were written so that translation into module language is mechanical. Thus, the decomposition in (i) is as in Theorems 6.1 and 6.3, while the decomposition in (ii) is as in Corollary 6.4. ∎

Call a decomposition of an R-module V as in Theorem 6.18 (ii) a **canonical decomposition** of V. If $V = \langle \alpha_1 \rangle \oplus \cdots \oplus \langle \alpha_s \rangle$ and if the order ideal of α_i is (r_i), then the **order** of V is defined to be (Πr_i).

Theorem 6.19 *Let V be a finite dimensional vector space over F and let $T: V \rightarrow V$ be a linear transformation. Then*

$$V = W_1 \oplus \cdots \oplus W_s,$$

† A much simpler proof was published by Ol'šanskii in 1982.

where each $W_i = \langle \alpha_i \rangle$ is a cyclic invariant subspace. Moreover, the cyclic invariant subspaces may be chosen to satisfy either of the following conditions. If the order ideal of α_i is $(f_i(x))$,

(i) each $f_i(x)$ is a power of an irreducible polynomial;

or

(ii) $f_i(x) | f_{i+1}(x)$ for $i = 1, \cdots, s-1$.

Proof   Recall that we may regard V as an $F[x]$-module V^T. Since V is finite dimensional, it has a basis $\{\beta_1, \cdots, \beta_n\}$, so that each vector $\gamma \in V$ may be written

$$\gamma = \Sigma c_i \beta_i, \qquad c_i \in F.$$

It follows that $\{\beta_1, \cdots, \beta_n\}$ generates V^T as an $F[x]$-module (since we are allowed to use any polynomials in $F[x]$ as coefficients, not merely constant ones, we may not need all the β_i to generate V^T as an $F[x]$-module). Further, every $\gamma \in V$ is annihilated by some polynomial (this follows, for example, because the $n+1$ vectors $\gamma, T\gamma, T^2\gamma, \cdots, T^n\gamma$ lying in the n-dimensional vector space V must be linearly dependent); therefore, every γ has finite order.

   We have verified that the hypotheses of Theorem 6.18 hold, so V^t may be decomposed into a direct sum of cyclic submodules. Since a submodule of V^T is just an invariant subspace (Exercise 6.24), the theorem follows. ∎

EXERCISES

**6.30.   A subspace W of V^T is a cyclic invariant subspace if and only if there is a vector $\alpha \in W$ such that $\{\alpha, T\alpha, T^2\alpha, \cdots, T^{k-1}\alpha\}$ is a basis of W for some $k \geq 1$ (compare Theorem 2.12).

6.31.   If W has a basis of the form

$$\{\alpha, T\alpha, T^2\alpha, \cdots, T^{k-1}\alpha\},$$

then adjoining $T^k\alpha$ to this set makes it dependent, and there is an equation

$$T^k\alpha = \sum_{i=0}^{k-1} c_i T^i \alpha.$$

Prove that the order ideal of α is generated by the polynomial

$$c(x) = x^k - c_{k-1}x^{k-1} - \cdots - c_0.$$

[HINT: $c(x)$ is a monic polynomial of degree k with $c(T) = 0$; by definition, $c(T)\alpha = 0$, and for $0 \leq i \leq k-1$, $c(T)T^i\alpha = T^i c(T)\alpha = 0$.]

   We remind the reader of the correspondence between linear transformations on V and matrices. Let V be m-dimensional and let $\{\varepsilon_1, \varepsilon_2, \cdots, \varepsilon_m\}$ be an ordered basis of V. If $T: V \to V$, then, for each i, $T\varepsilon_i$ is a linear combination of the ε_j:

$$T\varepsilon_i = \Sigma a_{ji}\varepsilon_j.$$

The matrix of T relative to the original basis of the ε_j is $A = [a_{ji}]$. Therefore, the coordinates of $T\varepsilon_1$ form the first *column* of A, the coordinates of $T\varepsilon_2$ form the second *column* of A, and so forth.

Definition Let $c(x) = x^k - c_{k-1}x^{k-1} - \cdots - c_0$. The **companion matrix** of $c(x)$ is the $k \times k$ matrix

$$\begin{bmatrix} 0 & 0 & 0 & \cdots & 0 & c_0 \\ 1 & 0 & 0 & \cdots & 0 & c_1 \\ 0 & 1 & 0 & \cdots & 0 & c_2 \\ 0 & 0 & 1 & \cdots & 0 & c_3 \\ \vdots & \vdots & \vdots & & \vdots & \vdots \\ 0 & 0 & 0 & \cdots & 1 & c_{k-1} \end{bmatrix}$$

(The companion matrix of $c(x) = x - a$ is the 1×1 matrix $[a]$.)

EXERCISES

**6.32. Let $T: V \to V$ and let W be a cyclic invariant subspace of V. If $\{\alpha, T\alpha, T^2\alpha, \cdots, T^{k-1}\alpha\}$ is a basis of W, the matrix of T on W is the $k \times k$ companion matrix of $c(x)$, where $c(x)$ generates the order ideal of α.

6.33. Let A be the companion matrix of $c(x)$. Prove that the characteristic polynomial of A is $c(x)$.

Definition Let A be a $k \times k$ matrix and let B be an $m \times m$ matrix. The **direct sum** of A and B is the $(k+m) \times (k+m)$ matrix

$$\begin{bmatrix} A & 0 \\ 0 & B \end{bmatrix}.$$

The analogous definition for the direct sum of a finite number of matrices may be supplied by the reader. Note that the direct sum of A and B is similar to the direct sum of B and A.

Theorem 6.20 *Every $n \times n$ matrix A over a field F is similar to a direct sum of companion matrices $C_1, \cdots, C_s$; if $f_i(x)$ is the characteristic polynomial of C_i, we may assume either*

(i) *each $f_i(x)$ is a power of an irreducible polynomial;*

or

(ii) *$f_i(x) | f_{i+1}(x)$, $i = 1, \cdots, s-1$.*

Proof Let V be the vector space of all n-tuples of elements in F; the **standard basis** of V is $\{\varepsilon_1, \cdots, \varepsilon_n\}$, where ε_i has ith coordinate 1 and all other coordinates 0. The matrix A defines a linear transformation $T: V \to V$ by $T(X) = AX$, where X is a column vector. Note that A is the matrix of T with respect to the standard basis because $A\varepsilon_i$ is the ith column of A. Given this T, we make V into an $F[x]$-module V^T. Now V is a direct sum of cyclic invariant subspaces W_i, by Theorem 6.19. If α_i is a generator of W_i, then Exercise 6.30 says that a new basis of V is

$$\{\alpha_1, T\alpha_1, T^2\alpha_1, \cdots; \alpha_2, T\alpha_2, T^2\alpha_2, \cdots; \cdots; \alpha_s, T\alpha_s, T^2\alpha_s, \cdots\}.$$

The matrix B of T with respect to this new basis is a direct sum of companion matrices, by Exercise 6.32. Finally, A and B are similar, for they represent the same linear transformation relative to different bases of V. ∎

A matrix B that is a direct sum of companion matrices as in Theorem 6.20(ii) is called a **rational canonical form**; the polynomials of the companion matrices are called the **invariant factors** of B. More generally, if a matrix A is similar to a rational canonical form B, then the invariant factors of A are defined to be the invariant factors of B. (The question whether A can be similar to different rational canonical forms will be answered shortly.)

Recall that the **minimum polynomial** of an $n \times n$ matrix A is the monic polynomial $m(x)$ of least degree for which $m(A) = 0$.

EXERCISES

6.34. (Cayley-Hamilton**)† Let A have invariant factors $f_1(x), \cdots, f_s(x)$, where $f_i(x) \mid f_{i+1}(x)$ for $i = 1, \cdots, s-1$. Prove that A satisfies its characteristic polynomial; more precisely, prove that the characteristic polynomial of A is $\Pi f_i(x)$ and that the minimum polynomial of A is $f_s(x)$. (Comparison with Exercise 6.7 shows that the characteristic polynomial is analogous to the order of a group and the minimum polynomial is analogous to the exponent.)

6.35. Give an example of two $n \times n$ matrices over $\mathbf{C}$ that are not similar but which have the same minimum polynomial and the same characteristic polynomial. (HINT: What is the analogous problem for abelian groups?)

6.36. If A is a companion matrix, its minimum polynomial is its characteristic polynomial.

This discussion has the disadvantage of not showing how to compute the invariant factors of a given matrix A. Details of the following procedure are discussed in the book of Albert, for example. If B is an $n \times n$ matrix with (polynomial) entries in

---

† Of course, usual proofs of the Cayley-Hamilton theorem are more elementary and do not use canonical forms.

$F[x]$, where F is a field, then one may use elementary row and column operations to put B in diagonal form D; moreover, the diagonal entries $d_1(x), \cdots, d_n(x)$ of D can be arranged so that $d_i(x)|d_{i+1}(x)$ for $1 \leq i < n$. In particular, for a matrix A with entries in F, the matrix $B = xE - A$ can be put in such a diagonal form, and the nonconstant diagonal entries are the invariant factors of A.

In Chapter 8, we shall study certain groups whose elements are nonsingular matrices. Since the order of an element is the same as the order of any of its conjugates, one can compute the order of a matrix by computing the order of its canonical form. Unfortunately, it is difficult to compute powers of companion matrices. If the field F is a large one, however, there is another canonical form whose powers are easily calculated.

Definition   A $k \times k$ **Jordan block** is a $k \times k$ matrix of the form

$$\begin{bmatrix} a & 0 & 0 & \cdots & 0 & 0 \\ 1 & a & 0 & \cdots & 0 & 0 \\ 0 & 1 & a & \cdots & 0 & 0 \\ 0 & 0 & 1 & \cdots & 0 & 0 \\ \vdots & \vdots & \vdots & \cdots & a & 0 \\ 0 & 0 & 0 & \cdots & 1 & a \end{bmatrix}$$

(A 1×1 Jordan block is a 1×1 matrix $[a]$.)

EXERCISES

6.37. (i) Prove that for every positive integer k,

$$\begin{bmatrix} b & 0 \\ 1 & b \end{bmatrix}^k = \begin{bmatrix} b^k & 0 \\ kb^{k-1} & b^k \end{bmatrix}.$$

(ii) Prove that for every positive integer k,

$$\begin{bmatrix} a & 0 & 0 \\ 1 & a & 0 \\ 0 & 1 & a \end{bmatrix}^k = \begin{bmatrix} a^k & 0 & 0 \\ ka^{k-1} & a^k & 0 \\ s(k)a^{k-2} & ka^{k-1} & a^k \end{bmatrix}$$

where $s(k) = k(k-1)/2$.

6.38. Let $T: W \to W$ and let $\{\alpha, T\alpha, \cdots, T^{k-1}\alpha\}$ be a basis of W, so that the matrix of T is the companion matrix of some polynomial $f(x)$. If

$$\beta_0 = \alpha, \ \beta_1 = (T - aE)\alpha, \cdots, \beta_{k-1} = (T - aE)^{k-1}\alpha,$$

prove that $\{\beta_0, \beta_1, \cdots, \beta_{k-1}\}$ is a basis of W (E is the identity transformation).

Theorem 6.21   *Let A be an $n \times n$ matrix over F, where F is a field that contains all the characteristic roots of A. Then A is similar to a direct sum of Jordan blocks.*

Proof   After looking at the primary decomposition in Theorem 6.20(i), we see that it suffices to prove that a companion matrix C whose characteristic polynomial $f(x)$ is a power of an irreducible polynomial is similar to a Jordan block.

The hypothesis on F tells us that $f(x) = (x - a)^k$ for some $a \in F$. Now C determines a transformation $T: W \to W$, where W is a vector space with a basis of the form $\{\alpha, T\alpha, \cdots, T^{k-1}\alpha\}$. Let us compute the matrix of T relative to the basis $\{\beta_0, \beta_1, \cdots, \beta_{k-1}\}$ of W that we examined in Exercise 6.38.

If $j + 1 \leq k$,

$$
\begin{aligned}
T\beta_j &= T(T - aE)^j\alpha \\
&= (T - aE)^j T\alpha \\
&= (T - aE)^j[aE + (T - aE)]\alpha \\
&= a(T - aE)^j\alpha + (T - aE)^{j+1}\alpha.
\end{aligned}
$$

If $j + 1 < k$, then $T\beta_j = a\beta_j + \beta_{j+1}$; if $j + 1 = k$, then $(T - aE)^{j+1} = (T - aE)^k = 0$, by the Cayley-Hamilton theorem. Therefore, $T\beta_{k-1} = a\beta_{k-1}$.

The matrix of T is thus a Jordan block J, and so C and J are similar. ∎

A direct sum of Jordan blocks is called a **Jordan canonical form**. Thus Theorem 6.21 says that if F contains enough elements, a square matrix is similar to a Jordan canonical form. For example, if F is the complex numbers (or any algebraically closed field), then every square matrix is similar to a Jordan canonical form. The polynomials (each a power of an irreducible), with multiplicities, corresponding to the Jordan blocks in a Jordan canonical form J are called **elementary divisors**. If a matrix A is similar to a Jordan canonical form J, then one defines the elementary divisors of A to be those of J; of course, it is necessary to know whether A can be similar to distinct Jordan canonical forms (see Theorem 6.26).

The uniqueness of the various canonical forms will follow from a translation of the fundamental theorem of finite abelian groups.

Definition   If V and W are R-modules, a function $f: V \to W$ is an **R-homomorphism** if

$$
f(v + v') = f(v) + f(v')
$$

and

$$
f(rv) = rf(v)
$$

for all $v, v' \in V$ and all $r \in R$. An R-homomorphism $f: V \to W$ is an **R-isomorphism** if it is a one-one correspondence; in this case, we say V and W are **R-isomorphic** and write $V \cong W$.

Example 7   If $R = \mathbf{Z}$, then V and W are merely abelian groups. Every homomorphism $f: V \to W$ is a $\mathbf{Z}$-homomorphism, for $f(mv) = mf(v)$ for every integer m.

Example 8   If R is a field, V and W are vector spaces and R-homomorphisms are linear transformations.

Example 9   If $R = F[x]$, then an R-homomorphism $f: V \to W$ is a linear transformation such that

$$f(\psi(x)v) = \psi(x)f(v)$$

for every polynomial $\psi(x) \in F[x]$.

EXERCISES

6.39.   Prove the first isomorphism theorem for modules (quotient modules have been defined on page 111). (HINT: Since modules are groups, the reader need check only that the group isomorphism in Theorem 2.18 is, in fact, an R-isomorphism.)

6.40.   Every cyclic R-module V is isomorphic to R/I, where I is the order ideal of a generator of V. Conclude that two cyclic modules are isomorphic if and only if they have generators with the same order ideal.

6.41.   Let R be a principal ideal domain and let $a, b \in R$ be relatively prime (gcd $= 1$). Then

$$R/(a) \oplus R/(b) \cong R/(ab).$$

In words, a direct sum of cyclic modules of relatively prime orders is cyclic.

Theorem 6.22   (Fundamental Theorem for Modules)   *Let R be a principal ideal domain and let V be a finitely generated R-module in which every element has finite order.*

(i)   *The order ideals (elementary divisors) corresponding to the summands occurring in a decomposition of V into a direct sum of primary cyclic submodules are uniquely determined by the module V.*

(ii)   *The order ideals (invariant factors) corresponding to the cyclic summands occurring in a canonical decomposition of V are uniquely determined by the module V.*

Proof   Translation of Theorem 6.10 and Corollary 6.11.   ∎

To apply this fundamental theorem to matrices, we continue our examination of $F[x]$-modules.

Lemma 6.23   *Let V and W be vector spaces over F and let $T: V \to V$ and $S: W \to W$ be linear transformations. A function $f: V^T \to W^S$ is an $F[x]$-homomorphism if and only if:*

(i)   *f is a linear transformation of the vector spaces V and W;*
(ii)   *$f(T(v)) = S(f(v))$ for all $v \in V$.*

Proof   If f is an $F[x]$-homomorphism, then $f(\psi(x)v) = \psi(x)f(v)$ for all $v \in V$ and all polynomials $\psi(x) \in F[x]$. In particular, if $\psi(x)$ is a constant polynomial, then we see that f is a linear transformation; if $\psi(x) = x$, then $f(xv) = xf(v)$. But the definition of scalar multiplication in V^T is $xv = T(v)$ and the definition of scalar multiplication in W^S is $xf(v) = S(f(v))$.

For the converse, we are told that

$$f(\psi(x)v) = \psi(x)f(v)$$

for all the polynomials $\psi(x) = $ constant and for the polynomial $\psi(x) = x$. It follows easily that the equation holds for all polynomials $\psi(x)$. ∎

Theorem 6.24   *If A and B are $n \times n$ matrices over a field F, then A is similar to B if and only if the corresponding $F[x]$-modules they determine are $F[x]$-isomorphic.*

Proof   Let V be the vector space of all n-tuples of elements in F. Define linear transformations $T: V \to V$ and $S: V \to V$ by $T(X) = AX$ and $S(X) = BX$, where X is a column vector in V. We let V^T denote V made into an $F[x]$-module by $x\alpha = T\alpha$ and we let V^S denote V made into an $F[x]$-module by $x\alpha = S\alpha$.

If A is similar to B, there is a nonsingular matrix P with $PAP^{-1} = B$, and P defines a linear transformation $f: V \to V$. We claim that f is even an $F[x]$-isomorphism between V^T and V^S. By Lemma 6.23, it suffices to prove that $f(T(v)) = S(f(v))$ for all $v \in V$, i.e., $fT = Sf$. In terms of matrices, this is the given equation $PA = BP$.

Suppose, conversely, that $f: V^T \to V^S$ is an $F[x]$-isomorphism. By Lemma 6.23, $Sf = fT$, or, since f is an isomorphism, $S = fTf^{-1}$. If P is the matrix corresponding to the linear transformation $f: V \to V$, then $B = PAP^{-1}$, i.e., A and B are similar. ∎

Theorem 6.25   *Two $n \times n$ matrices A and B over a field F are similar if and only if they have the same invariant factors. An $n \times n$ matrix is similar to exactly one rational canonical form.*

Proof   Only necessity needs proof. By the preceding theorem, A and B are similar if and only if the modules they determine are $F[x]$-isomorphic. It follows from part (ii) of the fundamental theorem for modules, Theorem 6.22, that the order ideals of cyclic summands in canonical decompositions are the same. But the monic generators $f_i(x)$ of these ideals are, by definition, the invariant factors.

If A is similar to two rational canonical forms B_1 and B_2, then B_1 and B_2 are similar, hence have the same invariant factors, hence are comprised of the same companion matrices. Therefore, $B_1 = B_2$. ∎

Theorem 6.26

(i) *If A is an $n \times n$ matrix over a field F that contains the characteristic roots of A, then any two Jordan canonical forms similar to A have the same elementary divisors.*

(ii) *Two $n \times n$ matrices A and B over a field F containing their characteristic roots are similar if and only if they have the same elementary divisors.*

Proof

(i) If A is similar to a Jordan canonical form J having elementary divisors $(x - a_i)^{e_i}$, then the $F[x]$-module associated to A (and to J) is a direct sum of cyclic modules $\langle \alpha_i \rangle$, where the order ideal of α_i is generated by $(x - a_i)^{e_i}$. But the fundamental theorem asserts the uniqueness of these polynomials and the multiplicity of each.

(ii) The first part shows that it makes sense to speak of the elementary divisors of the matrices A and B. The proof now proceeds as that of Theorem 6.25. ∎

The reader should note that two Jordan canonical forms are similar if and only if the Jordan blocks occurring in one are a permutation of the Jordan blocks in the other.

EXERCISES

6.42. If b and b' are nonzero elements of a field F, then $\begin{bmatrix} a & b \\ 0 & c \end{bmatrix}$ and $\begin{bmatrix} a & b' \\ 0 & c \end{bmatrix}$ are similar.

*6.43. Let A and B be $n \times n$ matrices with entries in a field F. Supposing that F is a subfield of a field K and that A and B are similar over K, prove that A and B are similar over F. (HINT: A rational canonical form over F is also a rational canonical form over K.)

6.44. (**Jordan decomposition**) (i) Recall that a matrix is **diagonalizable** if it is similar to a diagonal matrix. If F is an algebraically closed field, then every $n \times n$ matrix A over F may be written

$$A = D + N,$$

where D is diagonalizable, N is nilpotent, and $DN = ND$. (It may be shown that D and N are uniquely determined by A.)

(ii) Recall that a matrix U is called **unipotent** if $U - E$ is nilpotent. If F is algebraically closed, prove every nonsingular $n \times n$ matrix A over F may be written

$$A = DU,$$

where D is diagonalizable, U is unipotent, and $DU = UD$. [HINT: Define $U = E + ND^{-1}$.] (It may be shown that D and U are uniquely determined by A.)

THE REMAK-KRULL-SCHMIDT THEOREM

If a nonabelian group G is a direct product of normal subgroups, each of which cannot be factored further, is this factorization unique? The affirmative answer for a large class of groups (which contains all finite groups) is the main result of this section. Since we shall consider nonabelian groups, we return to the multiplicative notation.

Definition   A group G is **indecomposable** if $G \neq \{1\}$ and if $G \cong H \times K$, then either H or $K = \{1\}$.

EXERCISES

6.45.   Which of the following groups are indecomposable? (a) $\mathbf{Z}$; (b) the additive group of rationals; (c) S_n; (d) $\mathbf{Z}(21)$; (e) the multiplicative group of positive rationals; (f) $\mathbf{Z}(p^n)$, where p is prime; (g) the quaternions.

6.46.   Give necessary and sufficient conditions that a finite abelian group be indecomposable.

Definition   A homomorphism $f: G \to G$ is called an **endomorphism** of G; an isomorphism $f: G \to G$ is called an **automorphism** of G.

There are certain endomorphisms of G that arise quite naturally in the consideration of direct products, as we saw in Exercise 2.79.

Definition   Let $G = H_1 \times \cdots \times H_m$. The homomorphisms $\varepsilon_i: G \to H_i$ defined by $\varepsilon_i(h_1 h_2 \cdots h_m) = h_i$ are called **projections**.

Throughout this section, direct products are internal. Thus, each H_i is a normal subgroup of G and there is an inclusion map $\lambda_i: H_i \to G$. The composite $\lambda_i \varepsilon_i$ is then an endomorphism of G.

EXERCISES

6.47.   Let ε_i be a projection and λ_i the inclusion (as in the above paragraph). The endomorphism $\lambda_i \varepsilon_i$ is **idempotent**, i.e., $\lambda_i \varepsilon_i \circ \lambda_i \varepsilon_i = \lambda_i \varepsilon_i$.

6.48.   An endomorphism β of G is **normal** if

$$a\beta(x)a^{-1} = \beta(axa^{-1})$$

for every a and x in G. Prove that $\lambda_i \varepsilon_i$ is normal, where ε_i is a projection and λ_i is the corresponding inclusion.

**6.49.   The composite of normal endomorphisms is a normal endomorphism.

**6.50.   If $\beta: G \to G$ is a normal endomorphism and $H \lhd G$, then $\beta(H) \lhd G$.

We now introduce a new way of combining endomorphisms; unfortunately, the new function we get is not always a homomorphism.

Definition   If α and β are endomorphisms of G, then $\alpha + \beta \colon G \to G$ is the function defined by

$$(\alpha + \beta)(x) = \alpha(x)\beta(x) \quad \text{for every } x \in G.$$

EXERCISES

6.51.   If G is abelian and α and β are endomorphisms of G, then $\alpha + \beta$ is also an endomorphism of G.

6.52.   Let α and β be endomorphisms of S_3 defined as follows: α is conjugation by (123) and β is conjugation by (132). Show that $\alpha + \beta$ is not an endomorphism of S_3.

**6.53.   If $G = H_1 \times H_2 \times \cdots \times H_n$ has projections $\varepsilon_i \colon G \to H_i$ and inclusions $\lambda_i \colon H_i \to G$, then the sum of any k distinct $\lambda_i \varepsilon_i$ is a normal endomorphism of G. Moreover,

$$\lambda_1 \varepsilon_1 + \cdots + \lambda_n \varepsilon_n = 1_G.$$

We now consider a condition on a group G that will ensure it is a direct product of indecomposable groups (for there do exist groups without this property).

Definition   A group G has the ACC (**ascending chain condition**) if every increasing chain of normal subgroups of G stops; i.e., if

$$A_1 \subset A_2 \subset \cdots \subset A_n \subset A_{n+1} \subset \cdots$$

is a chain of normal subgroups of G, then there is an integer t for which $A_t = A_{t+1} = A_{t+2} = \cdots$.

A group G has the DCC (**descending chain condition**) if every decreasing chain of normal subgroups of G

$$B_1 \supset B_2 \supset \cdots \supset B_n \supset B_{n+1} \supset \cdots$$

stops.

A group G **has both chain conditions** if it has both chain conditions!

Every finite group has both chain conditions; $\mathbf{Z}$ has the ACC but not the DCC; in Chapter 10, we shall study a group $\mathbf{Z}(p^\infty)$ that has the DCC but not the ACC.

Lemma 6.27   *If G has either chain condition, then G is a direct product of a finite number of indecomposable groups.*

Proof   Let us call a group *good* if it satisfies the conclusion of this lemma; otherwise it is *bad*. Clearly, an indecomposable group is good. If A and B are good, so is $A \times B$. Thus, if a group is bad, it is not indecomposable and it has a proper, bad factor.

Suppose now that G is bad. Set $G = A_0$. By induction, we may assume there exist bad subgroups $A_0, \cdots, A_n$ such that A_i is a direct factor of A_{i-1} for $i = 1, \cdots, n$. Since A_n is bad, it has a proper, bad factor A_{n+1}. By induction, we obtain a strictly decreasing infinite chain of normal subgroups of G:

$$G = A_0 > A_1 > A_2 > \cdots.$$

If G has the DCC, we have reached a contradiction.

Now assume G has the ACC. Since each A_i is a direct factor of A_{i-1}, we have normal subgroups B_i with $A_{i-1} = A_i \times B_i$. There is thus an increasing chain of normal subgroups

$$B_1 < B_1 \times B_2 < B_1 \times B_2 \times B_3 < \cdots$$

and we have reached a contradiction in this case, too. ∎

There do exist indecomposable groups having neither chain condition, e.g., the additive group of rationals.

Lemma 6.28 *Let G have both chain conditions and let α be a normal endomorphism of G. Then α is one-one if and only if α is onto. (Thus, either property ensures that α is an automorphism.)*

Proof Suppose α is one-one and $g \notin \alpha(G)$. We prove, by induction, that $\alpha^n(g) \notin \alpha^{n+1}(G)$. Otherwise there would be an element $h \in G$ with $\alpha^n(g) = \alpha^{n+1}(h)$, so that $\alpha(\alpha^{n-1}(g)) = \alpha(\alpha^n(h))$. Since α is one-one, $\alpha^{n-1}(g) = \alpha^n(h) \in \alpha^n(G)$, which contradicts the inductive hypothesis. Thus, we have a strictly descending chain of subgroups

$$G \supset \alpha(G) \supset \alpha^2(G) \supset \cdots.$$

Since α is normal, each $\alpha^n(G)$ is a normal subgroup of G, and so the DCC is violated. Therefore, α is onto.

Suppose α is onto. Let $A_n = \text{kernel } \alpha^n$; each A_n is a normal subgroup of G because α^n is a homomorphism (the normality of α is here irrelevant). Thus, we have the ascending chain of normal subgroups:

$$\{1\} = A_0 \subset A_1 \subset A_2 \subset \cdots.$$

Since G satisfies the ACC, this chain stops. Let t be the smallest integer for which $A_t = A_{t+1} = \cdots$. We claim that $t = 0$, which will prove the theorem. If $t \geq 1$, there is an $x \in A_t$ with $x \notin A_{t-1}$, i.e., $\alpha^t(x) = 1$, but $\alpha^{t-1}(x) \neq 1$. Since α is onto, there is an element $g \in G$ with $\alpha(g) = x$. Hence, $1 = \alpha^t(x) = \alpha^{t+1}(g)$, so that $g \in A_{t+1} = A_t$. Therefore, $\alpha^t(g) = 1$; but $\alpha^t(g) = \alpha^{t-1}(\alpha(g)) = \alpha^{t-1}(x)$, so that $\alpha^{t-1}(x) = 1$, which is a contradiction. Thus α is one-one. ∎

Definition An endomorphism α of G is **nilpotent** if there is a positive integer k such that $\alpha^k = 0$, where 0 denotes the trivial endomorphism that sends every element of G into the identity.

Theorem 6.29 (Fitting's lemma, 1934) *Let G have both chain conditions and let α be a normal endomorphism of G. Then $G = K \times H$, where K and H are each invariant under α, $\alpha|K$ is nilpotent, and $\alpha|H$ is onto.*

Proof Let $K_n = \text{kernel } \alpha^n$ and $H_n = \text{image } \alpha^n$. As we observed above, there are two chains of normal subgroups:

$$G \supset H_1 \supset H_2 \supset \cdots \quad \text{and} \quad \{1\} \subset K_1 \subset K_2 \subset \cdots.$$

Since G has both chain conditions, each of these chains stops: the H_n after r steps and the K_n after s steps. Let t be the larger of r and s, so that $K_t = K_{t+1} = \cdots$ and $H_t = H_{t+1} = \cdots$; define $H = H_t$ and $K = K_t$. It is easy to check that H and K are each invariant under α.

Suppose $x \in H \cap K$. Now $x \in H$ implies $\alpha^t(g) = x$ for some $g \in G$, and $x \in K$ implies $\alpha^t(x) = 1$. Therefore, $\alpha^{2t}(g) = \alpha^t(x) = 1$, so that $g \in K_{2t} = K_t$. Hence, $\alpha^t(g) = 1$ and $x = 1$.

Let $g \in G$. Then $\alpha^t(g) \in H_t = H_{2t}$, so there is an $x \in G$ with $\alpha^{2t}(x) = \alpha^t(g)$. Applying α^t to $g\alpha^t(x^{-1})$ gives 1, so that $g\alpha^t(x^{-1}) \in K_t = K$. Hence, $g = [g\alpha^t(x^{-1})]\alpha^t(x) \in KH$, so that $G = K \times H$.

Now $\alpha(H) = \alpha(H_t) = \alpha(\alpha^t(G)) = \alpha^{t+1}(G) = H_{t+1} = H_t = H$, so that $\alpha|H$ is onto. Finally, let $x \in K$; then $\alpha^t(x) \in K \cap H = \{1\}$ so that $\alpha|K$ is nilpotent. ∎

Corollary 6.30 *Let G be an indecomposable group having both chain conditions. Any normal endomorphism α of G is either nilpotent or an automorphism.*

Proof By Theorem 6.29, $G = K \times H$ with $\alpha|K$ nilpotent and $\alpha|H$ onto. Since G is indecomposable, either $G = K$ or $G = H$. In the first case, α is nilpotent; in the second case, α is onto and hence is an automorphism, by Lemma 6.28. ∎

Theorem 6.31 *Let G be an indecomposable group having both chain conditions and let α_1 and α_2 be normal, nilpotent endomorphisms of G. If $\alpha_1 + \alpha_2$ is also an endomorphism of G, then it is nilpotent.*

Proof If $\alpha_1 + \alpha_2$ is an endomorphism of G, it is immediately seen to be normal, so that, by Corollary 6.30, it is either nilpotent or an automorphism. Suppose $\alpha_1 + \alpha_2$ is an automorphism and γ is its inverse; γ is easily seen to be normal. Set $\lambda_1 = \alpha_1\gamma$ and $\lambda_2 = \alpha_2\gamma$, so that $1_G = \lambda_1 + \lambda_2$, i.e., $\lambda_1(x)\lambda_2(x) = x$ for all $x \in G$. In particular, $\lambda_1(x^{-1})\lambda_2(x^{-1}) = x^{-1}$. If we take the inverse of both sides, we see that $\lambda_2(x)\lambda_1(x) = x$, and so $\lambda_1 + \lambda_2 = \lambda_2 + \lambda_1$. Now the equation $\lambda_1(\lambda_1 + \lambda_2) = (\lambda_1 + \lambda_2)\lambda_1$ implies that $\lambda_1\lambda_2 = \lambda_2\lambda_1$. It follows that the set of all endomorphisms of G obtained from λ_1 and λ_2 by sums and products forms an algebraic system† in which the binomial theorem holds: for any integer $m > 0$,

$$(\lambda_1 + \lambda_2)^m = \lambda_1^m + \binom{m}{1}\lambda_1^{m-1}\lambda_2 + \binom{m}{2}\lambda_1^{m-2}\lambda_2^2 + \cdots + \lambda_2^m.$$

† This system need not be a commutative ring, for additive inverses may not exist.

Since α_1 and α_2 are nilpotent, λ_1 and λ_2 are nilpotent (they cannot be automorphisms since they have nontrivial kernels). Therefore, there are positive integers r and s with $\lambda_1^r = 0$ and $\lambda_2^s = 0$. If we take m large enough ($m = r + s - 1$ will do), then we obtain $(\lambda_1 + \lambda_2)^m = 0$. Since $\lambda_1 + \lambda_2 = 1_G$, we have $1_G = 0$, which contradicts the fact that, as any indecomposable group, $G \neq \{1\}$. ∎

Corollary 6.32 *Let G be an indecomposable group having both chain conditions; let $\alpha_1, \cdots, \alpha_n$ be a set of normal, nilpotent endomorphisms such that every sum of distinct α_i is an endomorphism. Then $\alpha_1 + \cdots + \alpha_n$ is nilpotent.*

Proof Induction on n. ∎

Theorem 6.33 (Remak-Krull-Schmidt)† *Let G be a group having both chain conditions. If*

$$G = H_1 \times H_2 \times \cdots \times H_s$$

and

$$G = K_1 \times K_2 \times \cdots \times K_t$$

are two decompositions of G into indecomposable groups, then $s = t$ and there is a reindexing so that $H_i \cong K_i$ for all i. Moreover, given any r between 1 and s, the reindexing may be chosen so that

$$G = H_1 \times \cdots \times H_r \times K_{r+1} \times \cdots \times K_t.$$

REMARK Our conclusion is stronger than saying that the factors appearing in the two factorizations are determined up to isomorphism; we can even replace factors in one decomposition with factors of the other.

Proof We shall give the proof for the case $r = 1$, leaving the rest of the proof for the reader to finish by induction. Given the first decomposition, we must find a renumbering of the K so that $H_i \cong K_i$ for all i and

$$G = H_1 \times K_2 \times \cdots \times K_t.$$

Let $\varepsilon_i : G \to H_i$ and $\eta_j : G \to K_j$ be projections; let $\lambda_i : H_i \to G$ and $\mu_j : K_j \to G$ be inclusion maps. Then the maps $\lambda_i \varepsilon_i$ and $\mu_j \eta_j$ are endomorphisms of G.

Now Exercise 6.53 says that every partial sum of $\Sigma \, \mu_j \eta_j$ is a normal endomorphism of G. Hence any partial sum of

$$\Sigma \, \varepsilon_1 \mu_j \eta_j \lambda_1 = \varepsilon_1 \circ \Sigma \, \mu_j \eta_j \circ \lambda_1$$

is a normal endomorphism of H_1. Corollary 6.32 implies that not every

---

† This theorem was first stated in 1909 by Wedderburn, but his proof was incorrect. The first correct proof was given, for finite groups, by Remak (1911); a simpler proof was given by Schmidt (1912). The theorem was extended to abelian groups with operators by Krull (1925), and to nonabelian groups with operators (generalizing our Theorem 6.33) by Schmidt (1928).

$\varepsilon_1 \mu_j \eta_j \lambda_1$ is nilpotent. By Corollary 6.30, one of these maps must be an automorphism of H_1. We renumber so that $\varepsilon_1 \mu_1 \eta_1 \lambda_1$ is an automorphism; let γ be its inverse.

We shall now show that the map $\eta_1 \lambda_1 : H_1 \to K_1$ is an isomorphism. We do know that the following composite ψ is the identity:

$$H_1 \xrightarrow{\gamma} H_1 \xrightarrow{\lambda_1} G \xrightarrow{\eta_1} K_1 \xrightarrow{\mu_1} G \xrightarrow{\varepsilon_1} H_1.$$

Consider the normal composite $\theta \colon K_1 \to K_1$:

$$K_1 \xrightarrow{\mu_1} G \xrightarrow{\varepsilon_1} H_1 \xrightarrow{\gamma} H_1 \xrightarrow{\lambda_1} G \xrightarrow{\eta_1} K_1.$$

Using the fact that the composite $\psi \colon H_1 \to H_1$ above is the identity, it follows easily that θ is idempotent: $\theta \circ \theta = \theta$. Since K_1 is indecomposable with both chain conditions, θ is either nilpotent or an automorphism, by Corollary 6.30. Thus, either $\theta = 0$ or $\theta = 1_{K_1}$. But $\theta \neq 0$ because θ occurs in the composite $\psi \circ \psi = \psi = 1_{H_1}$. Therefore θ is the identity and $\eta_1 \lambda_1 : H_1 \to K_1$ is an isomorphism (with inverse $\gamma \varepsilon_1 \mu_1$). It follows that $\varepsilon_1 \mu_1 : K_1 \to H_1$ is an isomorphism, for γ is an isomorphism.

Now η_1 sends $K_2 \times \cdots \times K_t$ into 1, but $\eta_1 \lambda_1$ is an isomorphism on H_1. Therefore,

$$H_1 \cap (K_2 \times \cdots \times K_t) = \{1\}.$$

If we define $G^* = \langle H_1, K_2 \times \cdots \times K_t \rangle$, then

$$G^* = H_1 \times K_2 \times \cdots \times K_t.$$

If $x \in G$, then $x = k_1 k_2 \cdots k_t$, where each $k_j \in K_j$. Since $\varepsilon_1 \mu_1$ is an isomorphism, the map $\beta \colon G \to G$ defined by $\beta(x) = \varepsilon_1(k_1) k_2 \cdots k_t$ is one-one and has image G^*. By Lemma 6.28, β must be onto and so $G^* = G$. Finally,

$$K_2 \times \cdots \times K_t \cong G/H_1 \cong H_2 \times \cdots \times H_s,$$

so that the remainder of the theorem follows by induction on $\max \{s, t\}$. ∎

EXERCISES

6.54. Use the Remak-Krull-Schmidt theorem to prove the fundamental theorem of finite abelian groups.

6.55. If $K \lhd G$ and if both K and G/K have both chain conditions, prove that G has both chain conditions. Conclude that if K and H each have both chain conditions, so does $K \times H$. (HINT: Use the Dedekind law, Exercise 2.58.)

6.56. Let G have both chain conditions; if there is a group H with $G \times G \cong H \times H$, then $G \cong H$.

**6.57. Let G have both chain conditions. If $G \cong A \times B$ and $G \cong A \times C$, then $B \cong C$.

6.58. Let G be the additive group of $\mathbf{Z}[x]$. Prove that $G \times \mathbf{Z} \cong G \times \mathbf{Z} \times \mathbf{Z}$. Conclude that the "cancellation law" is not always valid.

*6.59. A subgroup H is **subnormal** in G if there is a normal series

$$G = G_0 \supset G_1 \supset \cdots \supset G_m \supset H \supset \{1\}.$$

Prove that a group G has a composition series if and only if G has both chain conditions on subnormal subgroups.

The Remak-Krull-Schmidt theorem holds for algebraic systems other than groups (with both chain conditions); in particular, it holds for R-modules having both chain conditions (where the phrase "normal subgroup" is replaced by "submodule" in the definitions of ACC and DCC). The proof of the Remak-Krull-Schmidt theorem given above is also a proof for R-modules if one further replaces "homomorphism" by "R-homomorphism" and "normal endomorphism" by "R-endomorphism". As an illustration of the value of this translation, we state Fitting's lemma for vector spaces (i.e., R-modules, where R is a field).

Fitting's Lemma for Vector Spaces   *If V is a finite-dimensional vector space and $T: V \to V$ is a linear transformation, then $V = W_1 \oplus W_2$, where W_1 and W_2 are invariant under T, $T|W_1$ is nonsingular, and $T|W_2$ is nilpotent.*

The matrix version of Fitting's lemma thus says: Every $n \times n$ matrix over a field F is similar to a matrix of the form

$$\begin{bmatrix} A & 0 \\ 0 & B \end{bmatrix},$$

where A is nilpotent and B is nonsingular.

The proofs of generalizations of the Remak-Krull-Schmidt theorem to more general situations, e.g., to groups with operators or to lattices, are easily accessible in the literature: see A. G. Kurosh,† *The Theory of Groups*, Vol. II (see Bibliography).

---

† This is the same Kuroš mentioned elsewhere (this alternative transliteration is used here because it appears so in the book cited).

chapter seven

Extensions

THE EXTENSION PROBLEM

If G is a group having a normal subgroup K, then we can "factor" G into the two groups K and G/K. The study of extensions involves the inverse question: Given K and G/K, to what extent may one recapture G?

Definition If K and Q are groups, an **extension** of K by Q is a group G such that:

 (i) G contains K as a normal subgroup;
(ii) $G/K \cong Q$.

(As a mnemonic, K denotes kernel and Q denotes quotient.)

In a heuristic sense, an extension G is a "product" of K and Q.

There is another description of extensions that is convenient.

Definition A sequence of groups and homomorphisms $1 \to K \to G \xrightarrow{\pi} Q \to 1$ is a **short exact sequence** if $K = \ker \pi$ and $Q = \operatorname{im} \pi$. (We may regard the map $K \to G$ as the inclusion.)

The first isomorphism theorem shows $K \lhd G$ and $Q \cong G/K$, i.e., G is an extension of K by Q.

Example 1   Both $\mathbf{Z}(6)$ and S_3 are extensions of $\mathbf{Z}(3)$ by $\mathbf{Z}(2)$.

Example 2   For any two groups K and Q, their direct product $K \times Q$ is an extension of K by Q (and also of Q by K).

Example 3   Every extension of a solvable group by a solvable group is itself solvable; an extension of a nilpotent group by a nilpotent group need not be nilpotent (but see Exercise 5.47).

The extension problem (formulated by Hölder) is, given K and Q, to determine all extensions of K by Q. We can better understand the Jordan-Hölder theorem in the light of this problem. Let G be a group with composition series

$$G = K_0 \supset K_1 \supset \cdots \supset K_{n-2} \supset K_{n-1} \supset K_n = \{1\}$$

and corresponding factor groups

$$K_0/K_1 = Q_1, \cdots, K_{n-1}/K_n = Q_n.$$

Now $K_{n-1} = Q_n$ since $K_n = \{1\}$, but something more interesting happens at the next stage: $K_{n-2}/K_{n-1} = Q_{n-1}$, so that K_{n-2} is an extension of K_{n-1} by Q_{n-1}. If we could solve the extension problem, we could recapture K_{n-2} from K_{n-1} and Q_{n-1}, i.e., from Q_n and Q_{n-1}. Once we have K_{n-2}, we can attack K_{n-3} in a similar manner, for $K_{n-3}/K_{n-2} = Q_{n-2}$. Thus, a solution of the extension problem allows us to recapture K_{n-3} from Q_n, Q_{n-1}, and Q_{n-2}. We continue climbing up the composition series until we reach $K_0 = G$; to do this, we need extensions and the factor groups $Q_1, \cdots, Q_n$. The group G is thus a "product" of the Q_i, and the Jordan-Hölder theorem says that the simple groups occurring as factors in this "factorization" of G are uniquely determined by G. We could thus survey all finite groups if we knew all finite simple groups and if we could solve the extension problem. In particular, we could survey all finite solvable groups if we could solve the extension problem.

A solution of the extension problem consists of determining all groups G with $G/K \cong Q$. But what does "determining" a group G mean? We gave two answers to this question in Chapter 1 when we considered "knowing" a group G. One answer is that a multiplication table for G can be constructed; a second answer is that the isomorphism class of G can be characterized. Schreier (1926) solved the extension problem in the first sense; given K and Q, all multiplication tables of extensions G of K by Q can be constructed. On the other hand, no solution is known in the second sense. For example, given K and Q, Schreier's solution does not allow us to calculate the number of nonisomorphic extensions of K by Q.

We shall see that, in essence, the extensions of K by Q themselves form an abelian group! The computation of one of these groups is the Schur-Zassenhaus lemma (Theorem 7.24). More general techniques for computing groups of extensions will be considered in Chapter 11, where we discuss homological algebra. There is a second group one may invent. Its elements are groups, and, whenever G is an extension of K by Q, then G is the product of K and Q. This construction is called the *Grothendieck group*, and it is examined in Chapter 10.

EXERCISES

7.1. Any two extensions of K by Q have the same number of elements, namely, $|K||Q|$.

7.2. There are exactly two nonisomorphic extensions of $\mathbf{Z}(3)$ by $\mathbf{Z}(2)$; there is exactly one extension of $\mathbf{Z}(2)$ by $\mathbf{Z}(3)$.

7.3. Every nonabelian group of order p^3, p prime, is an extension of $\mathbf{Z}(p)$ by $\mathbf{Z}(p) \times \mathbf{Z}(p)$.

7.4. If G is an extension of K by Q, must G contain a subgroup isomorphic to Q?

7.5. If $(a, b) = 1$ and if K and Q are abelian groups of orders a and b, respectively, prove that there is only one abelian extension of K by Q.

AUTOMORPHISM GROUPS

The coming construction is essential for a discussion of extensions; it is also of great intrinsic interest.

Definition The **automorphism group** of a group G, denoted by Aut(G), is the set of all automorphisms of G under the binary operation of composition.

It is easy to check that Aut(G) is a group; indeed, Aut(G) is a subgroup of S_G.

Definition An automorphism α of G is **inner** if it is conjugation by an element of G, i.e., $\alpha(x) = \gamma_a(x) = axa^{-1}$ for some $a \in G$; otherwise, α is **outer**.

Theorem 7.1

(i) *If H is a subgroup of a group G, then $C_G(H) \lhd N_G(H)$ and $N_G(H)/C_G(H)$ can be imbedded in* Aut(H).

(ii) *The set of all inner automorphisms of G, denoted by* Inn(G), *is a normal subgroup of* Aut(G) *and* $G/Z(G) \cong$ Inn(G).

Proof

(i) If $a \in G$, let γ_a denote conjugation by a, and let $v_a = \gamma_a | H$. The function $a \mapsto v_a$ is easily seen to be a homomorphism of $N_G(H)$ into Aut(H). The kernel of this map is $C_G(H)$, for the following statements are equivalent: a is in the kernel; v_a is the identity function on H; $aha^{-1} = h$ for all $h \in H$; $a \in C_G(H)$. It follows from the first isomorphism theorem that $C_G(H) \lhd N_G(H)$ and that $N_G(H)/C_G(H)$ is isomorphic to a subgroup of Aut(H).

(ii) If $H = G$, then $N_G(H) = G$, $C_G(H) = Z(G)$, and the map $a \mapsto v_a = \gamma_a$ has Inn(G) as its image. The isomorphism established above is now $G/Z(G) \cong$ Inn(G).

To prove Inn $(G) \lhd$ Aut (G), we must show that if $\alpha \in$ Aut (G), then $\alpha \gamma_a \alpha^{-1}$ is an inner automorphism. In fact, $\alpha \gamma_a \alpha^{-1} = \gamma_{\alpha(a)}$, as the reader should check. ∎

EXERCISES

*7.6. Prove that $\text{Aut}(S_3) \cong S_3$.

7.7. Prove that $\text{Aut}(V) \cong S_3$, where V is the 4-group. Conclude that non-isomorphic groups can have isomorphic automorphism groups.

*7.8. Let G be an elementary abelian group of order p^n, where p is prime. Prove that $\text{Aut}(G) \cong GL(n, p)$.

7.9. Let H and K be finite groups whose orders are relatively prime. Prove that $\text{Aut}(H \times K) \cong \text{Aut}(H) \times \text{Aut}(K)$.

*7.10. Let G be a finite group with $\text{Aut}(G) = \{1\}$. Prove that G has at most two elements. (The finiteness condition is unnecessary, but the reader may not possess the tools for dealing with the general case until Chapter 10.)

7.11. Prove that $\text{Aut}(Q) \cong S_4$, where Q is the group of quaternions. [HINT: Inn(Q) is isomorphic to the 4-group, and it is a subgroup of Aut(Q) equal to its own centralizer. Use Theorem 7.1 with $G = \text{Aut}(Q)$ and $H = \text{Inn}(Q)$.]

Definition A **unit** in a commutative ring R with 1 is an element having a multiplicative inverse in R. The **group of units**, denoted by $U(R)$, is the multiplicative group of all units in R.

EXERCISES

**7.12. Prove that $\text{Aut}(\mathbf{Z}(n)) \cong U(\mathbf{Z}/n\mathbf{Z})$ and hence has order $\varphi(n)$. [HINT: An integer t represents a unit in $\mathbf{Z}/n\mathbf{Z}$ if and only if $(t, n) = 1$.]

**7.13. If p is prime, then $\text{Aut}(\mathbf{Z}(p)) \cong \mathbf{Z}(p-1)$. (HINT: Use Theorem 2.16.)

**7.14. If p is prime, then $a \equiv b \pmod{p}$ implies
$$a^{p^{n-1}} \equiv b^{p^{n-1}} \pmod{p^n}.$$

Theorem 7.2

$$\text{Aut}(\mathbf{Z}(2^m)) \cong \begin{cases} \{1\} & \textit{if } m = 1 \\ \mathbf{Z}(2) & \textit{if } m = 2 \\ \mathbf{Z}(2) \times \mathbf{Z}(2^{m-2}) & \textit{if } m \geq 3. \end{cases}$$

If p is an odd prime,

$$\text{Aut}(\mathbf{Z}(p^n)) \cong \mathbf{Z}((p-1)p^{n-1}).$$

Proof By Exercise 7.12, $\text{Aut}(\mathbf{Z}(2^n)) \cong U(\mathbf{Z}/2^n\mathbf{Z})$ and has order $\varphi(2^n)$. Since $\varphi(2) = 1$ and $\varphi(4) = 2$, the theorem holds when $n \leq 2$. If $n \geq 3$, then this theorem is precisely Lemma 5.34.

Assume that p is odd and that $n \geq 2$. (We already know the theorem for $n = 1$.) Denote $U(\mathbf{Z}/p^n\mathbf{Z})$ by G. Now $|G| = \varphi(p^n) = (p-1)p^{n-1}$; we shall show that $G \cong \mathbf{Z}(p-1) \times \mathbf{Z}(p^{n-1})$.

Let $B = \{\bar{b} \in G: b \equiv 1 \ (\text{mod } p)\}$; note that B is a subgroup of G. If $1 \leq b < p^n$, then b has a unique expression in base p:

$$b = b_0 + b_1 p + \cdots + b_{n-1}p^{n-1}, \quad 0 \leq b_i \leq p-1.$$

It follows that $|B| = p^{n-1}$, for $b \in B$ if and only if $b_0 = 1$. We claim $B = \langle \overline{1+p} \rangle$. Since $|B| = p^{n-1}$, it suffices to show $(1+p)^{p^{n-2}} \not\equiv 1 \ (\text{mod } p^n)$, for then $\overline{1+p}$ must have order p^{n-1}. The incongruence is proved by induction on n, the case $n = 2$ being trivial.

Since $1 + p \equiv 1 \ (\text{mod } p)$, Exercise 7.14 gives

$$(1+p)^{p^{n-3}} \equiv 1 \ (\text{mod } p^{n-2}),$$

so that, by induction,

$$(1+p)^{p^{n-3}} = 1 + kp^{n-2} \qquad \text{where } k \not\equiv 0 \ (\text{mod } p).$$

Therefore

$$(1+p)^{p^{n-2}} = (1 + kp^{n-2})^p$$
$$= 1 + \binom{p}{1}kp^{n-2} + \cdots + k^p p^{p(n-2)}$$
$$\equiv 1 + kp^{n-1} \ (\text{mod } p^n)$$
$$\not\equiv 1 \ (\text{mod } p^n).$$

By the primary decomposition, $G = A \times B$, where A is a subgroup of order $p - 1$. Consider the map $f: G \to U(\mathbf{Z}/p\mathbf{Z})$ defined by $f(\bar{a}) = \text{cls } a$, where cls a means the congruence class of a modulo p. Clearly f is onto with kernel B, so that $G/B \cong U(\mathbf{Z}/p\mathbf{Z}) \cong \mathbf{Z}(p-1)$. Since $G/B \cong A$, we have $A \cong \mathbf{Z}(p-1)$. Therefore $G \cong \mathbf{Z}(p-1) \times \mathbf{Z}(p^{n-1}) \cong \mathbf{Z}((p-1)p^{n-1})$, by Exercise 2.74. ∎

EXERCISES

7.15. Give a complete set of invariants describing $U(\mathbf{Z}/n\mathbf{Z})$.

7.16. Let X be the set of all conjugacy classes of a group G. Every automorphism α of G induces a permutation of X. Conclude that $\text{Aut}(G)$ is a permutation group on X as well as a permutation group on G.

**7.17. Let G be a finite group and let C be a conjugacy class of G consisting of h elements of order t. If no other conjugacy class of G comprised of elements of order t has exactly h elements, then $\alpha(C) = C$ for every $\alpha \in \text{Aut}(G)$.

Theorem 7.1(ii) suggests the following class of groups.

Definition   A group G is **complete** if $Z(G) = \{1\}$ and every automorphism of G is inner.

If G is a complete group, then $\text{Aut}(G) \cong G$. An interesting generalization of Exercise 7.6 is that almost all the symmetric groups are complete.

Lemma 7.3   *If $\alpha: S_n \to S_n$ is an automorphism that preserves transpositions, i.e., if α sends every transposition into a transposition, then α is inner.*

Proof   Recall that two elements of S_n are conjugate if and only if they have the same cycle structure, so that every conjugation of S_n preserves transpositions.

If $\pi \in S_n$ and $\gamma: S_n \to S_n$, denote $\gamma(\pi)$ by π^γ.

We shall prove by induction on t that there exist conjugations $\beta_2, \cdots, \beta_t$ such that $\beta_t^{-1} \cdots \beta_2^{-1} \alpha$ fixes $(1\,2), \cdots, (1\,t)$. Now $(1\,2)^\alpha = (i\,j)$; define β_2 to be conjugation by $(1\,i)(2\,j)$ (where our notation sets, e.g., $(1\,i)$ = identity if $1 = i$). Our quick way of computing conjugations (Lemma 3.9) shows that $(1\,2)^\alpha = (1\,2)^{\beta_2}$, and so $\beta_2^{-1}\alpha$ fixes $(1\,2)$.

Let $\gamma = \beta_t^{-1} \cdots \beta_2^{-1}\alpha$ be given by the inductive hypothesis. Since γ preserves transpositions, $(1\ t+1)^\gamma = (l\,k)$. Now $\{1,2\} \cap \{l,k\} \neq \varnothing$, otherwise $[(1\,2)(1\,t+1)]^\gamma = (1\,2)(l\,k)$ has order 2 while $(1\,2)(1\,t+1)$ has order 3. Thus $(1\ t+1)^\gamma = (1\ k)$ or $(2\ k)$. We must have $k > t$, because γ is one-one. Define $\beta_{t+1}: S_n \to S_n$ to be conjugation by $(k\ t+1)$. Now β_{t+1} fixes $(1\,2), \cdots, (1\,t)$ and $(1\ t+1)^{\beta_{t+1}} = (1\ t+1)^\gamma$, so that $\beta_{t+1}^{-1}\gamma = \beta_{t+1}^{-1}\beta_t^{-1} \cdots \beta_2^{-1}\alpha$ fixes $(1\,2), \cdots, (1\ t+1)$.

It follows that $\beta_n^{-1} \cdots \beta_2^{-1}\alpha$ fixes $(1\,2), \cdots, (1\,n)$. Since these $n-1$ transpositions generate S_n, by Exercise 3.24, $\beta_n^{-1} \cdots \beta_2^{-1}\alpha$ is the identity, and $\alpha = \beta_2 \cdots \beta_n$ is inner. ∎

Theorem 7.4   *S_n is complete if $n \neq 2$ and $n \neq 6$.*

Proof   We remark that $S_2 \cong \mathbf{Z}(2)$ is not complete because it has a center. If $n \geq 3$, we have seen in Exercise 3.37 that S_n is centerless.

Let C_1 be the conjugacy class of S_n consisting of all transpositions; every element in C_1 has order 2. Since every permutation is a product of disjoint cycles, the only elements in S_n of order 2 are products of disjoint transpositions; let C_k be the conjugacy class consisting of all products of k disjoint transpositions. We shall show that if $n \neq 6$, the size of each C_k $(k \neq 1)$ is distinct from the size of C_1. Exercise 7.17 will then give $\alpha(C_1) = C_1$ for every $\alpha \in \text{Aut}(S_n)$, i.e., every $\alpha \in \text{Aut}(S_n)$ preserves transpositions. An application of Lemma 7.3 will then complete the proof.

There are $n(n-1)/2$ transpositions in S_n. How many k-tuples of disjoint transpositions are there? Answer:

$$\frac{n(n-1)}{2} \cdot \frac{(n-2)(n-3)}{2} \cdots \frac{(n-2k+2)(n-2k+1)}{2}.$$

How many distinct products of k disjoint transpositions are there? If τ_1, $\tau_2, \cdots, \tau_k$ are disjoint transpositions, they commute, and so the order in which they are written is irrelevant. There being $k!$ different orderings, the size of C_k is

$$\frac{1}{k!}\frac{1}{2^k} n(n-1)(n-2)\cdots(n-2k+1).$$

The problem is now reduced to the question: Can the size of C_k equal $n(n-1)/2$? Equivalently, does the following equation hold for $k > 1$?

(*) $(n-2)(n-3)\cdots(n-2k+1) = k!\,2^{k-1}.$

Since the left side is positive, we must have $n \geq 2k$. Therefore, for fixed n,

left side $\geq (2k-2)(2k-3)\cdots(2k-2k+1) = (2k-2)!.$

An easy induction shows that if $k \geq 4$, then

$$(2k-2)! > k!\,2^{k-1},$$

so that (*) can hold only if $k < 4$, regardless of the value of n. We may now assume that $k = 2$ or $k = 3$. It is easy to see that (*) never holds if $k = 2$, so that only the case $k = 3$ remains. Now, since $n \geq 2k$, we must have $n \geq 6$. If $n > 6$, the left side of (*) $\geq 5 \cdot 4 \cdot 3 \cdot 2 = 120$, while the right side of (*) $= 3!\,2^2 = 24$. (If $n = 6$ and $k = 3$, then (*) does hold.) We have shown that if $n \neq 6$, then there is no $k > 1$ for which (*) holds. Hence, if $n \neq 6$, the size of each C_k $(k \neq 1)$ is distinct from the size of C_1. ∎

Corollary 7.5 $\mathrm{Aut}(S_n) \cong S_n$ for $n \neq 2$ and $n \neq 6$.

We now show that S_6 is a genuine exception to Theorem 7.4.

Definition   A subgroup K of S_X, the symmetric group on a set X, is **transitive** if, for every pair of elements x, $y \in X$, there is a permutation $\sigma \in K$ with $\sigma(x) = y$.

We shall examine transitive subgroups more thoroughly in Chapter 9; here we need the following fact: If H is a subgroup of a group G, if X is the family of left cosets of H in G, and if $\rho: G \to S_X$ is the representation of Theorem 3.18 $[\rho(g): aH \mapsto gaH]$, then $\rho(G)$ is a transitive subgroup of S_X (given aH and bH in X, then $\rho(ba^{-1}): aH \mapsto bH$).

Lemma 7.6   *There exists a transitive subgroup K of S_6 having order 120; moreover, such a subgroup cannot contain a transposition.*

Proof   A Sylow 5-subgroup P of S_5 must have 6 conjugates (by the Sylow theorems), so the normalizer $N(P)$ of P in S_5 has index 6. The representation ρ of S_5 on the cosets of $N(P)$ is a homomorphism $\rho: S_5 \to S_6$ that must be one-one, for $\ker \rho \subset N(P)$ and, for $n \geq 4$, the only proper normal subgroup of S_n is the alternating group A_n. Therefore $K = \mathrm{im}\,\rho$ is a transitive subgroup of S_6 having order 120.

Now K contains an element α of order 5 which must be a 5-cycle, say, $\alpha = (12345)$. If $(ij) \in K$, then transitivity of K provides $\beta \in K$ with $\beta(j) = 6$. Therefore $\beta(ij)\beta^{-1} = (l6)$ for some $l \neq 6$. Conjugating $(l6)$ by the powers of α shows K contains $(16), (26), (36), (46), (56)$. But these transpositions generate S_6, by Exercise 3.24. ∎

If H is the subgroup composed of all $\beta \in S_6$ with $\beta(6) = 6$, then $H \cong S_5$ has order 120, but it is easy to see H is not transitive. Of course, H does contain transpositions.

Theorem 7.7 (Hölder, 1895)   *There exists an outer automorphism of S_6.*

Proof   Let K be a transitive subgroup of S_6 of order 120 and let the cosets of K be $\alpha_1 K, \cdots, \alpha_6 K$. If $\varphi: S_6 \to S_6$ is the representation of S_6 on the cosets of K, the argument in the lemma shows ker $\varphi = \{1\}$; thus φ is one-one, hence onto: $\varphi \in \text{Aut}(S_6)$. Were φ an inner automorphism, it would preserve the cycle structure of every permutation in S_6. But $\varphi((12))$ sends $\alpha_i K$ to $(12)\alpha_i K$ for each i. Can $(12)\alpha_i K = \alpha_i K$? If so, then the transposition $\alpha_i^{-1}(12)\alpha_i \in K$, contradicting the lemma. Therefore $\varphi((12))$ fixes no cosets, hence cannot be a transposition (which fixes all but two cosets). We conclude that φ is an outer automorphism. ∎

To see that the outer automorphism just constructed is essentially unique, we first refer the reader to the table on page 44 describing S_6. Recall that two permutations in S_6 lie in the same conjugacy class if and only if they have the same cycle structure.

Theorem 7.8 $\text{Aut}(S_6)/\text{Inn}(S_6) \cong \mathbf{Z}(2)$.

Proof   If $Y = \{C_1, C_2, \cdots, C_{11}\}$ is the set of conjugacy classes of S_6 and $\varphi \in \text{Aut}(S_6)$, then $\varphi \in S_Y$. Now φ is inner if and only if $\varphi(C_2) = C_2$, by Lemma 7.3; therefore, φ is outer if and only if φ interchanges C_2 and C_{10}, these being the only conjugacy classes having 15 elements. If φ and ψ are outer automorphisms, therefore, then $\varphi\psi^{-1}(C_2) = C_2$, $\varphi\psi^{-1} \in \text{Inn}(S_6)$, and $\text{Aut}(S_6)/\text{Inn}(S_6)$ has order 2. ∎

An explicit formula for an outer automorphism φ can be given (see G. Janusz and J. Rotman, *Amer. Math. Monthly*, 1982). In particular, such an automorphism φ interchanges C_2 and C_{10}, interchanges C_3 and C_{11}, and fixes each of C_6 and C_8 setwise; moreover, there exists an outer automorphism φ of order 2.

The next theorem provides another source of (possibly infinite) complete groups.

Theorem 7.9   *If G is a simple nonabelian group, then $\text{Aut}(G)$ is complete.*

Proof   Let $I = \text{Inn}(G) \lhd \text{Aut}(G) = A$. Since G is simple and not abelian, $Z(G) = \{1\}$, and so $G \cong I$. First of all, $Z(A) = \{1\}$. In fact, we show $C_A(I) = \{1\}$: if $\alpha \in C_A(I)$, then for every conjugation $\gamma_b, \gamma_b = \alpha\gamma_b\alpha^{-1} = \gamma_{\alpha(b)}$; hence $\alpha(b)b^{-1} \in Z(G) = \{1\}$ for all $b \in G$; thus $\alpha = 1$.

To see that every automorphism of A is inner, suppose $\sigma \in \mathrm{Aut}(A)$. Since $I \lhd A$, we have $\sigma(I) \lhd A$, so that $I \cap \sigma(I) \lhd \sigma(I)$. Simplicity of $\sigma(I)$ gives $I \cap \sigma(I) = \{1\}$ or $I \cap \sigma(I) = \sigma(I)$. In the first case,

$$[I, \sigma(I)] \subset I \cap \sigma(I) = \{1\},$$

which implies $\sigma(I) = \{1\}$, by paragraph one. This contradiction shows that $I \cap \sigma(I) = \sigma(I)$, i.e., $\sigma(I) \subset I$. Now, if $\gamma_b \in I$, then $\sigma(\gamma_b) \in I$; therefore,

$$\sigma(\gamma_b) = \gamma_{\alpha(b)}$$

for some $\alpha(b) \in G$, and one easily checks that $\alpha \in \mathrm{Aut}(G) = A$.

Let $\tau = \sigma \circ (\Gamma_\alpha)^{-1}$, where $\Gamma_\alpha : A \to A$ is conjugation by α. Observe that

$$\tau(\gamma_b) = \sigma \Gamma_\alpha^{-1}(\gamma_b) = \sigma(\alpha^{-1} \gamma_b \alpha)$$
$$= \sigma(\gamma_{\alpha^{-1}(b)})$$
$$= \gamma_{\alpha \alpha^{-1}(b)} = \gamma_b.$$

Thus, τ fixes everything in I.

We claim that τ is the identity on A. If not, $\tau(\beta) \neq \beta$ for some $\beta \in A$. For every $b \in G$,

$$\beta \gamma_b \beta^{-1} = \tau(\beta \gamma_b \beta^{-1}) \quad \text{(since } I \lhd A \text{ and } \tau \text{ fixes } I\text{)}$$
$$= \tau(\beta) \gamma_b \tau(\beta)^{-1} \quad \text{(since } \tau \text{ fixes } I\text{)}.$$

Hence $\tau(\beta)\beta^{-1} \in C_A(I) = \{1\}$ and $\tau(\beta) = \beta$, a contradiction. Therefore $\tau = 1$ and σ is the inner automorphism Γ_α. ■

It follows that $\mathrm{Aut}(\mathrm{Aut}(G)) \cong \mathrm{Aut}(G)$ whenever G is a nonabelian simple group. There is a beautiful theorem of Wielandt with a similar conclusion. For a centerless group G, we know that G can be imbedded in $\mathrm{Aut}(G)$; moreover, $\mathrm{Aut}(G)$ is also centerless. Thus, this process may be iterated,

$$G \subset \mathrm{Aut}(G) \subset \mathrm{Aut}(\mathrm{Aut}(G)) \subset \cdots,$$

to give the **automorphism tower** of G. Wielandt (1939) proved that if G is a finite centerless group, then its automorphism tower is constant from some point on.

Theorem 7.10 *Let $H \lhd G$, where H is a complete group. Then H is a direct factor of G, i.e., there exists a normal subgroup K of G with $G = H \times K$.*

Proof If $g \in G$, then $gxg^{-1} \in H$ for all $x \in H$, since $H \lhd G$. The function $x \mapsto gxg^{-1}$ is thus an automorphism of H. Since every automorphism of H is inner, there exists an element $\gamma = \gamma(g) \in H$ with $\gamma x \gamma^{-1} = gxg^{-1}$ for all $x \in H$. Let $K = C_G(H)$, the centralizer of H in G. Clearly K contains each $\gamma^{-1}g$. We claim that $G = H \times K$. First of all, $HK = G$, for if $g \in G$, then $g = \gamma(\gamma^{-1}g)$. Second, if $x \in H \cap K$, then $x \in Z(H) = \{1\}$, since H is centerless. Now $gKg^{-1} = \gamma(\gamma^{-1}g)K(g^{-1}\gamma)\gamma^{-1} \subset \gamma K \gamma^{-1}$, since $\gamma^{-1}g \in K$; furthermore, $\gamma K \gamma^{-1} \subset K$, by the definition of K. It follows that $K \lhd G$, and hence $G = H \times K$. ■

EXERCISES

7.18. If G is a complete group and $G' \neq G$, then G is not the commutator subgroup of any group containing it. Conclude that S_n, $n \neq 2, 6$, is never a commutator subgroup.

7.19. Prove the alternating groups A_n are never complete.

*7.20. (i) If G is a finite abelian group with more than two elements, then $\text{Aut}(G)$ is never of odd order. [HINT: If $x \mapsto x^{-1}$ is the identity, then G is elementary abelian of order 2^m and $\text{Aut}(G) \cong GL(m, 2)$.] The finiteness hypothesis is not necessary, but the reader must wait until Chapter 10 before seeing this.

(ii) If G is not abelian, then $\text{Aut}(G)$ is never cyclic. [HINT: Prove that $\text{Inn}(G)$ is not cyclic in this case.]

Conclude that a cyclic group of odd order is never $\text{Aut}(G)$ for any (finite) group G.

SEMIDIRECT PRODUCTS

A group G is the direct product of two normal subgroups K and Q in case $K \cap Q = \{1\}$ and $KQ = G$. A natural generalization of direct products that often arises is the situation in which only one of the subgroups is required to be normal.

Definition A group G is a **semidirect product** of K by Q, denoted by $G = K \rtimes Q$, if G contains subgroups K and Q such that:

(i) $K \triangleleft G$;
(ii) $KQ = G$;
(iii) $K \cap Q = \{1\}$.

Definition Let K and Q be (not necessarily normal) subgroups of a group G. Then Q is a **complement** of K if $K \cap Q = \{1\}$ and $KQ = G$.

Of course, $KQ = G$ implies $QK = G$ so that K is a complement of Q. An extension G of K by Q is a semidirect product of K by Q if and only if K has a complement (necessarily isomorphic to Q). The remark after the proof of Theorem 7.8 thus shows that $\text{Aut}(S_6)$ is a semidirect product of S_6 by $\mathbf{Z}(2)$. A subgroup K of G may not have a complement, and, even if it does, a complement need not be unique. In S_3, for example, every subgroup of order 2 serves as a complement to A_3. Indeed, it follows from the product formula that if $|G| = mn$, where $(m, n) = 1$, and if K is a subgroup of order m, then a subgroup Q of G is a complement of K if and only if $|Q| = n$.

EXERCISES

7.21. S_n is a semidirect product of A_n by $\mathbf{Z}(2)$.

7.22. Let G be a finite solvable group of order ab, where $(a, b) = 1$. If G contains a

normal subgroup K of order a, then G is a semidirect product of K by some
group Q of order b. (HINT: Use P. Hall's theorem, Theorem 5.23.)

*7.23. D_n is a semidirect product of $\mathbf{Z}(n)$ by $\mathbf{Z}(2)$.

 7.24.   If p is a prime, then $\mathbf{Z}(p^n)$ is not a semidirect product.

 7.25.   The quaternions Q is not a semidirect product.

*7.26.   Both S_3 and $\mathbf{Z}(6)$ are semidirect products of $\mathbf{Z}(3)$ by $\mathbf{Z}(2)$.

Exercise 7.26 is a bit jarring at first, for it says that, in contrast to a direct
product, a semidirect product of K by Q is not determined up to isomorphism by the
two subgroups. When we reflect on this, however, we see that a semidirect product
should depend on how K is normal in G.

Lemma 7.11   *Let G be a semidirect product of K by Q. There is a
homomorphism $\theta: Q \rightarrow \mathrm{Aut}(K)$ defined by*

$$\theta_x(k) = xkx^{-1}, \qquad all \ k \in K, \ x \in Q.$$

Moreover,

$$\left. \begin{array}{r} \theta_x(\theta_y(k)) = \theta_{xy}(k) \\ \theta_1(k) = k \end{array} \right\} \qquad k \in K, \ x, \ y, \ 1 \in Q.$$

Proof   Straightforward, using the normality of K.   ∎

The object of our study is to recapture a semidirect product G from K and Q, so
let us begin with only K, Q, and some homomorphism $\theta: Q \rightarrow \mathrm{Aut}(K)$.

Definition   Given K, Q, and $\theta: Q \rightarrow \mathrm{Aut}(K)$, then a semidirect product G of K by Q
realizes θ if

$$\theta_x(k) = xkx^{-1} = k^x \qquad \text{for all } k \in K.$$

In this language, Lemma 7.11 states that every semidirect product G of K by Q
determines some θ that G realizes. The intuitive meaning of θ is that it describes how
K is normal in G.

Definition   Let K, Q, and $\theta: Q \rightarrow \mathrm{Aut}(K)$ be given. Then $K \rtimes_\theta Q$ is the set of all
ordered pairs $(k, x) \in K \times Q$ under the binary operation

$$(k, x)(k_1, y) = (k\theta_x(k_1), xy).$$

Theorem 7.12   *Let K, Q, and $\theta: Q \rightarrow \mathrm{Aut}(K)$ be given; then $G = K \rtimes_\theta Q$ is a
semidirect product of K by Q that realizes θ.*

Proof We first prove that $G = K \rtimes_\theta Q$ is a group. Multiplication is associative:

$$[(k, x)(k_1, y)](k_2, z) \qquad\qquad (k, x)[(k_1, y)(k_2, z)]$$
$$= (k\theta_x(k_1), xy)(k_2, z) \qquad\qquad = (k, x)(k_1\theta_y(k_2), yz)$$
$$= (k\theta_x(k_1)\theta_{xy}(k_2), xyz); \qquad = (k\theta_x(k_1\theta_y(k_2)), xyz).$$

The formulas in Lemma 7.11 show the final entries in each column are equal. The identity element is $(1, 1)$, for

$$(1, 1)(k, x) = (1 \cdot \theta_1(k), 1x) = (k, x).$$

The inverse of (k, x) is $((\theta_{x^{-1}}(k))^{-1}, x^{-1})$, for

$$((\theta_{x^{-1}}(k))^{-1}, x^{-1})(k, x) = ((\theta_{x^{-1}}(k))^{-1}\theta_{x^{-1}}(k), x^{-1}x) = (1, 1).$$

Let us identify K with the subset of G consisting of all pairs of the form $(k, 1)$. Since the only "twist" occurs in the first coordinate, the map $\pi: G \to Q$ defined by $\pi(k, x) = x$ is a homomorphism. It is quickly checked that $\ker \pi = K$, so that K is a normal subgroup of G.

Identify Q with all pairs $(1, x)$. Then Q is a subgroup of G with $KQ = G$ and $K \cap Q = \{(1, 1)\}$. Therefore, G is a semidirect product of K by Q. To see that G realizes θ, compute:

$$(1, x)(k, 1)(1, x)^{-1} = (\theta_x(k), x)(1, x^{-1}) = (\theta_x(k), 1). \quad\blacksquare$$

Since $G = K \rtimes_\theta Q$ does realize θ, there can be no confusion if we write k^x instead of $\theta_x(k)$. The binary operation of $K \rtimes_\theta Q$ may be now written:

$$(k, x)(k_1, y) = (kk_1^x, xy).$$

Theorem 7.13 *If G is a semidirect product of K by Q, then $G \cong K \rtimes_\theta Q$ for some $\theta: Q \to \mathrm{Aut}(K)$.*

Proof As in Lemma 7.11, define $\theta_x(k) = xkx^{-1}$. Since $G = KQ$, each $g \in G$ has an expression $g = kx$, where $k \in K$ and $x \in Q$; this expression is unique because $K \cap Q = \{1\}$. Multiplication in G satisfies

$$(kx)(k_1 x_1) = k(xk_1 x^{-1})xx_1$$
$$= kk_1^x xx_1.$$

It is now easy to see that the map $K \rtimes_\theta Q \to G$ defined by $(k, x) \mapsto kx$ is an isomorphism. $\quad\blacksquare$

To illustrate how these results may be used, we construct the group T of order 12 defined in Exercise 4.39 and the nonabelian groups of order p^3 (also, see Exercise 7.44).

Example 4 The group T of order 12 is a semidirect product of $\mathbf{Z}(3)$ by $\mathbf{Z}(4)$: $T = \mathbf{Z}(3) \rtimes \mathbf{Z}(4)$.

Let $\mathbf{Z}(3) = \langle k \rangle$ and let $\mathbf{Z}(4) = \langle x \rangle$. Define $\theta\colon \mathbf{Z}(4) \to \mathrm{Aut}(\mathbf{Z}(3)) \cong \mathbf{Z}(2)$ by sending x into the generator. In detail,

$$k^x = k^2 \quad \text{and} \quad (k^2)^x = k,$$

while x^2 acts on $\langle k \rangle$ as the identity automorphism.

The group $\mathbf{Z}(3) \rtimes_\theta \mathbf{Z}(4)$ has order 12. If $a = (k^2, x^2)$ and $b = (1, x)$, then the reader may verify that

$$a^6 = 1 \quad \text{and} \quad b^2 = a^3 = (ab)^2,$$

which are the relations of T.

Example 5 Let p be a prime, $K = \langle k_1, k_2 \rangle$ be elementary abelian of order p^2, and $Q = \langle x \rangle$ be cyclic of order p. Define $\theta\colon Q \to \mathrm{Aut}(K) \cong GL(2, p)$ by

$$x^i \mapsto \begin{bmatrix} 1 & 0 \\ i & 1 \end{bmatrix}.$$

Thus, $k_1^x = k_1 k_2$ and $k_2^x = k_2$. The commutator $k_1^x k_1^{-1}$ is seen to be k_2. Therefore, $G = K \rtimes_\theta Q$ is a group of order p^3, $G = \langle k_1, k_2, x \rangle$, and these generators satisfy relations: $k_1^p = k_2^p = x^p = 1$, $k_2 = xk_1 x^{-1} k_1^{-1}$, and $[k_2, k_1] = 1 = [k_2, x]$.

If p is odd, we have the nonabelian group of order p^3 in which every element has order p; if $p = 2$, we have D_4. Note, using Exercise 7.23, that $D_4 \cong \mathbf{Z}(4) \rtimes_\psi \mathbf{Z}(2)$ and $D_4 \cong \mathbf{V} \rtimes_\theta \mathbf{Z}(2)$, where $\mathbf{V}$ is the 4-group; a group may thus have distinct factorizations as a semidirect product.

Example 6 Let $K = \langle k \rangle$ be cyclic of order p^2 and let $Q = \langle x \rangle$ be cyclic of order p. For odd prime p, Theorem 7.2 says $\mathrm{Aut}(K) \cong \mathbf{Z}(p-1) \times \mathbf{Z}(p)$ and a generator of $\mathbf{Z}(p)$ is α, where $\alpha(k) = k^{1+p}$. Define $\theta\colon Q \to K$ by $\theta(x) = \alpha$. The group $G = K \rtimes_\theta Q$ has order p^3, generators x, k, and relations $x^p = 1$, $k^{p^2} = 1$, and $xkx^{-1} = k^x = k^{1+p}$. We have constructed the second nonabelian group of order p^3 (see Exercise 4.41).

EXERCISES

7.27. $K \rtimes_\theta Q \cong K \times Q$, the direct product, if θ is **trivial** ($\theta_x = 1_K$ for all $x \epsilon Q$).

*7.28. If p and q are distinct primes, construct all semidirect products of $\mathbf{Z}(p)$ by $\mathbf{Z}(q)$. Compare your results with Theorem 4.14; the condition $q \mid p - 1$ should now be more understandable.

Definition The **holomorph** of K is $K \rtimes_\theta \mathrm{Aut}(K)$, where $\theta\colon \mathrm{Aut}(K) \to \mathrm{Aut}(K)$ is the identity map.

EXERCISES

7.29. Let G be the holomorph of K. Prove that every automorphism of K is the restriction of an inner automorphism of G.

7.30. For any group K, a **holomorphism** is a one-one correspondence $h: K \to K$ such that

$$h(xy^{-1}z) = h(x)h(y)^{-1}h(z) \qquad \text{for every } x, y, z \in K.$$

(Every automorphism of K is a holomorphism, as is every left or right translation.) Prove that $\text{Hol}(K)$, the set of all holomorphisms of K, is a subgroup of the symmetric group S_K, and that $\text{Hol}(K)$ is isomorphic to the holomorph of K.

7.31. Prove that $\text{Hol}(G)$ is isomorphic to the subgroup of S_G generated by $\text{Aut}(G)$, $\lambda(G)$, and $\rho(G)$, where $\lambda, \rho: G \to S_G$ are the left and right regular representations. Indeed, prove that $\langle \text{Aut}(G), \lambda(G) \rangle = \langle \text{Aut}(G), \rho(G) \rangle \cong \text{Hol}(G)$.

Definition A short exact sequence $1 \to K \to G \xrightarrow{\ \pi\ } Q \to 1$ **splits** if there exists a homomorphism $s: Q \to G$ with $\pi s = 1_Q$.

EXERCISE

**7.32. Prove that a short exact sequence $1 \to K \to G \xrightarrow{\ \pi\ } Q \to 1$ splits if and only if G is a semidirect product of K by Q_1, where Q_1 is a subgroup of G isomorphic to Q with $\pi(Q_1) = Q$, i.e., K has a complement isomorphic to Q via π.

Definition A **retraction** (or *projection*) is a homomorphism $\varphi: G \to G$ with $\varphi \circ \varphi = \varphi$. A **retract** of G is a subgroup Q of G with $Q = \text{im } \varphi$, where $\varphi: G \to G$ is a retraction.

EXERCISE

*7.33. If $\varphi: G \to G$ is a retraction, then $\ker \varphi$ is the normal subgroup of G generated by $\{g\varphi(g^{-1}): g \in G\}$. Moreover, a group G is a semidirect product of K by Q if and only if there is a retraction $\varphi: G \to G$ with $K = \ker \varphi$ and $Q = \text{im } \varphi$.

If $K \lhd G$ and Q is a subgroup of G, then we have seen that the following four statements are equivalent:

(i) G is a semidirect product of K by Q ($G = K \rtimes Q$);
(ii) there is a split short exact sequence

$$1 \to K \to G \to Q \to 1;$$

(iii) Q is a complement of K;

(iv) Q is a retract of G with kernel K.

Let A and Q be groups with Q finite. Define K to be the direct product of $|Q|$ copies of A. One can think of K either as "vectors" (a_x) whose coordinates a_x in A are indexed by Q or, more formally, as A^Q, all functions $f: Q \to A$ under pointwise multiplication:

$$(f_1 f_2)(x) = f_1(x) f_2(x) \qquad x \in Q.$$

For $y \in Q$ and $f \in A^Q$, define $f^y \in A^Q$ by

$$f^y(x) = f(xy) \qquad x \in Q.$$

If f is regarded as a vector with xth coordinate $a_x = f(x)$, then f^y is obtained from f by permuting its coordinates by translating the indices by y.

For example, if $Q = \{1, x, y, xy\}$ is the 4-group and $f = (a_1, a_x, a_y, a_{xy}) \in A \times A \times A \times A$, then $f^y = (a_y, a_{xy}, a_1, a_x)$. Finally, define $\theta: Q \to \text{Aut}(K)$ by $\theta_y(f) = f^y$.

Definition Let A and Q be groups, with Q finite. The **wreath product** of A by Q, denoted by $A \wr Q$ (or A wr Q), is the semidirect product

$$A \wr Q = A^Q \rtimes_\theta Q,$$

where θ is the homomorphism defined above.

EXERCISES

7.34. If A and Q are finite, then

$$|A \wr Q| = |A|^{|Q|} |Q|.$$

7.35. Wreath product is not associative: if Q and R are finite, then

$$(A \wr Q) \wr R \ncong A \wr (Q \wr R).$$

7.36. Prove that $D_4 \cong \mathbf{Z}(2) \wr \mathbf{Z}(2)$.

Theorem 7.14 (**Kaloujnine and Krasner, 1951**) *If K and Q are groups with Q finite, then $K \wr Q$ contains an isomorphic copy of every extension of K by Q.*

Proof Assume G is a group having a homomorphism $\pi: G \to Q$ that is onto and whose kernel is K; for $a \in G$, denote $\pi(a)$ by $\bar{a}$. For each $x \in Q$ choose an element $l(x) \in G$ with $\pi(l(x)) = x$.

Define $\varphi: G \to K \wr Q$ by $\varphi(a) = (\sigma_a, \bar{a})$, where $\sigma_a: Q \to K$ is defined by

$$\sigma_a(x) = l(x) a l(x\bar{a})^{-1}, \qquad x \in Q.$$

Note that $\pi(\sigma_a(x)) = 1$ so that $\sigma_a(x) \in K$. To check φ is a homomorphism, we must compare $(\sigma_{ab}, \overline{ab})$ with $(\sigma_a, \bar{a})(\sigma_b, \bar{b}) = (\sigma_a \sigma_b^{\bar{a}}, \overline{ab})$, where $\sigma_b^{\bar{a}}(x) = \sigma_b(x\bar{a})$.

But

$$(\sigma_a \sigma_b^{\bar{a}})(x) = \sigma_a(x)\sigma_b(x\bar{a}) = l(x)al(x\bar{a})^{-1}l(x\bar{a})bl(\overline{xab})^{-1} = \sigma_{ab}(x).$$

Finally, if $a \in \ker \varphi$, then $\bar{a} = 1$ and $\sigma_a(x) = 1 = l(x)al(x\bar{a})^{-1}$, which together imply $a = 1$. ∎

Corollary 7.15 *Let $\mathscr{A}$ be a class of finite groups closed under subgroups and semidirect products (if G, $H \in \mathscr{A}$ and S is a subgroup of G, then $S \in \mathscr{A}$ and $G \rtimes H \in \mathscr{A}$). Then $\mathscr{A}$ is closed under extensions.*

Proof Assume $1 \to K \to G \to Q \to 1$ is a short exact sequence with $K \in \mathscr{A}$ and $Q \in \mathscr{A}$. Since $\mathscr{A}$ is closed under semidirect products, it is closed under direct products, and $K^Q \in \mathscr{A}$; it follows that $K \wr Q \in \mathscr{A}$. Since $\mathscr{A}$ is closed under subgroups, the theorem gives $G \in \mathscr{A}$. ∎

Let k and m be positive integers. How many integers $\le m$ are divisible by k? The answer is t, where k, $2k$, $3k$, $\cdots$, $tk \le m$, while $(t+1)k > m$; thus, $t = [m/k]$, the greatest integer in m/k. If we let p be a prime, what is the largest power of p dividing $m!$? By our initial remarks, $[m/p]$ factors of $m!$ are divisible by p, $[m/p^2]$ factors are divisible by p^2, etc. Therefore, if $m! = p^N m'$, where $(m', p) = 1$, then

$$N = [m/p] + [m/p^2] + [m/p^3] + \cdots$$

(this sum is finite, for $[m/p^k] = 0$ as soon as $p^k > m$). In particular, if $m = p^n$, then $N = p^{n-1} + p^{n-2} + \cdots + p + 1$; we have computed the order of a Sylow p-subgroup of the symmetric group S_{p^n}.

Theorem 7.16 *If p is a prime, then a Sylow p-subgroup of the symmetric group S_{p^n} is the wreath product of $\mathbf{Z}(p)$ with itself n times.*†

Proof The proof is by induction on n, the case $n = 1$ holding because a Sylow p-subgroup of S_p is cyclic of order p. Now suppose $n > 1$. Let B_0 be the set of integers 1 through p^{n-1} and let $B_i = B_0 + ip^{n-1}$. Thus the set of integers 1 through p^n is the disjoint union $B_0 \cup B_1 \cup \cdots \cup B_{p-1}$. Let $\alpha \in S_{p^n}$ be the permutation that "adds p^{n-1} modulo p^n": if $b_0 \in B_0$,

$$\alpha(b_0 + ip^{n-1}) = \begin{cases} b_0 + (i+1)p^{n-1} & \text{if } i < p-1; \\ b_0 & \text{if } i = p-1. \end{cases}$$

Thus, $\alpha(B_0) = B_1$, $\alpha(B_1) = B_2$, $\cdots$, $\alpha(B_{p-1}) = B_0$. One thus sees that α has order p.

Let $H = H_0$ be a Sylow p-subgroup of $S_{B_0} \cong S_{p^{n-1}}$. Now $\alpha|B_i: B_i \to B_{i+1}$ (subscripts taken modulo p) is a one-one correspondence; hence (Exercise 3.2) it induces an isomorphism $S_{B_i} \cong S_{B_{i+1}}$. Let H_1 be the image of H_0, H_2 the image of H_1, etc. We have thus chosen one Sylow p-subgroup for each i.

Now S_{p^n} contains $K = H_0 \times H_1 \times \cdots \times H_{p-1}$ (disjoint permutations

---

† The cyclic groups are associated as follows: if W_n is the wreath product of n cyclic groups, then $W_{n+1} = W_n \wr \mathbf{Z}(p)$.

commute) and $\langle \alpha \rangle$. Moreover $K \cap \langle \alpha \rangle = \{1\}$, and if $(\beta_0, \beta_1, \cdots, \beta_{p-1}) \in K$, then Lemma 3.9 gives $(\beta_0, \beta_1, \cdots, \beta_{p-1})^\alpha = (\beta_1, \beta_2, \cdots, \beta_0)$. It follows easily that $\langle K, \alpha \rangle \cong H \wr \langle \alpha \rangle \cong H \wr \mathbf{Z}(p)$. By induction, H is a wreath product of $n-1$ copies of $\mathbf{Z}(p)$, so that $\langle K, \alpha \rangle$ is a wreath product of n copies of $\mathbf{Z}(p)$.

To check that $\langle K, \alpha \rangle \cong H \wr \mathbf{Z}(p)$ is a Sylow p-subgroup, it suffices to see that its order is $p^{N(n)}$, where $N(n) = p^{n-1} + p^{n-2} + \cdots + p + 1$. We know that $|H| = p^{N(n-1)}$; hence $|H \wr \mathbf{Z}(p)| = (p^{N(n-1)})^p p = p^{pN(n-1)+1} = p^{N(n)}$. ∎

Using Theorem 7.16, one may compute a Sylow p-subgroup of any S_n. Write n in base p:

$$n = a_0 + a_1 p + \cdots + a_t p^t \qquad \text{where } 0 \leq a_i \leq p-1.$$

Now partition the set of integers 1 through n into a_0 singletons, a_1 subsets with p elements, and so forth. On each of these subsets X, construct a Sylow p-subgroup of S_X. The direct product of all these subgroups is a Sylow p-subgroup of S_n (it is a subgroup because disjoint permutations commute; it is a Sylow subgroup because it has the right order). As an illustration, we compute a Sylow 2-subgroup of S_6 (this has been done by hand in Exercise 4.16). In base 2, $6 = 0 \cdot 1 + 1 \cdot 2 + 1 \cdot 4$. A Sylow 2-subgroup of S_2 is $\mathbf{Z}(2)$; a Sylow 2-subgroup of S_4 is $\mathbf{Z}(2) \wr \mathbf{Z}(2) \cong D_4$. Hence, a Sylow 2-subgroup of S_6 is isomorphic to $D_4 \times \mathbf{Z}(2)$.

FACTOR SETS

Since there are nonsimple groups that are not semidirect products, our survey of extensions is still incomplete. Notice the sort of survey we already have; if we know K, Q, and θ, then we know $K \rtimes_\theta Q$ in the sense that we can write a multiplication table for it.

In discussing extensions G of K by Q, it is convenient to use the multiplicative notation for Q and the additive notation for G and its subgroup K (this is one of the few instances in which one uses additive notation for a nonabelian group). Thus, if $k \in K$ and $g \in G$, the conjugate of k by g is written $g + k - g$.

Definition If $\pi: G \to Q$ is onto and $x \in Q$, then a **lifting** of x is an element $l(x) \in G$ with $\pi l(x) = x$.

Lemma 7.17 *Let G be an extension of K by Q, where K is abelian. There is a homomorphism $\theta: Q \to \mathrm{Aut}(K)$ such that*

$$\theta_x(k) = l(x) + k - l(x)$$

for every $k \in K$; moreover θ is independent of the choice of liftings $\{l(x): x \in Q\}$.

Proof If $a \in G$, let γ_a denote conjugation by a. Since $K \lhd G$, $\gamma_a|K$ is an automorphism of K, and the function $\mu: G \to \mathrm{Aut}(K)$ defined by $\mu(a) = \gamma_a|K$ is easily seen to be a homomorphism. If $a \in K$, then $\mu(a) = 1_K$, for K is abelian. There is thus a homomorphism $\mu_\#: G/K \to \mathrm{Aut}(K)$ defined by $\mu_\#(Ka) = \mu(a)$.

The first isomorphism theorem not only says $Q \cong G/K$; it says (Exercise 2.45) that for any choice of liftings $l(x)$, $x \in Q$, the map $\lambda: Q \to G/K$ defined by $\lambda(x) = Kl(x)$ is an isomorphism. [Note that if $\{l_1(x): x \in Q\}$ is another choice of liftings, then $l(x)l_1(x)^{-1} \in K$ for every $x \in Q$ and $Kl(x) = Kl_1(x)$; the isomorphism $\lambda: Q \to G/K$ is thus independent of the choice of liftings $l(x)$.] Let $\theta: Q \to \text{Aut}(K)$ be the composite $\mu_\# \lambda$. If $x \in Q$ and $l(x)$ is a lifting, then $\theta_x = \mu_\# \lambda(x) = \mu_\#(Kl(x)) = \mu(l(x)) \in \text{Aut}(K)$; therefore, if $k \in K$,

$$\theta_x(k) = \mu(l(x))(k) = l(x) + k - l(x). \quad \blacksquare$$

There is a version of Lemma 7.17 that provides a homomorphism θ when K is not assumed abelian [then $\theta: Q \to \text{Aut}(K)/\text{Inn}(K)$], but this general situation is rather complicated. From now until we discuss the Schur-Zassenhaus lemma, we assume that K is abelian.

We shall examine (not necessarily split) extensions given the data: abelian K, arbitrary Q, and $\theta: Q \to \text{Aut}(K)$. This study is called **cohomology of groups** because of an analogy with methods of cohomology theory in algebraic topology. Our aim is to write a multiplication table (rather, an addition table!) for any extension G arising from the data K, Q, and θ.

Once a homomorphism $\theta: Q \to \text{Aut}(K)$ is given, we abbreviate the expression $\theta_x(k)$ by xk. The following formulas are valid for all x, y, $1 \in Q$ and k, $k' \in K$:

$$x(k + k') = xk + xk';$$
$$(xy)k = x(yk);$$
$$1k = k.$$

Notation   We record that $\theta: Q \to \text{Aut}(K)$ allows one to write xk for all $x \in Q$ and $k \in K$ by writing $_\theta K$ instead of K.

[These formulas should remind the reader of the definition of a module. Indeed, a (possibly noncommutative) ring $\mathbf{Z}Q$ can be constructed, the *integral group ring* of Q, and the abelian group K is a (left) module admitting scalars from this ring. This remark is the basis of further examinations of group extensions; for more details we refer the reader to the book of S. Mac Lane, *Homology*, Chapter IV.]

Since the object of our study is to recapture G from K and Q, let us assume we begin with only K, Q, and θ, but that there is no G in sight.

Definition   Let K and Q be groups, K abelian, and $\theta: Q \to \text{Aut}(K)$ a homomorphism. An extension G of K by Q **realizes** θ if

$$xk = l(x) + k - l(x)$$

for every $k \in K$ and every lifting $l(x)$ of x.

In this language, Lemma 7.17 says that every extension G of an abelian K by Q determines a θ that G realizes. The intuitive meaning of θ is that it describes how K is normal in G.

The extension problem is now posed as follows: Given an abelian group K, a group Q, and a homomorphism $\theta: Q \to \text{Aut}(K)$, determine all extensions G of K by Q that realize θ. (For a discussion of the more complicated case when K is not abelian, the reader is again referred to the book of Mac Lane.)

Let G be an extension of K by Q and let $\pi: G \to Q$ be a homomorphism of G onto Q having kernel K. A choice of lifting $l(x)$ of each $x \in Q$ defines a function $l: Q \to G$ (which may not be a homomorphism) satisfying $\pi \circ l = 1_Q$. The range of such a function l is called a **transversal** of K in G (or a **complete set of coset representatives** of K in G), for it consists of exactly one representative from each coset of K.

Suppose K is abelian and G is an extension of K by Q realizing θ. We identify Q with G/K, and for each $x \in Q$, we choose a lifting $l(x) \in G$; for computational ease, choose $l(1) = 0$. Once this choice of transversal has been made, every element $g \in G$ has a unique expression of the form

$$g = k + l(x) \qquad x \in Q, k \in K$$

[after all, $l(x)$ is a representative of a coset of K in G, and G is the disjoint union of these cosets]. We have the following formulas:

(i) $l(x) + k = xk + l(x)$ for every $x \in Q$ and $k \in K$ (this being the statement that G realizes θ);
(ii) $l(x) + l(y) = f(x, y) + l(xy)$ for some $f(x, y) \in K$ [$l(x) + l(y)$ and $l(xy)$ are representatives of the same coset of K].

Definition   The function $f: Q \times Q \to K$ defined by formula (ii) is called a **factor set**†
of G.

Notice that a factor set depends on the choice of transversal. If G is a semidirect product, Theorem 7.12 shows that there is a transversal ℓ making $f(x, y) = 0$ for every x and y in Q, namely, $\ell(x) = (0, x)$, all $x \in Q$. In fact, it follows from formula (ii) that $\ell: Q \to G$ is a homomorphism if and only if the corresponding factor set is identically zero. Thus, a factor set may be fruitfully thought of as a "measure" of G's deviation from being a semidirect product, for it tells how the transversal ℓ fails to be a homomorphism.

Theorem 7.18   *Let K be abelian and let $\theta: Q \to \text{Aut}(K)$. A function $f: Q \times Q \to K$ is a factor set if and only if it satisfies the formulas:*

(i) $f(1, y) = 0 = f(x, 1)$ *for every* $x, y \in Q$;
(ii) $xf(y, z) - f(xy, z) + f(x, yz) - f(x, y) = 0$ *for every* $x, y, z \in Q$.

Proof   Suppose f is a factor set. This means that there is an extension G of K by Q realizing θ, a transversal l has been chosen, and f satisfies

$$l(x) + l(y) = f(x, y) + l(xy).$$

---

† The term *factor set* is a misnomer; factor function would be more suggestive. However, we conform to standard usage. Factor sets are often called **cocycles** to stress the analogy with cohomology theory. The initial Z in the forthcoming notation $Z^2(Q, {}_\theta K)$ abbreviates the German *Zykel*.

Since we have assumed that $l(1) = 0$, $0 + l(y) = f(1, y) + l(1\ y)$, so that $f(1, y) = 0$; a similar calculation shows that $f(x, 1) = 0$, so that condition (i) is established.

Formula (ii) arises from associativity; one only need calculate the consequences of $[l(x) + l(y)] + l(z) = l(x) + [l(y) + l(z)]$.

Suppose, conversely, that we have a function $f: Q \times Q \to K$ which satisfies (i) and (ii). We shall construct an extension G of K by Q realizing θ and we shall choose a transversal l such that f is the factor set determined by these data.

Let G be the set of all pairs $(k, x) \in K \times Q$ with the binary operation

$$(k, x) + (k', y) = (k + xk' + f(x, y), xy).$$

(Note the similarity to the construction of $K \rtimes_\theta Q$.)

The proof that G is a group is quite similar to the proof of Theorem 7.12; formula (ii) is needed to prove associativity; the identity is $(0, 1)$; the inverse of (k, x) is $(-x^{-1}k - x^{-1}f(x, x^{-1}), x^{-1})$.

That G is an extension of K by Q is also easy: identify K with all pairs of the form $(k, 1)$ and define $\pi: G \to Q$ by $\pi(k, x) = x$.

Does G realize θ? We must show that if $x \in Q$ and $k \in K$, then $xk = l(x) + k - l(x)$, where $l(x)$ is any lifting of x. We identify xk, as any element of K, with $(xk, 1)$; also, the definition of π yields $l(x) = (k', x)$ for some $k' \in K$. Compute:

$$(k', x) + (k, 1) - (k', x) = (k' + xk, x) + (-x^{-1}k' - x^{-1}f(x, x^{-1}), x^{-1})$$
$$= (k' + xk + x[-x^{-1}k' - x^{-1}f(x, x^{-1})] + f(x, x^{-1}), 1) = (xk, 1),$$

for K is abelian.

Finally, define a transversal l by $l(x) = (0, x)$. A straightforward computation shows that $l(x) + l(y) - l(xy) = (f(x, y), 1)$, as desired. ∎

Definition $Z^2(Q, {}_\theta K)$ is the set of all factor sets $f: Q \times Q \to K$.

EXERCISE

7.37.   Prove that $Z^2(Q, {}_\theta K)$ is an abelian group under the operation

$$(f + f')(x, y) = f(x, y) + f'(x, y).$$

We are near a solution of the extension problem. Given abelian K, Q, and θ, we can write a multiplication table for an extension G of K by Q realizing θ once we are given a factor set. Moreover, factor sets are concrete objects that are succinctly characterized by Theorem 7.18.

It is quite possible that different factors sets f and f' give rise to the same extension. Indeed, let us again assume that we have an extension G, and recall that a factor set f is defined by

$$l(x) + l(y) = f(x, y) + l(xy).$$

The factor set f thus depends on the transversal l. If we choose a different transversal l', we get a second factor set f' defined by

$$l'(x) + l'(y) = f'(x, y) + l'(xy),$$

but, quite clearly, f' still gives the group G.

Lemma 7.19 *Let G be an extension of K by Q realizing θ; let l and l' be transversals giving rise to factor sets f and f'. There exists a function $\alpha \colon Q \to K$ with $\alpha(1) = 0$ and*

$$f'(x, y) - f(x, y) = x\alpha(y) - \alpha(xy) + \alpha(x)$$

for every $x, y \in Q$.

Proof For each $x \in Q, l(x)$ and $l'(x)$ are just different representatives of the same coset of K in G. There is thus an element $\alpha(x) \in K$ with

$$l'(x) = \alpha(x) + l(x).$$

Since we are consistently lifting 1 to 0, $l'(1) = l(1) = 0$, so that $\alpha(1) = 0$. The main formula is derived as follows:

$$
\begin{aligned}
l'(x) + l'(y) &= \alpha(x) + l(x) + \alpha(y) + l(y) \\
&= \alpha(x) + x\alpha(y) + l(x) + l(y) \qquad \text{(since } G \text{ realizes } \theta) \\
&= \alpha(x) + x\alpha(y) + f(x, y) + l(xy) \\
&= \alpha(x) + x\alpha(y) + f(x, y) - \alpha(xy) + l'(xy).
\end{aligned}
$$

If follows that $f'(x, y) = \alpha(x) + x\alpha(y) + f(x, y) - \alpha(xy)$. This gives the desired formula, for each term lies in the abelian group K. ∎

Definition A **coboundary** is a function $g \colon Q \times Q \to K$ such that

$$g(x, y) = x\alpha(y) - \alpha(xy) + \alpha(x)$$

for some $\alpha \colon Q \to K$ with $\alpha(1) = 0$. The set of all coboundaries is denoted by $B^2(Q, {}_\theta K)$.

EXERCISE

7.38. $B^2(Q, {}_\theta K)$ is a subgroup of $Z^2(Q, {}_\theta K)$.

Lemma 7.19 says that factor sets arising from different transversals lie in the same coset of B^2 in Z^2. We have been led to the following quotient group and equivalence relation.

Definition If K is an abelian group and $\theta \colon Q \to \mathrm{Aut}(K)$ is a homomorphism, then $H^2(Q, {}_\theta K)$ is defined to be $Z^2(Q, {}_\theta K)/B^2(Q, {}_\theta K)$ and is called the **second cohomology group**.

Definition   Two extensions G and G' of K by Q realizing θ are **equivalent** if there are factor sets f of G' and f' of G' with $f' - f \in B^2(Q, {}_\theta K)$, i.e., the factor sets determine the same element of $H^2(Q, {}_\theta K)$.

Observe that we have defined a relation on groups, not merely on factor sets: If one chooses other factor sets f_0 of G and f'_0 of G', then $f'_0 - f_0 = (f'_0 - f') - (f_0 - f) + (f' - f) \in B^2(Q, {}_\theta K)$, since $f'_0 - f'$ and $f_0 - f$ are in $B^2(Q, {}_\theta K)$, by Lemma 7.19.

EXERCISES

*7.39.   Let G and G' be extensions of K by Q realizing θ, where K is abelian. We have the diagram

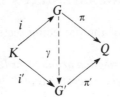

where $\pi: G \to Q$ are $\pi': G' \to Q$ are onto and $i: K \to G$ and $i': K \to G'$ are one-one. Prove that G is equivalent to G' if and only if there is an isomorphism $\gamma: G \to G'$ such that $\gamma \circ i = i'$ and $\pi' \circ \gamma = \pi$. (HINT: Use multiplication tables for G and G' to define γ.)

7.40.   Let K be abelian and let $\theta: Q \to \mathrm{Aut}(K)$. Prove that any two semidirect products of K by Q realizing θ are equivalent.

*7.41.   Let p be an odd prime. Give an example of isomorphic extensions of $\mathbf{Z}(p)$ by $\mathbf{Z}(p)$ that are not equivalent. (HINT: Let G be cyclic of order p^2 with generator x, let K be cyclic of order p with generator k, let $i: k \mapsto px$, and let $i': k \mapsto 2\,px$.)

We summarize the results of this section in the following theorem.

Theorem 7.20   *Let K be abelian and let $\theta: Q \to \mathrm{Aut}(K)$. The set E of all equivalence classes of extensions of K by Q realizing θ forms an abelian group isomorphic to $H^2(Q, {}_\theta K)$ whose identity element is the class of the semidirect product.*

Proof   If G is an extension of K by Q realizing θ, let $[G]$ denote its equivalence class, so that E is the set of all $[G]$. We now define a one-one correspondence $\varphi: H^2(Q, {}_\theta K) \to E$ as follows: $\varphi(f + B^2) = [G_f]$, where G_f is the extension determined by f that we constructed in Theorem 7.18. First of all, φ is well defined, for if both f and f' lie in the same coset of B^2, the definition of equivalence says that G_f and $G_{f'}$ are equivalent. Conversely, φ is one-one, for if $\varphi(f + B^2) = \varphi(f' + B^2)$, then G_f and $G_{f'}$ are equivalent, i.e., $f' - f \in B^2$.

Furthermore, φ is onto, for if G is an extension and f is a factor set of G, then $\varphi(f + B^2) = [G]$.

By Exercise 1.20, there is a unique addition defined on E making it a group and φ an isomorphism. The last part of the theorem follows from the fact that an extension is a semidirect product if and only if it has a factor set in B^2. ∎

In Chapter 11, we shall give an explicit construction of the sum of two classes in E when all groups are abelian.

Corollary 7.21   *Let K be abelian and let $\theta: Q \to \operatorname{Aut}(K)$. Every extension of K by Q realizing θ is a semidirect product if and only if $H^2(Q, {}_\theta K) = \{0\}$.*

THE SCHUR-ZASSENHAUS LEMMA

We now apply the results of the preceding section.

Theorem 7.22   *Let K be an abelian group of order m and let Q be a group of order n. If $(m, n) = 1$, then every extension G of K by Q is a semidirect product.*

Proof   By Corollary 7.21, it suffices to prove that every factor set f lies in $B^2(Q, {}_\theta K)$.

Define $\sigma: Q \to K$ by

$$\sigma(x) = \sum_{y \in Q} f(x, y);$$

σ is well defined since Q is finite and K is abelian. Now, if we sum the formula

$$xf(y, z) - f(xy, z) + f(x, yz) = f(x, y)$$

over all $z \in Q$, we obtain

$$x\sigma(y) - \sigma(xy) + \sigma(x) = nf(x, y)$$

(as z ranges over all elements of Q, so does yz for fixed y). Since $(m, n) = 1$, there are integers a and b with $am + bn = 1$. If we define $\alpha: Q \to K$ by $\alpha(x) = b\sigma(x)$, then $\alpha(1) = 0$ and

$$x\alpha(y) - \alpha(xy) + \alpha(x) = f(x, y).$$

This formula says that $f \in B^2(Q, {}_\theta K)$, as desired. ∎

Theorem 7.23   *Let K be an abelian group of order m, Q a group of order n, $(m, n) = 1$, and G an extension of K by Q. Then any two subgroups of G of order n (i.e., any two complements of K) are conjugate.*

Proof   Let Q_1 be a subgroup of G of order n. Now $K \cap Q_1 = \{1\}$, for if $a \in K \cap Q_1$, its order must divide m and n, and $(m, n) = 1$. Moreover, $|KQ_1| = |KQ_1| |K \cap Q_1| = |K| |Q_1| = |G|$, so that $KQ_1 = G$. It follows that G is a semidirect product of K by Q_1. If Q_2 is a second subgroup of G of order n, then

G is a semidirect product of K by Q_2 as well. By Exercise 7.32, Q_1 consists of liftings $l_1(x)$, and Q_2 consists of liftings $l_2(x)$, where l_1 and l_2 are homomorphisms; it follows that the factor sets f_1 and f_2 determined by the $l_1(x)$ and by the $l_2(x)$ are each zero. Now, if $l_1(x) = \alpha(x) + l_2(x)$, then

$$0 = f_1(x, y) - f_2(x, y) = x\alpha(y) - \alpha(xy) + \alpha(x).\dagger$$

Summing over all $y \in Q$ gives the equation:

$$0 = xk - k + n\alpha(x)$$

where

$$k = \sum_{y \in Q} \alpha(y).$$

This is an equation in K. If $am + bn = 1$ and $h = bk$,

$$-\alpha(x) = xh - h = -h + xh.$$

It follows that $-h + Q_1 + h = Q_2$, for $-h + l_1(x) + h = -h + xh + l_1(x)$ $= -\alpha(x) + l_1(x) = l_2(x) - l_1(x) + l_1(x) = l_2(x)$. ∎

Theorem 7.24 (**Schur-Zassenhaus lemma, 1937**)†† *If K and Q are finite groups of orders m and n, respectively, and if $(m, n) = 1$, then every extension G of K by Q is a semidirect product.*

Proof As we saw at the beginning of the proof of Theorem 7.23, it suffices to prove that G contains a subgroup of order n.

We perform an induction on m, noting that if $m = 1$, the theorem is trivial. Suppose K contains a proper subgroup T that is also normal in G. Then $K/T \lhd G/T$ and $(G/T)/(K/T) \cong G/K \cong Q$. If $|K/T| = m'$, then $m' < m$ and $|G/T| = m'n$. By induction, G/T contains a subgroup N/T of order n. Now $|N| = n|T|$ and $(n, |T|) = 1$, for $|T|$ divides $|K| = m$. Since $T \lhd N$ and $|T| < m$, the inductive hypothesis gives a subgroup of N, and hence of G, of order n.

We may now assume that K is a minimal normal subgroup of G. If p is a prime dividing m and if P is a Sylow p-subgroup of K, then the Frattini argument gives $G = KN_G(P)$. There are isomorphisms

$$Q \cong G/K = KN_G(P)/K \cong N_G(P)/K \cap N_G(P) = N_G(P)/N_K(P).$$

Therefore $|N_K(P)|n = |N_K(P)||Q| = |N_G(P)|$. If $N_G(P)$ is a proper subgroup of G, then $|N_K(P)| < m$, and so $N_G(P)$ contains a subgroup of order n, by induction. Therefore, we may assume that $N_G(P) = G$, i.e., that $P \lhd G$.

Since $K \supset P$ and K is a minimal normal subgroup of G, $K = P$. Now $Z(P)$ is a characteristic subgroup of P, so that $P \lhd G$ implies $Z(P) \lhd G$. Again, minimality implies $Z(P) = P$, for a finite p-group has a nontrivial center. But now P is abelian, and the proof is completed by Theorem 7.22. ∎

---

† A function $\alpha: Q \to K$ such that $\alpha(xy) = \alpha(x) + x\alpha(y)$ (as this one is) is called a **crossed homomorphism** or a **derivation**.
†† Schur (1904) first proved this theorem in the special case Q cyclic.

The following generalization of Theorem 7.23 is true: Under the hypotheses of the Schur-Zassenhaus lemma, any two subgroups of order n are conjugate. One proves this theorem by first assuming that either K or Q is solvable. Since $(m, n) = 1$, one of these groups must have odd order, and hence one of K or Q must be solvable, by the theorem of Feit and Thompson that every group of odd order is solvable.

Had we known Theorems 7.22 and 7.23, we could have applied them in the proof of P. Hall's theorem (Theorem 5.23), for the second case in that proof is the situation in which the normal subgroup is abelian.

EXERCISES

7.42. Use the Schur-Zassenhaus lemma and Exercise 7.28 to reclassify all groups of order pq, where p and q are distinct primes.

7.43. Prove that every group of order p^2q, where $p > q$ are primes, has a normal Sylow p-subgroup and classify all such groups.

*7.44. Using Corollary 4.18, reclassify all groups of order 12.

7.45. Use factor sets to prove the existence of the generalized quaternions Q_n. (HINT: Use Exercise 4.51.)

7.46. Let $p_1 < p_2 \cdots < p_t$ be primes such that $(p_i, p_j - 1) = 1$ for all $i < j$. Prove that every group G of order $n = p_1 p_2 \cdots p_t$ is cyclic. (HINT: Use Exercise 4.46, which states that G must contain a normal Sylow p_t-subgroup.) We remark that such n may be characterized by the condition $(n, \varphi(n)) = 1$.

We sketch a proof that there exist sets of primes $\{p_1, p_2, \cdots, p_t\}$ for arbitrarily large t that satisfy the hypothesis of the last exercise. Let $p_1 = 3$, and suppose there are $p_1 < p_2 < \cdots < p_t$ with $(p_i, p_j - 1) = 1$ for all $i < j$. We quote a theorem of Dirichlet: If $(a, b) = 1$, then the arithmetic progression a, $a+b$, $a+2b$, $\cdots$, $a+mb$, $\cdots$ contains infinitely many primes. Since $(2, p_1 p_2 \cdots p_t) = 1$, there is a positive integer m such that $p_{t+1} = 2 + mp_1 p_2 \cdots p_t$ is prime; $\{p_1, p_2, \cdots, p_{t+1}\}$ satisfies the desired conditions.

The reader may test his group-theoretic muscles on the following problem: For which integers n is every group of order n abelian? The number $5929 = 7^2 11^2$ should be considered before any false conjectures are made.

TRANSFER AND BURNSIDE'S THEOREM

We have seen two theorems which prove a given group G is a semidirect product: P. Hall's theorem (Theorem 5.23) for solvable groups and the Schur-Zassenhaus lemma. In each theorem, one starts with a normal subgroup K of G and constructs a complement Q (where $Q \cong G/K$). We now aim for a companion theorem, due to Burnside, that begins with a subgroup Q of G (a Sylow subgroup) and in certain cases

constructs a normal complement K. It is natural to seek a homomorphism whose
kernel is K; such a homomorphism is called the "transfer".

Lemma 7.25 *Let Q be a subgroup of finite index in G and let $\{x_1, \cdots, x_n\}$ and
$\{y_1, \cdots, y_n\}$ be left transversals† of Q in G. For fixed $a \in G$ and each i with
$1 \le i \le n$, there is a unique j and a unique $q_i \in Q$ with*

$$ay_i = x_j q_i.$$

Moreover, the function σ on $\{1, \cdots, n\}$ defined by $\sigma(i) = j$ is a permutation.

Proof Since the left cosets of Q partition G, there is a unique coset $x_j Q$
containing ay_i; the first statement follows. Assume $\sigma(i) = \sigma(k) = j$. Then
$ay_i = x_j q_i$ and $ay_k = x_j q_k$; thus $ay_i q_i^{-1} = ay_k q_k^{-1}$, $y_i^{-1} y_k = q_i^{-1} q_k \in Q$,
$y_i Q = y_k Q$, and $i = k$. Therefore σ is one-one, hence is a permutation. ∎

This lemma will be used in two cases. The first has $x_i = y_i$ for all i, i.e., the
transversals coincide. In this case

$$ax_i = x_{\sigma i} q_i,$$

where $\sigma \in S_n$ and $q_i \in Q$. The second case has two transversals, but we set $a = 1$. Now

$$y_j = x_{\alpha j} c_j,$$

where $\alpha \in S_n$ and $c_j \in Q$.

Definition If Q is a subgroup of finite index in a group G, then the **transfer** is the
function $T: G \to Q/Q'$ defined by

$$T(a) = \Pi \, q_i Q',$$

where $\{x_1, \cdots, x_n\}$ is a left transversal of Q in G and $ax_i = x_j q_i$.

REMARK The transfer $T: G \to Q/Q'$ is often denoted by $V_{G \to Q}$, the letter V abbreviating
the original German term *Verlagerung*.

Theorem 7.26 *If Q is a subgroup of finite index in a group G, the transfer
$T: G \to Q/Q'$ is a homomorphism whose definition is independent of the choice of
left transversal of Q in G.*

Proof Let $\{x_1, \cdots, x_n\}$ and $\{y_1, \cdots, y_n\}$ be left transversals of Q in G. By
Lemma 7.25, we have the following equations for $a \in G$:

$$
\begin{aligned}
ax_i &= x_{\sigma i} q_i, & \sigma \in S_n, \ q_i \in Q; \\
ay_i &= y_{\tau i} c_i, & \tau \in S_n, \ c_i \in Q; \\
y_i &= x_{\alpha i} u_i, & \alpha \in S_n, \ u_i \in Q.
\end{aligned}
$$

Now

$$ay_i = ax_{\alpha i} u_i = x_{\sigma \alpha i} q_{\alpha i} u_i.$$

† A **left transversal** of Q in G is a subset of G consisting of exactly one element from each left coset of Q in
G; a **right transversal** is defined similarly.

Defining j by $\alpha j = \sigma \alpha i$, we have $y_j = x_{\sigma \alpha i} u_j$, whence

$$ay_i = y_j u_j^{-1} q_{\alpha i} u_i.$$

The uniqueness assertion of Lemma 7.25 and the definition of j give

$$c_i = u_j^{-1} q_{\alpha i} u_i = u_{\alpha^{-1} \sigma \alpha i}^{-1} q_{\alpha i} u_i.$$

In the abelian group Q/Q', we may rearrange factors; thus

$$\Pi c_i Q' = \Pi u_{\alpha^{-1} \sigma \alpha i}^{-1} q_{\alpha i} u_i Q' = \Pi q_{\alpha i} Q',$$

because $\alpha^{-1} \sigma \alpha \in S_n$, and so the inverse of each u_i occurs and cancels u_i. Finally, $\Pi q_{\alpha i} Q' = \Pi q_i Q'$ since $\alpha \in S_n$. We have shown that $T(a)$ is independent of the choice of transversal.

Let $a, b \in G$ and let $\{x_1, \cdots, x_n\}$ be a left transversal of Q. Thus, $ax_i = x_{\sigma i} q_i$ and $bx_i = x_{\tau i} p_i$, where $q_i, p_i \in Q$. Then

$$abx_i = ax_{\tau i} p_i = x_{\sigma \tau i} q_{\tau i} p_i.$$

Therefore

$$T(ab) = \Pi q_{\tau i} p_i Q' = (\Pi q_{\tau i} Q')(\Pi p_i Q') = (\Pi q_i Q')(\Pi p_i Q') = T(a)T(b). \quad \blacksquare$$

Suppose a subgroup Q of finite index in a group G has a (not necessarily normal) complement K. Since $G = KQ$ and $K \cap Q = \{1\}$, the subgroup $K = \{x_1, \cdots, x_n\}$ is a left transversal of Q in G. If $a \in K$, then $ax_i \in K$ for all i: $ax_i = x_{\sigma i}$. But the general formula is $ax_i = x_{\sigma i} q_i$, so that each $q_i = 1$. We conclude that if $a \in K$, then $T(a) = 1$, i.e., $K \subset \ker T$. If Q is abelian, then $Q' = 1$ and we may identify Q/Q' with Q; thus $\mathrm{im}\, T \subset Q$ in this case. These remarks indicate we are on the right track in the search for a normal complement K: the candidate for K is $\ker T$.

The following formula for the transfer is useful: $T(a)$ is essentially a product of conjugates of certain powers of a.

Lemma 7.27   *Let Q be a subgroup of finite index in G and let $\{x_1, \cdots, x_n\}$ be a left transversal of Q in G. For $a \in G$, there exist elements $g_1, \cdots, g_m$ of G and positive integers $n_1, \cdots, n_m$ such that*

   (i) *each $g_i \in \{x_1, \cdots, x_n\}$;*

   (ii) *$g_i^{-1} a^{n_i} g_i \in Q$;*

   (iii) *$\Sigma n_i = n = [G:Q]$;*

   (iv) *$T(a) = \Pi(g_i^{-1} a^{n_i} g_i)Q'$.*

REMARK   The elements g_i and integers n_i depend on a.

Proof   We know $ax_i = x_{\sigma i} q_i$ where $\sigma \in S_n$ and $q_i \in Q$. Write σ as a product of disjoint cycles (including one 1-cycle for each fixed point): $\sigma = \alpha_1 \cdots \alpha_m$. Suppose $\alpha_i = (j_1, j_2, \cdots, j_r)$. Then

$$ax_{j_1} = x_{\sigma j_1} q_{j_1} = x_{j_2} q_{j_1}, \qquad ax_{j_2} = x_{j_3} q_{j_2}, \cdots, ax_{j_r} = x_{j_1} q_{j_r}.$$

It follows that

$$x_{j_1}^{-1} a^r x_{j_1} = q_{j_r} \cdots q_{j_1} \in Q.$$

Define $g_i = x_{j_1}$ and $n_i = r$; all the conclusions are now immediate. ∎

Theorem 7.28 *If Q is a subgroup of index n in G and if $Q \subset Z(G)$, then $T(a) = a^n$ for all $a \in G$.*

Proof First of all, $Q \subset Z(G)$ implies Q is an abelian normal subgroup of G. That Q is abelian allows us to regard the transfer as a homomorphism $T: G \to Q$. If $a \in G$ and $g^{-1} a^r g \in Q$, then normality of Q gives $a^r = g(g^{-1} a^r g) g^{-1} \in Q$. But $Q \subset Z(G)$ now gives $g^{-1} a^r g = a^r$. The result now follows from formulas (iii) and (iv) of Lemma 7.27. ∎

Corollary 7.29 If a group G has a subgroup Q of index n with $Q \subset Z(G)$, then $a \mapsto a^n$ is a homomorphism.

Proof We have just seen that $a \mapsto a^n$ is the transfer. ∎

The reader should try to prove this last corollary without using the transfer; I do not know a more elementary proof.

EXERCISES

7.47. Let Q be a subgroup of finite index in G and let $\{y_1, \cdots, y_n\}$ be a *right* transversal of Q in G. For $a \in G$, $y_i a = p_i y_{\tau i}$ for $p_i \in Q$ and $\tau \in S_n$. Prove that $R: G \to Q/Q'$ defined by $R(a) = \Pi p_i Q'$ is the transfer, i.e., $R(a) = T(a)$ for all $a \in G$. (HINT: If $\{y_1, \cdots, y_n\}$ is a right transversal of Q, then $\{y_1^{-1}, \cdots, y_n^{-1}\}$ is a left transversal of Q.)

7.48. If $T: G \to Q/Q'$ is the transfer, then $G' \subset \ker T$ and T induces a homomorphism $\bar{T}: G/G' \to Q/Q'$, namely $aG' \mapsto T(a)$.

7.49. Prove that transfer is transitive: if $P \subset Q \subset G$ are subgroups of finite index and $T: G \to Q/Q'$, $U: G \to P/P'$, and $V: Q \to P/P'$ are transfers, then $\bar{U} = \bar{V}\bar{T}$.

7.50. If Q has index n in G and K is a subgroup of G with $G = KQ$ and $Q \subset C_G(K)$, then $T(a) = a^n Q'$ for all $a \in G$. [HINT: There is a transversal of Q contained in K. If $a \in K$ and $g \in K$, then $g^{-1} a^r g \in Q \subset C_G(K)$ implies $a^r \in g C_G(K) g^{-1} = C_G(gKg^{-1}) = C_G(K)$.]

7.51. If G is a finite group of order mn, where $(m, n) = 1$, and if $Q \subseteq Z(G)$ is a subgroup of order m (hence index n), then $K = \ker T$ is a normal complement of Q (where $T: G \to Q/Q'$ is the transfer). Conclude that $G = K \times Q$.

7.52. Let G be a group having no elements of finite order and having a cyclic subgroup of finite index. Prove that G is cyclic.

Lemma 7.30 *Let Q be a Sylow subgroup of a finite group G. If $a, b \in C_G(Q)$ are conjugate in G, then they are conjugate in $N_G(Q)$.*

Proof Let $b = g^{-1}ag$ for $g \in G$; then $b \in g^{-1}C_G(Q)g = C_G(g^{-1}Qg)$. Since Q and $g^{-1}Qg$ are contained in $C_G(b)$, both are Sylow subgroups of $C_G(b)$. The Sylow theorems provide $c \in C_G(b)$ with $Q = c^{-1}g^{-1}Qgc$. Clearly $gc \in N_G(Q)$ and $c^{-1}g^{-1}agc = c^{-1}bc = b$. ∎

Theorem 7.31 (Burnside, 1900) *Let G be a finite group and let Q be a Sylow subgroup contained in the center of its normalizer; then Q has a normal complement K (and K is a characteristic subgroup of G).*

Proof First of all, $Q \subset Z(N_G(Q))$ implies Q is abelian. Therefore we may regard the transfer $T: G \to Q$. Let us compute $T(a)$ for $a \in Q$. By Lemma 7.27, $T(a) = \Pi g_i^{-1} a^{n_i} g_i$. Now if a^r and $g^{-1}a^r g$ lie in Q, they are conjugate elements in $C_G(Q)$ [Q abelian implies $Q \subset C_G(Q)$]. By Lemma 7.30, there is $c \in N_G(Q)$ with $g^{-1}a^r g = c^{-1}a^r c$. But $Q \subset Z(N_G(Q))$ implies $c^{-1}a^r c = a^r$. Hence, if $n = [G:Q]$, then $T(a) = a^n$ for $a \in Q$. If $|Q| = q$, then $(n, q) = 1$ (because Q is a Sylow subgroup) and there are integers α, β with $1 = \alpha n + \beta q$. Therefore, when $a \in Q$, $a = a^{\alpha n}a^{\beta q} = (a^\alpha)^n$, and we conclude that $T: G \to Q$ is onto: if $x \in Q$, then $T(x^\alpha) = x^{\alpha n} = x$. The first isomorphism theorem gives $G/K \cong Q$, where $K = \ker T$. It follows that $G = KQ$ and $K \cap Q = \{1\}$ [for $|K| = n$, hence $(|K|, |Q|) = 1$]. Therefore K is a normal complement of Q. That K is characteristic follows from Exercise 5.37. ∎

We remark that Q abelian and $N_G(Q) = C_G(Q)$ implies $Q \subset Z(N_G(Q))$.
There are interesting consequences of Burnside's theorem.

Theorem 7.32 *Let Q be a Sylow p-subgroup of a finite group G, where p is the smallest prime divisor of $|G|$. If Q is cyclic, then Q has a normal complement.*

Proof By Theorem 7.1, there is an imbedding $N_G(Q)/C_G(Q) \to \operatorname{Aut}(Q)$. Obviously $|N/C|$ divides $|G|$. Now Q is cyclic of order p^m, say, so that Theorem 7.2 gives $|\operatorname{Aut}(Q)| = p^{m-1}(p-1)$; since Q is a Sylow p-subgroup and $Q \subset C = C_G(Q)$, p does not divide $|N/C|$; therefore $|N/C|$ divides $p-1$. But $(p-1, |G|) = 1$ because p is the smallest prime divisor of $|G|$. Hence $|N/C| = 1$ and $N_G(Q) = C_G(Q)$. As remarked above, this implies $Q \subset Z(N_G(Q))$ and Burnside's theorem applies. ∎

Theorem 7.33 (Hölder, 1895) *If every Sylow subgroup of a finite group G is cyclic, then G is solvable.*

Proof If p is the smallest prime divisor of $|G|$, then Theorem 7.32 provides a normal complement K to a Sylow p-subgroup Q. Thus, $G/K \cong Q$. By induction, K is solvable, and hence G is solvable, by Theorem 5.14. ∎

Corollary 7.34 (Frobenius, 1893)† *Every group G of square-free order is solvable.*

Proof Every Sylow subgroup of G must be cyclic. ∎

Corollary 7.35 *A nonabelian simple group cannot have a cyclic Sylow 2-subgroup.*

REMARK   This result was proved in Exercise 3.45.

Corollary 7.36 *Let p be the smallest prime divisor of $|G|$, where G is a nonabelian simple group. Either p^3 divides $|G|$ or 12 divides $|G|$.*

Proof Let Q be a Sylow p-subgroup of G. By Corollary 7.35, Q cannot be cyclic. Therefore $|Q| \geq p^2$ and, if $|Q| = p^2$, then Q is elementary abelian. Because Q is a two-dimensional vector space over $\mathbb{Z}/p\mathbb{Z}$, $|\text{Aut}(Q)|$ $= (p^2 - 1)(p^2 - p) = p(p + 1)(p - 1)^2$. Now $N_G(Q)/C_G(Q)$ is imbedded in $\text{Aut}(Q)$ so that $|N/C|$ divides $|\text{Aut}(Q)|$. Moreover, $|N/C| \neq 1$ lest Burnside's theorem apply, so p does not divide $|N/C|$ because $Q \subset C = C_G(Q)$. Since p is the smallest prime divisor of $|G|$, we have $|N/C|$ dividing $p + 1$. If p is odd, this is impossible, for the smallest prime divisor of $|N/C| \geq p + 2$. We may thus assume $p = 2$. Now $|N/C| \neq 1$ and divides $p + 1 = 3$. Therefore $|G|$ is divisible by $p^2 \cdot 3 = 12$. ∎

The simple group A_5 has order 60, and 60 is divisible by 12 but not by 8. In 1958, Suzuki discovered an infinite family of nonabelian simple groups whose order is not divisible by 3 (their orders are, of course, divisible by 8). The magnificent result of Feit and Thompson asserts that every nonabelian simple group has even order.

EXERCISES

7.53. If G is a nonabelian group with $|G| \leq 100$ and $|G| \neq 60$, then G is not simple. (The next order of a nonabelian simple group is 168.) (See Exercise 4.48.)

7.54. If p and q are primes, then every group of order $p^2 q^2$ is solvable.

chapter eight

Some Simple
Linear Groups

FINITE FIELDS

The Jordan-Hölder theorem tells us that once we know extensions and simple groups, we know all groups that possess composition series. There are several families of simple groups (in addition to the cyclic groups of prime order and the large alternating groups), and our main concern in this chapter is the exhibition of one of these families, the projective unimodular groups. These are essentially groups of matrices of determinant 1 whose entries are allowed to lie in (almost) any field. Since these groups are finite only when the underlying field is finite, we begin our discussion by examining the finite fields.

Definition   A **prime field** is a field having no proper subfields.

Theorem 8.1   *Every prime field K is isomorphic to either $\mathbf{Z}/p\mathbf{Z}$ or the rationals, $\mathbf{Q}$.*

Proof   If 1 is the unit element of K, let $R = \{n \cdot 1 : n \in \mathbf{Z}\} \subset K$; it is easily checked that R is a domain. Further, the map $f : \mathbf{Z} \to R$ defined by $f(n) = n \cdot 1$ is a ring homomorphism of $\mathbf{Z}$ onto R. By the first isomorphism theorem for rings, $R \cong \mathbf{Z}/I$, where I is the kernel of f. Since R is a domain, I is a prime ideal. If $I = (p)$, then $R \cong \mathbf{Z}/(p) = \mathbf{Z}/p\mathbf{Z}$. In this case, R is a field, and so $R = K$. If

$I = (0)$, then $R \cong \mathbf{Z}$; it follows that K contains a subfield S isomorphic to the rationals, for K must contain the multiplicative inverse of each nonzero element in R. Since K is a prime field, $K = S$. ∎

Theorem 8.2 *Every field F contains a unique prime field K.*

Proof Let K be the intersection of all subfields of F; recall that every subfield of F contains 1, so that $K \neq \{0\}$. The reader may now prove that K is a prime field that is visibly the only prime field contained in F. ∎

Definition A field F with prime field K has **characteristic p** if $K \cong \mathbf{Z}/p\mathbf{Z}$; otherwise, F has **characteristic 0** (in which case $K \cong \mathbf{Q}$).

Observe that if F has characteristic p, then $pa = 0$ for every $a \in F$.

Corollary 8.3 *If F is a finite field, then F has exactly p^n elements for some prime p.*

Proof Since F is finite, it must have characteristic p for some prime p. Moreover, F is a finite-dimensional vector space over $\mathbf{Z}/p\mathbf{Z}$. If F has dimension n, then one may choose a basis of F and count exactly p^n vectors $(k_1, k_2, \cdots, k_n)$, where $k_i \in \mathbf{Z}/p\mathbf{Z}$. ∎

We remark that there do exist infinite fields of characteristic p, e.g., all rational functions over $\mathbf{Z}/p\mathbf{Z}$ (a rational function is a quotient of two polynomials).

The next question is whether there are any finite fields aside from $\mathbf{Z}/p\mathbf{Z}$. To exhibit them, we generalize the method of adjoining a root of a polynomial to a field. In Chapter 5, this process was simple because we could work within the complex numbers. We must be wily here, however, for there is no such larger field available to us at the outset.

Lemma 8.4 *Let K be a field and let $\varphi(x)$ be an irreducible polynomial in $K[x]$. There exists a field F containing a root of $\varphi(x)$ and a subfield isomorphic to K.*

Proof Let $R = K[x]$ and let I be the ideal $(\varphi(x))$. Since $\varphi(x)$ is irreducible, R/I is a field, by Corollary 6.17.

Set $F = R/I$. If $\varphi(x) = \Sigma \, k_i x^i$, where $k_i \in K$, then

$$\begin{aligned}
\varphi(x + I) &= \Sigma \, k_i (x + I)^i \\
&= \Sigma \, k_i (x^i + I) \\
&= \Sigma \, (k_i x^i + I) \\
&= (\Sigma \, k_i x^i) + I \\
&= \varphi(x) + I.
\end{aligned}$$

Since $\varphi(x) \in I$, $\varphi(x) + I = I$, the zero element of F. Therefore, the element $x + I$ in F is a root of $\varphi(x)$.

Finally, the reader may verify that $\{k + I : k \in K\}$ is a subfield of F that is isomorphic to K. ∎

Theorem 8.5   *If K is a field and $f(x) \in K[x]$, there is a field F containing K over which $f(x)$ is a product of linear factors.*

Proof   We perform an induction on the degree d of $f(x)$. If $d = 1$, then K itself is the desired field. If $d > 1$, then $f(x) = \varphi(x)g(x)$, where $\varphi(x)$ is an irreducible polynomial. By Lemma 8.4, there is a field F_0 containing K and a root α of $\varphi(x)$. In $F_0[x]$, $\varphi(x) = (x - \alpha)\psi(x)$ and so $f(x) = (x - \alpha)\psi(x)g(x)$. Since $\psi(x)g(x)$ has degree less than d, the inductive hypothesis provides a field F containing F_0 (hence containing K and α) in which $\psi(x)g(x)$ (hence $f(x)$) is a product of linear factors. ∎

EXERCISES

**8.1.   If F is a field of characteristic $p > 0$, then for every $a, b \in G$,

$$(a + b)^{p^k} = a^{p^k} + b^{p^k} \qquad \text{for all } k > 0.$$

**8.2.   If F is a finite field with exactly q elements, then every element in F is a root of $x^q - x$. (HINT: Use Lagrange's theorem on the group of nonzero elements of F.)

Definition   Let F be a field and let

$$f(x) = a_n x^n + a_{n-1} x^{n-1} + \cdots + a_0 \in F[x].$$

The **derivative** of $f(x)$, denoted by $f'(x)$, is the polynomial

$$f'(x) = n a_n x^{n-1} + (n-1) a_{n-1} x^{n-2} + \cdots + a_1 \in F[x].$$

EXERCISES

8.3.   Prove the usual formulas of calculus for the derivatives of sums and products of polynomials in $F[x]$.

**8.4.   Let $f(x) \in F[x]$. Then $f(x)$ and $f'(x)$ have a nonconstant common factor if and only if $f(x)$ has repeated roots. [HINT: Using Theorem 8.5, one may assume $f(x)$ and $f'(x)$ are products of linear factors.]

Theorem 8.6   *Let p be a prime and n be a positive integer. There exists a field F with exactly p^n elements.*

Proof   Let $q = p^n$, and consider the polynomial

$$f(x) = x^q - x \in (\mathbf{Z}/p\mathbf{Z})[x].$$

By Theorem 8.5, there is a field F_0 containing $\mathbf{Z}/p\mathbf{Z}$ in which $f(x)$ is a product of linear factors. Note that since F_0 contains $\mathbf{Z}/p\mathbf{Z}$, it has characteristic p.

Let F be the subset of F_0 consisting of all the roots of $f(x)$. Since $f(x)$ has degree q, $|F| \leq q$. To prove that $|F| = q$, it suffices to prove that all the roots of $f(x)$ in F_0 are distinct, i.e., that $f(x)$ has no repeated roots. Now $f'(x) = qx^{q-1} - 1 = -1$, since $q = p^n$ and F_0 has characteristic p. Now apply Exercise 8.4.

Let us prove F is a field. If a and b are roots of $f(x)$, then, using Exercise 8.1, $(a - b)^q = a^q - b^q = a - b$, i.e., $a - b$ is a root of $f(x)$. Hence, F is a group under addition. Further, $(ab)^q = a^q b^q = ab$, so that ab is also a root of $f(x)$; thus, F is a commutative ring. Finally, if $f(a) = 0$ and $a \neq 0$, then $a^{q-1} = 1$; the inverse of a is thus a^{q-2}. Therefore, F is a field having precisely $q = p^n$ elements. ∎

Theorem 8.7 (**E. H. Moore, 1893**) *Any two fields having exactly p^n elements are isomorphic.*

Proof If $|F| = p^n = q$, then, by Exercise 8.2, every element in F is a root of $f(x) = x^q - x$. In the language of Chapter 5, F is a root field of $f(x)$. The reader may now adapt the proof of Lemma 5.5 to prove its analog: Any two root fields of $f(x)$ over $\mathbf{Z}/p\mathbf{Z}$ are isomorphic. The result follows. ∎

It was Galois who discovered these finite fields, and they are called **Galois fields** in his honor. The field with p^n elements is thus denoted by $GF(p^n)$.

Notation If K is a field, denote the multiplicative group of its nonzero elements by $K^{\#}$.

We proved in Theorem 2.16 that $K^{\#}$ is cyclic when K is finite.

Definition An element $\rho \in GF(q)$ (where $q = p^n$) is a **primitive element** if ρ is a generator of the cyclic group $GF(q)^{\#}$.

Recall that a field K containing a subfield F may be regarded as a vector space over F; the dimension of K so viewed is denoted by $[K:F]$.

Lemma 8.8 *If ρ is a primitive element of $GF(p^n)$, then ρ is a root of an irreducible polynomial $g(x)$ in $(\mathbf{Z}/p\mathbf{Z})[x]$ of degree n.*

Proof Write K for $GF(p^n)$ and let its prime field be denoted by F ($F \cong \mathbf{Z}/p\mathbf{Z}$). There is an irreducible polynomial $g(x) \in F[x]$ having ρ as a root ($g(x)$ is a factor of $x^{p^n} - x$); let the degree of $g(x)$ be d. By Exercise 5.5, $[F(\rho): F] = d$, where $F(\rho)$ is the subfield of K obtained by adjoining ρ to F. But $F(\rho) = K$ because ρ is a primitive element, and so $d = [F(\rho):F] = [K:F] = n$. ∎

It follows that for every positive integer n there exists an irreducible polynomial in $(\mathbf{Z}/p\mathbf{Z})[x]$ of degree n.

Theorem 8.9   *For every prime p, the group $Aut(GF(p^n))$ of all automorphisms of the field $GF(p^n)$ is cyclic of order n.*

Proof   Since $GF(p^n)$ is obtained from $\mathbf{Z}/p\mathbf{Z} = GF(p)$ by adjoining a primitive element ρ, an automorphism σ of $GF(p^n)$ is completely determined by its value on ρ. If $g(x)$ is the irreducible polynomial in $(\mathbf{Z}/p\mathbf{Z})[x]$ having ρ as a root, then $\sigma(\rho)$ is also a root of $g(x)$. By the lemma, $g(x)$ has degree n; there are thus at most n possible values of $\sigma(\rho)$ and hence at most n automorphisms σ. But Exercise 8.1 exhibits some automorphisms of $GF(p^n)$, namely $\sigma: a \mapsto a^p$ and its powers. If $1 \le k < n$, then $\sigma^k = 1$ implies $a = \sigma^k(a) = a^{p^k}$ for every $a \in GF(p^n)$; in particular, $1 = \rho^{p^k-1}$, and this contradicts the fact that ρ has order $p^n - 1$ (since it generates $GF(p^n)^{\#}$). We conclude that σ has order n and that $\langle\sigma\rangle = Aut(GF(p^n))$.   ∎

THE GENERAL LINEAR GROUP

Groups of nonsingular matrices are as natural an object of study as groups of permutations. In investigating the structure of these groups, we shall discover a new family of simple groups.

Definition   Let K be a field. The **general linear group** $GL(m, K)$ is the multiplicative group of all nonsingular $m \times m$ matrices over K. When $K = GF(q)$, we may write $GL(m, q)$ instead of $GL(m, K)$.

Theorem 8.10 $|GL(m, q)| = (q^m - 1)(q^m - q) \cdots (q^m - q^{m-1})$.

Proof   Let V be an m-dimensional vector space over $GF(q)$, and let $\{\alpha_1, \alpha_2, \cdots, \alpha_m\}$ be an ordered basis. Regarding $GL(m, q)$ as linear transformations on V, we can exhibit a one-one correspondence between $GL(m, q)$ and the family of all ordered bases of V. If $T \in GL(m, q)$, then $\{T\alpha_1, T\alpha_2, \cdots, T\alpha_m\}$ is an ordered basis of V (because T is nonsingular); if $\{\beta_1, \beta_2, \cdots, \beta_m\}$ is an ordered basis of V, there is a unique $T \in GL(m, q)$ with $T\alpha_i = \beta_i$ for all i.

An ordered basis of V consists of vectors $\{\beta_1, \beta_2, \cdots, \beta_m\}$. Since there are q^m vectors in V, there are $q^m - 1$ choices for β_1 (the zero vector not being a candidate). Having chosen β_1, the only restriction on β_2 is that it not lie in the subspace spanned by β_1; there are thus $q^m - q$ choices for β_2. More generally, having chosen an independent set $\{\beta_1, \beta_2, \cdots, \beta_i\}$, the only restriction on β_{i+1} is that it not lie in the subspace spanned by $\{\beta_1, \beta_2, \cdots, \beta_i\}$; there are thus $q^m - q^i$ choices for β_{i+1}. Therefore, there are exactly $(q^m - 1)(q^m - q) \cdots (q^m - q^{m-1})$ ordered bases of V.   ∎

Definition   If t is a nonnegative integer and ρ is a primitive element of $GF(q)$, then

$$M(t) = \{A \in GL(m, q): \det A = \text{power of } \rho^t\}.$$

Lemma 8.11   *Let t be a divisor of $q - 1$. If $\Omega = |GL(m, q)|$, then $M(t)$ is a normal subgroup of $GL(m, q)$ of order Ω/t.*

Proof We use the correspondence theorem in the setting

$$\det: GL(m, q) \to GF(q)^{\#}.$$

If t is a divisor of $q - 1 = |GF(q)^{\#}|$, then the cyclic subgroup $\langle \rho^t \rangle$ of $GF(q)^{\#}$ has order $(q - 1)/t$, and hence index t. Furthermore, $\langle \rho^t \rangle \lhd GF(q)^{\#}$, for the latter group is abelian. Since $M(t)$ is the subgroup of $GL(m, q)$ that corresponds to $\langle \rho^t \rangle$, we have $M(t) \lhd GL(m, q)$, of index t, and of order Ω/t.   ∎

Theorem 8.12 *Let* $q - 1 = p_1 p_2 \cdots p_k$, *where the* p *are* (*not necessarily distinct*) *primes. The following normal series is the beginning of a composition series:*

$$GL(m, q) = M(1) \supset M(p_1) \supset M(p_1 p_2) \supset \cdots \supset M(q - 1).$$

Proof We have already seen above that each of the terms in this series is normal in $GL(m, q)$. Furthermore, if $\Omega = |GL(m, q)|$,

$$\left| \frac{M(p_1 p_2 \cdots p_i)}{M(p_1 p_2 \cdots p_{i+1})} \right| = \frac{\Omega/p_1 p_2 \cdots p_i}{\Omega/p_1 p_2 \cdots p_{i+1}} = p_{i+1}.$$

Since the factor groups have prime order, they are simple.   ∎

The last subgroup in the above chain, $M(q - 1)$, is of special interest; it consists of all matrices of determinant $\rho^{q-1} = 1$.

Definition   A matrix is **unimodular** if it has determinant 1.

Definition   Let K be a field. The **special linear group** $SL(m, K)$ is the multiplicative group of all $m \times m$ unimodular matrices over K. When $K = GF(q)$, we may write $SL(m, q)$ instead of $SL(m, K)$.

One would discover the subgroup SL without recalling determinants, for SL is the commutator subgroup of GL, as we shall see later (Theorem 8.24).

The following elementary matrices are introduced in order to analyze the structure of the subgroup $SL(m, K)$.

Definition   Let λ be a nonzero element of K and $i \neq j$ integers between 1 and m. A **transvection**† $B_{ij}(\lambda)$ is the $m \times m$ matrix differing from the identity matrix E in that it has λ as its ijth entry.

Note that every transvection is unimodular. Also, $B_{ij}(\lambda)A$ is the matrix obtained from a matrix A by adding λ times its jth row to its ith row.

Lemma 8.13   *If* $A \in GL(m, K)$, *then* $A = UD(\mu)$, *where* U *is a product of transvections,* $D(\mu)$ *is the diagonal matrix with diagonal entries* $\{1, 1, \cdots, 1, \mu\}$, *and* $\mu = \det A$.

† Most authors call every matrix similar to some $B_{ij}(\lambda)$ a transvection.

Proof We prove by induction on t that A can be transformed, by a sequence of operations that add a multiple of one row to another row, into a matrix of the form

$$A' = \begin{bmatrix} E_t & * \\ 0 & C \end{bmatrix},$$

where E_t is a $t \times t$ identity matrix.

Since A is nonsingular, not every entry a_{i1} in the first column is 0. Adding some row to the second if necessary, we may assume $a_{21} \neq 0$. Now add $a_{21}^{-1}(1 - a_{11})$ times the second row to the first row to get 1 in the 1, 1 position. We may now clean out the remainder of the first column so that all other entries in it are 0.

Suppose A has been transformed into A'. Note that A' is nonsingular, so that the matrix C is also nonsingular. As before, we may insert 1 in the $t + 1$, $t + 1$ position using row operations involving only the rows of C (assuming that C has at least two rows); it follows that the first t columns of A' are not altered by this insertion. Adding on a suitable multiple of row $t + 1$ to the other rows cleans out column $t + 1$ so that all its entries are 0 except the 1 in position $t + 1$, $t + 1$.

This process may be continued until A has the form

$$\begin{bmatrix} E_{m-1} & * \\ 0 & \mu \end{bmatrix},$$

where $\mu \neq 0$. Adding suitable multiples of the last row to the other rows cleans out the last column, leaving $D(\mu)$.

In terms of matrix multiplication, we have shown that $PA = D(\mu)$, where P is a product of transvections. Since the inverse of $B_{ij}(\lambda)$ is $B_{ij}(-\lambda)$, which is again a transvection, $A = P^{-1}D(\mu)$ is the desired factorization. ∎

Theorem 8.14 $GL(m, K)$ *is a semidirect product of* $SL(m, K)$ *by* $K^{\#}$; $SL(m, K)$ *is generated by transvections.*

Proof First of all, $SL(m, K) \lhd GL(m, K)$. If Δ is the set of all matrices $D(\mu)$ = diagonal $\{1, \cdots, 1, \mu\}$, then Δ is a subgroup of $GL(m, K)$ isomorphic to $K^{\#}$. Since $\det(D(\mu)) = \mu$, $\Delta \cap SL(m, K) = \{E\}$, so that GL is a semidirect product of SL by $K^{\#}$.

If $A = UD(\mu)$ is a factorization of an element A of GL as in the lemma, then $\det A = \mu$. It follows that if A is unimodular, $D(\mu) = D(1) = E$, and so $A = U$ is a product of transvections. ∎

Notation If K is a field, let $Z_1(m, K)$ denote the group of all $m \times m$ scalar matrices kE, $k \in K$, with $k^m = 1$. If $K = GF(q)$, we may write $Z_1(m, q)$ instead of $Z_1(m, K)$.

Theorem 8.15 *The center of SL* (m, K) *is* $Z_1(m, K)$.

Proof   It suffices to prove that A in the center of $SL(m, K)$ implies A is a scalar matrix kE; that $k^m = 1$ will then follow from det $A = 1$.

Let us consider $SL(m, K)$ as transformations on a vector space over K with basis $\{e_1, \cdots, e_m\}$. The transvection $B_{ij}(1)$ is that transformation B which sends e_j into $e_i + e_j$ and which fixes the other basis elements. Let A be the transformation sending e_l into $\Sigma a_{kl} e_k$.

If $l \neq j$, then $ABe_l = Ae_l$. On the other hand,

$$BAe_l = B(\Sigma a_{kl} e_k) = \Sigma a_{kl} Be_k$$
$$= \left(\sum_{k \neq j} a_{kl} e_k\right) + a_{jl}(e_i + e_j)$$
$$= Ae_l + a_{jl} e_i.$$

If $AB = BA$, then $a_{jl} = 0$ whenever $j \neq l$; therefore A is diagonal: $Ae_l = a_{ll}e_l$. An easy computation shows that $ABe_j = a_{ii}e_i + a_{jj}e_j$ while $BAe_j = a_{jj}(e_i + e_j)$. Hence $a_{ii} = a_{jj}$. Therefore, if A commutes with all $B_{ij}(1)$, then A is scalar. ∎

Theorem 8.16 $|Z_1(m, q)| = d$, where $d = (m, q-1)$.

Proof   If $G = \langle \rho \rangle$ is a cyclic group of order n and d is a divisor of n, then it is easy to see $\{x \in G : x^d = 1\}$ is a subgroup of order d, namely, $\langle \rho^{n/d} \rangle$. Now the homomorphism $Z_1(m, q) \to GF(q)^{\#}$, given by $kE \mapsto k$, is one-one. Therefore, it suffices to show $k^m = 1$ if and only if $k^d = 1$, where $d = (m, q-1)$. Now $m = dc$ for some integer c, so $k^d = 1$ implies $k^m = k^{dc} = 1$. Conversely, there are integers a and b with $d = am + b(q-1)$. Thus

$$k^d = k^{am + b(q-1)} = k^{ma} k^{(q-1)b} = k^{ma},$$

and so $k^m = 1$ implies $k^d = 1$. ∎

EXERCISE

8.5.   Let $H \lhd SL(2, K)$ and let $A \in H$. If A is similar to $\begin{bmatrix} a & b \\ c & d \end{bmatrix}$, then there is a nonzero $\mu \in K$ such that H contains $\begin{bmatrix} a & \mu^{-1}b \\ \mu c & d \end{bmatrix}$. (HINT: Use the factorization of Lemma 8.13.)

PSL(2, *K*)

Our preceding discussion allows us to extend the normal series of Theorem 8.12 one step further:

$$GL(m, K) \supset \cdots \supset M(q-1) = SL(m, K) \supset Z_1(m, K).$$

Moreover, $Z_1(m, K)$ is an abelian group, so its composition factors are no secret if it is finite. Let us investigate the last factor group of this series.

Definition The **projective unimodular group** $PSL(m, K)$ is the group $SL(m, K)/Z_1(m, K)$. When $K = GF(q)$, we may write $PSL(m, q)$ instead of $PSL(m, K)$.

We shall see in Chapter 9 that these groups are intimately related to projective geometry.

Theorem 8.17 If $d = (m, q - 1)$, then

$$|PSL(m, q)| = \frac{(q^m - 1)(q^m - q) \cdots (q^m - q^{m-1})}{d(q-1)}.$$

Proof If $\Omega = |GL(m, q)|$, then $|SL(m, q)| = \Omega/q - 1$, by Lemma 8.11. The theorem now follows from Theorems 8.10 and 8.16. ∎

For the remainder of this section, we concentrate on the case $m = 2$ with the aim of proving that the groups $PSL(2, q)$ are simple when $q > 3$.

Lemma 8.18 *If a normal subgroup H of $SL(2, q)$ contains a transvection $B_{ij}(\mu)$, then $H = SL(2, q)$.*

Proof By Theorem 8.14, it suffices to prove that H contains every transvection.† Denote $GF(q)$ by K.

If we conjugate $B_{12}(\mu)$ by a unimodular matrix, we have

$$\begin{bmatrix} a & b \\ c & d \end{bmatrix}\begin{bmatrix} 1 & \mu \\ 0 & 1 \end{bmatrix}\begin{bmatrix} d & -b \\ -c & a \end{bmatrix} = \begin{bmatrix} 1 - \mu ac & \mu a^2 \\ -\mu c^2 & 1 + \mu ac \end{bmatrix}.$$

In particular, this conjugate is $B_{12}(\mu a^2)$ if $c = 0$, and it is $B_{21}(-\mu c^2)$ if $a = 0$. Furthermore, these matrices lie in H since H is normal.

The map $k \mapsto k^2$ is an endomorphism of the abelian group $K^\#$ whose kernel consists of all k with $k^2 = 1$. Since K is a field, the polynomial $x^2 - 1$ has at most two roots, and so the kernel has order 1 or order 2 (it has order 1 when K has characteristic 2). It follows that at least half the elements of $K^\#$ are squares.

Let

$$\Gamma = \{\lambda \in K: B_{12}(\lambda) \in H\} \cup \{0\}.$$

It is easy to see that Γ is a subgroup of K (where we consider K only as an additive group). Moreover, we know that Γ contains 0 and all elements of the form μa^2. Therefore, Γ contains more than half the elements of K, and so $\Gamma = K$, by Lagrange's theorem. Hence, H contains all transvections of the form $B_{12}(\lambda)$, and a similar argument shows that H contains all transvections of the form $B_{21}(\lambda)$. ∎

† Any two transvections are similar, hence are conjugate in GL; they need not be conjugate in SL.

The transvections thus play the same role in the study of the special linear groups as the 3-cycles play in the study of the alternating groups. We now prove the main theorem of this section.

Theorem 8.19 (Jordan-Moore)† *The groups PSL*(2, *q*) *are simple if and only if q* > 3.

Proof First of all, Theorem 8.17 gives

$$|PSL(2,q)| = \begin{cases} (q+1)(q^2-q) & \text{if } q = 2^n; \\ \tfrac{1}{2}(q+1)(q^2-q) & \text{if } q = p^n, p \text{ an odd prime.} \end{cases}$$

Therefore, $|PSL(2,2)| = 6$ and $|PSL(2,3)| = 12$, so that these groups cannot be simple.

Let H be a normal subgroup of $SL(2, q)$ that contains a matrix not in $Z_1(m, q)$. By the correspondence theorem, it suffices to prove that $H = SL(2, q)$.

Suppose H contains a matrix

$$A = \begin{bmatrix} r & 0 \\ s & t \end{bmatrix},$$

where $r \neq \pm 1$. If

$$S = \begin{bmatrix} 1 & 0 \\ 1 & 1 \end{bmatrix},$$

then H also contains

$$SAS^{-1}A^{-1} = \begin{bmatrix} 1 & 0 \\ 1-t^2 & 1 \end{bmatrix}.$$

Since $\det A = 1 = rt$, $t \neq \pm 1$ and $1 - t^2 \neq 0$. This last matrix is thus a transvection, and so $H = SL(2, q)$, by Lemma 8.18.

To complete the proof, we have only to produce a matrix in H whose top row is $[r, 0]$ where $r \neq \pm 1$. Let M be a matrix in H that is not in $Z_1(m, q)$. Now M is similar to either a diagonal matrix or a matrix

$$\begin{bmatrix} 0 & -1 \\ 1 & x \end{bmatrix},$$

for the only rational canonical forms for 2×2 matrices are a direct sum of two 1×1 companion matrices, i.e., a diagonal matrix, or a 2×2 companion matrix (which has the above form because M is unimodular). In the first event, Exercise 8.5 shows we have our desired matrix (since $M \notin Z_1(m, q)$, and M is

† Jordan (1870) proved this theorem for q prime. In 1893, after F. Cole had discovered the simple group of order 504, E. H. Moore recognized this group as $PSL(2, 8)$ and proved the simplicity of $PSL(2, q)$ for all prime powers $q > 3$.

unimodular). In the second event, H contains, by Exercise 8.5, a matrix

$$C = \begin{bmatrix} 0 & -\mu^{-1} \\ \mu & x \end{bmatrix}.$$

Setting

$$T = \begin{bmatrix} \alpha^{-1} & 0 \\ 0 & \alpha \end{bmatrix},$$

H contains the matrix U defined by

$$U = TCT^{-1}C^{-1} = \begin{bmatrix} \alpha^{-2} & 0 \\ \mu x(\alpha^2 - 1) & \alpha^2 \end{bmatrix}.$$

We are done if $\alpha^{-2} \neq \pm 1$, i.e., if $\alpha^4 \neq 1$. If $q > 5$, such a nonzero α does exist in $GF(q)$, for the polynomial $z^4 - 1$ has at most four roots in a field. If $q = 4$, then every $\alpha \in GF(4)$ satisfies the equation $\alpha^4 = \alpha$, so $\alpha \neq 1$ implies $\alpha^4 \neq 1$.

   Only the case $q = 5$ remains. There are two possibilities. Assume the entry x in C is nonzero. Choose $\alpha \in GF(5) (\cong \mathbf{Z}/5\mathbf{Z})$ so that $\alpha^2 - 1 \neq 0$; note that $\alpha^2 = -1$. The lower left corner of U, $\lambda = \mu x(\alpha^2 - 1)$, is nonzero, and so

$$U = \begin{bmatrix} -1 & 0 \\ \lambda & -1 \end{bmatrix}.$$

Hence, $U^2 = B_{21}(-2\lambda) \in H$; therefore H contains a transvection, and we are done. If the entry x is zero, then

$$C = \begin{bmatrix} 0 & -\mu^{-1} \\ \mu & 0 \end{bmatrix} \in H.$$

Therefore H contains

$$\begin{bmatrix} 1 & z \\ 0 & 1 \end{bmatrix} \begin{bmatrix} 0 & -\mu^{-1} \\ \mu & 0 \end{bmatrix} \begin{bmatrix} 1 & -z \\ 0 & 1 \end{bmatrix} = \begin{bmatrix} \mu z & (-\mu z^2 - \mu^{-1}) \\ * & * \end{bmatrix}.$$

If we choose $z = 2\mu^{-1}$, then the top row of this last matrix is $[2, 0]$, as desired. ∎

   We have exhibited an infinite family of simple groups. Are any of its members distinct from simple groups we already know? Using the formula of Theorem 8.17, we see that both $PSL(2, 4)$ and $PSL(2, 5)$ have order 60. By Exercise 4.49, any two simple groups of order 60 are isomorphic; therefore

$$PSL(2, 4) \cong A_5 \cong PSL(2, 5).\dagger$$

If $q = 7$, however, we do get a new simple group, for $|PSL(2, 7)| = 168$, which is neither prime nor $\frac{1}{2}n!$. If we take $q = 8$, we see that there is a simple group of order 504; if we take $q = 11$, we have a simple group of order 660.

---

† The only other isomorphisms involving A_n and PSL are $PSL(2, 9) \cong A_6$, $PSL(2, 7) \cong PSL(3, 2)$, and Theorem 9.67: $PSL(4, 2) \cong A_8$.

EXERCISES

8.6. If K is a field of characteristic $\neq 2$, then

$$x = \left(\frac{x+1}{2}\right)^2 - \left(\frac{x-1}{2}\right)^2 \qquad \text{for every } x \in K.$$

Use this remark to prove that $PSL(2, K)$ is simple for every, possibly infinite, field K of characteristic not 2. (HINT: The finiteness of K was used only in the consideration of the subgroup Γ in the proof of Lemma 8.18.) It is known that $PSL(2, K)$ is simple for every infinite field K.

8.7. If $G = \bigcup_{n=1}^{\infty} G_n$, where $G_n \subset G_{n+1}$ for all n and every G_n is simple, then G is simple. Conclude that $PSL(2, K)$ is simple, where $K = \bigcup_{n=1}^{\infty} GF(2^{n!})$ is the algebraic closure of $GF(2)$.

8.8. Show $SL(2, 3) \ncong S_4$. Prove that $PSL(2, 3) \cong A_4$. (HINT: Use Theorem 4.17.)

8.9. What is the Sylow 2-subgroup of $SL(2, 3)$?

8.10. What are the composition factors of $GL(2, 7)$?

8.11. Prove that the commutator subgroup of $GL(2, q)$ is $SL(2, q)$ when $q > 3$. What are the commutator subgroups of $GL(2, 3)$ and $GL(2, 2)$?

8.12. Let $SO(2, \mathbf{R})$ denote the group of all 2×2 real matrices of the form

$$\begin{bmatrix} \cos \alpha & \sin \alpha \\ -\sin \alpha & \cos \alpha \end{bmatrix}.$$

Prove that $SO(2, \mathbf{R})$ is not isomorphic to a quotient group of $GL(2, \mathbf{R})$.

PSL(*m, K*)

We shall prove in this section that $PSL(m, K)$ is simple for every field K and all $m \geq 3$. As a consequence of this, we shall be able to exhibit two nonisomorphic simple groups having the same finite order. The proof we present is due to E. Artin and is much more elegant than matrix manipulation.

Notation If V is a vector space over a field K, then $GL(V)$ denotes the group of all nonsingular linear transformations on V. The subgroup of all transformations of determinant 1 is denoted by $SL(V)$.

Of course, if $\dim V = m$, then choosing a basis of V allows one to define isomorphisms $GL(V) \cong GL(m, K)$ and $SL(V) \cong SL(m, K)$.

Definition A **hyperplane** H in V is a subspace of dimension $m-1$ (where $\dim V = m$).

If H is a hyperplane in V and $\alpha \in V$, $\alpha \notin H$, then every vector $\beta \in V$ has a unique expression of the form

$$\beta = \mu\alpha + \gamma, \qquad \mu \in K, \quad \gamma \in H$$

(under the natural map $V \to V/H \cong K$, $\alpha \mapsto \alpha + H \neq 0$, so that the scalar multiples of α form a transversal of H in V).

Lemma 8.20 *Let H be a hyperplane in V and let $A \in GL(V)$ fix H (pointwise). If $\alpha \in V$, $\alpha \notin H$, then*

$$A(\alpha) = \mu\alpha + \gamma, \qquad \mu \in K, \quad \gamma \in H.$$

Moreover, given any $\beta \in V$,

$$A(\beta) = \mu\beta + \gamma', \qquad \gamma' \in H.$$

Proof The remarks above show that, as any vector in V, $A(\alpha)$ has an expression as displayed. If $\beta \in V$, then

$$\beta = b\alpha + \gamma'', \qquad b \in K, \quad \gamma'' \in H.$$

Hence,

$$
\begin{aligned}
A(\beta) &= bA(\alpha) + \gamma'' && \text{(for } A \text{ fixes } H) \\
&= b(\mu\alpha + \gamma) + \gamma'' \\
(*) \qquad &= \mu(b\alpha + \gamma'') + [(1-\mu)\gamma'' + b\gamma] \\
&= \mu\beta + \gamma' && \text{where } \gamma' \in H. \quad \blacksquare
\end{aligned}
$$

The scalar $\mu = \mu(A)$ in Lemma 8.20 is thus uniquely determined by any A that fixes a hyperplane pointwise.

Definition Let $A \in GL(V)$ fix a hyperplane H pointwise and let $\mu = \mu(A)$. If $\mu \neq 1$, then A is called a **dilatation**; if $\mu = 1$ and $A \neq E$, then A is called a **transvection**.

The next theorem shows that if a transformation A is a transvection, then there is a matrix representing A that is a transvection $B_{ij}(\lambda)$.

Theorem 8.21 *Let $A \in GL(V)$ fix a hyperplane H. If A is a dilatation, then A may be represented (relative to a suitable basis of V) by a matrix $D(\mu)$ = diagonal $\{1, \cdots, 1, \mu\}$; if A is a transvection, then A may be represented by $B_{21}(1)$. Moreover, a transvection has no characteristic vectors outside of H.*

Proof Let us first establish notation. Choose $\alpha \in V$, $\alpha \notin H$, and let

$$A(\alpha) = \mu\alpha + \gamma, \qquad \mu \in K, \quad \gamma \in H.$$

Of course, every nonzero vector in H is a characteristic vector of A; are there any others? If $\beta \in V$, then

$$\beta = b\alpha + \gamma'', \qquad b \in K, \quad \gamma'' \in H.$$

Equation (*) shows that

$$(**) \qquad A(\beta) = c\beta \quad \text{if and only if } \mu = c \text{ and } b\gamma = (\mu - 1)\gamma''.$$

If A is a dilatation, then $\mu - 1 \neq 0$ and we may solve for

$$\gamma'' = b(\mu - 1)^{-1}\gamma.$$

Therefore, any choice of $b \neq 0$ gives a characteristic vector β outside H. If $\{\tau_1, \cdots, \tau_{m-1}\}$ is a basis of H and if β is a characteristic vector outside H, then $\{\tau_1, \cdots, \tau_{m-1}, \beta\}$ is a basis of V and, relative to this basis, A has matrix $D(\mu)$.

Assume now that A is a transvection. Since $\mu = 1$, condition (**) shows that A has a characteristic vector outside H if and only if $\gamma = 0$. This implies that $A = E$, contradicting the proviso in the definition of transvection excluding E. Therefore A has no characteristic vectors outside H. Let τ_1 be any vector outside H and define $\tau_2 = A\tau_1 - \tau_1$. Note that $A\tau_1 = \tau_1 + \gamma_0$, where $\gamma_0 \in H$ ($\mu = 1$ since A is a transvection); therefore $\tau_2 = \gamma_0 \in H$. Now $\{\tau_1, \tau_2\}$ is independent, lest τ_1 be characteristic. Extend $\{\tau_1, \tau_2\}$ to a basis $\{\tau_1, \cdots, \tau_m\}$ of V, where $\{\tau_3, \cdots, \tau_m\}$ is chosen from H. Relative to this basis, A has matrix $B_{21}(1)$, for $A\tau_1 = \tau_1 + \tau_2$ while $A\tau_i = \tau_i$ for $i \neq 1$. ∎

Corollary 8.22   *All matrices $B_{ij}(\lambda)$, where $i \neq j$ and $\lambda \neq 0$, are conjugate in $GL(m, K)$.*

Proof   It suffices to prove that $B_{ij}(\lambda)$ also represents the transvection A. Choose τ_i outside H and define $\tau_j = \lambda^{-1}(A\tau_i - \tau_i)$; for $l \neq i, j$, choose $\tau_l \in H$ so that $\{\tau_1, \cdots, \tau_m\}$ is a basis of V. Relative to this basis, A has matrix $B_{ij}(\lambda)$. ∎

Corollary 8.23   *$GL(V)$ is generated by all dilatations and transvections, and $SL(V)$ is generated by all transvections.*

Proof   Theorem 8.21 allows us to translate Lemma 8.13 from matrices to linear transformations. ∎

Let us look at transvections more closely. If A is a transvection of V fixing a hyperplane H and if $\alpha \notin H$, then

$$A(\alpha) = \alpha + \gamma, \qquad \gamma \in H.$$

If $\beta \in V$, then

(***) $$\beta = b\alpha + \gamma', \qquad b \in K, \quad \gamma' \in H,$$

and equation (*) in the midst of Lemma 8.20 gives

$$A(\beta) = \beta + b\gamma.$$

The function $f: V \to K$ defined by $\beta \mapsto b$ is, from (***), K-linear (it is a **functional** on V) and has kernel H. Thus, A determines a functional f and a vector $\gamma \in H = \ker f$ so that A satisfies the formula

$$A(\beta) = \beta + f(\beta)\gamma, \qquad \text{all } \beta \in V.$$

Notation   Given a functional f and $\gamma \in \ker f$, define $T_{f,\gamma}: V \to V$ by

$$T_{f,\gamma}(\beta) = \beta + f(\beta)\gamma.$$

It is easy to see $T_{f,\gamma}$ is a transvection when $f \not\equiv 0$ and $\gamma \neq 0$. Moreover, every transvection equals $T_{f,\gamma}$ for some f and γ.

EXERCISES

**8.13.   If f, g are functionals and $f(\gamma) = g(\gamma) = f(\gamma') = 0$, then

$$T_{f,\gamma} \circ T_{f,\gamma'} = T_{f,\gamma+\gamma'} \quad \text{and} \quad T_{f,\gamma} \circ T_{g,\gamma} = T_{f+g,\gamma}.$$

**8.14. $T_{af,\gamma} = T_{f,a\gamma}$ for all $a \in K$.

**8.15. $T_{f,\gamma} = T_{g,\delta}$ if and only if there is a scalar $a \in K$ with $g = af$ and $\gamma = a\delta$.

**8.16.   If $S \in GL(V)$, then

$$ST_{f,\gamma}S^{-1} = T_{fS^{-1},S\gamma}.$$

Exercise 8.16 says that a conjugate of a transvection is again a transvection. This is precisely why linear transformations are preferable to matrices, for a conjugate of $B_{ij}(\lambda)$ need not be a matrix of the same form.

Theorem 8.24   *The commutator subgroup of $GL(V)$ is $SL(V)$ unless V is a two-dimensional vector space over $GF(2)$.*

Proof   We know that det: $GL \to K^{\#}$ has kernel SL, so that $GL/SL \cong K^{\#}$; since $K^{\#}$ is abelian, $(GL)' \subset SL$.

For the reverse inclusion, let $\pi: GL \to GL/(GL)'$ be the natural map. Since all transvections are conjugate in GL, by Corollary 8.22, $\pi(T) = \pi(T')$ for any two transvections T and T'; let d denote their common value. If we avoid the exceptional case in the statement, every hyperplane H contains nonzero vectors γ, γ' (not necessarily distinct) such that $\gamma + \gamma'$ is also nonzero. Choose a hyperplane H, nonzero vectors $\gamma, \gamma', \gamma + \gamma' \epsilon H$, and a functional f having kernel H. By Exercise 8.13, we have in GL

$$T_{f,\gamma} \circ T_{f,\gamma'} = T_{f,\gamma+\gamma'}.$$

Since $\gamma, \gamma', \gamma + \gamma'$ are all nonzero, each of these terms is a transvection (they are not E). Applying π to this equation gives $d^2 = d$ in $K^{\#}$, whence $d = 1$. Thus, every transvection is in ker $\pi = (GL)'$. Since SL is generated by transvections, $SL \subset (GL)'$, as desired.

Finally, when V is a two-dimensional vector space over $GF(2)$, then $GL(V)$ is a genuine exception:

$$GL(V) = SL(V) \cong S_3,$$

while $(S_3)' = A_3$, a proper subgroup.   ∎

To this point, we have allowed $m = \dim V \geq 2$; the assumption $m \geq 3$ enters crucially in the next result.

Theorem 8.25   *If $m \geq 3$, then all transvections are conjugate in $SL(V)$.*

REMARK   We know transvections are conjugate in GL; we claim they are conjugate even in SL when $m \geq 3$.

Proof Let $T_{f, \gamma}$ and $T_{f', \gamma'}$ be transvections, and let H and H' be the hyperplanes fixed by each. Choose vectors β, $\beta' \in V$ with

$$f(\beta) = 1 = f'(\beta').$$

We claim there is a transformation S in $GL(V)$ with

$$S(\gamma) = \gamma', \quad S(H) = H', \quad \text{and} \quad S(\beta) = \beta'.$$

There is a basis of H one of whose terms is γ, say $\{\gamma, \gamma_2, \cdots, \gamma_{m-1}\}$; similarly, there is a basis $\{\gamma', \gamma'_2, \cdots, \gamma'_{m-1}\}$ of H' containing γ'. Since H and H' are hyperplanes and $\beta \notin H$, $\beta' \notin H'$, then $\{\beta, \gamma, \gamma_2, \cdots, \gamma_{m-1}\}$ and $\{\beta', \gamma', \gamma'_2, \cdots, \gamma'_{m-1}\}$ are bases of V. Define $S: V \to V$ to be the transformation taking the first of these ordered bases onto the second.

If $m \geq 3$, we claim that we can further choose S of determinant 1. Suppose $\det S = d$. Since $m \geq 3$, the first basis of V constructed above contains at least one other vector besides β and γ. Redefine S so that $S(\gamma_{m-1}) = d^{-1} \gamma'_{m-1}$. Relative to the basis $\{\beta, \gamma, \gamma_2, \cdots, \gamma_{m-1}\}$, the new transformation has as its matrix that of the original S with the last column multiplied by d^{-1}. This new transformation thus has determinant 1 as well as the other properties of S.

We claim $S T_{f, \gamma} S^{-1} = T_{f', \gamma'}$, which will complete the proof. By Exercise 8.16, we know $S T_{f, \gamma} S^{-1} = T_{f S^{-1}, S\gamma}$. Now $S\gamma = \gamma'$; to see that $f S^{-1} = f'$, it suffices to check whether these maps agree on a basis of V; they do agree on $\{\beta', \gamma', \gamma'_2, \cdots, \gamma'_{m-1}\}$. ∎

EXERCISE

8.17. Show that $\begin{bmatrix} 1 & 1 \\ 0 & 1 \end{bmatrix}$ and $\begin{bmatrix} 1 & -1 \\ 0 & 1 \end{bmatrix}$ are not conjugate in $SL(2, 3)$. Generalize to any field K in which -1 is not a square.

We need a bit more information about transvections before we can prove the main theorem.

Notation If H is a hyperplane, then $\mathbf{T}(H)$ is the set consisting of the identity together with all transvections fixing H.

[$\mathbf{T}(H)$ is the analog of the subgroup Γ that arose in the 2×2 case of Lemma 8.18.]

Notation For a vector space V over a field K, let $Sc_1(V)$ denote the group of all scalar transformations of determinant 1 and write $PSL(V) = SL(V)/Sc_1(V)$.

If V is an m-dimensional vector space over a field K, a choice of basis of V enables one to define an isomorphism $PSL(V) \cong PSL(m, K)$.

Lemma 8.26   *Let H be a hyperplane of V.*

(i) *There is a functional f with $H = \ker f$ such that*

$$\mathbf{T}(H) = \{T_{f,\gamma} : \gamma \in H\};$$

(ii) $\mathbf{T}(H)$ *is an abelian subgroup of $SL(m, K)$; in fact, $\mathbf{T}(H) \cong H$;*

(iii) $C_{SL}(\mathbf{T}(H)) = Sc_1(V) \cdot \mathbf{T}(H)$.

Proof

(i) Assume that $T_{f,\gamma}$ and $T_{f',\gamma'}$ are in $\mathbf{T}(H)$ and that $f \not\equiv 0$. Let $\{\alpha_1, \cdots, \alpha_m\}$ be a basis of V whose first $m-1$ terms lie in H. If $f(\alpha_m) = a$ and $f'(\alpha_m) = a'$, then set $b = a'a^{-1}$; it is easy to check that $f' = bf$. By Exercise 8.14,

$$T_{f',\gamma'} = T_{bf,\gamma'} = T_{f,b\gamma'}.$$

(ii) One verifies quickly that $\mathbf{T}(H)$ is a subgroup of SL and that $T_{f,\gamma} \mapsto \gamma$ is an isomorphism $\mathbf{T}(H) \to H$ (where we choose the same functional f for all T's, by (i)).

(iii) Since $\mathbf{T}(H)$ is abelian, we have $C_{SL}(\mathbf{T}(H)) \supset Sc_1(V) \cdot \mathbf{T}(H)$. For the reverse inclusion, suppose $A \in SL$ and

$$AT_{f,\gamma}A^{-1} = T_{f,\gamma}, \qquad \text{all } T_{f,\gamma} \in \mathbf{T}(H).$$

By Exercise 8.16,

$$T_{fA^{-1}, A\gamma} = T_{f,\gamma};$$

by Exercise 8.15, there is a scalar $a \in K$ with $fA^{-1} = af$ and $\gamma = aA\gamma$. Therefore

$$A(\gamma) = a^{-1}\gamma, \qquad \text{all } \gamma \in H$$

and

$$fA = a^{-1}f.$$

If $\{\alpha_1, \cdots, \alpha_m\}$ is a basis of V whose first $m-1$ terms lie in H, then the last two equations give, for $\beta \in V$,

$$A(\beta) = a^{-1}\beta + \gamma', \qquad \gamma' \in H.$$

This says that A is a scalar multiple of a transvection, as desired. ∎

Theorem 8.27 (Jordan-Dickson)†   *The groups $PSL(V)$ are simple whenever V is a vector space of dimension ≥ 3 over an arbitrary field K.*

Proof   We show that if $G \lhd SL$ and $G \not\subseteq Sc_1(V)$, then $G = SL$. By Theorem 8.25, it suffices to prove G contains a transvection. Choose $A \in G$ with $A \notin Sc_1(V)$. There is thus a transvection T that does not commute with A:

$$B = (T^{-1}AT)A^{-1} \neq E.$$

Note that $B \in G$ since G is normal. Now

$$B = T^{-1}(ATA^{-1}) = T_1 T_2,$$

---

† Jordan (1870) proved this theorem for $K = GF(p)$; Dickson (1897) proved the theorem for every finite field K, four years after Moore had done the case $m = 2$.

where T_i is a transvection, $i = 1, 2$. If $T_i = T_{f_i, \gamma_i}$ and $H_i = \ker f_i$, then, for all $\beta \in V$,

$$T_i(\beta) = \beta + f_i(\beta)\gamma_i, \qquad \text{where } \gamma_i \in H_i, \quad i = 1, 2.$$

Let W be the subspace of V spanned by γ_1 and γ_2, so that dim $W \leq 2$. Since dim $V \geq 3$, there is a hyperplane H of V containing W. Now

$(\#)$ $$B(H) \subset H,$$

for if $\eta \in H$, then

$$B(\eta) = T_1 T_2(\eta) = T_2(\eta) + f_1(T_2(\eta))\gamma_1$$
$$= \eta + f_2(\eta)\gamma_2 + f_1(T_2(\eta))\gamma_1 \in H + W = H.$$

Next, we claim $H_1 \cap H_2 \neq \{0\}$. If $H_1 = H_2$, this is surely true. If $H_1 \neq H_2$, then $H_1 + H_2 = V$ (hyperplanes are maximal subspaces) and $\dim(H_1 + H_2) = m$. It follows from

$$\dim(H_1 \cap H_2) + \dim(H_1 + H_2) = \dim H_1 + \dim H_2$$

that $\dim(H_1 \cap H_2) = m - 2 \geq 1$, whence $H_1 \cap H_2 \neq \{0\}$. Choose $\zeta \in H_1 \cap H_2$, $\zeta \neq 0$. Then

$(\#\#)$ $$B(\zeta) = T_1 T_2(\zeta) = \zeta.$$

We may assume B is not a transvection (or we are done, since $B \in G$). Therefore $B \notin \mathbf{T}(H)$, which is wholly comprised of transvections. If $B = aS$, where $a \in K$ and $S \in \mathbf{T}(H)$, then $(\#\#)$ says that ζ is a characteristic vector of S. By Theorem 8.21, $\zeta \in H$ and so $S(\zeta) = \zeta$. Thus

$$\zeta = B(\zeta) = aS(\zeta) = a\zeta$$

and $a = 1$; hence $B = S \in \mathbf{T}(H)$, a contradiction. Therefore

$$B \notin Sc_1(V) \cdot \mathbf{T}(H) = C_{SL}(\mathbf{T}(H)),$$

by Lemma 8.26(iii). There is thus a transvection $U \in \mathbf{T}(H)$ with the commutator C defined by

$$C = UBU^{-1}B^{-1} \neq E.$$

Clearly $C = (UBU^{-1})B^{-1} \in G$. Also, if $\eta \in H$,

$$C(\eta) = UBU^{-1}B^{-1}(\eta) = UB(B^{-1}(\eta)) = \eta,$$

since $B^{-1}(\eta) \in H$, by $(\#)$, and $U \in \mathbf{T}(H)$ fixes H. Therefore C fixes H; hence, C is either a dilatation or a transvection. It is not a dilatation, for det $C = 1$. Hence C is a transvection in G, and the proof is complete. ∎

We shall give different proofs of Theorems 8.19 and 8.27 in Chapter 9 (see Theorem 9.50).

EXERCISE

8.18 Let $O(n, K)$ denote the $n \times n$ **orthogonal group** over a field K consisting of all $A \in GL(n, K)$ such that $AA^t = E$, where A^t is the transpose of A. If $n \geq 3$, prove that $O(n, K)$ is not isomorphic to a quotient group of $GL(n, K)$. Compare with Exercise 8.12.

Observe that $|PSL(3, 4)| = 20{,}160 = \frac{1}{2}8!$, so that $PSL(3, 4)$ and A_8 have the same order. We now show these groups are not isomorphic.

Lemma 8.28 (Compare Exercise 3.32) *Let H be a normal subgroup of prime index in a finite group G, let $h \in H$, and assume there is an element not in H that commutes with h. If an element $k \in H$ is conjugate to h in G, then k is conjugate to h in H.*

Proof In light of Theorem 3.8, it suffices to prove $[H:C_H(h)] = [G:C_G(h)]$. Using the product formula,

$$[H:C_H(h)] = [H:C_G(h) \cap H] = |HC_G(h)|/|C_G(h)| = [G:C_G(h)]$$

[the last equation results from $HC_G(h) = G$ because $HC_G(h)$ is a subgroup $(H \lhd G)$ properly containing the maximal subgroup H of G ($[G:H]$ is prime)]. ∎

Theorem 8.29 (Schottenfels, 1900) A_8 *and* $PSL(3, 4)$ *are not isomorphic. Therefore, there are nonisomorphic finite simple groups having the same order.*

Proof The elements $(12)(34)$ and $(12)(34)(56)(78)$ have order 2, are even, and are not conjugate in A_8 (these elements are not even conjugate in S_8). In contrast, we shall show that all elements of order 2 in $PSL(3, 4)$ are conjugate, and this will prove the theorem.

A nonscalar matrix $A \in SL(3, 4)$ corresponds to an element of order 2 in $PSL(3, 4)$ if and only if A^2 is scalar. Since A^2 is scalar if and only if $(P^{-1}AP)^2$ is scalar, we may assume A is a rational canonical form. If A is diagonal (i.e., a direct sum of three 1×1 companion matrices), then A^2 scalar implies A is scalar $[a^2 = b^2 = c^2$ implies $a = b = c$ because $GF(4)$ has characteristic 2]; if A is a 3×3 companion matrix and $A^2 = \gamma E$, then A satisfies $x^2 - \gamma = 0$; but the minimum polynomial of a companion matrix coincides with its characteristic polynomial, which in this case has degree 3. We conclude that A has the form

$$A = \begin{bmatrix} a & 0 & 0 \\ 0 & 0 & b \\ 0 & 1 & c \end{bmatrix}.$$

Now det $A = 1 = ab$ ($-1 = 1$ here) forces $b = a^{-1}$, and A^2 scalar forces $c = 0$. Thus,

$$A = \begin{bmatrix} a & 0 & 0 \\ 0 & 0 & a^{-1} \\ 0 & 1 & 0 \end{bmatrix}.$$

There are only three such matrices; if α is a primitive element of $GF(4)$, they are:

$$A = \begin{bmatrix} 1 & 0 & 0 \\ 0 & 0 & 1 \\ 0 & 1 & 0 \end{bmatrix}; \qquad B = \begin{bmatrix} \alpha & 0 & 0 \\ 0 & 0 & \alpha^2 \\ 0 & 1 & 0 \end{bmatrix}; \qquad C = \begin{bmatrix} \alpha^2 & 0 & 0 \\ 0 & 0 & \alpha \\ 0 & 1 & 0 \end{bmatrix}$$

Note that $A^2 = E$, $B^2 = \alpha^2 E$, and $C^2 = \alpha E$. It follows that if $M \in SL(3, 4)$ has order 2, then M is similar to A, i.e., there is $P \in GL(3, 4)$ with $M = P^{-1}AP$. In particular, $\alpha^2 B$ and αC have order 2, so there are P, $Q \in GL(3, 4)$ with

(*) $\qquad\qquad P^{-1}AP = \alpha^2 B \quad$ and $\quad Q^{-1}AQ = \alpha C$.

Now $SL(3, 4)$ is a normal subgroup of index 3 in $GL(3, 4)$ [for $GL(3, 4)/SL(3, 4) \cong GF(4)^{\#}$, the multiplicative group of nonzero elements of $GF(4)$], and the matrix

$$\begin{bmatrix} \alpha & 0 & 0 \\ 0 & 1 & 0 \\ 0 & 0 & 1 \end{bmatrix}$$

commutes with A. The lemma shows we may assume P and Q lie in $SL(3, 4)$. Finally, equations (*) show that A, B, and C become conjugate in $PSL(3, 4)$. We have proved that all elements of order 2 in $PSL(3, 4)$ are conjugate, and this shows $A_8 \not\cong PSL(3, 4)$. ∎

Another way to prove this last theorem is to observe that A_8 contains an element of order 15, namely, (123)(45678), but $PSL(3, 4)$ contains no such element.

There are infinitely many integers n for which there exist two nonisomorphic simple groups of order n (see Artin, p. 210), but there exist no integers n for which there exist three nonisomorphic simple groups of order n. For discussion of other simple groups, the reader is referred to the books of Artin, Carter, Dieudonné, Gorenstein (1982), and Jacobson cited in the bibliography.

chapter nine

Permutations and the Mathieu Groups

The Mathieu groups are five remarkable simple groups discovered by Mathieu in 1861 and 1873. This chapter is devoted to proving their existence and displaying some of their properties. Along the way, some interesting scenery is visible.

G-SETS

In Chapter 3, we defined permutation groups, orbits, and stabilizers. With the representation Theorems 3.18 and 3.21 in mind, we generalize these notions. Unless indicated otherwise, we assume throughout this chapter that all groups and sets are finite and all vector spaces are finite-dimensional.

Definition If X is a set and G a group, then X is a **G-set** if there is a function $G \times X \to X$, denoted by $(g, x) \mapsto gx$, such that

    (i) $1x = x$;
    (ii) $g(hx) = (gh)x$

for every $x \in X$ and every $g, h \in G$. One says G **acts** on X; if $|X| = n$, then n is called the **degree** of X.

Note that if X is a G-set and H is a subgroup of G, then X is also an H-set (just restrict the given function from $G \times X$ to $H \times X$).

Theorem 9.1 *Every G-set X defines a homomorphism $\varphi: G \to S_X$ (its **action**), namely, $\varphi(g): x \mapsto gx$. Conversely, every homomorphism $\psi: G \to S_X$ equips X with the structure of a G-set.*

Proof Assume X is a G-set. If $g \in G$, we claim $\varphi(g)$ is a permutation of X, for $g^{-1}(gx) = (g^{-1}g)x = 1x = x$; similarly $g(g^{-1}x) = x$, so the inverse of $\varphi(g)$ is $\varphi(g^{-1})$. That φ is a homomorphism is immediate from axiom (ii). For the converse, given ψ, define $gx = \psi(g)(x)$. ∎

This theorem thus shows that G-sets are just permutation representations of G, i.e., homomorphisms into symmetric groups.

Let us denote the set of all functions from a set X to itself by X^X. One sees easily that X^X under composition is a semigroup with 1; of course, $S_X \subset X^X$. If G is a group and $\theta: G \to X^X$ is a homomorphism (of semigroups) with $\theta(1) = 1_X$, then im $\theta \subset S_X$ [because $1_X = \theta(1) = \theta(gg^{-1}) = \theta(g)\theta(g^{-1})$ and, similarly, $1_X = \theta(g^{-1})\theta(g)$]. It follows that such a homomorphism θ makes X into a G-set.

Definition A G-set X is **faithful** if its action $\varphi: G \to S_X$ is one-one.

If X is a faithful G-set, we may identify G with a subgroup of S_X via φ, and G becomes a permutation group on X. Cayley's theorem asserts every group G of order n is a faithful G-set of degree n; because $|S_n| = n!$ is so much larger than n, faithful G-sets of smaller degree are much more valuable.

Example 1 Assume that X has some "structure" and that Aut(X), the set of all permutations of X preserving the structure, forms a group under composition. Then X is a faithful G-set for every subgroup G of Aut(X).

Example 2 If X is a G-set and $gx = x$, where $g \in G$ and $x \in X$, one says g **fixes** x. If each $g \in G$ fixes every $x \in X$, then X is called a **trivial** G-set.

Example 3 If X is a G-set, then $N = \{g \in G: gx = x \text{ for all } x \in X\}$ is a normal subgroup of G ($N = \ker \varphi$), and X becomes a faithful (G/N)-set if one defines $\bar{g}x = gx$, where $\bar{g}$ is the coset gN. Therefore, if X is a nontrivial G-set and G is simple, then X is faithful.

Example 4 If H is a subgroup of G and $G /\!/ H$ is the set of all left cosets of H in G, then X is a G-set (Theorem 3.18) that may not be faithful.

Example 5 If H is a subgroup of G and X is the set of all conjugates of H in G, then X is a G-set (Theorem 3.21) that may not be faithful.

Example 6 If V is a vector space, then $GL(V)$ acts faithfully on V and also on $V^{\#} = V - \{0\}$. If G is a group, then Aut(G) acts faithfully on G and also on $G^{\#} = G - \{1\}$.

Definition If X is a G-set, the **G-orbit** of a point $x \in X$ is the subset of X

$$Gx = \{gx: g \in G\}.$$

When no confusion can arise, we will write **orbit** instead of G-orbit.

Definition If X is a G-set and $x_1, \cdots, x_k \in X$, then the **stabilizer** $G_{x_1, \cdots, x_k}$ is the subgroup

$$G_{x_1, \cdots, x_k} = \{g \in G: gx_i = x_i, \quad i = 1, \cdots, k\}.$$

That $G_{x_1, \cdots, x_k}$ is the set of all $g \in G$ fixing each x_i, $i = 1, \cdots, k$, may also be phrased

$$G_{x_1, \cdots, x_k} = \bigcap_{i=1}^{k} G_{x_i}.$$

It follows that if X is a G-set and $x, y \in X$, then X is also a G_x-set and a G_y-set and

$$(G_x)_y = G_{x, y} = (G_y)_x.$$

We have already seen that if X is a G-set and $x \in X$, the index $[G:G_x]$ is the number of elements in the orbit Gx (Theorem 3.22); moreover, Theorem 3.26, the Cauchy-Frobenius theorem, gives a formula for the number of orbits in X.

Let us illustrate these ideas by giving a proof, due to McKay, of Cauchy's Theorem 4.2. Assume G is a finite group and p is a prime. Define

$$X = \{(a_1, \cdots, a_p) \in G \times G \times \cdots \times G: a_1 a_2 \cdots a_p = 1\}.$$

Note that $|X| = |G|^{p-1}$, for having chosen the first $p - 1$ coordinates arbitrarily, we must set $a_p = (a_1 a_2 \cdots a_{p-1})^{-1}$. Theorem 1.4 shows $(a_1, \cdots, a_p) \in X$ implies $(a_p, a_1, \cdots, a_{p-1}) \in X$. It follows that X is a $\mathbf{Z}(p)$-set, where $g \in \mathbf{Z}(p)$ acts by cyclically permuting the coordinates. By Corollary 3.23, the cardinal of each $\mathbf{Z}(p)$-orbit is 1 or p. Now a $\mathbf{Z}(p)$-orbit with just one element is a p-tuple having all its coordinates equal, say, $a_i = a$ for all i; in other words, such orbits correspond to elements $a \in G$ with $a^p = 1$. Clearly $(1, 1, \cdots, 1)$ is such an orbit; were this the only such orbit, we would have

$$|X| = |G|^{p-1} = 1 + kp$$

for some integer k, i.e., $|G|^{p-1} \equiv 1 \bmod p$. If $|G|$ is divisible by p, we have a contradiction, whence X has another orbit with one element, and this shows G has an element of order p. (As A. Mann remarked to me, if $|G|$ is not divisible by p, one has another proof of Fermat's theorem.)

Let us now examine the structure of G-sets.

Definition A G-set X is **transitive†** if, for each $x, y \in X$, there exists $g \in G$ with $gx = y$.

---

† When a transitive G-set X is given, one often abuses language and says that the group G is transitive; this remark applies to other adjectives as well.

Clearly, a G-set X is transitive if and only if it has only one orbit; moreover, if X is nontrivial, then $|X| > 1$.

Theorem 9.2 *Every G-set X is partitioned into its G-orbits, each of which is a transitive G-set. Conversely, if a G-set X is partitioned into transitive G-sets $\{X_i : i \in I\}$, then the X_i are the G-orbits of X.*

Proof Each G-orbit Gx is visibly a transitive G-set. Defining $x \sim y$ (for $x, y \in X$) to mean $gx = y$ for some $g \in G$, it is easy to see $\sim$ is an equivalence relation on X whose equivalence classes are precisely the G-orbits.

To prove the converse, it suffices to show each $X_i = Gx_i$, where $x_i \in X_i$. Now $Gx_i \subset X_i$ because X_i is a G-set. For the reverse inclusion, if $y \in X_i$, then transitivity gives $y = gx_i$ for some $g \in G$. Therefore $y \in Gx_i$ and $X \subset Gx_i$. ∎

This unique decomposition permits one to focus on transitive G-sets.

EXERCISES

**9.1. Show that S_n cannot act transitively on a set X with t elements, where $n \geq 5$ and $2 < t < n$. (HINT: Consider the representation $S_n \rightarrow S_t$ in light of Example 3.) Conclude that every orbit of an S_n-set with more than two elements has at least n elements.

**9.2. Prove that if $n \geq 5$ and $2 < t < n$, then S_n has no subgroup of index t.

9.3. (i) If H is a subgroup of G and $G/\!/H$ is the set of left cosets of H in G, then the representation of Theorem 3.18 makes $G/\!/H$ a transitive G-set of degree $[G:H]$. In particular, when $H = \{1\}$, the (left) regular representation of the Cayley theorem makes G a faithful transitive G-set.
(ii) If H is a subgroup of G and Y is the set of conjugates of H in G, then the conjugation representation of Theorem 3.21 makes Y a transitive G-set of degree $[G: N_G(H)]$.

**9.4. If X is a G-set and H is a subgroup of G, then every G-orbit is a disjoint union of H-orbits.

9.5. If G is a group and $G^{\#} = G - \{1\}$ is a transitive $\mathrm{Aut}(G)$-set, then G has no proper characteristic subgroups. (See Theorem 5.20.)

9.6. Assume G is a finite group for which $G^{\#}$ is a transitive $\mathrm{Aut}(G)$-set. Show that G is an elementary abelian p-group for some prime p. [The group G of rationals is an infinite group with $G^{\#}$ a transitive $\mathrm{Aut}(G)$-set.]

9.7. If X is a transitive G-set and N is a normal subgroup of G, regard X as an N-set and prove every two stabilizers N_x and N_y, for $x, y \in X$, have the same order.

*9.8. Let G be a group with $n < |G|$. Prove that G is isomorphic to a transitive subgroup of S_n if and only if G contains a subgroup H of index n such that neither H nor any proper subgroup of H is normal in G. (HINT: For sufficiency, use Theorem 3.18; for necessity, choose H to be the stabilizer of any symbol.)

The basic philosophy here is that the existence of a G-set of a special type may force constraints on G. For example, Exercise 9.8 shows that if there exists a faithful transitive G-set of degree n, then G has a subgroup of index n (with certain added properties). It follows immediately that $|G|$ is divisible by n; we give a second proof of this fact (see also Theorem 3.22).

Theorem 9.3 *If X is a transitive G-set of degree n and $x \in X$, then*

$$|G| = n|G_x|.$$

If X is a faithful G-set, then $|G_x|$ is a divisor of $(n-1)!$.

Proof We know $|G| = [G:G_x]|G_x|$. But $[G:G_x]$ is the cardinal of the orbit Gx, and $Gx = X$ because G acts transitively on X. The second statement follows because G is (isomorphic to) a subgroup of S_n. ∎

We may now give a swift proof of Exercise 3.30 when G is finite. If G has only two conjugacy classes, then every two nonidentity elements of G are conjugate. Thus, G acts transitively (by conjugation) on $G^{\#}$, and Theorem 9.3 asserts $|G| - 1$ divides $|G|$; therefore $|G| = 2$.

Theorem 9.4 *Let X be a transitive G-set and let $x, y \in X$.*

(i) *If $tx = y$ for some $t \in G$, then $G_y = tG_x t^{-1}$.*
(ii) *X has the same number of G_x-orbits as of G_y-orbits.*

Proof

(i) If q fixes x, then $tgt^{-1}y = tgx = tx = y$ and tgt^{-1} fixes y. The result follows easily.
(ii) Since G acts transitively, there is $t \in G$ with $tx = y$. Let the G_x-orbits be $\{G_x a_i : i \in I\}$, where $a_i \in X$. Define $b_i = ta_i \in X$. We claim the sets $G_y b_i$, $i \in I$, are the G_y-orbits of X. Clearly, each of these is a transitive G_y-set. Further,

$$G_y b_i = tG_x t^{-1} b_i = tG_x a_i.$$

Since t acts as a permutation of X, it carries any partition of X to another partition; therefore the subsets $G_y b_i$ partition X, and Theorem 9.2 completes the proof. ∎

EXERCISE

9.9.   Let X be a G-set, and let H be a subgroup of G. If the H-orbits of X are $\{o_i, i \in I\}$, then the orbits of gHg^{-1} are $\{go_i, i \in I\}$. Use this result to give a second proof of Theorem 9.4(ii).

Definition If X is a transitive G-set, then the **rank** of X is the number of G_x-orbits of the G_x-set X.

Note that Theorem 9.4 shows the definition of rank is independent of the choice of stabilizer G_x. There are several characterizations of rank. We prove one, but only state another interesting one:

$$|G| \operatorname{rank} X = \sum_{g \in G} F(g)^2,$$

where $F(g)$ is the number of $x \in X$ fixed by g.

Theorem 9.5 *Let X be a transitive G-set and let $x \in X$. Then rank X is the number of G_x-G_x double cosets in G.*

Proof Define π: $\{G_x\text{-orbits of } X\} \to \{G_x\text{-}G_x \text{ double cosets of } G\}$ by $\pi(G_x y) = G_x g G_x$, where $gx = y$. Now π is well defined, for if $hx = y$, then $gx = hx$ and $G_x g G_x = G_x h G_x$. Conversely, π is one-one, for if $G_x g G_x = \pi(G_x y) = \pi(G_x z) = G_x h G_x$ (where $gx = y$ and $hx = z$), then $gx = hx$, $y = z$, and $G_x y = G_x z$. Finally, π is onto, for if $g \in G$ and $gx = y$, then $\pi(G_x y) = G_x g G_x$. ∎

Observe that rank $X \geq 2$, for G_x is always a proper subgroup of G (because G acts transitively). Let us consider the minimal case (in which G_x acts transitively on $X - \{x\}$).

Definition Let X be a G-set of degree n and let $k \leq n$ be a positive integer. One says X is *k*-**transitive** if, for every pair of k-tuples $(x_1, \cdots, x_k)$ and $(y_1, \cdots, y_k)$ having distinct entries in X, there is $g \in G$ with $gx_i = y_i$, for $i = 1, \cdots, k$.

Of course, 1-transitivity is ordinary transitivity, and if $k > 1$, then k-transitivity implies $(k - 1)$-transitivity. A k-transitive G-set X is called **doubly transitive** if $k = 2$, **triply transitive** if $k = 3, \cdots$, and **multiply transitive** if $k > 1$. Examples of multiply transitive G-sets will be given later.

Theorem 9.6 *Every multiply transitive G-set X has rank 2, and, if $x \in X$ and $g \notin G_x$, then $G = G_x \cup G_x g G_x$.*

Proof Since G acts k-transitively on X for some $k > 1$, G_x acts transitively on $X - \{x\}$: $X - \{x\}$ has just one G_x-orbit. Hence X has rank 2, and Theorem 9.5 shows there are only two G_x-G_x double cosets. ∎

EXERCISES

**9.10. Let X be a transitive G-set. Show that X is k-transitive, where $k \geq 2$, if and only if, for every $x \in X$, the G_x-set $X - \{x\}$ is $(k - 1)$-transitive.

9.11. Let X be a k-transitive G-set. Show that stabilizers of k distinct elements of X are conjugate subgroups of G.

Theorem 9.7 *If X is a k-transitive G-set of degree n, then*

$$|G| = n(n-1)\cdots(n-k+1)|G_{x_1,\cdots,x_k}|$$

for every choice of k distinct elements $x_1, \cdots, x_k$ in X. If G acts faithfully, then $|G_{x_1,\cdots,x_k}|$ is a divisor of $(n-k)!$.

Proof   Choose $x_1 \in X$. By Theorem 9.3, $|G| = n|G_{x_1}|$. But Exercise 9.10 shows G_{x_1} acts $(k-1)$-transitively on $X - \{x_1\}$. Therefore, if $x_2, \cdots, x_k$ are distinct elements of $X - \{x_1\}$, induction gives

$$|G_{x_1}| = (n-1)\cdots(n-k+1)|G_{x_1, x_2, \cdots, x_k}|.$$

If G acts faithfully, the last statement follows from G being imbedded in S_n.   ∎

Definition   A k-transitive G-set X is **sharply k-transitive** if only the identity of G fixes k distinct elements of X.

Corollary 9.8   *The following are equivalent for a faithful k-transitive G-set X of degree n:*

(i) *X is sharply k-transitive;*
(ii) *if $(x_1, \cdots, x_k)$ and $(y_1, \cdots, y_k)$ are k-tuples with distinct entries in X, there is a unique $g \in G$ with $gx_i = y_i$ for $i = 1, \cdots, k$;*
(iii) *$|G| = n(n-1)\cdots(n-k+1)$;*
(iv) *every stabilizer of k distinct elements in X is $\{1\}$.*

If $k \geq 2$, these conditions are equivalent to

(v) *for every $x \in X$, the G_x-set $X - \{x\}$ is sharply $(k-1)$-transitive.*

Proof   The straightforward arguments are left as an exercise.   ∎

Sharp k-transitivity is especially interesting when $k = 1$ and $k = 2$.

Definition   A sharply 1-transitive G-set X is called **regular.**

Thus, a faithful G-set is regular if it is transitive and only the identity of G has any fixed points. The reader should check that the (left) regular representation of the Cayley theorem makes G into a regular G-set.

EXERCISES

9.12.   If G is abelian, then a faithful transitive G-set X is regular. (HINT: If $x, y \in X$, then $G_x = G_y$.)

9.13.   If X is a sharply k-transitive G-set, then X is not $(k+1)$-transitive. (In particular, a regular G-set is not doubly transitive.)

Our discussion of sharply 2-transitive G-sets begins with a technical definition.

Definition If X is a G-set, the **Frobenius kernel** N of G is the subset

$$N = \{1\} \cup \{g \in G : g \text{ has no fixed points}\}.$$

In general, the Frobenius kernel is not a subgroup of G.

Lemma 9.9 *If X is a faithful sharply 2-transitive G-set of degree n, then the Frobenius kernel N of G has exactly n elements.*

Proof The hypothesis gives $|G| = n(n-1)$ and $G_{x,y} = \{1\}$ for distinct $x, y \in X$. For each $x \in X$, the stabilizer G_x has order $n - 1$, whence $|G_x^\#| = n - 2$. If $x \neq y$, then $G_x \cap G_y = G_{x,y} = \{1\}$, so $\{G_x^\# : x \in X\}$ is a disjoint family and $|\bigcup_{x \in X} G_x^\#| = n(n-2)$. But N is just the complement of this union, so $|N| = n(n-1) - n(n-2) = n$. ∎

We shall now classify the groups G occurring in Lemma 9.9 in the special case when the degree n is odd. The next theorem illustrates the basic philosophy: knowledge of permutation representations of a group G can yield important information about G.

Theorem 9.10 *Let X be a faithful sharply 2-transitive G-set of odd degree n.*

(i) *Each G_x contains a unique element of order 2.*

(ii) *G_x has a center of even order and a Sylow 2-subgroup of G is either cyclic or generalized quaternion.*

(iii) *The Frobenius kernel N of G is a normal subgroup of G.*

(iv) *n is a power of an odd prime p.*

(v) *N is an elementary abelian p-group.*

(vi) *G is a semidirect product of N by G_x.*

REMARK   See Exercise 9.19 for another property of N.

Proof

(i) Since $|G| = n(n-1)$ is even, G contains an element g of order 2. Now g, being a permutation of the n points in X, has a cycle decomposition; by Exercise 3.14, g is a product of disjoint transpositions, say, $g = \tau_1 \cdots \tau_m$. Because $|X| = n$ is odd, g must fix some $x \in X$, i.e., $g \in G_x$; moreover, because X is sharply 2-transitive, g can fix nothing else, whence $m = \frac{1}{2}(n-1)$. If h is another element of order 2 in G, then $h \in G_y$ for some $y \in X$ (perhaps $y = x$) and h is also a product of disjoint transpositions: $h = \tau_1' \cdots \tau_m'$. Note that g and h can have no factors in common: if $\tau_i = \tau_j'$, then we may assume $\tau_i = \tau_m$ and $\tau_j' = \tau_1'$ (because disjoint cycles commute), $gh = gh^{-1} = 1$ (because gh fixes the two elements moved by τ_m), and $g = h$.

Assume G_x has t elements of order 2. As $\{G_x^\# : x \in X\}$ is a disjoint family of subsets, G contains nt such elements, each of which involves

$m = \frac{1}{2}(n-1)$ transpositions; thus, there exist $\frac{1}{2}nt(n-1)$ distinct transpositions. Because $S_X \cong S_n$ contains only $\frac{1}{2}n(n-1)$ transpositions, we
must have $t = 1$.

(ii) If $g, h \in G_x$ and g has order 2, then hgh^{-1} also has order 2, whence
$hgh^{-1} = g$ and g commutes with h. The second statement is immediate
from Theorem 5.36.

(iii) If T is the set of all elements in G of order 2, we claim $TT \subset N$. Otherwise,
there exist $g, h \in T$ with $gh \neq 1$ and gh fixing some point $y \in X$. Define
$z = hy$. Both g and h have the transposition (y, z) as a factor: $hy = z$ and
$hz = h^2y = y$; $gz = ghy = y$ and $gy = hy = z$ (since $g = g^{-1}$, $ghy = y$
implies $gy = hy$). From part (i), we conclude that $g = h$, giving the
contradiction $gh = 1$.

For fixed $g \in T$, the functions $T \to N$ defined by $h \mapsto gh$ and $h \mapsto hg$ are
one-one. By part (i), we know $|T| = n$; by Lemma 9.9, we know $|N| = n$. It
follows that the two functions $T \to N$ are onto, i.e., $gT = N = Tg$.

The Frobenius kernel N contains 1 and is obviously closed under
inverses and conjugation by elements in G. To see N is closed under
multiplication, choose $g \in T$ and observe

$$NN = (Tg)(gT) = Tg^2T = TT \subset N.$$

(iv) Choose an element $h \in N$ of prime order p (p is odd since it divides
$|N| = n$). Assume $f \in G$ commutes with h, i.e., $hfh^{-1} = f$. If $f \in G_x$ for some
$x \in X$, then

$$f \in G_x \cap hG_xh^{-1} = G_x \cap G_{hx} = G_{x,hx} = \{1\}$$

because $hx \neq x$. Thus, if $f \neq 1$, we have $f \notin \bigcup_{x \in X} G_x$, that is, $f \in N$. We
conclude that $C_G(h)$, the centralizer of h, is contained in N, and so
$[G:C_G(h)] = [G:N][N:C_G(h)] \geq n-1$. But $[G:C_G(h)]$ is the number
of conjugates of h, all of which lie in the normal subgroup N of order n. It
follows that $N^{\#}$ consists precisely of the conjugates of h, whence N is a
p-group of exponent p. Therefore $n = |N|$ is a power of p.

(v) If $g \in G$ has order 2, then conjugation by g is an automorphism of N
satisfying the conditions of Exercise 1.38. Thus, N is abelian, and, since
every $h \in N^{\#}$ has order p, N is a vector space over $GF(p)$, i.e., N is an
elementary abelian p-group.

(vi) Since the orders of N and G_x are consecutive integers, they are relatively
prime, and the last statement follows at once from the Schur-Zassenhaus
lemma. ∎

Here is the proper context for this last theorem. Let X be a faithful transitive
G-set such that each $g \in G^{\#}$ has at most one fixed point. If no $g \in G^{\#}$ has a fixed
point, then X is just a regular G-set; if some $g \in G^{\#}$ has a fixed point (i.e., stabilizers of
points are nontrivial), then G is called a **Frobenius group**. It is easy to see that if there
exists a faithful sharply 2-transitive G-set of any (not necessarily odd) degree, then G
is a Frobenius group. In 1901, Frobenius proved that Frobenius kernels of Frobenius
groups are normal subgroups; in 1959, Thompson proved these Frobenius kernels

are nilpotent groups. Every Frobenius group G is a semidirect product of its Frobenius kernel by G_x (here G_x is called a **Frobenius complement**). Every Frobenius complement is a group in which every Sylow subgroup is either cyclic or generalized quaternion. Such groups are usually solvable (Theorem 7.33); however, there are nonsolvable Frobenius complements, e.g., $SL(2,5)$.

Here are some obvious examples of highly transitive G-sets of degree n.

Theorem 9.11 *For every n, the symmetric group S_n acts sharply n-transitively on $X = \{1, \cdots, n\}$; for every $n \geq 3$, the alternating group A_n acts sharply $(n-2)$-transitively on X.*

Proof The first statement is obvious since S_n contains every permutation of X.

We show by induction on $n \geq 3$ that A_n acts $(n-2)$-transitively on X. When $n = 3$, then $A_3 = \langle (123) \rangle$ acts transitively on $X = \{1, 2, 3\}$. If $n > 3$, note that the stabilizer of n, $(A_n)_n$, is A_{n-1}, which acts $(n-3)$-transitively on $\{1, \cdots, n-1\}$ by induction. Exercise 9.10 completes the induction. Finally, this action is sharp, for $|A_n| = \frac{1}{2}n! = n(n-1) \cdots (n-(n-2)+1)$. ∎

Aside from the actions of the symmetric and alternating groups (regarded as "trivial"), it is known no k-transitive G-sets exist for $k > 5$.

We continue the analysis of transitive G-sets.

Definition A **block** of a G-set X is a subset B of X such that, for each $g \in G$, either $gB = B$ or $gB \cap B = \varnothing$ (of course, $gB = \{gb: b \in B\}$).

Clearly $B = \varnothing$ and $B = X$ are blocks, as is every one-point subset of X; these are called **trivial** blocks, and any other block is called **nontrivial**.

Definition A transitive G-set X is **primitive** if it contains no nontrivial blocks; a G-set containing a nontrivial block is called **imprimitive**.

Example 7 Let X be the 4 vertices of a square construed as a D_4-set in the usual way.

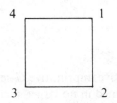

One may check that $\{1, 3\}$ and $\{2, 4\}$ are blocks, so the transitive D_4-set X is imprimitive.

Example 8 Viewing a group G as an $\mathrm{Aut}(G)$-set, one sees that any conjugacy class B of G is a block. If G is not abelian, therefore, then G is an imprimitive $\mathrm{Aut}(G)$-set.

It is usually difficult to see whether a G-set X is primitive; various examples will appear as we discuss primitivity.

Theorem 9.12   *Let B be a nontrivial block in a transitive G-set X of degree n.*

(i) *If $g \in G$, then gB is a block.*
(ii) *There are elements $g_1, g_2, \cdots, g_m$ of G such that*
    $Y = \{B, g_1B, g_2B, \cdots, g_mB\}$ *is a partition of X.*
(iii) *$|B|$ divides $|X|$.*
(iv) *Y is a transitive G-set of degree $n/|B|$.*

Proof

(i) Assume $gB \cap hgB \neq \varnothing$ for some $h \in G$. It follows that $B \cap g^{-1}hgB \neq \varnothing$ whence $g^{-1}hgB = B$ and $hgB = gB$.
(ii) Choose $b \in B$ and $x_1 \notin B$. Since G acts transitively, there is $g_1 \in G$ with $g_1 b = x$. That B is a block implies B and g_1B are disjoint. If $X = B \cup g_1B$, we are done. Otherwise, choose $x_2 \notin B \cup g_1B$ and choose $g_2 \in G$ with $g_2 b = x_2$. Because B and g_1B are blocks, it is easy to see g_2B meets neither of them. The proof is finished by iteration of this procedure.
(iii) If $g \in G$, then $|B| = |gB|$. The result now follows from (ii).
(iv) For any $g \in G$, observe that gg_iB meets some g_jB, and so $gg_iB = g_jB$. It follows easily that $\varphi \colon G \to S_Y$ defined by $\varphi(g) \colon g_iB \mapsto gg_iB$ makes Y a transitive G-set visibly having degree $|Y| = n/|B|$. ∎

Definition   The set Y in Theorem 9.12(ii) is called an **imprimitive system** (generated by B).

Corollary 9.13   *A transitive G-set of prime degree is primitive.*

For a prime p, let $\mathrm{Aff}(1, p)$ be the group under composition of all $f \colon GF(p) \to GF(p)$ of the form $f(x) = ax + b$, where $a, b \in GF(p)$ and $a \neq 0$. Burnside (1900) proved that if there exists a faithful transitive G-set of prime degree, then either G is doubly transitive or G is a subgroup of $\mathrm{Aff}(1, p)$ (compare Theorem 9.27).

EXERCISES

*9.14.   Assume X is a transitive imprimitive G-set and B is a maximal nontrivial block, i.e., B is contained in no larger such. If Y is the imprimitive system generated by B, show that Y is a primitive G-set.

**9.15.   Let X be a G-set with $x, y \in X$. If H is a subgroup of G, then $Hx \cap Hy \neq \varnothing$ implies $Hx = Hy$. If $H \lhd G$ (so that $gH = Hg$), the subsets Hx are blocks of X.

9.16.   (i) Let X be a transitive G-set. If $x \in X$ and A is a subset of X, then define B as the intersection of those gA, $g \in G$, containing x. Prove that B is a block.

(ii)   Let X be a primitive G-set and let A be a nonempty proper subset of X. If x, y are distinct elements of X, there exists $g \in G$ with $x \in gA$ and $y \notin gA$. [HINT: The block B in part (i) must be $\{x\}$.]

Theorem 9.14   *Every multiply transitive G-set X is primitive.*

Proof   Assume X has a nontrivial block B. Choose distinct elements, $x, y \in B$ and $z \notin B$. There is $g \in G$ with $gx = x$ and $gy = z$, and so B and gB are distinct and intersect, a contradiction.   ∎

Here is a characterization of primitive G-sets.

Theorem 9.15   *Assume X is a transitive G-set. Then X is primitive if and only if, for each $x \in X$, the stabilizer G_x is a maximal subgroup of G.*

Proof†   Assume X is primitive and there exists a subgroup H with $G_x \subsetneqq H \subsetneqq G$. Define $B = Hx$; we claim B is a block. If $g \in G$ and $B \cap gB \neq \varnothing$, then $hx = gh'x$ for some $h, h' \in H$. Thus, $h^{-1}gh' = k \in G_x \subset H$ and $g \in H$; hence $gB = gHx = Hx = B$. It remains to show B is nontrivial to reach a contradiction. Clearly $B = Hx \neq \varnothing$. If $Hx = X$, choose $g \in G$, $g \notin H$; then $gx = hx$ for some $h \in H$, whence $h^{-1}g \in G_x \subset H$ and $g \in H$, contradicting $H \neq G$. Finally, Hx is not a one-point set because $G_x \subsetneqq H$.

Assume every G_x is a maximal (proper) subgroup, yet there exists a nontrivial block B in X. Define a subgroup H of G:

$$H = \{g \in G: gB = B\}.$$

Choose $x \in B$. As B is a block, $G_x \subset H$. Since B is nontrivial, there exists $y \in B$ with $y \neq x$. Transitivity provides $g \in G$ with $gx = y$. Plainly $g \notin G_x$ yet $g \in H$ (for $B \cap gB \neq \varnothing$); hence $G_x \subsetneqq H$. Finally, Theorem 9.12(ii) shows $H \neq G$, otherwise $B = X$. Therefore B cannot exist and X is primitive.   ∎

Assume X is a transitive G-set and $H \lhd G$. In general, X need not be H-transitive. For example, if V is a vector space, then $V^{\#} = V - \{0\}$ is a transitive $GL(V)$-set; if H is the center of $GL(V)$, it is easy to see H does not act transitively on $V^{\#}$.

Theorem 9.16   *If X is a faithful primitive G-set, $H \lhd G$, and $H \neq \{1\}$, then X is H-transitive.*

Proof   We have seen in Exercise 9.15 that every subset Hx, $x \in X$, is a block. Since G acts primitively, either each $Hx = X$ (and X is H-transitive) or $Hx = \{x\}$ for every $x \in X$, contradicting faithfulness and $H \neq \{1\}$.   ∎

Since $GL(V)$ acts faithfully on $V^{\#}$, the example just before Theorem 9.16 shows that $V^{\#}$ is not a primitive $GL(V)$-set.

† We shall give another proof of this theorem after Theorem 9.19.

EXERCISES

9.17. Prove there exist groups G having no faithful primitive G-sets. (HINT: If G does have such a G-set, then its Frattini subgroup $\Phi(G) = \{1\}$.)

*9.18. If G is a solvable group and X is a faithful primitive G-set, then X has degree a power of a prime. (HINT: Apply Theorem 9.16 to a minimal normal subgroup of G.)

We now use these results to obtain simplicity criteria.

Theorem 9.17 *Assume X is a faithful primitive G-set with simple stabilizer G_x. Then either G is simple or X is a regular H-set for every normal subgroup $H \neq \{1\}$.*

Proof Assume $H \neq \{1\}$ is a normal subgroup of G. By Theorem 9.16, X is H-transitive. Therefore, either H acts regularly or $H \cap G_x \neq \{1\}$ for some $x \in X$. But $H \cap G_x \lhd G_x$, so simplicity of G_x gives $H \cap G_x = G_x$, i.e., $G_x \subset H$. By Theorem 9.15, either $H = G_x$ or $H = G$, and the first case cannot occur because H acts transitively. Thus, G is simple. ∎

Definition A normal subgroup H of G for which a G-set X is a regular H-set is called a **regular normal subgroup** of G.

We remark that all regular normal subgroups have the same order, namely, $|X|$.

EXERCISE

*9.19. Let X be a faithful sharply 2-transitive G-set of odd degree. Show that the Frobenius kernel of G is a regular normal subgroup.

Definition If X and Y are G-sets, a function $\theta: X \to Y$ is a **G-map** if

$$\theta(gx) = g\theta(x)$$

for all $g \in G$ and $x \in X$. If θ is also a one-one correspondence, then θ is a **G-isomorphism** and X and Y are **isomorphic** G-sets, denoted by $X \cong Y$.

Usually there is no confusion in saying X is a G-set and not displaying the action of G, but because we now wish to compare two such, let us denote a G-set by (X, φ), where $\varphi: G \to S_X$ is the action of G on X.

Example 9 If G is a group and λ and ρ are the left and right regular representations of G $[\lambda(g): x \mapsto gx$ and $\rho(g): x \mapsto xg^{-1}]$, then (G, λ) and (G, ρ) are isomorphic G-sets. Define $\theta: G \to G$ by $\theta(x) = x^{-1}$. Now

$$\theta(\lambda(g)x) = \theta(gx) = x^{-1}g^{-1} = \theta(x)g^{-1} = \rho(g)\theta(x).$$

Since θ is a one-one correspondence, it is a G-isomorphism.

Example 10   If X and Y are G-sets, their cartesian product $X \times Y$ is a G-set under "diagonal action": $g(x, y) = (gx, gy)$. This construction allows one to give an analog for G-sets of the Chinese Remainder Theorem for rings.

Let G be a finite group and let H and K be subgroups with $HK = G$. Then $G//(H \cap K)$ [all left cosets of $H \cap K$ in G on which G acts by left translation] and $(G//H) \times (G//K)$, the latter with diagonal action, are isomorphic G-sets.

Define $\theta\colon G//(H \cap K) \to (G//H) \times (G//K)$ by $x(H \cap K) \mapsto (xH, xK)$. It is straightforward to verify that θ is a well defined one-one G-map. Since $HK = G$, the product formula shows $[G\colon H \cap K] = [G\colon H][G\colon K]$, and so θ must be onto; therefore, θ is a G-isomorphism.

EXERCISES

9.20.   If X and Y are isomorphic G-sets and X has any of the following properties, then so does Y: faithful; k-transitive ($k \geq 1$); sharply k-transitive ($k \geq 1$); primitive.

*9.21.   Let H be a subgroup of G and let $X = G//H$ be the set of all left cosets of H made into a G-set by left translation. Show that X is a primitive G-set if and only if H is a maximal subgroup of G. (HINT: If B is a block in X containing H, then the union of the cosets in B is a subgroup.)

Of course, different ways of viewing a G-set may be valuable. The next theorem, however, shows that Theorem 3.18 describes all transitive G-sets up to isomorphism.

Theorem 9.18   *Every transitive G-set X is isomorphic to the set $Y = G//H$ of all left cosets of a subgroup H of G on which G acts by left translation.*

Proof   Let $X = \{x_1, \cdots, x_n\}$, let $H = G_{x_1}$, and for each i choose $g_i \in G$ with $g_i x_1 = x_i$ (which is possible because G acts transitively). The routine argument that $\theta\colon X \to Y$ defined by $\theta(x_i) = g_i H$ is a one-one correspondence is left to the reader (recall that $n = |\text{orbit of } x_1| = [G\colon H]$). Is θ a G-isomorphism? If $g \in G$, then $gx_i = x_j$ for some j and

$$\theta(gx_i) = \theta(x_j) = g_j H.$$

On the other hand,

$$g\theta(x_i) = gg_i H.$$

But $gg_i x_1 = gx_i = x_j = g_j x_1$ implies $g_j^{-1} gg_i \in G_{x_1} = H$, hence $g_j H = gg_i H$. ∎

Theorem 9.19
(i) *If H and K are subgroups of a group G, then $G//H$ and $G//K$ (with G-actions left translation) are G-isomorphic if and only if H and K are conjugate subgroups.*

(ii) *Two transitive G-sets (X, φ) and (X, ψ) are G-isomorphic if and only if stabilizers of points in each are conjugate subgroups of G.*

Proof
(i) Assume $\theta: G//H \to G//K$ is a G-isomorphism. There exists $g \in G$ with $\theta(H) = gK$. If $h \in H$, then

$$gK = \theta(H) = \theta(hH) = h\theta(H) = hgK.$$

Therefore, $g^{-1}hg \in K$ and $g^{-1}Hg \subset K$. Since $\theta(g^{-1}H) = g^{-1}\theta(H)$ $= g^{-1}gK = K$, we see that $\theta^{-1}(K) = g^{-1}H$. The argument above now gives $gKg^{-1} \subset H$, whence $g^{-1}Hg = K$.

For the converse, choose $g \in G$ with $g^{-1}Hg = K$. Observe that the following are equivalent for $a, b \in G$: $aH = bH$; $a^{-1}b \in H$; $g^{-1}a^{-1}bg \in g^{-1}Hg = K$; $agK = bgK$. We conclude that the function $\theta: G//H \to G//K$ given by $\theta(aH) = agK$ is well defined and one-one. Clearly θ is onto, for $b \in G$ implies $bK = \theta(bg^{-1}H)$. Finally, θ is a G-map, for $\theta(abH) = (ab)gK$ and $a\theta(bH) = a(bgK)$.

(ii) Let H and K be stabilizers of points in (X, φ) and (X, ψ), respectively. By Theorem 9.18, (X, φ) is G-isomorphic to $G//H$ and (X, ψ) is G-isomorphic to $G//K$. The result now follows easily from part (i). ■

It is now easy to exhibit nonisomorphic transitive G-sets of the same degree: choose two nonconjugate subgroups of G having the same index.

Here is a second proof of Theorem 9.15 using these ideas: A transitive G-set X is primitive if and only if each stabilizer G_x is a maximal subgroup of G. We know $X \cong Y$, where Y is the set of all left cosets of G_x on which G acts by translation. By Exercise 9.21, G_x is a maximal subgroup of G if and only if Y is primitive.

EXERCISES

9.22. If G is nilpotent, a faithful primitive G-set must have prime degree. (HINT: Theorem 9.18 and Exercise 9.21.)

**9.23. If G is solvable, the degree of a faithful primitive G-set is a prime power. (HINT: Exercise 9.18.) Conclude that the index of a maximal subgroup of a solvable group is a prime power.

Lemma 9.20   *Let X be a transitive G-set and let H be a regular normal subgroup of G. Fix $x \in X$ and let G_x act on $H^\#$ by conjugation. Then the G_x-sets $H^\#$ and $X - \{x\}$ are isomorphic.*

Proof   Define $\theta: H^\# \to X - \{x\}$ by $\theta(h) = hx$. If $\theta(h) = \theta(k)$, then $h^{-1}k \in H_x = \{1\}$ (since H acts regularly), and so θ is one-one. Regularity also gives $|H| = |X|$, hence $|H^\#| = |X - \{x\}|$ and θ must be onto. Finally, we show θ is a G-map. If $g \in G_x$ and $h \in H^\#$, denote the action of g on h by $g * h$: thus, $g * h = ghg^{-1}$. Therefore,

$$\theta(g * h) = \theta(ghg^{-1}) = ghg^{-1}x = ghx$$

because $g^{-1} \in G_x$; on the other hand,

$$g\theta(h) = g(hx). \quad \blacksquare$$

Theorem 9.21 *If X is a k-transitive G-set ($k \geq 2$) of degree n and if G has a regular normal subgroup H, then $k \leq 4$. Moreover,*

(i) *if $k = 2$, then H is an elementary abelian p-group for some prime p and $n = p^m$;*

(ii) *if $k = 3$, then $H \cong \mathbf{Z}(3)$ and $n = 3$ or H is an elementary abelian 2-group and $n = 2^m$;*

(iii) *if $k = 4$, then $H \cong \mathbf{V}$ and $n = 4$.*

Proof If X is a k-transitive G-set with $k \geq 2$, then for fixed $x \in X$ we have $X - \{x\}$ a $(k-1)$-transitive G_x-set. By Lemma 9.20, $H^\#$ is a $(k-1)$-transitive G_x-set, where G_x acts on $H^\#$ by conjugation.

(i) Assume $k = 2$. Since conjugation is an (inner) automorphism, it follows easily that all elements of $H^\#$ have the same order which must be a prime p. Thus H is a p-group; since $Z(H)$ is a nontrivial characteristic subgroup, H must be abelian. Finally, $|H| = n$ because H is regular.

(ii) Assume $k = 3$. Thus $H^\#$ is a multiply transitive, hence primitive, G_x-set. If $h \in H^\#$, it is easy to see $\{h, h^{-1}\}$ is a block. By primitivity, either $\{h, h^{-1}\} = H^\#$ has two elements, whence $H \cong \mathbf{Z}(3)$ and $n = 3$, or $\{h, h^{-1}\} = \{h\}$, i.e., h has order 2. Since 3-transitivity implies 2-transitivity, the result of case (i) holds; thus H is an elementary abelian 2-group and $n = |H| = 2^m$.

(iii) Assume $k = 4$. In this case, $k - 1 = 3$ and $|H|^\# \geq 3$ [which excludes $\mathbf{Z}(3)$ and $\mathbf{Z}(2)$]. It follows that H contains a copy of $\mathbf{V}$, say, $\{1, h, k, hk\}$. Now $G_{x,h}$ acts 2-transitively, hence primitively, on $H^\# - \{h\}$. But one sees easily that $\{k, hk\}$ is now a block, whence $H^\# - \{h\} = \{k, hk\}$. We conclude that $H \cong \mathbf{V}$ and $n = |H| = 4$.

Finally, we cannot have $k \geq 5$ for $n = 4 < k$. $\quad \blacksquare$

Of course, the case $k = 4$ actually occurs when $G = S_4$ and $H = \mathbf{V}$. The case $k = 2$ should be compared with Theorem 9.10.

Corollary 9.22 *Let X be a faithful k-transitive G-set, where $k \geq 2$, and assume G_x is simple (for some $x \in X$).*

(i) *If $k \geq 4$, then G is simple.*

(ii) *If $k = 3$ and $|X|$ is not a power of 2, either $G \cong S_3$ or G is simple.*

(iii) *If $k = 2$ and $|X|$ is not a prime power, then G is simple.*

Proof By Theorem 9.17, either G is simple or G contains a regular normal subgroup H. If H exists, Theorem 9.21 implies $k \leq 4$ and, if $k = 4$, that $H \cong \mathbf{V}$ and $|X| = 4$. Now the only 4-transitive subgroup of S_4 is S_4 itself, but the stabilizer of a point is S_3 which is not simple. This proves (i). The other two statements are also immediate consequences of Theorem 9.21 [note that the stabilizer of a point of an S_3-set is the simple group $S_2 \cong \mathbf{Z}(2)$ so that S_3 is a genuine exception in part (ii)]. ∎

Of course, the assumption that the stabilizer of a point is simple does not usually hold; even so, we shall see that Corollary 9.22 is useful.

As an application, here is a second proof of the simplicity of the large alternating groups.

Theorem 9.23 A_n *is simple for* $n \geq 5$.

Proof We do an induction on $n \geq 5$. If $n = 5$, the result is Lemma 3.12. Assume $n \geq 6$. By Theorem 9.11, A_n acts k-transitively on $X = \{1, 2, \cdots, n\}$, where $k = n - 2 \geq 4$. Now $(A_n)_n$, the stabilizer of n, is just A_{n-1} (for it consists of all the even permutations of $\{1, 2, \cdots, n-1\}$) and hence is simple, by induction. Therefore A_n is simple, by Corollary 9.22(i). ∎

Here is another criterion for simplicity, preceded by an easy lemma. Later, we shall use it to give a second proof of simplicity of the large PSL's.

Lemma 9.24 *Let X be a G-set and let N be a subgroup of G acting transitively on X. For each $x \in X$, we have $G = NG_x$.*

Proof If $g \in G$, transitivity of N gives $n \in N$ with $nx = gx$. Hence $n^{-1}g = h \in G_x$ and $g = nh \in NG_x$. ∎

This lemma is the heart of the Frattini argument (Theorem 4.12): If $K \lhd G$ and P is a Sylow p-subgroup of K, then $G = KN_G(P)$. If X denotes the set of all the conjugates of P in K, then normality of K implies G acts on X by conjugation [whence the stabilizer of P is $N_G(P)$] and the Sylow theorems imply K acts transitively on X.

Theorem 9.25 **(Iwasawa, 1941)** *Let X be a G-set. Assume there is $x \in X$ and a subgroup H of G_x such that*

(i) *X is a faithful primitive G-set;*
(ii) *H is an abelian normal subgroup of G_x;*
(iii) *the conjugates $\{gHg^{-1} : g \in G\}$ generate G;*
(iv) *G is **perfect**, i.e., $G = G'$.*

Then G is a simple group.

Proof Assume $N \neq \{1\}$ is a normal subgroup of G. If $g \in G$, condition (iii) gives $g = \prod g_i h_i g_i^{-1}$, where $g_i \in G$ and $h_i \in H$. Now N acts transitively on X, by Theorem 9.16, so that Lemma 9.24 applies to factor each $g_i = n_i s_i$ for

$n_i \in N$ and $s_i \in G_x$. Therefore, normality of H in G_x gives

$$g = \Pi \, n_i s_i h_i s_i^{-1} n_i^{-1} \in NHN \subset HN,$$

and we have $G = HN$. Since H is abelian, $G/N = HN/N \cong H/H \cap N$ is abelian. Therefore $N \supset G' = G$, by condition (iv).† $\blacksquare$

AFFINE GEOMETRY

In this section, we consider groups acting on vector spaces and affine spaces. All vector spaces considered are finite-dimensional.

Notation If K is a field, the (n-dimensional) vector space of all n-tuples over K is denoted by $V(n, K)$. If $K = GF(q)$, we may write $V(n, q)$ instead of $V(n, K)$.

There are two groups acting on a vector space V that come to mind at once: $GL(V) \cong GL(n, K)$ and $SL(V) \cong SL(n, K)$; moreover, each of these acts on $V^{\#} = V - \{0\}$ as well.

Theorem 9.26 *Let V be a vector space of dimension n over a field K. Then $V^{\#}$ is a transitive $GL(V)$-set that is regular when $n = 1$. If $n \geq 2$, then $V^{\#}$ is doubly transitive if and only if $K = GF(2)$.*

Proof That $GL(V)$ acts transitively on $V^{\#}$ follows from the facts: (i) every nonzero vector is part of a basis; (ii) there exists a nonsingular linear transformation taking any basis of V onto any other basis of V. If $n = 1$, only the identity fixes a nonzero vector, so $GL(V)$ acts regularly.

Assume $n \geq 2$. If $K \neq GF(2)$, there exists a dependent set containing two nonzero vectors, namely $\{x, \lambda x\}$, where $x \in V^{\#}$ and $\lambda \in K$, $\lambda \neq 0, 1$. If $\{y, z\}$ is an independent set in V, there is no $g \in GL(V)$ with $g(x) = y$ and $g(\lambda x) = z$, so that $V^{\#}$ is not a doubly transitive $GL(V)$-set. If $K = GF(2)$, however, any pair of nonzero vectors forms an independent set, hence can be extended to a basis, and fact (ii) cited in the first paragraph applies. $\blacksquare$

Here is another group acting on vector spaces.

Definition If V is a vector space and $y \in V$, then **translation** by y is the function $t_y \colon V \to V$ defined by

$$t_y(x) = x + y$$

for all $x \in V$. Let $\mathrm{Tr}(V)$ denote the group of all translations of V under composition.

Definition If V is a vector space, the set of all $a \colon V \to V$ of the form $a = t_{x_0} g$, i.e.,

$$a(x) = g(x) + x_0,$$

† One need assume only that H is solvable, for the proof would then show G/N solvable; this together with $G = G'$ implies $G = N$ (if G is perfect, so is any quotient group, and $\{1\}$ is the only group both solvable and perfect).

where $g \in GL(V)$ and $x, x_0 \in V$, forms a group under composition, denoted by Aff(V), called the **affine group**. When $V = V(n, K)$ we may write Aff(n, K), and when $K = GF(q)$, we may write Aff(n, q).

EXERCISES

**9.24. Tr(V) is an abelian normal subgroup of Aff(V), and Tr(V) is isomorphic to the additive group V.

9.25. Choose a basis of V. To the function $a: V \to V$ with $a(x) = g(x) + x_0$, assign the $(n + 1) \times (n + 1)$ matrix

$$M(a) = \begin{bmatrix} A & B \\ 0 & 1 \end{bmatrix},$$

where A is the matrix of g and B is the column whose entries are the coordinates of x_0. Prove that $M: \text{Aff}(V) \to GL(n + 1, K)$ is a homomorphism that is one-one and whose image is the subgroup of all matrices in $GL(n + 1, K)$ whose bottom row is $[0, 0, \cdots, 0, 1]$.

Theorem 9.27 *A vector space V of dimension n over a field K is a doubly transitive* Aff(V)-*set that is sharply* 2-*transitive when $n = 1$.*

Proof First of all, it is easy to see Tr(V), hence Aff(V), acts transitively on V. Choose $w \in V^\#$. If u and v are distinct vectors in V, it suffices to find $a \in \text{Aff}(V)$ with $a(w) = u$ and $a(0) = v$. Now a must have the form: $a(x) = g(x) + x_0$. Define $x_0 = v$ and choose $g \in GL(V)$ with $g(w) = u - v$ (such a g exists because $w \neq 0$ and $u - v \neq 0$). Therefore, Aff(V) acts doubly transitively on V. Finally, if $n = 1$, the chosen g is unique and the action is sharp. ∎

It is clear that V is never a triply transitive Aff(V)-set.

We hesitated to define affine isomorphisms [even though Aff(V) consists of certain such, called affinities] because we had not yet defined affine spaces. Before giving the definition, we remark that the adjective "affine" means "finite" and its significance will become clear when we discuss projective spaces.

Let us now consider "linear" subsets of a vector space V. Clearly we must not restrict attention to subspaces of V; were the only lines one-dimensional subspaces, all lines would pass through the origin. Here is the obvious definition.

Definition Let V be a vector space over a field K and let S be a subspace of dimension m. An **affine m-subset** of V is a subset of the form $S + v$ for some $v \in V$. The **dimension** of $S + v$ is defined to be m, the dimension of S.

There are special names for certain affine m-subsets: if $m = 0, 1$, or 2, then affine m-subsets are called **points**, **lines**, or **planes**, respectively. If V has dimension n, an affine $(n - 1)$-subset is called a **hyperplane**.

One may view an affine subset $S + v$ in two ways: it is a translate of a subspace $[S + v = t_v(S)]$; it is a coset of the subspace S (with representative v). The usual facts are assembled in the next lemma.

Lemma 9.28 *Let V be a vector space of dimension $n \geq 2$ over a field K.*

(i) *Two distinct lines $L_i = Ku_i + v_i$, $i = 1, 2$, are either disjoint or intersect in a unique point.*

(ii) *Two distinct points $u, v \in V$ lie on a unique line L, namely, $L = K(u - v) + v$.*

(iii) *(Parallelogram law) If $\{u, v\}$ is independent, then*

$$\{u + v\} = (Ku + v) \cap (Kv + u).$$

(iv) *Two distinct hyperplanes $H + x$ and $J + y$ are disjoint if and only if $H = J$. In particular, if $n = 2$, then distinct lines $Ku_1 + v_1$ and $Ku_2 + v_2$ are disjoint if and only if $Ku_1 = Ku_2$.*

Proof

(i) Assume $z \in L_1 \cap L_2$. As lines are cosets, Exercise 2.12 says $L_1 \cap L_2 = (Ku_1 \cap Ku_2) + z$. Now $Ku_1 = Ku_2$ cannot occur lest the distinct lines L_1 and L_2 be distinct cosets of Ku_1, hence disjoint. Therefore $Ku_1 \cap Ku_2 = \{0\}$ and $L_1 \cap L_2 = \{z\}$.

(ii) It is clear $K(u - v) + v$ is a line containing u and v; a second such line intersects this one in at least two points, contradicting (i).

(iii) The statement is suggested by the "parallelogram law" depicting addition of vectors in the plane.

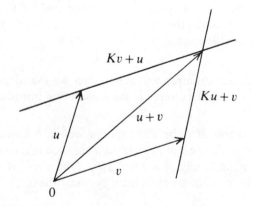

To prove (iii), first note that independence of $\{u, v\}$ implies the lines $Ku + v$ and $Kv + u$ are distinct. Visibly, $u + v \in (Ku + v) \cap (Kv + u)$, so the result follows from (i).

(iv) If $H = J$, then distinct hyperplanes are distinct cosets of H, hence are disjoint. If $H \neq J$, then $H + J = V$ (because H and J are hyperplanes). Hence $y - x = h + j$ for $h \in H$ and $j \in J$. Therefore $h + x = -j + y \in (H + x) \cap (J + y)$. The last remark follows because a hyperplane in a two-dimensional space is a line. ∎

It is necessary to focus attention on the affine subsets of a vector space (its "geometry"), ignoring the presence of vector addition and scalar multiplication. We thus define an affine space to be a "model" of this aspect of a vector space.

Definition Let A be a set for which there is an integer n and, for each m with $0 \leq m \leq n$, a family $L_m(A)$ of subsets of A. An **affine n-space over K** is an ordered pair (A, α) such that

(i) $\alpha: V \to A$ is a one-one correspondence from an n-dimensional vector space V over K;

(ii) a subset S of V is an affine m-subset if and only if $\alpha(S) \in L_m(A)$.

We call V the **associated vector space** and n the **dimension** of (A, α). A subset S in $L_m(A)$ is called an **affine m-subspace**.

In brief, one thinks of $L_0(A)$ as the points in A, $L_1(A)$ as the lines in A, and so forth.

Example 11 Every vector space V is an affine space (associated to itself via $\alpha = 1_V$) in which $L_m(V) = \{$all affine m-subsets$\}$.

Example 12 Let V be a vector space and let $L_m(V) = \{$all affine m-subsets$\}$. For fixed $x \in V$, let $t_x: V \to V$ be translation by x. Then (V, t_x) is an affine space with associated vector space V.

Example 13 Let V be a vector space, S a subspace, and $x \in V$. Define $A = S + x$, define $L_m(A) = \{T + x: T$ is an affine m-subset of $S\}$, and define $\alpha: S \to A$ by $\alpha(s) = s + x$. Then (A, α) is an affine space.

One more important example will arise when we examine projective spaces (indeed, this as yet undefined example is the reason one introduces affine spaces).

Definition Let (A, α) and (B, β) be affine spaces over K. A function $h: A \to B$ is an **affine isomorphism** if h is a one-one correspondence such that a subset S of A lies in $L_m(A)$ for some m if and only if $h(S) \in L_m(B)$. One says (A, α) and (B, β) are **isomorphic** if there is an affine isomorphism between them.

Definition All affine isomorphisms of (A, α) to itself (called **affine automorphisms**) form a group under composition, denoted by $\mathrm{Aut}(A)$, called the **automorphism group**.

If (A, α) is an affine space with associated vector space V, and if V is construed as the affine space $(V, 1_V)$ of Example 11, then it is easy to see the function $\mathrm{Aut}(A) \to \mathrm{Aut}(V)$ defined by $h \mapsto \alpha^{-1} h \alpha$ is a group isomorphism. Therefore, up to isomorphism, the group $\mathrm{Aut}(A)$ does not depend on the function α.

Notation   If $V = V(n, K)$, we may write Aut(n, K) instead of Aut(V); when $K = GF(q)$, we may write Aut(n, q).

Here are some obvious examples of affine isomorphisms.

Definition   Let (A, α) and (B, β) be affine spaces over K with associated vector spaces V and W, respectively. A function $f: A \rightarrow B$ is an **affinity** if $f = \beta^{-1} g \alpha$ for some nonsingular linear transformation $g: V \rightarrow W$. All affinities of (A, α) to itself form a group under composition, denoted by $GL(A)$.

Three remarks are necessary. First, every affinity is an affine isomorphism, whence $GL(A)$ is a subgroup of Aut(A). Second, the function $GL(A) \rightarrow GL(V)$ defined by $f \rightarrow \alpha^{-1} f \alpha$ is a group isomorphism, so that the notation $GL(A)$ agrees with the usual notation when both are defined, i.e., when A is a vector space. Third, one may use Examples 11 and 12 to construe every $a \in \text{Aff}(V)$ as an affinity (V, t_{-y}) $\rightarrow (V, 1_v)$; of course, t_{-y} depends on a.

Theorem 9.29   *Two affine spaces over K are isomorphic if and only if they have the same dimension.*

Proof   Let (A, α) and (B, β) be affine spaces over K with associated vector spaces V and W, respectively. It is obvious that isomorphic affine spaces have the same dimension, for V and W are isomorphic. Conversely, if (A, α) and (B, β) have dimension n, then V and W have dimension n, and there is a nonsingular linear transformation $g: V \rightarrow W$. The function $f: A \rightarrow B$ defined by $f = \beta g \alpha^{-1}$ is an affine isomorphism, even an affinity, and (A, α) and (B, β) are isomorphic.   ∎

We may now return to vector spaces, for every affine space is isomorphic to a vector space (viewed as an affine space over itself). To see affine isomorphisms that are neither linear transformations nor translations, we need a short digression.

Assume $\sigma: K \rightarrow K$ is an automorphism of a field K (thus, σ is a permutation with $\sigma(\lambda + \mu) = \sigma(\lambda) + \sigma(\mu)$, $\sigma(\lambda \mu) = \sigma(\lambda) \sigma(\mu)$, and $\sigma(1) = 1$). If $K = GF(q) = GF(p^r)$, we have seen in Theorem 8.9 that Aut(K) is a cyclic group of order r with generator $\sigma: \lambda \mapsto \lambda^p$. We remark that the real field **R** has no automorphisms other than the identity [if $\sigma: \mathbf{R} \rightarrow \mathbf{R}$ is an automorphism, it fixes every rational and is necessarily continuous (σ preserves the relation $\lambda < \mu$ because a real number is positive if and only if it is a square)]. The complex field **C** does have automorphisms, e.g., complex conjugation.

Definition   Let V and W be vector spaces over a field K. A function $g: V \rightarrow V$ is a **semilinear transformation**† if there exists $\sigma \in \text{Aut}(K)$ such that

$$g(x + y) = g(x) + g(y)$$

---

† In the special case $K = \mathbf{C}$ and $\sigma =$ complex conjugation, f is sometimes called "conjugate linear" or "sesquilinear".

and

$$g(\lambda x) = \sigma(\lambda)g(x)$$

for every $x, y \in V$ and $\lambda \in K$.

Of course, every linear transformation is semilinear (with σ the identity).

Example 14 Choose a basis of V so that each $x \in V$ has coordinates: $x = (x_1, \cdots, x_n)$. If $\sigma \in \text{Aut}(K)$, define $\sigma_*: V \to V$ by

$$\sigma_*(x_1, \cdots, x_n) = (\sigma(x_1), \cdots, \sigma(x_n));$$

it is a simple matter to check that σ_* is semilinear.

Definition A semilinear transformation is **nonsingular** if it is a one-one correspondence.

EXERCISES

9.26. If $f: V \to W$ is semilinear with $f(\lambda x) = \sigma(\lambda)f(x)$, then the automorphism σ is uniquely determined by f.

9.27. The semilinear transformation σ_* of Example 14 is nonsingular and it depends on the choice of basis of V.

*9.28. The composite of semilinear transformations is semilinear (if f and g are semilinear with field automorphisms σ and τ respectively, then gf is semilinear with field automorphism $\tau\sigma$). If a semilinear transformation is nonsingular, then its inverse is also semilinear.

Definition $\Gamma L(V)$ is the group of all nonsingular semilinear transformations of V to itself under composition. If $V = V(n, K)$, we may write $\Gamma L(V) = \Gamma L(n, K)$, and when $K = GF(q)$ we may write $\Gamma L(n, K) = \Gamma L(n, q)$.

That $\Gamma L(V)$ is a group is immediate from Exercise 9.28.

The point of this digression is that, obviously, every nonsingular semilinear transformation is an affine isomorphism. It follows that if $h: V \to W$ is a one-one correspondence of the form

$$h(x) = g(x) + w_0,$$

where $x, w_0 \in V$ and g is nonsingular semilinear, then h is an affine isomorphism. The remarkable fact is that every affine isomorphism has this form (Theorem 9.31).

Recall that the subspace of a vector space V generated by a subset X is denoted by $\langle X \rangle$.

Lemma 9.30 *Let V and Y be vector spaces over a field K, let $h: V \to Y$ be an affine isomorphism with $h(0) = 0$, and let W be a subspace of V with basis $\{u, v\}$.*

(i) $h(Ku) = Kh(u);$
(ii) $\{h(u), h(v)\}$ is independent;
(iii) $h(W) = \langle h(u), h(v) \rangle;$
(iv) $h|W: W \to h(W)$ is an affine isomorphism.

Proof

(i) Ku is the line containing u and 0; thus $h(Ku)$ is the line containing $h(u)$ and $h(0) = 0$, namely, $Kh(u)$.

(ii) Were $\{h(u), h(v)\}$ dependent, then the points $h(u)$, $h(v)$, 0 would be collinear; applying the affine isomorphism h^{-1} gives the contradiction $u, v, 0$ collinear.

(iii)† For each ordered pair $\lambda, \mu \in K$ (not both 0), let $L(\lambda, \mu)$ denote the line containing λu and μv. It is easy to see

$$W = \bigcup_{\lambda, \mu} L(\lambda, \mu);$$

therefore

$$h(W) = \bigcup_{\lambda, \mu} h(L(\lambda, \mu)).$$

Consider a fixed $L(\lambda, \mu)$. By (i), $h(\lambda u) = \alpha h(u)$ and $h(\mu v) = \beta h(v)$ for some $\alpha, \beta \in K$. Therefore $h(\lambda u)$ and $h(\mu v)$, hence $h(L(\lambda, \mu))$, lie inside $\langle h(u), h(v) \rangle$. We conclude that $h(W) = h(\langle u, v \rangle) \subset \langle h(u), h(v) \rangle$. For the reverse inclusion, apply h^{-1} to $\langle h(u), h(v) \rangle$ [noting that, by (ii), $\{h(u), h(v)\}$ is independent] and obtain $h^{-1}(\langle h(u), h(v) \rangle) \subset \langle u, v \rangle = W$; hence $\langle h(u), h(v) \rangle \subset h(\langle u, v \rangle) = h(W)$.

(iv) This is immediate since we now know $h(W)$ is a vector space having the same dimension as W. ∎

Theorem 9.31 *Assume S and T are isomorphic vector spaces of dimension at least 2 over a field K. Then every affine isomorphism $h: S \to T$ has the form $h = t_y g$, i.e.,*

$$h(x) = g(x) + y,$$

where $x \in S$, $y \in T$, and $g: S \to T$ is a nonsingular semilinear transformation.

REMARK One must assume dim $S \geq 2$, for any one-one correspondence between one-dimensional spaces is an affine isomorphism.

Proof Composing with the translation $x \mapsto x - h(0)$, we may assume $h(0) = 0$, and it now suffices to prove h is semilinear. Because h is a one-one correspondence, it preserves intersections; in particular, $h(L_1 \cap L_2) = h(L_1) \cap h(L_2)$ if L_1 and L_2 are lines.

Assume $\{u, v\}$ is independent; we claim

$$h(Ku + v) = Kh(u) + h(v).$$

---

† The definition of affine isomorphism says $h(W)$ is an affine 2-subset of V; hence $h(W) = \langle hu, hv \rangle$ since it contains 0. We give this longer proof which assumes only that h preserves lines.

By Lemma 9.30(iii), both $h(Ku+v)$ and $h(Ku)$ are lines contained in $\langle h(u), h(v) \rangle$; indeed, they are disjoint because $Ku+v$ and Ku are disjoint. Since $h(Ku) = Kh(u)$, by Lemma 9.30(i), we may apply Lemma 9.28(iv) to $\langle h(u), h(v) \rangle$ to obtain

$$h(Ku+v) = Kh(u) + y$$

for some y. In particular, there are scalars α, $\beta \in K$ with $h(v) = \alpha h(u) + y$ and $h(\lambda u + v) = \beta h(u) + y$, where $\lambda \in K$ and β depends on λ. Thus,

$$\begin{aligned} h(\lambda u + v) &= \beta h(u) + h(v) - \alpha h(u) \\ &= (\beta - \alpha)h(u) + h(v) \in Kh(u) + h(v). \end{aligned}$$

Therefore $h(Ku+v) \subset Kh(u) + h(v)$ and equality follows because both are lines.

We first prove $h(u+v) = h(u) + h(v)$ when $\{u, v\}$ is independent. By Lemma 9.28 (iii), we know $\{u + v\} = (Ku+v) \cap (Kv+u)$. Because h preserves intersections,

$$\begin{aligned} \{h(u+v)\} &= h(Ku+v) \cap h(Kv+u) \\ &= [Kh(u)+h(v)] \cap [Kh(v)+h(u)] = \{h(u)+h(v)\}, \end{aligned}$$

the last equality because $\{h(u), h(v)\}$ is independent [Lemma 9.30(ii)]. It remains to evaluate $h(\lambda u + \mu u)$; we do this in two steps. Choose w so that $\{u, w\}$ is independent. Then $\{u+w, -u\}$ is independent and

$$h(w) = h((u+w) - u) = h(u+w) + h(-u) = h(u) + h(w) + h(-u).$$

It follows that $h(-u) = -h(u)$. Consequently, if $\lambda + \mu = 0$, we have $h(\lambda u + \mu u) = 0 = h(\lambda u) + h(\mu u)$. Finally, assume $\lambda + \mu \neq 0$. One checks that $\{\lambda u + w, \mu u - w\}$ is independent and so

$$\begin{aligned} h(\lambda u + \mu u) &= h(\lambda u + w) + h(\mu u - w) \\ &= h(\lambda u) + h(w) + h(\mu u) + h(-w) = h(\lambda u) + h(\mu u). \end{aligned}$$

We have proved that h is additive.

If $u \neq 0$ and $\lambda \in K$, then $h(Ku) = Kh(u)$ implies there is $\sigma_u(\lambda) \in K$ with $h(\lambda u) = \sigma_u(\lambda)h(u)$. Plainly $\sigma_u \colon K \to K$ is a permutation [because $h(Ku) = Kh(u)$] with $\sigma_u(1) = 1$. Additivity of h implies

$$\begin{aligned} \sigma_u(\lambda + \mu)h(u) &= h((\lambda + \mu)u) = h(\lambda u + \mu u) \\ &= h(\lambda u) + h(\mu u) = [\sigma_u(\lambda) + \sigma_u(\mu)]h(u); \end{aligned}$$

since $h(u) \neq 0$, we see that σ_u is additive.

Next we show that σ_u does not depend on u. Choose w so that $\{u, w\}$ is independent. Now

$$h(\lambda u + \lambda w) = h(\lambda u) + h(\lambda w) = \sigma_u(\lambda)h(u) + \sigma_w(\lambda)h(w).$$

On the other hand,

$$h(\lambda u + \lambda w) = \sigma_{u+w}(\lambda)h(u+w) = \sigma_{u+w}(\lambda)[h(u) + h(w)].$$

Equating coefficients,

$$\sigma_u(\lambda) = \sigma_{u+w}(\lambda) = \sigma_w(\lambda).$$

Lastly, if $\mu \in K^{\#}$, then $\{\mu u, w\}$ is independent and we have, with μu in place of u, that $\sigma_{\mu u}(\lambda) = \sigma_w(\lambda)$.

It remains to show $\sigma\colon K \to K$ is multiplicative (the subscript may now be omitted). But

$$h(\lambda \mu u) = \sigma(\lambda \mu) h(u),$$

and also

$$h(\lambda \mu u) = \sigma(\lambda) h(\mu u) = \sigma(\lambda)\sigma(\mu)h(u).$$

Therefore h is semilinear. ∎

The astute reader has observed that the proof used the fact that $h(L)$ is a line whenever L is a line, but not that $h(S)$ is an affine subset for any higher dimensional affine subset S. The hypothesis of Theorem 9.31 can thus be weakened.

Corollary 9.32   Let (A, α) and (B, β) be affine spaces over K of dimension ≥ 2 with associated vector spaces V and W, respectively. If $f\colon A \to B$ is an affine isomorphism, then

$$f = \beta t_z g \alpha^{-1},$$

where $g\colon V \to W$ is a nonsingular semilinear transformation and $t_z\colon W \to W$ is translation by $z = \beta^{-1} f\alpha(0)$.

Proof   If $g'\colon V \to W$ is defined by $g' = \beta^{-1} f\alpha$, then g' is an affine isomorphism; if $g = t_{-z}g'$, then $g\colon V \to W$ is an affine isomorphism with $g(0) = 0$. By Theorem 9.31, g is a nonsingular semilinear transformation. Therefore,

$$g = t_{-z}g' = t_{-z}\beta^{-1}f\alpha$$

and the result follows. ∎

Theorem 9.33   Let V be a vector space over a field K.

(i) $\text{Aff}(V) \subset \text{Aut}(V)$, and $\text{Aut}(V)$ acts doubly transitively on V.

(ii) If $\text{Aut}(K) = \{1\}$ (e.g., $K = GF(p)$ for prime p, $K = \mathbf{Q}$ or $K = \mathbf{R}$), then $\text{Aff}(V) = \text{Aut}(V)$.

(iii) $\text{Aut}(V)$ is a semidirect product of $\text{Tr}(V)$ by $\Gamma L(V)$.

(iv) If $V = V(n, q)$, then $|\text{Aut}(n, q)| = q^n |\Gamma L(n, q)|$.

Proof

(i) That $\text{Aff}(V) \subset \text{Aut}(V)$ is obvious from their definitions; the second statement results from the double transitivity of the action of $\text{Aff}(V)$ (Theorem 9.27).

(ii) If $\text{Aut}(K) = \{1\}$, every semilinear transformation is linear.

(iii) If $h \in \text{Aut}(V)$, then h has the form $h(x) = g(x) + x_0$; define

π: Aut$(V) \to \Gamma L(V)$ by π: $h \mapsto g$. It is easy to see that ker $\pi = $ Tr(V) and that π fixes $\Gamma L(V)$ pointwise, whence π is a retraction (Exercise 7.33) and Aut(V) is a semidirect product as described.

(iv) By Exercise 9.24, Tr$(V) \cong V$, so when $V = V(n,q)$, $|$Aut$(n,q)|$ $= |$Tr$(V)| \, |\Gamma L(n,q)| = q^n |\Gamma L(n,q)|$. ∎

Theorem 9.34 *If V is a vector space over a field K, then $\Gamma L(V)$ is a semidirect product of $GL(V)$ by* Aut(K). *If $V = V(n,q)$ and $q = p^r$, then*

$$|\Gamma L(n,q)| = r|GL(n,q)|.$$

Proof Each nonsingular semilinear transformation h determines a unique automorphism σ of K. The function θ: $\Gamma L(V) \to$ Aut(K) defined by $h \mapsto \sigma$ is a homomorphism onto Aut(K) having kernel $GL(V)$. As in Example 14, choose a basis of V and, for each $\sigma \in$ Aut(K), consider the semilinear transformation σ_*. It is easy to see the set of all such σ_* is a subgroup of $\Gamma L(V)$ intersecting $GL(V)$ trivially. Therefore $\Gamma L(V)$ is a semidirect product. When $V = V(n,q)$, then

$$|\Gamma L(n,q)| = |\text{Aut}\,(GF(p^r))|\,|GL(n,q)| = r|GL(n,q)|,$$

by Theorem 8.9. ∎

We remark that $|$Aut$(n,q)| = rq^n|GL(n,q)|$; since $|GL(n,q)|$ has been computed in Theorem 8.10, we have explicit formulas for $|$Aut$(n,q)|$ and $|\Gamma L(n,q)|$.

EXERCISES

9.29. Show that Aff(V) is a semidirect product of Tr(V) by $GL(V)$ and that $|$Aff$(n,q)| = q^n|GL(n,q)|$.

9.30. Let (V, α) and (W, β) be affine spaces over K in which V and W are vector spaces over K. Show that every affine isomorphism f: $V \to W$ has the form $f = t_w g t_v$, where $w \in W$, $v \in V$ and g: $V \to W$ is a nonsingular semilinear transformation.

9.31. Choose a basis of V and, for $\sigma \in$ Aut(K), let σ_*: $V \to V$ be the semilinear transformation of Example 14. If $g \in GL(V)$, let $[\lambda_{ij}]$ be the matrix of g with respect to the chosen basis. Show that $\sigma_* g \sigma_*^{-1}$ is linear and has matrix $[\sigma(\lambda_{ij})]$. Conclude that det$(\sigma_* g \sigma_*^{-1}) = \sigma(\det g)$.

9.32. Show that $SL(V) \lhd \Gamma L(V)$.

9.33. Show that $Z(n, K) \lhd \Gamma L(n, k)$, where $Z(n, K)$ is the center of $GL(n, K)$ consisting of all nonzero scalar matrices.

9.34. Prove that if $q = p^r$, then Aut$(1, q)$ is a solvable group of order $rp^r(p^r - 1)$ and that $GF(q)$ is a doubly transitive Aut$(1, q)$-set. [Huppert (1957) has shown that if G is solvable and there exists a faithful doubly transitive G-set, then, with only finitely many exceptions, G is a subgroup of some Aut$(1, q)$.]

PROJECTIVE GEOMETRY

We now turn from affine geometry to projective geometry. The reader is aware that projective geometry involves adjoining "points at infinity" to the "finite points" of an affine space (whence the adjective "affine"). The need for projective geometry was first felt by artists who were obliged to understand perspective in order to paint replicas of three-dimensional space on two-dimensional canvas. Indeed, if one regards a viewer's eyes as a vertex, then the problem of drawing in perspective amounts to analyzing conical projections from this vertex onto planes (see Pedoe, p. 2); this is the origin of the adjective "projective".

The definition of projective space will involve a construction. After some elementary discussion, we shall see (Theorem 9.39) that the construction does yield an enlargement of affine space.†

Let V be a vector space over a field K. Define an equivalence relation on $V^{\#} = V - \{0\}$ by $x \sim y$ if there exists $\lambda \in K^{\#}$ with $y = \lambda x$. If $x \in V^{\#}$, denote its equivalence class by $[x]$.

Definition   If V is a vector space over K of dimension $n + 1$, the set of equivalence classes

$$P(V) = \{[x]: x \in V^{\#}\}$$

is called a **projective n-space**; one says $P(V)$ has (projective) **dimension** n.

As with affine space, projective space has lost the algebraic operations possessed by vector spaces: one can neither add points of $P(V)$ nor multiply them by scalars. These operations can be restored to an affine space (A, α) via the function α. This is not possible with projective space, and one must be content with an algebraic vestige: "projective subspaces".

Definition   Let $\varphi: V^{\#} \to P(V)$ be the natural map $x \mapsto [x]$. A **projective m-subspace** is a subset S of $P(V)$ of the form $\varphi(W^{\#})$, where W is an $(m + 1)$-dimensional subspace of V. One says S has **dimension** m.

Here, too, there are special names for certain subspaces: if $m = 0$, 1, or 2, a projective m-subspace is a **projective point**, **projective line**, or **projective plane**, respectively. If $P(V)$ has dimension n, a projective $(n-1)$-subspace is called a **projective hyperplane**.

The reason for lowering dimension in passing from V to $P(V)$ is now apparent: a line in V (through the origin) becomes a projective point; a plane in V (through the origin) becomes a projective line, and so forth.

The following easy result is fundamental.

Theorem 9.35   *Assume V is a vector space over a field K and $\dim V = n \geq 2$.*

(i) *If $x, y \in V$, then $[x] \neq [y]$ if and only if $\{x, y\}$ is independent.*

---

† Our exposition is influenced by that in Gruenberg and Weir.

(ii) *Every two distinct points in $P(V)$ lie on a unique projective line.*
(iii) *If Ω is a projective hyperplane in $P(V)$ and L is a projective line not contained in Ω, then Ω and L intersect in a unique point.*

Proof

(i) If $[x] \neq [y]$, then $x \neq \lambda y$ for $\lambda \in K^{\#}$ and $\{x, y\}$ is independent; the converse is just as obvious.

(ii) Let $[x] \neq [y]$ be points in $P(V)$ and choose $x, y \in V^{\#}$ representing each. A projective line L containing $[x]$ and $[y]$ must have the form $\varphi(W^{\#})$, where W is a two-dimensional subspace of V containing x and y. The set $\{x, y\}$ is independent, by (i), so that $\langle x, y \rangle$ is two-dimensional. This proves the existence and uniqueness of L.

(iii) Let $L = \varphi(W^{\#})$ and $\Omega = \varphi(H^{\#})$, where dim $W = 2$ and dim $H = n - 1$. Since L is not contained in Ω, the subspace W is not contained in H. Therefore $W + H = V$ and

$$\dim(W \cap H) = \dim W + \dim H - \dim(W + H)$$
$$= 2 + (n - 1) - n = 1.$$

Hence $\varphi((W \cap H)^{\#})$ is a projective point. ∎

Here are the analogs of affine isomorphisms (one could call them projective isomorphisms).

Definition If V and W are vector spaces over K, a **collineation** $\theta: P(V) \to P(W)$ is a one-one correspondence such that a subset S of $P(V)$ is a projective m-subspace if and only if $\theta(S)$ is a projective m-subspace of $P(W)$. Two projective spaces are **isomorphic** if there is a collineation between them.

Notation If $h: V \to W$ is a nonsingular semilinear transformation, then h induces a collineation, denoted by $P(h): P(V) \to P(W)$, defined by $[x] \mapsto [h(x)]$.

Definition A collineation of the form $P(h)$ as described above is called a **projectivity** if h is a nonsingular *linear* transformation.

Thus, projectivities are analogous to affinities in that each arises from linear transformations. We shall see that the only other collineations arise from semilinear transformations (Theorem 9.41).

Theorem 9.36 *Two projective spaces $P(V)$ and $P(W)$ are isomorphic if and only if they have the same dimension.*

Proof Necessity is obvious. For sufficiency, equality of the dimensions of $P(V)$ and $P(W)$ implies equality of the dimensions of V and W. There is thus a (linear) isomorphism $h: V \to W$ inducing a collineation $P(h)$ (which is even a projectivity) between $P(V)$ and $P(W)$. ∎

In view of this theorem, we may introduce notation for "the" projective n-space over a field K.

Notation   If $V = V(n+1, K)$, write $P^n(K)$ for $P(V)$. When $K = GF(q)$, we may write $P^n(q)$ for $P^n(K)$.

Theorem 9.37

(i) *For every $n \geq 0$ and every prime power q,*

$$|P^n(q)| = q^n + q^{n-1} + \cdots + q + 1.$$

In particular, every projective line has exactly $q+1$ points.

(ii) *The number of projective lines in the projective plane $P^2(q)$ is the same as the number of points in $P^2(q)$, namely, $q^2 + q + 1$.*

Proof

(i)   To compute $|P^n(q)|$, note that if $V = V(n+1, q)$, then $|V^{\#}| = q^{n+1} - 1$. Since $V^{\#}$ is partitioned into classes of the form $[x]$ each of which has $q-1$ elements ($[x] = \{\lambda x : \lambda \neq 0\}$), we have

$$|P^n(q)| = (q^{n+1} - 1)/(q - 1) = q^n + q^{n-1} + \cdots + q + 1.$$

Setting $n = 1$ in this formula gives the number of points on a projective line.

(ii)† Take a line L and a point $x \notin L$. For each point $y \in L$, there is a line joining x and y, so we have exhibited $q + 1$ lines through x. There can be no other lines through x, for every line through x meets L [Theorem 9.35(iii)].

   For each of the $q + 1$ points on L, there are q lines through it distinct from L. This exhibits $(q+1)q + 1$ lines in $P^2(q)$ (the extra "one" counts L). There can be no other lines, since every two lines meet.   ∎

We shall use the next elementary lemma about vector spaces to prove the basic theorem relating affine and projective spaces which asserts a projective space arises from an affine space by adjoining a "hyperplane at infinity".

Lemma 9.38   *Assume H is a subspace of a vector space V and that $y \notin H$. If T is a subspace of H with $T + u \subset H$ for some $u \in H$, then*

(i) $\dim \langle T + u + y \rangle = 1 + \dim T;$
(ii) $\langle T + u + y \rangle \cap (H + y) = T + u + y.$

Proof

(i) Since $u \in H$ and $y \notin H$, we have $u + y \notin H$ and $\langle T + u + y \rangle = T \oplus \langle u + y \rangle$.
(ii) Clearly $T + u + y$ lies in the left side. For the reverse inclusion, assume $t + \lambda(u + y) = h + y$, where $t \in T$ and $h \in H$. Then $(1 - \lambda)y = t + \lambda u - h \in H$. Since $y \notin H$, we must have $\lambda = 1$ and so $h + y = t + u + y \in T + u + y$.   ∎

---

† There is a more elegant proof based on a duality present in projective planes. Also see the footnote to Theorem 9.58.

Theorem 9.39   *Let $P(V)$ be a projective n-space over K. Assume Ω is a projective hyperplane in $P(V)$, and let H be the linear subspace of V with $\varphi(H^{\#}) = \Omega$. If $A = P(V) - \Omega$, the complement of Ω, then there exists $\alpha\colon H \to A$ so that (A, α) is an affine n-space over K.*

Moreover, every (affine) line L in A lies in a unique projective line L^* in $P(V)$, and L^* intersects Ω in a unique point.

Proof   Recall that the natural map $\varphi\colon V^{\#} \to P(V)$ is defined by $x \mapsto [x]$. Assume B is a nonempty subset of V such that $b \in B$ implies $\lambda b \in B$ whenever $\lambda \in K^{\#}$ (e.g., $B = W^{\#}$ has this property for any linear subspace W of V); we claim $\varphi(B \cap C) = \varphi(B) \cap \varphi(C)$ for every subset C of V. It is trivial to see the left side is contained in the right side. For the reverse inclusion, assume $[b] = [c] \in \varphi(B) \cap \varphi(C)$; then $c = \lambda b \in B \cap C$ and $[c] \in \varphi(B \cap C)$.

Choose $y \in V$, $y \notin H$. Define $\alpha\colon H \to A$ by

$$\alpha(h) = \varphi(h + y) = [h + y]$$

[note that $h + y \neq 0$ since $y \notin H$, so that $\varphi(h + y)$ is defined; moreover, $\varphi(h + y) \in A$ since $\varphi(h + y) \notin \Omega = \varphi(H^{\#})$].

Now α is one-one, for no two elements of the form $h + y$ (with $h \in H$) are equivalent. To see that α is onto, take $[x] \in A$. Since H is a linear hyperplane in V, we have $x = h + \lambda y$ for some $h \in H$ and $\lambda \in K$; moreover, $\lambda \neq 0$ lest $x \in H$ and $[x] \in \Omega$. Therefore $x \sim \lambda^{-1} x = \lambda^{-1} h + y$ and $[x] = \alpha(\lambda^{-1} h)$.

For $0 \leq m \leq n$, define

$$L_m(A) = \{ A \cap \varphi(W^{\#})\colon W \text{ is an } (m+1)\text{-dimensional linear subspace of } V \}.$$

Thus, a subset is in $L_m(A)$ if it is the intersection of A with a projective m-subspace of $P(V)$. It suffices to show a subset S of H is an affine m-subset if and only if $\alpha(S) \in L_m(A)$.

Assume $S = T + u$ is an affine m-subset of H (so $u \in H$ and T is an m-dimensional linear subspace of H). By Lemma 9.38,

$$T + u + y = \langle T + u + y \rangle \cap (H + y).$$

Using the first paragraph of this proof and the observation $0 \notin H + y$, we see

$$\begin{aligned}
\alpha(S) = \alpha(T + u) = \varphi(T + u + y) &= \varphi(\langle T + u + y \rangle \cap (H + y)) \\
&= \varphi(\langle T + u + y \rangle^{\#}) \cap \varphi(H + y) \\
&= \varphi(\langle T + u + y \rangle^{\#}) \cap A.
\end{aligned}$$

Since $\dim \langle T + u + y \rangle = m + 1$, it follows that $\alpha(S) \in L_m(A)$.

Assume S is a nonempty subset of H with $\alpha(S) = \varphi(S + y) \in L_m(A)$. By definition of $L_m(A)$, there is an $(m + 1)$-dimensional linear subspace W of V with

$$\varphi(S + y) = \varphi(W^{\#}) \cap A.$$

But $\varphi(W^{\#}) \cap A = \varphi(W^{\#}) \cap \varphi(H + y) = \varphi(W^{\#} \cap (H + y)) = \varphi(W \cap (H + y))$, by the first paragraph and the observation that $0 \notin H + y$. Since α is one-one,

$\varphi|(H+y)$ is one-one and $\varphi(S+y) = \varphi(W \cap (H+y))$ implies

$$S + y = W \cap (H+y).$$

We deduce that $W \not\subset H$ lest the right-hand side be empty. Also, Exercise 2.12 gives $W \cap (H+y) = (W \cap H) + z$, where $z = u + y \in H + y$. Therefore $S + y = (W \cap H) + u + y$ and $S = (W \cap H) + u$. We have shown that S is an affine subset of H; it remains to compute its dimension. Since $W \not\subset H$ and H is a linear hyperplane of V, we have $V = W + H$. Therefore

$$\begin{aligned} \dim (W \cap H) &= \dim W + \dim H - \dim (W + H) \\ &= (m+1) + n - (n+1) = m, \end{aligned}$$

and S is an affine m-subset of H, as desired.

Let L be a line in A, and choose distinct points $[u + y]$ and $[v + y]$ on L. Thus, $L = \alpha(\langle u - v \rangle + v) = \varphi(\langle u - v \rangle + v + y)$. By Theorem 9.35(ii), there is a unique projective line L^* containing $[u + y]$ and $[v + y]$, namely, $L^* = \varphi(W^\#)$, where $W = \langle u + y, \ v + y \rangle$. Since $\langle u - v \rangle + v + y \subset \langle u + y, v + y \rangle$, it follows that $L \subset L^*$. Visibly L^* is not contained in Ω, so Theorem 9.35(iii) says $L^* \cap \Omega$ is a unique point. ∎

EXERCISE

9.35. The function $\alpha: H \to V$ constructed in Theorem 9.39 depends on the choice of $y \notin H$. Show that if $y' \notin H$ and $\alpha': H \to V$ is the corresponding function, then the affine spaces (A, α) and (A, α') are isomorphic.

There is another way to view Theorem 9.39, using **homogeneous coordinates**. Assume a basis of V has been chosen so that each $x \in V$ has coordinates $x = (\lambda_0, \lambda_1, \cdots, \lambda_n)$. Thus, $[x] \in P(V)$ has a family of coordinates $(\lambda \lambda_0, \lambda \lambda_1, \cdots, \lambda \lambda_n)$ for $\lambda \neq 0$. If $\lambda_0 = 0$, then all 0th coordinates in $[x]$ are 0; similarly, if one $\lambda_0 \neq 0$, all 0th coordinates in $[x]$ are nonzero. Define $\Omega = \{[x] \in P(V): \lambda_0 = 0\}$; one can show Ω is a projective hyperplane. Define $A = \{[x] \in P(V): \lambda_0 \neq 0\}$. Each $[x] \in A$ has a unique set of coordinates of the form $(1, \lambda_0^{-1} \lambda_1, \cdots, \lambda_0^{-1} \lambda_n)$, and one may make A into an affine space by defining $\alpha: V(n, K) \to A$ by $(\mu_1, \cdots, \mu_n) \mapsto [(1, \mu_1, \cdots, \mu_n)]$. In the future, we shall write $[\lambda_0, \cdots, \lambda_n]$ instead of $[(\lambda_0, \cdots, \lambda_n)]$.

Before discussing Theorem 9.39 further, let us prove an immediate consequence of it.

Theorem 9.40 *Let V and W be vector spaces of dimension ≥ 2 and let Ω be a projective hyperplane in $P(V)$ with $A = P(V) - \Omega$. If θ_1 and θ_2 are collineations $P(V) \to P(W)$ agreeing on A (i.e., $\theta_1|A = \theta_2|A$), then $\theta_1 = \theta_2$.*

Proof Let L be an affine line in A and let L^* be the projective line it determines. Now $\theta_i(L^* \cap \Omega) = \theta_i(L^*) \cap \theta_i(\Omega)$ for $i = 1, 2$. Since $L^* \cap \Omega$ is a point of Ω (and $\theta_i(L^*) \cap \theta_i(\Omega)$ is a point of $\theta_i(\Omega)$), θ_1 and θ_2 agree on all points of Ω of the form $L^* \cap \Omega$. But every point $[u]$ of Ω has this form:

choose $[x] \notin \Omega$, let Λ be the projective line determined by $[x]$ and $[u]$, define $L = \Lambda - \{[u]\}$, and observe that $L^* = \Lambda$. As θ_1 and θ_2 agree on A by hypothesis, they agree everywhere. ∎

We illustrate Theorem 9.39 in the smallest interesting case. Let $V = V(3, 2)$ so that (A, α) is an affine plane over $GF(2)$. Draw the 4 points of A and the 6 lines between them.

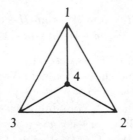

There are 3 pairs of parallel lines (e.g., $\{1, 2\}$ and $\{3, 4\}$), so $P(V)$ requires 3 points at infinity. By Theorem 9.37(i), $|P^2(2)| = 2^2 + 2 + 1 = 7$. The picture of $P^2(2)$ is

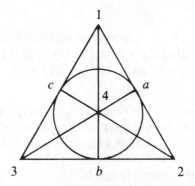

There are now 7 lines instead of 6, each having 3 points instead of 2. [Theorem 9.37(ii) asserts that the number of lines is the same as the number of points.] We have adjoined one "infinite" point to each of the old lines, thus forcing parallel lines to meet. The set of infinite points $\{a, b, c\}$ is a new line, the "line at infinity". Let us return to group theory.

Recall that if $g: V \to W$ is a nonsingular semilinear transformation, then g induces a collineation $P(g): P(V) \to P(W)$ by $[x] \mapsto [g(x)]$. We now show that, with an obvious exception, every collineation arises in this way.

Theorem 9.41   (Fundamental Theorem of Projective Geometry)   *If V and W are vector spaces over K of dimension ≥ 3, then every collineation $\theta: P(V) \to P(W)$ has the form $\theta = P(f)$ for some nonsingular semilinear transformation $f: V \to W$.*

REMARKS
  (i) One needs dim $V \geq 3$, otherwise $P(V)$ and $P(W)$ are projective lines and every one-one correspondence between them is a collineation.
 (ii) There is a stronger version of this theorem in which the field K is replaced by a division ring. This version does not affect finite groups, for Wedderburn's theorem asserts that every finite division ring is a field.

Proof We set up notation. Let $\varphi\colon V \to P(V)$ and $\psi\colon W \to P(W)$ be the natural maps. Choose a projective hyperplane Ω in $P(V)$; since θ is a collineation, $\theta(\Omega)$ is a projective hyperplane in $P(W)$. Let H be the linear hyperplane in V with $\varphi(H^{\#}) = \Omega$, let M be the linear hyperplane in W with $\psi(M^{\#}) = \theta(\Omega)$; choose $x \in V$ with $x \notin H$ and choose $y \in W$ with $[y] = \theta([x])$. In this notation, $P(g) = \psi g \varphi^{-1}$ whenever $g\colon V \to W$ is a semilinear transformation.

There are affine spaces (A, α) and (B, β), where $A = P(V) - \Omega$, $B = P(W) - \theta(\Omega)$, $\alpha = \varphi t_x\colon H \to A$, and $\beta = \psi t_y\colon M \to B$ (see the proof of Theorem 9.39). The restriction $\theta|A$, denoted by f, is an affine isomorphism $f\colon A \to B$ (because θ is a collineation). By Corollary 9.32, we have $f = \beta t_z g \alpha^{-1}$, where $g\colon H \to M$ is a nonsingular semilinear transformation and $z = \beta^{-1} f \alpha(0)$. Compute:

$$z = \beta^{-1} f \alpha(0) = t_{-y} \psi^{-1} f \varphi t_x(0)$$
$$= t_{-y} \psi^{-1} f[x]$$
$$= t_{-y} \psi^{-1}[y] = y - y = 0$$

(remember that ψ is one-one on $M + y$). Therefore $t_z = t_0$ is the identity and $\theta|A = f = \beta g \alpha^{-1}$.

We may write $V = \langle x \rangle \oplus H$ because H is a linear hyperplane and $x \notin H$. Define $\tilde{g}\colon V \to W$ by

$$\tilde{g}(\lambda x + u) = \sigma(\lambda) y + g(u),$$

where $\lambda \in K$, $u \in H$, and $\sigma \in \mathrm{Aut}(K)$ is determined by g. It is easy to see that $\tilde{g}$ is a semilinear transformation extending g; it is nonsingular because g is nonsingular and y is not in the linear hyperplane M.

Let us compute:

$$t_y \tilde{g} t_{-x}(\lambda x + u) = t_y \tilde{g}((\lambda - 1)x + u)$$
$$= \sigma(\lambda - 1)y + g(u) + y$$
$$= \sigma(\lambda)y + g(u) = \tilde{g}(\lambda x + u).$$

Therefore,

$$\beta \tilde{g} \alpha^{-1} = \psi t_y \tilde{g} t_{-x} \varphi^{-1} = \psi \tilde{g} \varphi^{-1} = P(\tilde{g}).$$

Since $\theta|A = f = \beta g \alpha^{-1} = P(\tilde{g})|A$, Theorem 9.40 gives $\theta = P(\tilde{g})$. ∎

Notation If V is a vector space over K, then $Sc(V)$ is the group of all nonzero scalar transformations $x \mapsto \lambda x$ with $\lambda \in K^{\#}$.

Theorem 9.42   *If* dim $V \geq 3$, *then* $\Gamma L(V)/Sc(V)$ *is isomorphic to the group of all collineations of* $P(V)$. *If* dim $V = 2$, *then* $\Gamma L(V)/Sc(V)$ *is a subgroup of the symmetric group on* $P(V)$, *the latter being the group of all collineations of* $P(V)$.

Proof   First of all, every permutation of $P^1(K)$ is a collineation, so the collineation group is the symmetric group when dim $V = 2$.

When dim $V \geq 2$, define a homomorphism π from $\Gamma L(V)$ to the group of collineations of $P(V)$ by $\pi(g) = P(g)$. Note that π is onto when dim $V \geq 3$, by Theorem 9.41. To compute ker π, assume $g \in \Gamma L(V)$ with $P(g) = $ identity, i.e., $[g(x)] = [x]$ for every $[x] \in P(V)$. Thus, $g(x) = \lambda_x x$ for all $x \in V$, where $\lambda_x \in K$.

The proof is completed, as in Theorem 9.31, by showing λ_x does not depend on x. We may choose $w \in V$ so that $\{x, w\}$ is independent, for dim $V \geq 2$. As g is additive,

$$g(x + w) = g(x) + g(w) = \lambda_x x + \lambda_w w;$$

on the other hand

$$g(x + w) = \lambda_{x+w}(x + w).$$

Equating coefficients gives $\lambda_x = \lambda_{x+w} = \lambda_w$. If $\mu \in K^{\#}$, then $\{\mu x, w\}$ is independent and $\lambda_{\mu x} = \lambda_w$. Therefore g is the scalar transformation $x \mapsto \lambda x$, where λ is the common value of the λ_x. ∎

Definition   For a vector space V, write

$$P\Gamma L(V) = \Gamma L(V)/Sc(V).$$

When $V = V(n, K)$, write $P\Gamma L(V) = P\Gamma L(n, K)$ and, when $K = GF(q)$, write $P\Gamma L(n, K) = P\Gamma L(n, q)$.

Corollary 9.43   *If* $n \geq 2$, $P\Gamma L(n + 1, K)$ *is the group of all collineations of* $P^n(K)$. *The group* $P\Gamma L(2, K)$ *is a subgroup of the symmetric group on* $P^1(K)$.

Now recall that a projectivity is a collineation of the form $P(g)$, where g is a nonsingular *linear* transformation.

Definition   For a vector space V, write

$$PGL(V) = GL(V)/Sc(V).$$

When $V = V(n, K)$, write $PGL(V) = PGL(n, K)$ and, when $K = GF(q)$, write $PGL(n, K) = PGL(n, q)$.

Corollary 9.44   *For all* $n \geq 1$, $PGL(n + 1, K)$ *is the group of all projectivities of* $P^n(K)$.

Proof   Let π_1 be the restriction to $GL(n + 1, K)$ of the homomorphism π of Theorem 9.42; thus, $\pi_1(g) = P(g)$. The image of π_1 is the group of projectivities and ker $\pi_1 = $ ker $\pi = Sc(V)$. ∎

The groups PSL discussed in Chapter 8 fit into the notational scheme above, and

$$PSL(V) \subset PGL(V) \subset P\Gamma L(V)$$

because $PSL(V) = SL(V)/Sc_1(V)$, where $Sc_1(V) = Sc(V) \cap SL(V)$, and so $PSL(V) \cong SL(V)Sc(V)/Sc(V)$. Moreover, for all $n \geq 1$, each of these groups acts faithfully on $P^n(K)$, by Theorem 9.42, where $V = V(n+1, K)$.

We have seen that $P\Gamma L(n+1, K)$, hence each of its subgroups, acts faithfully on $P^n(K)$ for all $n \geq 1$. Let us use this observation to prove that $P\Gamma L(2, 4) \cong S_5$ and $PGL(2, 4) \cong A_5$. From Theorems 9.34 and 8.10, it follows that $|P\Gamma L(2, 4)| = 120$ and $|PGL(2, 4)| = 60$. Since both groups act faithfully on $P^1(4)$, a set with 5 elements, we conclude each is a subgroup of S_5, from which the result follows.

Theorem 9.45 *For all $n \geq 1$ and every field K, $PSL(n+1, K)$ (and hence each of the larger groups $PGL(n+1, K)$ and $P\Gamma L(n+1, K)$) acts 2-transitively on $P^n(K)$.*

Proof Regard $P^n(K) = P(V)$, where $V = V(n+1, K)$. If $[x_1], [x_2]$ and $[y_1]$, $[y_2]$ are ordered pairs of distinct points in $P(V)$, then $\{x_1, x_2\}$ and $\{y_1, y_2\}$ are independent subsets of V. There are bases $\{x_1, x_2, \cdots, x_{n+1}\}$ and $\{y_1, y_2, \cdots, y_{n+1}\}$ of V, and the linear transformation g with $g(x_i) = y_i$, for all i, lies in $GL(n+1, K)$. Hence $P(g)([x_i]) = [y_i]$ for $i = 1, 2$, and $PGL(n+1, K)$ acts 2-transitively on $P^n(K)$.

Assume det $g = \lambda$. Define $h \in GL(n+1, K)$ by $h(x_1) = \lambda^{-1}y_1$ and $h(x_i) = y_i$ for $i \geq 2$. Then det $h = 1$, i.e., $h \in SL(n+1, K)$, and $P(h)([x_1]) = [\lambda^{-1}y_1] = [y_1]$ and $P(h)([x_2]) = [y_2]$. Thus $PSL(n+1, K)$ also acts 2-transitively on $P^n(K)$. ■

EXERCISES

**9.36. Show $PSL(n+1, K)$ acts faithfully and transitively on the set of all projective lines in $P^n(K)$. (HINT: Two points determine a line.)

9.37. Let $h: P(V) \to P(W)$ be a collineation. If dim $V \geq 3$ and there exists a projective line L in $P(V)$ such that $h|L$ is a projectivity, then h is a projectivity.

9.38. Using a permutation representation on $P^1(5)$, prove that $PGL(2, 5) \cong S_5$ and $PSL(2, 5) \cong A_5$. (HINT: $|P^1(5)| = 6$.)

It appears $P\Gamma L(2, K)$ is less important than $P\Gamma L(m, K)$ for $m \geq 3$ because it is not the full collineation group of its projective space. We shall now see this group has other features to interest us.

Definition Let K be a field and $\sigma \in \text{Aut}(K)$. A **semilinear fractional transfor-**

mation is a function $f\colon K \cup \{\infty\} \to K \cup \{\infty\}$ of the form

$$f(\lambda) = \frac{a\sigma(\lambda) + b}{c\sigma(\lambda) + d},$$

where $ad - bc \neq 0$; if $c = 0$, define $f(\infty) = \infty$; if $c \neq 0$, define $f(\infty) = ac^{-1}$. If σ is the identity, then f is called a **linear fractional transformation**.

Definition   All semilinear fractional transformations form a group under composition, denoted by $\Gamma LF(K)$; all linear fractional transformations form a subgroup, denoted by $LF(K)$. When $K = GF(q)$, we may write $\Gamma LF(K) = \Gamma LF(q)$ and $LF(K) = LF(q)$.

Theorem 9.46   *For every field K,*

$$P\Gamma L(2, K) \cong \Gamma LF(K) \quad and \quad PGL(2, K) \cong LF(K).$$

Proof   After choosing a basis of $V(2, K)$ one sees, using Theorem 9.34, that each $h \in \Gamma L(2, K)$ has a unique factorization $h = g\sigma_*$, where $\sigma \in \mathrm{Aut}(K)$ and g is nonsingular with matrix

$$g = \begin{bmatrix} a & b \\ c & d \end{bmatrix}.$$

Define $\psi\colon \Gamma L(2, K) \to \Gamma LF(K)$ by $\psi(g\sigma_*)\colon \lambda \mapsto (a\sigma(\lambda) + b)/(c\sigma(\lambda) + d)$. One checks easily that ψ is a homomorphism onto and that $\ker \psi$ consists of nonzero scalar matrices. The second isomorphism is just the restriction of the first one. ∎

   Let us display the groups that have arisen from a vector space V over K of dimension m:

$$SL(V) \subset GL(V) \subset \Gamma L(V) \subset \mathrm{Aut}(V)$$
$$\cup \qquad \cup \qquad\qquad\qquad \cup$$
$$Sc_1(V) \subset Sc(V) \qquad\longrightarrow \mathrm{Aff}(V)$$
$$\cup$$
$$\mathrm{Tr}(V);$$

$$PSL(V) \subset PGL(V) \subset P\Gamma L(V).$$

Recall that a semidirect product of a group A by a group B may be denoted by $A \rtimes B$ and that $\mathrm{Aut}(K)$ is the group of field automorphisms of K [not to be confused with $\mathrm{Aut}(V)$ when V is one-dimensional]. Here are isomorphisms involving these groups:

$$Z_1(m, K) \cong Sc_1(V) = Sc(V) \cap SL(V);$$
$$SL(m, K) \cong SL(V);$$
$$GL(m, K) \cong GL(V) \cong SL(V) \rtimes K^{\#};$$
$$\Gamma L(m, K) \cong \Gamma L(V) \cong GL(V) \rtimes \mathrm{Aut}(K);$$
$$\mathrm{Aut}(m, K) \cong \mathrm{Aut}(V) \cong \Gamma L(V) \rtimes \mathrm{Tr}(V);$$
$$\mathrm{Aff}(m, K) \cong \mathrm{Aff}(V) \cong GL(V) \rtimes \mathrm{Tr}(V).$$

When dim $V = m = 2$, there are two more isomorphisms:

$$P\Gamma L(V) \cong P\Gamma L(2, K) \cong \Gamma LF(K);$$
$$PGL(V) \cong PGL(2, K) \cong LF(K).$$

Suppose (X, ρ_1) is a G-set, (Y, ρ_2) is an H-set, and $\psi:G \to H$ is an isomorphism; when is it reasonable to identify these sets equipped with their respective group actions? Recall (Exercise 3.2) that a one-one correspondence $\theta: X \to Y$ induces an isomorphism θ_* of symmetric groups $S_X \to S_Y$ by $\alpha \mapsto \theta\alpha\theta^{-1}$. Since actions are just homomorphisms into symmetric groups, let us agree to identify X and Y if $\theta_*\rho_1 = \rho_2\psi$.

For each $g \in G$, this says $\theta\rho_1(g)\theta^{-1} = \theta_*\rho_1(g) = \rho_2\psi(g)$; for each $g \in G$ and $x \in X$, this says

$$\theta\rho_1(g)\theta^{-1}(x) = \rho_2\psi(g)(x).$$

(When $G = H$ and the map ψ is the identity, this definition becomes that of G-isomorphism.)

Consider the $\Gamma LF(K)$-set $K \cup \{\infty\}$ with action $\lambda \mapsto (a\sigma(\lambda) + b)/(c\sigma(\lambda) + d)$ and $\infty \mapsto \infty$ or ac^{-1} depending on whether $c = 0$ or $c \neq 0$; consider also the $P\Gamma L(V)$-set $P^1(K)$, where V is a two-dimensional vector space over K, with action $P(h)[v] = [h(v)]$. We claim the general situation above obtains here. Choosing a basis of V equips each $[v] \in P^1(K)$ with homogeneous coordinates $[v] = [\lambda, \mu]$ and factors each nonsingular semilinear transformation $h = g\sigma_*$, where

$$g = \begin{bmatrix} a & b \\ c & d \end{bmatrix}$$

and

$$\sigma_*[\lambda, \mu] = [\sigma(\lambda), \sigma(\mu)].$$

The action of $P\Gamma L(2, K)$ on $P^1(K)$ is thus given by

$$P(h)\begin{bmatrix} \lambda \\ \mu \end{bmatrix} = \begin{bmatrix} \begin{bmatrix} a & b \\ c & d \end{bmatrix}\begin{bmatrix} \sigma(\lambda) \\ \sigma(\mu) \end{bmatrix} \end{bmatrix} = \begin{bmatrix} a\sigma(\lambda) + b\sigma(\mu) \\ c\sigma(\lambda) + d\sigma(\mu) \end{bmatrix}.$$

Now define $\theta: P^1(K) \to K \cup \{\infty\}$ by $\theta[1, 0] = \infty$ and $\theta[\lambda, 1] = \lambda$. If $\lambda \in K$ and $c\sigma(\lambda) + d \neq 0$, then

$$\theta g\sigma_*\theta^{-1}(\lambda) = \theta g\sigma_*[\lambda, 1]$$
$$= \theta[a\sigma(\lambda) + b, c\sigma(\lambda) + d]$$
$$= \theta[(a\sigma(\lambda) + b)/(c\sigma(\lambda) + d), 1]$$
$$= (a\sigma(\lambda) + b)/(c\sigma(\lambda) + d).$$

A similar equation holds when $c\sigma(\lambda) + d = 0$ or when λ is replaced by ∞. Therefore, we may identify $P^1(K)$ and $K \cup \{\infty\}$ as sets with group action.

Theorem 9.47   *For every field* K, $P^1(K)$ *is a faithful, sharply* 3-*transitive* $PGL(2, K)$-*set.*

Proof   As in Theorem 9.46 and the subsequent discussion, we may consider $PGL(2, K) \cong LF(K)$ acting on $K \cup \{\infty\}$ as linear fractional transformations. Considering K as a one-dimensional vector space over itself, $\text{Aff}(K) = \{\lambda \mapsto a\lambda + b\}$ [the subgroup of $LF(K)$ consisting of all "numerators"] is the stabilizer of ∞. But Theorem 9.27 says $\text{Aff}(K)$ acts sharply 2-transitively on K, so that Exercise 9.10 and Corollary 9.8(v) will give the result once we show $LF(K)$ acts transitively on $K \cup \{\infty\}$. This is easy: if $\lambda \in K$, then $f(x) = x + \lambda$ sends 0 to λ; if $\lambda = \infty$, then $f(x) = 1/x$ sends 0 to ∞. ∎

Observe that $|P^1(q)| = q + 1$ implies $|PGL(2, q)| = (q+1)q(q-1)$, a formula we have known since Chapter 8; observe also that $P\Gamma L(2, q)$ acts 3-transitively on $P^1(q)$ (because its subgroup PGL does), but the action of $P\Gamma L$ is not sharp.

We now display a second family of sharply 3-transitive G-sets. If $h \in \Gamma LF(q)$, then $h(\lambda) = (a\sigma(\lambda) + b)/(c\sigma(\lambda) + d)$, where $\sigma \in \text{Aut}(GF(q))$ and $ad - bc \neq 0$. Multiplying numerator and denominator by $\mu \in GF(q)^{\#}$ does not change h, but its "determinant" does change to $\mu^2(ad - bc)$. If q is a power of 2, every element in $GF(q)$ is a square; however, if q is a power of an odd prime p [and if α is a primitive element of $GF(q)$], then the nonzero squares form a subgroup of index 2 in $GF(q)^{\#}$, namely, $\langle \alpha^2 \rangle$. In this second case, it thus makes sense to say det h is or is not a square.

A second ingredient in the definition is an analog of complex conjugation: an automorphism σ of $GF(q)$ having order 2. It is easy to see that such a σ exists (uniquely) when $q = p^{2n}$, in which case $\sigma(\lambda) = \lambda^{p^n}$.

Definition   Let $q = p^{2n}$, where p is an odd prime, and let $\sigma \in \text{Aut}(GF(q))$ have order 2 [i.e., $\sigma(\lambda) = \lambda^{p^n}$]. Define $Sh(q)$ as the subset $A \cup B$ of $\Gamma LF(q)$, where

$$A = \{h\colon \lambda \mapsto (a\lambda + b)/(c\lambda + d)\colon ad - bc \text{ is a square}\}$$

and

$$B = \{h\colon \lambda \mapsto (a\sigma(\lambda) + b)/(c\sigma(\lambda) + d)\colon ad - bc \text{ is not a square}\}.$$

The reader may check quickly that $Sh(q)$ is, in fact, a subgroup of $\Gamma LF(q)$ and, as such, acts faithfully on $P^1(q)$; moreover, A is a subgroup of $Sh(q)$ of index 2 and B is the other coset of A [Exercise 9.43 shows $Sh(q)$ is not a semidirect product of A by $\mathbf{Z}(2)$]. The next theorem explains our notation: "Sh" abbreviates "sharp".

Theorem 9.48   *If p is an odd prime and $q = p^{2n}$, then $P^1(q)$ is a faithful, sharply* 3-*transitive* $Sh(q)$-*set.*

Proof   Let $K = GF(q)$ and identify $P^1(q)$ with $K \cup \{\infty\}$; let us write G for $Sh(q)$. Now $G_{\infty} = A_{\infty} \cup B_{\infty}$, where

$$A_{\infty} = \{h\colon \lambda \mapsto a\lambda + b\colon a \in (K^{\#})^2\}$$

and

$$B_{\infty} = \{h\colon \lambda \mapsto a\sigma(\lambda) + b\colon a \notin (K^{\#})^2\}.$$

If a and b are distinct elements of K with $a \neq 0$, then a short computation shows there exists $h \in G$ with $h(0) = b$ and $h(1) = a$ (h will lie either in A_∞ or B_∞, depending on whether a is or is not a square). It follows that G_∞ acts doubly transitively on K. But in each of A_∞ and B_∞, there are q choices for b and $\frac{1}{2}(q-1)$ choices for a, so that $|G_\infty| = q(q-1)$, and the action of G_∞ on K is sharp. Finally, G acts transitively on $K \cup \{\infty\}$, for $\lambda \mapsto -1/\lambda$ is in G (the negative sign gives determinant 1) and it interchanges 0 and ∞. We conclude, by Corollary 9.8(v), that $G = Sh(q)$ acts sharply 3-transitively on $K \cup \{\infty\}$. ∎

Zassenhaus (1936) proved that the actions of $PGL(2, q)$ and $Sh(q)$ on $P^1(q)$ give the only faithful sharply 3-transitive G-sets. The first family is defined for all prime powers q; the second is defined for all even powers of odd primes. It is not obvious that $PGL(2, q) \not\cong Sh(q)$ when $q = p^{2n}$ and p is odd, though this is true. Here is the smallest instance of this fact.

Definition The group $Sh(9)$ is usually denoted by M_{10} and is called the **Mathieu group** of degree 10.

Theorem 9.49 $PGL(2, 9)$ and M_{10} are nonisomorphic groups of order 720, each of which acts sharply 3-transitively on $P^1(9)$.

Proof We have already seen each group acts on $P^1(9) = GF(9) \cup \{\infty\}$ as described, whence the order of each group is $10 \cdot 9 \cdot 8 = 720 = 16 \cdot 45$. Let $G = LF(9) \cong PGL(2,9)$ acting on $GF(9) \cup \{\infty\}$. Now $G_{\infty,0} = \{h: \lambda \mapsto a\lambda$ with $a \neq 0\} \cong GF(9)^\# \cong \mathbf{Z}(8)$; indeed, a generator of $G_{\infty,0}$ is $g: \lambda \mapsto \alpha\lambda$, where α is a primitive element of $GF(9)$. If $\tau(\lambda) = \lambda^{-1}$, then τ is an element of order 2 in G, and one checks that $\tau g \tau^{-1} = g^{-1}$. It follows that $\langle G_{\infty,0}, \tau \rangle$ is a dihedral group of order 16 and is a Sylow 2-subgroup of G.

Consider $H = M_{10}$ acting on $GF(9) \cup \{\infty\}$. Since $q = 3^2$, the automorphism σ is just $\sigma(\lambda) = \lambda^3$. Now $H_{\infty,0} = A_{\infty,0} \cup B_{\infty,0} = \{h: \lambda \mapsto a^2\lambda$ with $a \neq 0\} \cup \{h: \lambda \mapsto a\lambda^3$ with a not a square$\}$. One checks that this nonabelian subgroup of order 8 has only one element of order 2, so that $H_{\infty,0} \cong Q$, the quaternions. Since dihedral groups have no quaternion subgroups, it follows that the Sylow 2-subgroups of G and H are not isomorphic, *a fortiori*, G and H are not isomorphic. ∎

EXERCISES

In the following exercises, α denotes a primitive element of $GF(9)$, so that $\alpha^8 = 1$.

9.39. Prove that a Sylow 3-subgroup of M_{10} is elementary abelian of order 9.

**9.40. Prove that $(M_{10})_\infty$ is a group of order 72 having a normal Sylow 3-subgroup. Conclude that $(M_{10})_\infty$ is a semidirect product of $\mathbf{Z}(3) \times \mathbf{Z}(3)$ by Q. (HINT: $\lambda \mapsto \lambda + b$ has order 3 for every $b \in GF(9)^\#$; the inverse of

$\lambda \mapsto \alpha^{2i}\lambda + b$ is $\lambda \mapsto \alpha^{6i}\lambda - \alpha^{6i}b$, and the inverse of $\lambda \mapsto \alpha^{2i+1}\lambda^3 + b$ is $\lambda \mapsto \alpha^{2i+5}\lambda^3 - \alpha^{2i+5}b^3$.)

**9.41.   There are exactly 8 elements in $(M_{10})_\infty$ of order 3 and they are conjugate to one another in $(M_{10})_\infty$.

9.42.   Prove that the subgroup A of $Sh(q)$ is isomorphic to $PSL(2, q)$. Conclude that $M_{10} = Sh(9)$ is neither simple nor solvable. [HINT: If $h(\lambda) = (a\lambda + b)/(c\lambda + d)$ and $ad - bc = \mu^2$, multiply numerator and denominator by μ^{-1}.]

*9.43.   Show that M_{10} is not a semidirect product of A by $\mathbf{Z}(2)$. (HINT: B contains no element of order 2.)

*9.44.   Show that $M_{10} = \langle \sigma_1, \sigma_2, \sigma_3, \sigma_4, \sigma_5 \rangle$, where $\sigma_1(\lambda) = -\lambda^{-1}$, $\sigma_2(\lambda) = \lambda + 1$, $\sigma_3(\lambda) = \lambda + \alpha$, $\sigma_4(\lambda) = \alpha^2 \lambda$, and $\sigma_5(\lambda) = \alpha\lambda^3$.

**9.45. $GF(9)$ may be regarded as a two-dimensional vector space over $GF(3)$ with $\{1, \alpha\}$ as a basis. Verify the following coordinates [any root of an irreducible quadratic with coefficients in $GF(3)$ will serve as α; we further assume $\alpha^2 + \alpha = 1$]:

$$
\begin{aligned}
1 &= (1, 0); & \alpha^4 &= (-1, 0); \\
\alpha &= (0, 1); & \alpha^5 &= (0, -1); \\
\alpha^2 &= (1, -1); & \alpha^6 &= (-1, 1); \\
\alpha^3 &= (-1, -1); & \alpha^7 &= (1, 1).
\end{aligned}
$$

**9.46.   Prove that M_{10} consists of even permutations of $P^1(9)$. (HINT: Write each of the permutations σ_i, $1 \le i \le 5$, as a product of disjoint cycles.)

This section ends with a second proof of the simplicity of the PSL's. We remind the reader of a definition from Chapter 8. Let $V = V(n, q)$ and let $f: V \to GF(q)$ be a functional. If $x \in \ker f$, then a *transvection* $T_{f,x}$ is a function $V \to V$ defined by

$$T_{f,x}(v) = v + f(v)x$$

for all $v \in V$. Recall that every transvection lies in $SL(V)$ and that the set of them generates $SL(V)$.

Theorem 9.50 $PSL(n, q)$ is simple if $(n, q) \ne (2, 2)$ and $(n, q) \ne (2, 3)$.

Proof   We use Iwasawa's Theorem 9.25. If $G = PSL(n, q)$ and $P^{n-1}(q) = P(V)$, where $V = V(n, q)$, we know $P(V)$ is a faithful 2-transitive, hence primitive, G-set; thus, condition (i) of Theorem 9.25 holds.

Choose $x \in V$ and define a subset H of the stabilizer $G_{[x]}$ by

$$H = \{P(T_{f,x}): f \text{ is a functional with } f(x) = 0\}.$$

Applying P to the formula $T_{f,x}T_{g,x} = T_{f+g,x}$ of Exercise 8.13 shows H is an abelian subgroup of $G_{[x]}$. Recall Exercise 8.16: If S is a nonsingular linear transformation [which we choose to be in $SL(V)$],

$$ST_{f,x}S^{-1} = T_{fS^{-1},Sx}.$$

Now $P(S) \in G_{[x]}$ if and only if $Sx = \lambda x$ for some $\lambda \in GF(q)$. But $T_{g,\lambda x} = T_{\lambda g,x}$ (Exercise 8.14) and this shows H is a normal subgroup of $G_{[x]}$.

Since the transvections generate $SL(V)$, it suffices for condition (iii) to show $P(T_{g,y})$ is a conjugate of some $P(T_{f,x})$. Choose $S \in SL(V)$ with $Sx = y$. Then $P(S)P(T_{f,x})P(S)^{-1} = P(T_{fS^{-1},y})$. But, as f varies over all functionals with $f(x) = 0$, fS^{-1} varies over all functionals g with $g(y) = 0$.

It remains to prove G is perfect. Suppose we could show some transvection T is a commutator: $T = [A, B]$ for some $A, B \in SL(V)$. For any other transvection T', Corollary 8.22 gives T and T' conjugate in $GL(V)$, say, $T' = T^U$ for some $U \in GL(V)$. Since conjugation by U is a homomorphism, $T' = [A, B]^U = [A^U, B^U]$; since $SL(V) \lhd GL(V)$, both A^U and B^U lie in $SL(V)$. We conclude that every transvection would be a commutator and, since the transvections generate $SL(V)$, that $SL(V)$ and its image $PSL(V) = G$ would be perfect.

Assume $n \geq 3$, and let $\{e_1, \cdots, e_n\}$ be a basis of V. Let T be the transvection $T = T_{f,x}$ with $x = -e_2 - e_1$ and f the functional selecting the third coordinate: $f(\Sigma \lambda_i e_i) = \lambda_3$ (thus $Te_i = e_i$ for $i \neq 3$ and $Te_3 = e_3 - e_2 - e_1$). Define $A = T_{f,-e_1}$ (so $Ae_i = e_i$ for $i \neq 3$ and $Ae_3 = e_3 - e_1$), and define B by $B(e_1) = -e_2$, $B(e_2) = e_1$, and $B(e_i) = e_i$ for $i \geq 3$. Note that A, $B \in SL(V)$, and a routine calculation gives $T = [A, B] = ABA^{-1}B^{-1}$.

Finally, assume $n = 2$ and $q > 3$. There exists $\lambda \in GF(q)$ with $\lambda^2 \neq 1$. But

$$\begin{bmatrix} \lambda & 0 \\ 0 & \lambda^{-1} \end{bmatrix} \begin{bmatrix} 1 & 1 \\ 0 & 1 \end{bmatrix} \begin{bmatrix} \lambda^{-1} & 0 \\ 0 & \lambda \end{bmatrix} \begin{bmatrix} 1 & -1 \\ 0 & 1 \end{bmatrix} = \begin{bmatrix} 1 & \lambda^2 - 1 \\ 0 & 1 \end{bmatrix},$$

so the last transvection is exhibited as a commutator.

All the conditions of Theorem 9.25 have been verified, and we conclude $PSL(n, q)$ is simple with two exceptions. ∎

MATHIEU GROUPS

Examples of doubly and triply transitive G-sets have already been given. One regards the G-sets obtained from the symmetric and alternating groups as trivial, not because it is so easy to exhibit them, but because the real question is how transitive must a subgroup of S_n be before one can conclude it is either S_n or A_n? The Mathieu groups are the only other groups G providing examples of faithful 4-transitive G-sets (two of them are actually 5-transitive), and we shall discuss them in this section.

In 1873, Jordan proved there do not exist *sharply* 6-transitive G-sets (unless G is symmetric or alternating). In 1981, the classification of finite nonabelian simple groups was completed†: such a group is either an alternating group, a group of "Lie type" (projective unimodular groups are the easiest examples), or one of twenty-six "sporadic" groups (five of which are the Mathieu groups). This classification can be

† This theorem, one of the highest achievements of mathematics, is the culmination of work of about 100 mathematicians between 1950 and 1980 and consists of several hundred journal articles totaling some 5000 pages!

used to prove that no nontrivial 6-transitive G-sets exist; indeed, it yields a classification of all faithful doubly transitive G-sets, as well as the proof that the Mathieu groups give the only faithful 4-transitive groups.

A very easy technique is available for lowering transitivity. If X is a transitive G-set and $x \in X$, then the G_x-set $X - \{x\}$ is (sharply) $(k-1)$-transitive if and only if X is a (sharply) k-transitive G-set (Exercise 9.10 and Corollary 9.8). Can we reverse this process, starting with G_x and constructing G?

All G-sets in this section are faithful, and we shall call the groups G **permutation groups** from now on. Indeed, we finally succumb to the irresistible urge of applying to groups those adjectives heretofore reserved for G-sets, e.g., we will speak of a multiply transitive group G of degree n meaning there is some set $X = \{x_1, \cdots, x_n\}$ that is a multiply transitive G-set.

Definition   Let G be a permutation group on X and let $\tilde{X} = X \cup \{\infty\}$, where ∞ is a point not in X. A transitive permutation group $\tilde{G}$ on $\tilde{X}$ is a **transitive extension** of G if the stabilizer $\tilde{G}_\infty$ is G.

The next exercise shows that transitive extensions may not exist (actually, they rarely exist).

EXERCISE

9.47.   Show that the 4-group **V** has no transitive extension. (HINT: If $h \in S_5$ has order 5, then $\langle \mathbf{V}, h \rangle \supset A_5$.)

Theorem 9.51 (Witt, 1938)   *Let G be a multiply transitive permutation group acting on a set X. Suppose there is an element $\infty \notin X$, a permutation h of $\tilde{X} = X \cup \{\infty\}$, an element $x \in X$, and an element $g \in G$ such that*

(i) $g \notin G_x$;
(ii) $h(\infty) \in X$;
(iii) $h^2 \in G$ *and* $(gh)^3 \in G$;
(iv) $hG_x h = G_x$.

Then $\tilde{G} = \langle G, h \rangle$ is a transitive extension of G.

Proof   As G acts transitively on X, condition (ii) shows $\tilde{G}$ acts transitively on $\tilde{X}$. Suppose we show $G \cup GhG$ is a group (as Theorem 9.6 predicts); then it is clear that if $\tilde{G} = \langle G, h \rangle$, then $\tilde{G} = G \cup GhG$ and $\tilde{G}_\infty = G$ (since everything in G fixes ∞, each element in the double coset GhG moves ∞).

By Exercise 2.8, we need see only whether $G \cup GhG$ is closed under multiplication. Since $GG = G$, it suffices to show $(GhG)(GhG) = GhGhG \subset G \cup GhG$, and this will follow if we show $hGh \subset G \cup GhG$.

Since G acts multiply transitively on X, Theorem 9.6 gives $G = G_x \cup G_x g G_x$ (because $g \notin G_x$). Before computing, note that $h^2 = \gamma_1 \in G$ implies $h\gamma_1^{-1} = h^{-1} = \gamma_1^{-1}h$ and   that $(gh)^3 = \gamma_2 \in G$ implies hgh

$= g^{-1}h^{-1}g^{-1}\gamma_2$. Now

$$\begin{aligned}
hGh &= h(G_x \cup G_x g G_x)h \\
&= hG_x h \cup hG_x g G_x h \\
&= hG_x h \cup (hG_x h)h^{-1}gh^{-1}(hG_x h) \\
&= G_x \cup G_x h^{-1}gh^{-1}G_x \quad \text{[condition (iv)]} \\
&= G_x \cup G_x(\gamma_1^{-1}h)g(h\gamma_1^{-1})G_x \\
&= G_x \cup G_x\gamma_1^{-1}(g^{-1}h^{-1}g^{-1}\gamma_2)\gamma_1^{-1}G_x \\
&\subset G \cup Gh^{-1}G \\
&= G \cup G\gamma_1^{-1}hG = G \cup GhG. \quad \blacksquare
\end{aligned}$$

The conditions in the theorem yield information about the cycle structure of the permutation h. If $h(\infty) = a \in X$, then $h^2 \in G = \tilde{G}_\infty$ implies $h(a) = h^2(\infty) = \infty$. Therefore $h = (\infty, a)h'$, where $h' \in G_a$ is disjoint from (∞, a). Similarly, one may see that gh has a factor involving ∞ that is a 3-cycle.

There are five more Mathieu groups besides M_{10}, namely, $M_{11}, M_{12}, M_{22}, M_{23}, M_{24}$; the subscripts indicate the degree of each group's usual representation as a permutation group. We present the results of successful searches for building transitive extensions by the method of Theorem 9.51. The reader will understand these constructions better once the relation between the Mathieu groups and Steiner systems is seen.

Theorem 9.52 *There exists a sharply* 4-*transitive group* M_{11} *of degree* 11 *and order* $7920 = 11 \cdot 10 \cdot 9 \cdot 8 = 2^4 \cdot 3^2 \cdot 5 \cdot 11$ *such that the stabilizer of a point is* M_{10}.

Proof We know M_{10} acts (sharply) 3-transitively on $X = GF(9) \cup \{\infty\}$; we construct a transitive extension of M_{10} acting on $\tilde{X} = X \cup \{\omega\}$. For α a primitive element of $GF(9)$, define

$$x = \infty,$$
$$g = (0, \infty)(\alpha, \alpha^7)(\alpha^2, \alpha^6)(\alpha^3, \alpha^5),$$

and

$$h = (\omega, \infty)(\alpha, \alpha^2)(\alpha^3, \alpha^7)(\alpha^5, \alpha^6).$$

Note that $g \in M_{10}$ for $g(\lambda) = \lambda^{-1}$ [det $g = -1 = \alpha^4$, a square in $GF(9)$]. Using Exercise 9.45, one may check that $(\alpha, \alpha^2)(\alpha^3, \alpha^7)(\alpha^5, \alpha^6)$ is the permutation of $GF(9)$ given by $\lambda \mapsto \alpha^2\lambda + \alpha\lambda^3$.

It is evident that $g \notin (M_{10})_\infty$ [because $g(\infty) = 0$] and that $h(\omega) = \infty \in X$. Now $h^2 = 1$, and $gh = (\omega, 0, \infty)(\alpha, \alpha^6, \alpha^3)(\alpha^2, \alpha^7, \alpha^5)$ has order 3. To satisfy the last condition of Theorem 9.51, observe that if $f \in (M_{10})_\infty$, then

$$hfh(\infty) = hf(\omega) = h(\omega) = \infty.$$

Thus $h(M_{10})_\infty h = (M_{10})_\infty$ if each $hfh \in M_{10}$. Writing $(M_{10})_\infty = A_\infty \cup B_\infty$ (as in the definition of M_{10}), either $f = \alpha^{2i}\lambda + b$ or $f = \alpha^{2i+1}\lambda^3 + b$, where $i \geq 0$ and

$b \in GF(9)$. In the first case (computing with $\alpha^2\lambda + \alpha\lambda^3$),

$$hfh(\lambda) = (\alpha^{2i+4} + \alpha^{6i+4})\lambda + (\alpha^{2i+3} + \alpha^{6i+7})\lambda^3 + \alpha^2 b + \alpha b^3.$$

The coefficients of λ and λ^3 are $\alpha^{2i+4}(1+\alpha^{4i})$ and $\alpha^{2i+3}(1+\alpha^{4i+4})$, respectively. When i is even, the second coefficient is 0 and the first coefficient is $2\alpha^{2i+4}$; but $2 = -1 = \alpha^4$, so this coefficient is a square and $hfh \in A_\infty$. When i is odd, the first coefficient is 0 and the second coefficient is $2\alpha^{2i+3} = \alpha^{2i+7}$ (which is a nonsquare), whence $hfh \in B_\infty$. The second case ($f(\lambda) = \alpha^{2i+1}\lambda^3 + b$) is similar. The reader may calculate here that

$$hfh(\lambda) = \alpha^{2i+6}(1+\alpha^{4i})\lambda + \alpha^{2i+1}(1+\alpha^{4i+4})\lambda^3 + \alpha^2 b + \alpha b^3,$$

an expression of the same form as that treated in the first case.

It follows from Theorem 9.8(v) that $M_{11} = \langle M_{10}, h \rangle$ acts sharply 4-transitively on $\tilde{X}$ (for M_{10} acts sharply 3-transitively on X), and so $|M_{11}| = 11 \cdot 10 \cdot 9 \cdot 8 = 7920$. ∎

This procedure may be repeated; again the difficult part is discovering a good permutation to adjoin.

Theorem 9.53   *There exists a sharply 5-transitive group M_{12} of degree 12 and order $12 \cdot 11 \cdot 10 \cdot 9 \cdot 8 = 95,040 = 2^6 \cdot 3^3 \cdot 5 \cdot 11$ such that the stabilizer of a point is M_{11}.*

Proof   We know M_{11} acts (sharply) 4-transitively on $Y = \{GF(9), \infty, \omega\}$; we construct a transitive extension of M_{11} acting on $\tilde{Y} = Y \cup \{\Omega\}$. For α a primitive element of $GF(9)$, define

$$x = \omega,$$
$$h = (\omega, \infty)(\alpha, \alpha^2)(\alpha^3, \alpha^7)(\alpha^5, \alpha^6),$$

and

$$k = (\omega, \Omega)(\alpha, \alpha^3)(\alpha^2, \alpha^6)(\alpha^5, \alpha^7).$$

[Note that $h \in M_{11}$ is the same h occurring in the previous theorem and that the "other factor" $(\alpha, \alpha^3)(\alpha^2, \alpha^6)(\alpha^5, \alpha^7)$ of k is the permutation of $GF(9)$ given by $\lambda \mapsto \lambda^3$.] Clearly $k(\Omega) = \omega \in Y$ and $h \notin (M_{11})_\omega = M_{10}$. Next, $k^2 = 1$ and $hk = (\omega, \Omega, \infty)(\alpha, \alpha^7, \alpha^6)(\alpha^2, \alpha^5, \alpha^3)$ has order 3. To satisfy the last condition of Theorem 9.51, observe first that if $f \in (M_{11})_\omega = M_{10} = A \cup B$, then kfk also fixes ω. Finally $kfk \in M_{11}$: if $f(\lambda) = (a\lambda+b)/(c\lambda+d) \in A$, then $kfk(\lambda) = (a^3\lambda + b^3)/(c^3\lambda + d^3)$ has determinant $a^3d^3 - b^3c^3 = (ad-bc)^3$, which is a square because $ad - bc$ is; a similar argument holds when $f \in B$. Thus $kM_{10}k = M_{10}$.

It follows that $M_{12} = \langle M_{11}, k \rangle$ acts sharply 5-transitively on $\tilde{Y}$ (since M_{11} acts sharply 4-transitively on Y) and $|M_{12}| = 12 \cdot 11 \cdot 10 \cdot 9 \cdot 8 = 95,040$. ∎

EXERCISES

9.48. Let σ_6 and σ_7 be the permutations of $GF(9)$ defined by $\sigma_6(\lambda) = \alpha^2\lambda + \alpha\lambda^3$ and $\sigma_7(\lambda) = \lambda^3$ [so that $h = (\omega, \infty)\sigma_6$ and $k = (\omega, \Omega)\sigma_7$]. Regarding $GF(9)$ as a vector space over $GF(3)$, prove that σ_6 and σ_7 are linear transformations.

9.49. Prove that $GL(2, 3) \cong \langle \sigma_4, \sigma_5, \sigma_6, \sigma_7 \rangle$, where $\sigma_4(\lambda) = \alpha^2\lambda$ and $\sigma_5(\lambda) = \alpha\lambda^3$ (recall σ_4 and σ_5 are two of the generators of M_{10} given in Exercise 9.44). (HINT: Using the coordinates displayed in Exercise 9.45,

$$\sigma_4 = \begin{bmatrix} 1 & -1 \\ -1 & -1 \end{bmatrix}; \quad \sigma_5 = \begin{bmatrix} 0 & -1 \\ 1 & 0 \end{bmatrix}; \quad \sigma_6 = \begin{bmatrix} 1 & 1 \\ 0 & -1 \end{bmatrix}; \quad \sigma_7 = \begin{bmatrix} 1 & -1 \\ 0 & -1 \end{bmatrix}.)$$

*9.50. Let $W = \{g \in M_{12}: g(\{\infty, \omega, \Omega\}) = \{\infty, \omega, \Omega\}\}$. Show there is a homomorphism of W onto S_3 with kernel $(M_{12})_{\infty,\omega,\Omega}$. Conclude that $|W| = 6 \cdot 72$.

**9.51. Prove that Aut(2, 3), the group of all affine automorphisms of $GF(9)$ viewed as an affine plane over $GF(3)$, is isomorphic to the subgroup W of M_{12} in Exercise 9.50.

The theorem of Jordan mentioned at the beginning of this section states that S_4, S_5, A_6, and M_{11} are the only sharply 4-transitive groups, that S_5, S_6, A_7, and M_{12} are the only sharply 5-transitive groups, and that S_k, S_{k+1}, and A_{k+2} are the only sharply k-transitive groups for $k \geq 6$. We remind the reader that Zassenhaus classified all sharply 3-transitive groups [such groups are either $PGL(2, q)$ or $Sh(p^{2n})$ for odd p], while Thompson completed the classification of sharply 2-transitive groups (they are certain Frobenius groups). The classification of all sharply 1-transitive groups, i.e., of all regular groups, is, by Cayley's theorem, equivalent to the classification of all groups.

The "large" Mathieu groups are constructed as a sequence of transitive extensions beginning with $PSL(3, 4)$ [which acts doubly transitively on $P^2(4)$]. Since $|P^2(4)| = 4^2 + 4 + 1 = 21$, one begins with a permutation group of degree 21; recall from Chapter 8 that $|PSL(3, 4)| = 20,160$. If we describe the points of $P^2(4)$ by homogeneous coordinates $[\lambda, \mu, \nu]$, where $\lambda, \mu, \nu \in GF(4)$, then the action of $PSL(3, 4)$ on a point is just matrix multiplication on the column vector $[\lambda, \mu, \nu]$.

EXERCISES

**9.52. Define $f_1: P^2(4) \to P^2(4)$ by

$$f_1[\lambda, \mu, \nu] = [\lambda^2 + \mu\nu, \mu^2, \nu^2].$$

Using the fact that $\lambda \mapsto \lambda^2$ is an automorphism of $GF(4)$, prove that f_1 is a permutation of $P^2(4)$ of order 2 that fixes $[1, 0, 0]$.

**9.53. For β a primitive element of $GF(4)$, define $f_2: P^2(4) \to P^2(4)$ by

$$f_2[\lambda, \mu, \nu] = [\lambda^2, \mu^2, \beta\nu^2].$$

Show that f_2 is a permutation of $P^2(4)$ of order 2 that fixes $[1,0,0]$.

9.54. Define $f_3\colon P^2(4) \to P^2(4)$ by

$$f_3[\lambda,\mu,\nu] = [\lambda^2,\mu^2,\nu^2].$$

Show that f_3 is a permutation of $P^2(4)$ of order 2 that fixes $[1,0,0]$.

9.55. Show that $\langle PSL(3,4),\ f_2,\ f_3 \rangle = P\Gamma L(3,4)$. [HINT: $PSL(3,4)$ is a normal subgroup of $P\Gamma L(3,4)$ of index 6, and the quotient group is isomorphic to S_3. Indeed, if $\sigma(\lambda) = \lambda^2$ is the unique nontrivial automorphism of $GF(4)$, then $f_3 = \sigma_*$ and

$$f_2 = \begin{bmatrix} 1 & 0 & 0 \\ 0 & 1 & 0 \\ 0 & 0 & \beta \end{bmatrix} \sigma_* .]$$

Theorem 9.54 *There exists a 3-transitive group M_{22} of degree 22 and order* $443{,}520 = 22\cdot21\cdot20\cdot48 = 2^7\cdot3^2\cdot5\cdot7\cdot11$ *such that the stabilizer of a point is* $PSL(3,4)$.

Proof We show that $G = PSL(3,4)$ acting on $X = P^2(4)$ has a transitive extension. Using the notation of Theorem 9.51, let

$$x = [1,0,0],$$
$$g[\lambda,\mu,\nu] = [\mu,\lambda,\nu],$$

and

$$h_1 = (\infty, [1,0,0])f_1.$$

In matrix form,

$$g = \begin{bmatrix} 0 & 1 & 0 \\ 1 & 0 & 0 \\ 0 & 0 & 1 \end{bmatrix},$$

so that $\det g = -1 = 1 \in GF(4)$ and g represents an element in $PSL(3,4)$. It is plain that g does not fix $x = [1,0,0]$, and, using Exercise 9.52, that $h_1^2 = 1$. The following computation shows that $(gh_1)^3 = 1$. For $[\lambda,\mu,\nu] \neq [1,0,0]$, $[0,1,0]$, and ∞, one easily sees

$$(gh_1)^3[\lambda,\mu,\nu] = [\lambda\nu + \mu^2(\nu^3+1),\ \mu\nu + \lambda^2(\nu^3+1),\ \nu^2].$$

If $\nu \neq 0$, then $\nu^3 = 1$ and $\nu^3 + 1 = 0$, so that the right side is $[\lambda\nu,\mu\nu,\nu^2]$ $= [\lambda,\mu,\nu]$ (by definition of homogeneous coordinates). If $\nu = 0$ and $\lambda\mu \neq 0$, then the right side is $[\mu^2,\lambda^2,0] = [(\lambda\mu)\mu^2,(\lambda\mu)\lambda^2,0] = [\lambda,\mu,0]$, as desired. The remaining cases $[1,0,0]$, $[0,1,0]$, and ∞ may be quickly handled.

Finally, assume $k \in G_x$. Thus $k \in PSL(3,4)$ is represented by a matrix

$$k = \begin{bmatrix} 1 & * & * \\ 0 & a & b \\ 0 & c & d \end{bmatrix}$$

(because k fixes $x = [1, 0, 0]$) and $\det k = ad - bc = 1$. The reader may calculate that $h_1 k h_1$ is represented by the matrix

$$h_1 k h_1 = \begin{bmatrix} 1 & * & * \\ 0 & a^2 & b^2 \\ 0 & c^2 & d^2 \end{bmatrix}$$

which fixes $[1, 0, 0]$ and whose determinant is $a^2 d^2 - b^2 c^2 = (ad - bc)^2 = 1$. Thus $h_1 G_x h_1 = G_x$, and the theorem is proved. ∎

Theorem 9.55 *There exists a 4-transitive group M_{23} of degree 23 and order* $10,200,960 = 23 \cdot 22 \cdot 21 \cdot 20 \cdot 48 = 2^7 \cdot 3^2 \cdot 5 \cdot 7 \cdot 11 \cdot 23$ *such that the stabilizer of a point is M_{22}.*

Proof The proof follows that of the previous theorem, and so we only provide the necessary information. To $P^2(4) \cup \{\infty\}$ adjoin a new symbol ω. Let

$$x = \infty,$$
$$g = (\infty, [1, 0, 0]) f_1 = \text{the former } h_1,$$

and

$$h_2 = (\omega, \infty) f_2.$$

The reader may use Theorem 9.51 to show $M_{23} = \langle M_{22}, h_2 \rangle$ is a transitive extension of M_{22}. ∎

Theorem 9.56 *There exists a 5-transitive group M_{24} of degree 24 and order* $244,823,040 = 24 \cdot 23 \cdot 22 \cdot 21 \cdot 20 \cdot 48 = 2^{10} \cdot 3^3 \cdot 5 \cdot 7 \cdot 11 \cdot 23$ *such that the stabilizer of a point is M_{23}.*

Proof To $P^2(4) \cup \{\infty, \omega\}$ adjoin a new symbol Ω. Define

$$x = \omega,$$
$$g = (\omega, \infty) f_2 = \text{the former } h_2,$$

and

$$h_3 = (\Omega, \omega) f_3;$$

a proof using Theorem 9.51 shows $M_{24} = \langle M_{23}, h_3 \rangle$ is a transitive extension of M_{23}. ∎

EXERCISE

**9.56. Show that $\langle PSL(3, 4), h_2, h_3 \rangle$ is a subgroup of M_{24} isomorphic to $P\Gamma L(3, 4)$. (HINT: See Exercise 9.55.)

It is evident that the Mathieu groups M_{22}, M_{23}, and M_{24} are not sharply k-transitive for $k = 3, 4, 5$, respectively.

Theorem 9.56 (Miller, 1900)   *The Mathieu groups M_{22}, M_{23}, and M_{24} are simple.*

Proof   Since M_{22} is 3-transitive of degree 22 (and 22 is not a power of 2) and since the stabilizer of a point is the simple group $PSL\,(3, 4)$, Corollary 9.22(ii) shows M_{22} is simple. The group M_{23} is 4-transitive and the stabilizer of a point is the simple group M_{22}, so simplicity of M_{23} follows from Corollary 9.22(i). Finally, M_{24} is 5-transitive and the stabilizer of a point is the simple group M_{23}, so Corollary 9.22(i) shows M_{24} is simple.   ∎

The proof above does not apply to M_{11} because the stabilizer of a point, M_{10}, is not simple.

Theorem 9.57†   *The Mathieu groups M_{11} and M_{12} are simple.*

Proof   It is clear, using Corollary 9.22(i), that the simplicity of M_{11} implies the simplicity of M_{12}.

Assume H is a normal subgroup of M_{11} with $H \neq \{1\}$. By Theorem 9.16, H is transitive of degree 11, and so $|H|$ is divisible by 11. Let P be a Sylow 11-subgroup of H. Since 11^2 does not divide $|M_{11}|$, P is a Sylow 11-subgroup of M_{11} and is cyclic of order 11.

We claim $P \neq N_H(P)$. Otherwise $P = N_H(P)$ and P, being abelian, is plainly in the center of its normalizer. Burnside's Theorem 7.31 provides a normal complement Q to P in H. Therefore $(11, |Q|) = 1$ and Q is even characteristic in H; as $H \lhd M_{11}$, we have $Q \lhd M_{11}$. Theorem 9.16 shows Q is transitive of degree 11, whence $|Q|$ is divisible by 11, a contradiction..

We now compute $N_M(P)$ (when M_{11} and S_{11} occur in this proof as subscripts, we omit the subscript 11). In S_{11}, there are $11!/11 = 10!$ 11-cycles, hence 9! cyclic subgroups of order 11 (each of which consists of the identity and 10 11-cycles). Therefore $[S_{11} : N_S(P)] = 9!$ and $|N_S(P)| = 110$. There is an element τ of order 2 inverting an 11-cycle σ: if $\sigma = (1, 2, 3, \cdots, 11)$, then $\sigma^{-1} = (11, 10, 9, \cdots, 1)$ and $\tau\sigma\tau = \sigma^{-1}$, where $\tau = (1, 11)\,(2, 10)\,(3, 9)\,(4, 8)\,(5, 7)$; note that τ is odd. Since $M_{11} \subset A_{11}$ (Exercise 9.46 shows $M_{10} \subset A_{10}$ and $M_{11} = \langle M_{10}, h \rangle$ where h is even), $N_M(P) = N_S(P) \cap M_{11} \subset N_S(P) \cap A_{11}$; it follows from τ being odd that $|N_M(P)| = 11$ or 55. Finally, observe that $P \subset N_H(P) \subset N_M(P)$, and we have already seen that $P \neq N_H(P)$, so that

$$N_H(P) = N_M(P)$$

(and, incidentally, both have order 55). The Frattini argument gives $M_{11} = HN_M(P) = HN_H(P) = H$, since $N_H(P) \subset H$, and this shows M_{11} is simple.   ∎

The Mathieu groups were the first examples of "sporadic" simple groups, i.e., simple groups that do not appear to belong to some infinite class of simple groups as do, for example, $\mathbf{Z}(p)$, A_n, or $PSL\,(n, q)$.

---

† F. Cole (1896) proved the simplicity of M_{11} and G. A. Miller (1899) proved the simplicity of M_{12}.

STEINER SYSTEMS

Suppose X is a set, possibly endowed with some "structure", and suppose all structure preserving permutations of X form a group, Aut (X), under composition. Of course, every subgroup of Aut (X) acts faithfully on X. For example, if X is a set with no structure, Aut(X) is the symmetric group S_X; if X is a vector space, affine space, or projective space, then Aut(X) is, respectively, GL, the group of affine automorphisms, or $P\Gamma L$. A Steiner system (defined below) is a set endowed with a certain combinatorial structure one may regard as a generalized plane geometry, for its main feature is the existence of certain subsets (called blocks[†]) which may be regarded as generalized lines. The situation is described more precisely by integral parameters.

Definition Let $1 < t < k < v$ be integers. A **Steiner system** of **type** $S(t, k, v)$ is an ordered pair (X, B), where X is a set with v elements and B is a family of subsets of X (called **blocks**), each having k elements, such that every t elements of X lie in a unique block.

Example 15 Let X be the affine plane $V(2, q)$ and define blocks to be the lines in X. Plainly X has q^2 points and there are exactly q points on every line. Because every 2 points lie on a unique line, the affine plane gives a Steiner system of type $S(2, q, q^2)$.

Example 16 Let X be the projective plane $P^2(q)$ and define blocks to be the projective lines. By Theorem 9.37(i), X has $q^2 + q + 1$ points and there are exactly $q + 1$ points on every line. Because every 2 points lie on a unique line, the projective plane $P^2(q)$ gives a Steiner system of type $S(2, q+1, q^2+q+1)$.

Given parameters $1 < t < k < v$, the problem whether there exists a Steiner system of type $S(t, k, v)$ is unsolved. For example, one defines a **projective plane** of **order n** to be a Steiner system of type $S(2, n+1, n^2+n+1)$. It is conjectured that n must be a prime power, but it is still unknown whether there exists a projective plane of order 10.

Strict inequalities $1 < t < k < v$ are assumed for the parameters of a Steiner system of type $S(t, k, v)$ to eliminate uninteresting cases. If $t = 1$, every point lies in a unique block, and the blocks partition X (into subsets having k elements); if $t = k$, every subset having t elements is a block; if $k = v$, there is only one block. In the first case, all "lines" (blocks) are parallel; in the second case, there are too many blocks; in the third case, there are not enough blocks.

EXERCISES

9.57. Let X be an m-dimensional vector space over $GF(2)$ and let B be the family of all planes (affine 2-subsets) of X. Show that if $m \geq 3$, then (X, B) is a

[†] These blocks are not related to the blocks arising in the discussion of primitivity.

Steiner system of type $S(3, 4, 2^m)$. (HINT: Three distinct points cannot be collinear.)

*9.58. Let (X, B) be a Steiner system of type $S(t, k, v)$ with $t \geq 3$. Choose $x \in X$. Define $X' = X - \{x\}$ and define B' as the family of all $\beta - \{x\}$, where β is a block in B containing x. Prove that (X', B') is a Steiner system of type $S(t-1, k-1, v-1)$.

Definition   The Steiner system (X', B') of Exercise 9.58 is called a **contraction** of (X, B).

Here is a familiar method based on counting the elements of a set in two different ways. Let A and B be finite sets and let S be a subset of $A \times B$. For each $a \in A$, define

$$S(a, \quad) = \{b \in B: (a, b) \in S\};$$

for each $b \in B$, define

$$S(\quad, b) = \{a \in A: (a, b) \in S\}.$$

Clearly

$$\sum_{a \in A} |S(a, \quad)| = \sum_{b \in B} |S(\quad, b)|,$$

for both sides equal $|S|$. We deduce the **counting principle**: If $|S(a, \quad)| = m$ for all $a \in A$ and if $|S(\quad, b)| = n$ for all $b \in B$, then

$$m|A| = n|B|.$$

Theorem 9.58   *Let (X, B) be a Steiner system of type $S(t, k, v)$. If b is the number of blocks (i.e., $b = |B|$), then*

$$b = \frac{v(v-1)(v-2) \cdots (v-t+1)}{k(k-1)(k-2) \cdots (k-t+1)};$$

if r is the number of blocks containing a point $x \in X$, then r is independent of x and

$$r = \frac{(v-1)(v-2) \cdots (v-t+1)\dagger}{(k-1)(k-2) \cdots (k-t+1)}.$$

Proof   Let A be the family of all subsets of t (distinct) points in X; thus $|A| = v(v-1) \cdots (v-t+1)/t!$. Let B be the set of blocks, and let $S \subset A \times B$ consist of all $(\{x_1, \cdots, x_t\}, \beta)$ with $\{x_1, \cdots, x_t\} \subset \beta$. Since every set of t distinct points lies in a unique block, $|S(\{x_1, \cdots, x_t\}, \quad)| = 1$; since each block β has k points, $|S(\quad, \beta)| = k(k-1) \cdots (k-t+1)/t!$. The counting principle now gives the desired formula for b.

---

† In the special case of a projective plane, we have $(t, k, v) = (2, q+1, q^2+q+1)$ and $b = v$; this gives a second proof of Theorem 9.37(ii). Moreover, one sees $r = q+1$ in this case, proving that the number of lines through a given point is equal to the number of points on a line.

The second formula follows from the first: the number r of blocks containing x is the number of blocks in the contraction (X', B') [where $X' = X - \{x\}$ and which is of type $S(t-1, k-1, v-1)$]. We thus see that r does not depend on x. ∎

REMARK How many blocks in a Steiner system of type $S(t, k, v)$ contain a pair of points x, y? The contracted system (X', B') obtained by discarding x (see Exercise 9.58) is of type $S(t-1, k-1, v-1)$, and the number r' of blocks in (X', B') containing y is the same as the number of blocks in (X, B) containing $\{x, y\}$. We can compute r' as in the theorem (using the parameters of the contracted system):

$$r' = \frac{(v-2)(v-3) \cdots (v-t+1)}{(k-2)(k-3) \cdots (k-t+1)}.$$

Similarly one may count the number of blocks in (X, B) containing p points, where $1 \le p < t$, obtaining

$$\frac{(v-p)(v-p-1) \cdots (v-t+1)}{(k-p)(k-p-1) \cdots (k-t+1)}.$$

Definition Let (X, B) and (X_1, B_1) be Steiner systems. An **isomorphism** from (X, B) to (X_1, B_1) is a one-one correspondence $f: X \to X_1$ such that $\beta \in B$ implies $f(\beta) \in B_1$. If $(X, B) = (X_1, B_1)$, an isomorphism is called an **automorphism**.

For certain parameters t, k, v, there is a unique (to isomorphism) Steiner system of type $S(t, k, v)$, but there may exist nonisomorphic Steiner systems of the same type. For example, it is known there are nonisomorphic Steiner systems of types $S(2, 3, 13)$ and $S(2, 10, 91) = S(2, 9+1, 9^2 + 9 + 1)$.

Theorem 9.59 *All the automorphisms of a Steiner system (X, B) form a subgroup of S_X.*

Proof The only point needing discussion is whether the inverse of an automorphism h is itself an automorphism. But S_X is finite, so $h^{-1} = h^m$ for some $m > 0$, and it is obvious that h^m is an automorphism when h is. ∎

Notation The group of automorphisms of a Steiner system (X, B) is denoted by $\mathrm{Aut}(X, B)$.

Theorem 9.60 *If (X, B) is a Steiner system, then $\mathrm{Aut}(X, B)$ acts faithfully on B.*

Proof Assume $\varphi \in \mathrm{Aut}(X, B)$ and $\varphi(\beta) = \beta$ for all $\beta \in B$; we must show $\varphi = 1_X$. If $x \in X$, let **star**(x) denote the family of all blocks containing x; thus $|\mathrm{star}(x)| = r$. Since φ is an automorphism, $\varphi(\mathrm{star}(x)) = \mathrm{star}(\varphi(x))$; since φ fixes every block, $\varphi(\mathrm{star}(x)) = \mathrm{star}(x)$. Therefore, $\mathrm{star}(x) = \mathrm{star}(\varphi(x))$, and the number of blocks containing x, namely r, is the same as the number r' of blocks containing $\{x, \varphi(x)\}$. If $x \ne \varphi(x)$, the remark after Theorem 9.58 shows

$$r' = \frac{(v-2)(v-3) \cdots (v-t+1)}{(k-2)(k-3) \cdots (k-t+1)}.$$

The equation $r = r'$ yields $v - 1 = k - 1$, and this contradicts the inequality $k < v$. Therefore $\varphi = 1_X$. ∎

REMARK   It follows from $r \neq r'$ that if x and y are distinct, then there exists a block β with $x \in \beta$ and $y \notin \beta$; hence $\{x\} = \bigcap_{\beta \in \mathrm{star}(x)} \beta$.

We will now see that multiply transitive groups may determine Steiner systems.

Notation   If X is a G-set and U is a subgroup of G, then

$$F(U) = \{x \in X : gx = x \text{ for all } g \in U\}.$$

EXERCISE

**9.59.   If U is a subgroup of G and $g \in G$, denote its conjugate gUg^{-1} by U^g. If X is a G-set, show that $F(U^g) = gF(U)$ and $|F(U^g)| = |F(U)|$.

Definition   If $U \subset H \subset G$ are subgroups, then H **controls fusion** of U if, whenever $U^g \subset H$ for some $g \in G$, there exists $h \in H$ with $U^g = U^h$.

It is clear that H controls fusion of itself ($U = H$) and that H controls fusion of any of its Sylow p-subgroups (any prime p) (for any subgroup of H isomorphic to a Sylow subgroup is a Sylow subgroup).

Theorem 9.61   *Let G be a t-transitive permutation group on X (where $t \geq 2$ and $|X| = n$), let $H = G_{x_1, \cdots, x_t}$ be the stabilizer of t points $x_1, \cdots, x_t$ in X, and let H control fusion of a given subgroup U of H.*

(i) *$N_G(U)$ acts t-transitively on $F(U)$.*
(ii) *(**Carmichael-Witt**)† Let $m = |F(U)|$ and assume $m > t$. If $U \neq \{1\}$ and $U \lhd H$, then (X, B) is a Steiner system of type $S(t, m, n)$, where*

$$B = \{gF(U) : g \in G\}.$$

Proof

(i) Observe that $\{x_1, \cdots, x_t\} \subset F(U)$ because $U \subset H = G_{x_1, \cdots, x_t}$; thus $m = |F(U)| \geq t$. Let $y_1, \cdots, y_t$ be distinct elements of $F(U)$. Since G acts t-transitively on X, there is $g \in G$ with $gy_i = x_i$ for $i = 1, \cdots, t$. It follows that $gUg^{-1} \subset H$ (for gug^{-1} fixes each x_i for every $u \in U$). As H controls fusion of U, there is $h \in H$ with $gUg^{-1} = hUh^{-1}$. Therefore $h^{-1}g \in N_G(U)$ and $(h^{-1}g)y_i = h^{-1}x_i = x_i$ for each i.

---

† This theorem was proved by Witt (1938). However, Carmichael (1931) proved the following theorem. Let X be a t-transitive G-set of degree v. Assume there is a subset β of X (with $t < |\beta| = k < v$) such that for all $g \in G$ and all $x_1, \cdots, x_t$ in β, then $g(\{x_1, \cdots x_t\}) \subset \beta$ implies $g(\beta) = \beta$. If B is defined by $B = \{g\beta : g \in G\}$, then (X, B) is a Steiner system of type $S(t, k, v)$.

(ii) First, the parameters satisfy the necessary inequalities $1 < t < m < n$ (the last because $U \neq \{1\}$). It is clear that each subset $gF(U)$ has cardinal m.

Let $y_1, \cdots, y_t$ be distinct elements of X. There is $g \in G$ with $gx_i = y_i$, $i = 1, \cdots, t$, and thus $\{y_1, \cdots, y_t\} \subset gF(U)$. It remains to show $gF(U)$ is the unique subset of this form containing the y_i. Note that $g^{-1}y_i \in F(U)$ for every i. By (i), there is $h \in N_G(U)$ with $hg^{-1}y_i = x_i$ for all i. Similarly, if $\{y_1, \cdots, y_t\} \subset \gamma F(U)$ for some $\gamma \in G$, there is $\eta \in N_G(U)$ with $\eta\gamma^{-1}y_i = x_i$ for all i. Therefore $hg^{-1}\gamma\eta^{-1}$ fixes each x_i, and hence lies in H: there is $v \in H$ with $hg^{-1}\gamma\eta^{-1} = v$. But $U \triangleleft H$ implies $H \subset N_G(U)$, and so $g^{-1}\gamma = h^{-1}v\eta \in N_G(U)$. Thus $U^\gamma = U^g$ and $\gamma F(U) = gF(U)$, by Exercise 9.59. We have shown (X, B) is a Steiner system of type $S(t, m, n)$. ∎

According to the first part of this theorem, there is a homomorphism $\varphi: N_G(U) \to S_{F(U)}$, namely, the action of $N_G(U)$ on $F(U)$. By Theorem 9.1, whenever $g \in N_G(U)$, then $\varphi(g): x \mapsto gx$ for all $x \in F(U)$ and $\ker \varphi = K \cap N_G(U)$, where K is the stabilizer of the points in $F(U)$. Therefore $\varphi(g)$ is just the restriction of g to $F(U)$ and $\ker \varphi$ consists of those $g \in N_G(U)$ fixing every element of $F(U)$.

Lemma 9.62 *Let $H \subset M_{24}$ be the stabilizer of the five points*

$$\infty, \omega, \Omega, [1, 0, 0], \text{ and } [0, 1, 0].$$

(i) $|H| = 48$.

(ii) *H has a normal Sylow 2-subgroup U that is elementary abelian of order 16.*

(iii) *$F(U) = L \cup \{\infty, \omega, \Omega\}$, where L is the projective line $v = 0$, and so $|F(U)| = 8$.*

(iv) *Only the identity element of M_{24} fixes more than 8 points.*

Proof

(i) Immediate from Theorem 9.7.

(ii) Clearly a Sylow 2-subgroup U of H has order 16. Consider the group $\tilde{H}$ of all matrices over $GF(4)$ of the form

$$A = \lambda \begin{bmatrix} 1 & 0 & a \\ 0 & c & b \\ 0 & 0 & c^{-1} \end{bmatrix}$$

where λ and c are not 0. There are 3 choices for each of λ and c and 4 choices for each of a and b, so that $|\tilde{H}| = 3 \cdot 48$. Plainly the group obtained from $\tilde{H}$ by dividing out scalars has order 48, lies in $PSL(3, 4) \subset M_{24}$, and fixes the five listed points; this group is thus the subgroup H. If $\tilde{U}$ is the subgroup of $\tilde{H}$ consisting of all matrices A with $c = 1$, then $\tilde{U}$ has order $3 \cdot 16$ and corresponds to a subgroup U of H of order 16. Now $\tilde{U} \triangleleft \tilde{H}$, being the kernel of the map

$$A \mapsto \begin{bmatrix} \lambda & 0 & 0 \\ 0 & \lambda c & 0 \\ 0 & 0 & \lambda c^{-1} \end{bmatrix}.$$

Finally, every element in U has order 2, so that U is abelian and elementary.

(iii) Assume $[\lambda, \mu, v] \in F(U)$. For all $a, b \in GF(4)$,

$$\begin{bmatrix} 1 & 0 & a \\ 0 & 1 & b \\ 0 & 0 & 1 \end{bmatrix} \begin{bmatrix} \lambda \\ \mu \\ v \end{bmatrix} = \begin{bmatrix} \lambda + av \\ \mu + bv \\ v \end{bmatrix} = \begin{bmatrix} \xi\lambda \\ \xi\mu \\ \xi v \end{bmatrix}$$

for some $\xi \in GF(4)^{\#}$. If both λ and μ are 0, then $v \neq 0$ and $a = b = 0$; but a and b are not constrained. Therefore at least one of λ, μ is not zero. If $v \neq 0$, then $\xi = 1$ and both a and b are forced to be 0. We conclude that $v = 0$. All such points do lie in $F(U)$, and they form a projective line over $GF(4)$ with $4 + 1 = 5$ points. Since $F(U)$ also contains ∞, ω, and Ω, we have $|F(U)| = 8$.

(iv) We show that an element $h \in H^{\#}$ can fix at most 3 points in addition to the five points on the list. Consider the equation for $\xi \in GF(4)^{\#}$:

$$h\begin{bmatrix} \lambda \\ \mu \\ v \end{bmatrix} = \begin{bmatrix} 1 & 0 & a \\ 0 & c & b \\ 0 & 0 & c^{-1} \end{bmatrix} \begin{bmatrix} \lambda \\ \mu \\ v \end{bmatrix} = \begin{bmatrix} \lambda + av \\ c\mu + bv \\ c^{-1}v \end{bmatrix} = \begin{bmatrix} \xi\lambda \\ \xi\mu \\ \xi v \end{bmatrix}.$$

If $v = 0$, we may assume $\lambda \neq 0$ (since $[0, 1, 0]$ is on the list). Thus $\lambda = \lambda + av = \xi\lambda$ implies $\xi = 1$ which, coupled with $c\mu = c\mu + bv = \xi\mu$, gives $c = 1$. It follows that $h \in U$, by a computation as in (iii), and we are done. If $v \neq 0$, then $c^{-1}v = \xi v$ implies $\xi = c^{-1}$, and we may assume $c \neq 1$ lest we return to the case $h \in U$. But now the equations yield $\lambda = ac(1-c)^{-1}v$ and $\mu = bc(1-c^2)^{-1}v$, and so there is only one additional fixed point, namely, $[ac(1-c)^{-1}, bc(1-c^2)^{-1}, 1]$, and h fixes 6 (< 8) points. ∎

Theorem 9.63   *Neither M_{12} nor M_{24} has a transitive extension.*

Proof   To show M_{12} has no transitive extension, we show there is no sharply 6-transitive group G of degree 13. Such a group G would have order $13 \cdot 12 \cdot 11 \cdot 10 \cdot 9 \cdot 8$. If $g \in G$ has order 5, then g is a product of two disjoint 5-cycles (it cannot be a 5-cycle lest it fix $8 > 6$ points). Denote the fixed points of g by $\{a, b, c\}$. Regard G as a 3-transitive group and let H be the stabilizer of $\{a, b, c\}$. Now $\langle g \rangle$ is a Sylow 5-subgroup of H (it is even a Sylow 5-subgroup of G) so that Theorem 9.61(i) shows $N = N_G(\langle g \rangle)$ acts 3-transitively on $F(\langle g \rangle) = \{a, b, c\}$. This says there is a homomorphism φ of N onto S_3. We claim $C = C_G(\langle g \rangle) \nsubseteq \ker \varphi$; otherwise $S_3 = \operatorname{im} \varphi \subset \operatorname{im} \varphi_*$, where $\varphi_*: N/C \to S_3$ is defined by $\varphi_*(\sigma C) = \varphi(\sigma)$ for $\sigma \in N$. But Theorem 7.1 shows $N/C \hookrightarrow \operatorname{Aut}(\langle g \rangle)$, and this last group is abelian, forcing N/C and S_3 to be abelian, a contradiction. Now $C \triangleleft N$ implies $\varphi(C) \triangleleft \varphi(N) = S_3$; as $\varphi(C) \neq \{1\}$, $\varphi(C) = A_3 \cong \mathbf{Z}(3)$ whence 3 divides $|C|$. Cauchy's theorem provides an element $h \in C = C_G(\langle g \rangle)$ of order 3.

The element gh has order 15 (because g and h commute). Since G has degree 13, gh cannot be a 15-cycle and its factorization into disjoint cycles (of lengths > 1) must involve only 5-cycles and 3-cycles. The only possibilities for the cycle structure of gh are $(5, 5, 3)$, $(5, 3, 3)$, or $(5, 3)$. In any case $(gh)^5$ has order 3 (hence is not 1) and fixes more than 6 points, a contradiction.

A transitive extension G of M_{24} would have degree 25 and order $25 \cdot 24 \cdot 23 \cdot 22 \cdot 21 \cdot 20 \cdot 48$. If $g \in G$ has order 11, then g is a product of two disjoint 11-cycles [it cannot be an 11-cycle lest it fix more than 9 points, contradicting Lemma 9.62 (iv)]. Just as above, there is an element $h \in G$ of order 3 commuting with g, and gh has order 33. Since G has degree 25, gh cannot be a 33-cycle and its factorization into disjoint cycles (of lengths > 1) must involve only 11-cycles and 3-cycles. The only possibilities for the cycle structure of gh are $(11, 11, 3)$ or one 11-cycle and several 3-cycles. In either case, $(gh)^{11}$ has order 3 (hence is not 1) and fixes more than 9 elements, contradicting Lemma 9.62 (iv). ∎

Let us return to multiply transitive groups whose existence is known.

Theorem 9.64

(i) If $X = P^2(4) \cup \{\infty, \omega, \Omega\}$ is regarded as an M_{24}-set and $B = \{gF(U): g \in M_{24}\}$ (where U is the Sylow 2-subgroup of the stabilizer of 5 points), then (X, B) is a Steiner system of type $S(5, 8, 24)$.

(ii) If $gF(U)$ contains $\{\infty, \omega, \Omega\}$, then its remaining 5 points form a projective line. Conversely, for every projective line L', there is $g \in PSL(3, 4) \subset M_{24}$ with $gF(U) = L' \cup \{\infty, \omega, \Omega\}$.

Proof

(i) Lemma 9.62 shows the hypothesis of Theorem 9.61 is satisfied, so (X, B) is a Steiner system. Since M_{24} is 5-transitive and $|F(U)| = 8$, this system is of type $S(5, 8, 24)$.

(ii) The number r'' of blocks containing $\{\infty, \omega, \Omega\}$ is, by the remark after Theorem 9.58, the number r for the (doubly) contracted Steiner system of type $S(3, 6, 22)$. Using the formula of Theorem 9.58 with these parameters, we compute $r'' = 21$. If L is the projective line $v = 0$ (so $L \subset F(U)$) and if $g \in PSL(3, 4) = (M_{24})_{\infty, \omega, \Omega}$, then $gF(U) = g(L) \cup \{\infty, \omega, \Omega\}$. But $PSL(3, 4)$ acts transitively on the lines of $P^2(4)$ (Exercise 9.36) and $P^2(4)$ has exactly 21 lines [Theorem 9.37 (ii)]. Therefore, the 21 blocks containing $\{\infty, \omega, \Omega\}$ are as described. ∎

Corollary 9.65 *A Steiner system of type $S(5, 8, 24)$ exists.*

The next result (and subsequent ones relating Mathieu groups to automorphism groups of Steiner systems) are due to Carmichael (1931) and Witt (1938).

Theorem 9.66 M_{24} *is isomorphic to* Aut(X, B), *where* (X, B) *is a Steiner system of type* $S(5, 8, 24)$.

Proof The previous theorem shows (X, B) is a Steiner system of type $S(5, 8, 24)$, where $X = P^2(4) \cup \{\infty, \omega, \Omega\}$ and $B = \{gF(U): g \in M_{24}\}$ (recall that U is the Sylow 2-subgroup of the stabilizer of 5 points).

Every $f \in M_{24}$ is a permutation of X that carries blocks to blocks; therefore, $M_{24} \subset \text{Aut}(X, B)$. Assume $\varphi \in \text{Aut}(X, B)$. Multiplying by an element of M_{24} if necessary, we may assume φ fixes $\{\infty, \omega, \Omega\}$ and hence that $\varphi: P^2(4) \to P^2(4)$. Using Theorem 9.64 (ii), we see φ is a collineation of $P^2(4)$, for it carries projective lines to projective lines. Now Exercise 9.56 shows M_{24} contains a copy of $P\Gamma L(3, 4)$, the group of all collineations. There thus exists $g \in M_{24}$ with $g|P^2(4) = \varphi|P^2(4)$; since $M_{24} \subset \text{Aut}(X, B)$, it follows that $\varphi g^{-1} \in \text{Aut}(X, B)$. But φg^{-1} can permute only ∞, ω, Ω; because every block has 8 elements and 5 elements determine a block, it follows that φg^{-1} fixes every block. Finally, Theorem 9.60 forces $\varphi g^{-1} = 1$, whence $\varphi = g \in M_{24}$. ∎

It is a fact that there is a unique (to isomorphism) Steiner system of type $S(5, 8, 24)$; therefore, a quick definition of M_{24} is that it is the automorphism group of this system.

We interrupt this discussion, for we may now give a short proof that $PSL(4, 2) \cong A_8$, a fact we mentioned in Chapter 8. An elementary abelian group of order 16 (e.g., the subgroup U) is a 4-dimensional vector space over $GF(2)$ and $\text{Aut}(U) \cong GL(4, 2) = SL(4, 2) = PSL(4, 2)$, whence $|\text{Aut}(U)| = |GL(4, 2)| = 8!/2$, by Theorem 8.10.

Theorem 9.67 $PSL(4, 2) \cong A_8$.

Proof Theorem 9.61(i) asserts $N(U)$ (in M_{24}) acts 5-transitively on $F(U)$, a set with 8 elements. Therefore $|N(U)| = 8 \cdot 7 \cdot 6 \cdot 5 \cdot 4 \cdot s$, where s is the order of the stabilizer of five points in $N(U)$. It follows that if we identify the symmetric group on $F(U)$ with S_8, then $[S_8 : N(U)] = 8!/|N(U)| = t \leq 6$. Now Exercise 9.2 gives $t = 1$ or 2, i.e., $N(U) = S_8$ or A_8. There is a homomorphism $\varphi: N(U) \to \text{Aut}(U)$ given by $g \mapsto \gamma_g$, where γ_g is conjugation by g. Now $\ker \varphi \neq A_8$ or S_8 (if $g \in H \subset N(U)$ has order 3, then $\gamma_g \neq 1$) and so φ is one-one. Since $|\text{Aut}(U)| = 8!/2$, it follows that $A_8 = N(U) \cong \text{Aut}(U) \cong PSL(4, 2)$. ∎

Theorem 9.68 M_{23} is isomorphic to $\text{Aut}(X', B')$, where (X', B') is a Steiner system of type $S(4, 7, 23)$.

REMARK It is known that a Steiner system with these parameters is unique to isomorphism.

Proof If $X' = X - \{\Omega\}$, $\beta' = F(U) - \{\Omega\}$, and $B' = \{g\beta': g \in M_{23}\}$, it is easy to see (X', B') is contracted from the system (X, B) of M_{24}, hence is a Steiner system of type $S(4, 7, 23)$.

One sees at once that $M_{23} \subset \text{Aut}(X', B')$. For the reverse inclusion, let $\varphi \in \text{Aut}(X', B')$ and regard φ as a permutation of X with $\varphi(\Omega) = \Omega$. Multiplying by an element of M_{23} if necessary, we may assume φ fixes ∞ and ω. Since (X', B') is a contraction of (X, B), a block β' in B' containing $\{\infty, \omega\}$ has the form $\beta' = L \cup \{\infty, \omega\}$, where L is a projective line in $P^2(4)$. As in the proof of Theorem 9.66, $\varphi | P^2(4)$ preserves lines, hence is a collineation. Since M_{24} contains a copy of $P\Gamma L(3, 4)$, there is $g \in M_{24}$ with $\varphi | P^2(4) = g | P^2(4)$.

Let $B_\Omega = \{\beta \in B : \Omega \in \beta\}$. Since $\varphi \in \text{Aut}(X', B')$ and $B' = \{\beta - \{\Omega\} : \beta \in B_\Omega\}$, it follows that $\varphi(\beta) \in B$ for all $\beta \in B_\Omega$. Now $g \in M_{24} = \text{Aut}(X, B)$, so $g(\beta)$ is a block for every $\beta \in B_\Omega$ (indeed, for every $\beta \in B$). But the blocks $\varphi(\beta)$ and $g(\beta)$ have at least five points in common [since φ and g agree on $P^2(4)$], so $\varphi(\beta) = g(\beta)$ for all $\beta \in B_\Omega$ (there is a unique block containing four given points). Hence, by the remark after Theorem 9.60,

$$\{\Omega\} = \{\varphi(\Omega)\} = \varphi\left(\bigcap_{\beta \in B_\Omega} \beta\right) = \bigcap_{\beta \in B_\Omega} \varphi(\beta) = \bigcap_{\beta \in B_\Omega} g(\beta) = \{g(\Omega)\},$$

i.e., $g(\Omega) = \Omega$. Therefore $g \in (M_{24})_\Omega = M_{23}$. The argument now ends as that of Theorem 9.66: $\varphi g^{-1} \in \text{Aut}(X', B')$ since $M_{23} \subset \text{Aut}(X', B')$; φg^{-1} fixes B'; finally, $\varphi = g \in M_{23}$. ∎

The next result is mildly surprising.

Theorem 9.69 M_{22} *is a subgroup of index 2 in* $\text{Aut}(X'', B'')$, *where* (X'', B'') *is a Steiner system of type* $S(3, 6, 22)$.

REMARK   It is known that a Steiner system with these parameters is unique to isomorphism.

Proof If $X'' = X - \{\Omega, \omega\}$, $\beta'' = F(U) - \{\Omega, \omega\}$, and $B'' = \{g\beta'' : g \in M_{22}\}$, it is easy to see (X'', B'') is doubly contracted from the system (X, B) of M_{24}, hence is a Steiner system of type $S(3, 6, 22)$.

One sees at once that $M_{22} \subset \text{Aut}(X'', B'')$. Assume $\varphi \in \text{Aut}(X'', B'')$ and regard φ as a permutation of X fixing Ω and ω. As in the proof of Theorem 9.68, we may assume $\varphi(\infty) = \infty$ and $\varphi | P^2(4)$ is a collineation. There thus exists $g \in M_{24}$ with $g | P^2(4) = \varphi | P^2(4)$. Moreover, the proof of Theorem 9.68 gives $g(\omega) = \omega$. Therefore, φg^{-1} is a permutation of X fixing $P^2(4) \cup \{\omega\}$. If φg^{-1} fixes Ω, then $\varphi g^{-1} = 1$ and $\varphi = g$; since φ (hence g) fixes Ω and ω, we have $\varphi \in (M_{24})_{\Omega,\omega} = M_{22}$. The other possibility is $\varphi g^{-1} = (\infty, \Omega)$.

Let us first show $[\text{Aut}(X'', B'') : M_{22}] \leq 2$. If $\varphi_1, \varphi_2 \in \text{Aut}(X'', B'') - M_{22}$, we have just seen that we may assume each fixes ∞, ω, and Ω and that $\varphi_i = (\infty, \Omega)g_i$, where $g_i \in M_{24}$ and $i = 1, 2$. But $\varphi_1^{-1}\varphi_2 = g_1^{-1}g_2 \in (M_{24})_{\Omega,\omega} = M_{22}$, and this shows there are at most two cosets of M_{22}.

Recall the definitions of the elements h_2 and h_3 in M_{24}: $h_2 = (\omega, \infty)f_2$, where f_2 acts on $P^2(4)$ and fixes Ω; $h_3 = (\omega, \Omega)f_3$, where f_3 acts on $P^2(4)$ and fixes ∞. Define $g = h_3 h_2 h_3 = (\Omega, \infty)f_3 f_2 f_3$, and define $\varphi : X'' \to X''$ to be the function fixing ∞ and with $\varphi | P^2(4) = f_3 f_2 f_3$. By Exercise 9.55, $\varphi | P^2(4)$ is a collineation; since φ fixes ∞, it follows that $\varphi \in \text{Aut}(X'', B'')$. On the other hand, $\varphi \notin M_{22}$, lest $\varphi g^{-1} = (\Omega, \infty) \in M_{24}$, contradicting Lemma 9.62 (iv). Therefore $[\text{Aut}(X'', B'') : M_{22}] = 2$. ∎

Corollary 9.70 M_{22} *has an outer automorphism of order 2 and* $\text{Aut}(X'', B'') \cong M_{22} \rtimes \mathbf{Z}(2)$.

Proof The automorphism $\varphi \in \text{Aut}(X'', B'')$ outside M_{22} constructed above has order 2, for f_3 and f_2 have order 2 (Exercises 9.54 and 9.53), and hence the conjugate $\psi = f_3 f_2 f_3$ also has order 2. It follows that $\text{Aut}(X'', B'')$ is a semidirect product of M_{22} by $\mathbf{Z}(2)$. Now observe that this semidirect product is not a direct product: ψ does not commute with $h_1 = (\infty, [1, 0, 0]) f_1$ in M_{22} (if $[\lambda, \mu, \nu] \in P^2(4)$, then $h_1([\lambda, \mu, \nu]) = [\lambda^2 + \mu\nu, \mu^2, \nu^2]$ and $\psi h_1 \psi([\lambda, \mu, \nu])$ $= [\lambda + \beta\mu\nu, \mu^2, \beta\nu^2]$). It follows from Theorem 7.10 that M_{22} is not a complete group, i.e., M_{22} has an outer automorphism. ∎

The "small" Mathieu groups M_{11} and M_{12} are also intimately related to Steiner systems, but the construction of Theorem 9.61 does not apply because the actions of the groups are sharp.

Lemma 9.71 *Consider* $X = GF(9) \cup \{\infty, \omega, \Omega\}$ *as an* M_{12}*-set. There is an isomorphic copy* T *of* S_6 *in* M_{12} *acting 6-transitively on*

$$\beta_0 = \{\infty, \omega, \Omega, 1, -1, 0\}.$$

Proof For each permutation σ of the five elements $\infty, \omega, \Omega, 1, -1$, sharp 5-transitivity provides a unique $\sigma' \in M_{12}$ with $\sigma' | (\beta_0 - \{0\}) = \sigma$, and the function $\sigma \mapsto \sigma'$ is an isomorphism of S_5 with a subgroup S of M_{12}.

Let us regard X as an S-set; what are its orbits? Obviously, one orbit is $\{\infty, \omega, \Omega, 1, -1\}$, and we now investigate the set Y of the other seven points. Choose $g \in S$ of order 3 permuting $\{\infty, \omega, \Omega\}$ (so that g fixes 1 and -1). It follows that g is a product of three disjoint 3-cycles (fewer 3-cycles would fix too many points of X), and the orbits of $\langle g \rangle$ on the 7-point set Y have sizes $(3, 3, 1)$. By Exercise 9.4, the S-orbits of Y have possible sizes $(3, 3, 1)$, $(6, 1)$, or $(3, 4)$ (S cannot act transitively on Y because 7 does not divide $|S| = 120$); by Exercise 9.1, only one of these possibilities can occur, namely, two orbits having sizes 6 and 1, respectively.

What is the S-orbit of size 1; what point is fixed by every element of S? Let $\sigma \in S$ correspond to the transposition $(1, -1)$: σ fixes ∞, ω, Ω and $\sigma(\lambda) = -\lambda$ for $\lambda \in GF(9)$. The only other fixed point of σ is 0, and so $\{0\}$ must be the 1-point S-orbit.

Consider $\gamma \in M_{12}$, where γ fixes ∞, ω, Ω and $\gamma(\lambda) = \lambda + 1$ for $\lambda \in GF(9)$; thus, γ is an element of order 3 in $(M_{12})_{\infty, \omega, \Omega} = (M_{10})_\infty$. Write $H = (M_{10})_\infty$. By Exercise 9.41, all subgroups in H of order 3 are conjugate in H, so H controls fusion of $\langle \gamma \rangle$. Regard M_{12} as 3-transitive with H the stabilizer of 3 points. Theorem 9.61 shows $N(\gamma)$, the normalizer of $\langle \gamma \rangle$ in M_{12}, acts 3-transitively on $F(\langle \gamma \rangle) = \{\infty, \omega, \Omega\}$ (elements of order 3 can fix at most 3 points). This action is a homomorphism $\varphi: N(\gamma) \to S_3$ (permuting ∞, ω, Ω) that must be onto and whose kernel is $N(\gamma) \cap (M_{12})_{\infty, \omega, \Omega} = N(\gamma) \cap (M_{10})_\infty$. Using the hint for Exercise 9.40, one calculates easily that $\ker \varphi = \langle \gamma \rangle$; we conclude that $|N(\gamma)| = 18$. But we can exhibit 18 elements normalizing $\langle \gamma \rangle$. Recall from the definition of M_{11} and M_{12} the elements

$$h = (\infty, \omega)(\alpha^2 \lambda + \alpha \lambda^3) \quad \text{and} \quad k = (\Omega, \omega)\lambda^3.$$

It is easy to see that h and k commute with γ and that $\langle h, k \rangle \cong S_3$. Therefore $\langle \gamma, h, k \rangle = \langle \gamma \rangle \times \langle h, k \rangle$ is a subgroup of order 18 centralizing, hence normalizing $\langle \gamma \rangle$, and

$$N(\gamma) = \langle \gamma, h, k \rangle.$$

Define $T = \langle S, N(\gamma) \rangle$. Since S is a subgroup of T, the possible sizes of the T-orbits of X are $(6, 6)$, $(5, 7)$, $(1, 11)$, and 12. Now $(5, 7)$ is ruled out because 7 does not divide $|T|$ (7 does not even divide $|M_{12}|$); the possibility $(1, 11)$ cannot occur because the common fixed point 0 of S is moved by $\gamma \in T$; there cannot be one orbit of size 12 because T does not act transitively on X: if $t \in S$, then $t1 \in \{\infty, \omega, \Omega, 1, -1\}$ and if $t \in N(\gamma)$, then $t1 \in \{0, 1, -1\}$ [use the explicit description of $N(\gamma)$]. We conclude that T has two orbits of size 6, one of which is $\beta_0 = \{\infty, \omega, \Omega, 1, -1, 0\}$.

Finally, T acts transitively on β_0 and the stabilizer of 0 is S. Since $S(\cong S_5)$ acts sharply 5-transitively on $\beta_0 - \{0\}$, T acts sharply 6-transitively on β_0 and $T \cong S_6$. ∎

Theorem 9.72 *If $X = GF(9) \cup \{\infty, \omega, \Omega\}$ is regarded as an M_{12}-set and $B = \{g\beta_0 : g \in M_{12}\}$, where $\beta_0 = \{\infty, \omega, \Omega, 1, -1, 0\}$, then (X, B) is a Steiner system of type $S(5, 6, 12)$.*

Proof It is plain that every block $g\beta_0$ has six elements. If $\{x_1, \cdots, x_5\} \subset X$, then 5-transitivity provides $g \in M_{12}$ taking the 5-tuple $(\infty, \omega, \Omega, 1, -1)$ to $(x_1, \cdots, x_5)$ and $\{x_1, \cdots, x_5\} \subset g\beta_0$. It remains to prove uniqueness of a block containing five given points.

If $g_1\beta_0$ and $g_2\beta_0$ have five common points, then so do $g_1^{-1}g_1\beta_0 = \beta_0$ and $g_1^{-1}g_2\beta_0$; it thus suffices to prove that whenever $g \in M_{12}$ and β_0 and $g\beta_0$ have five common points $y_1, \cdots, y_5$, then $\beta_0 = g\beta_0$. Now $y_i = gx_i$, where $x_i \in \beta_0$ and $i = 1, \cdots, 5$. Write $\beta_0 = \{x_1, \cdots, x_5, x_6\} = \{y_1, \cdots, y_5, y_6\}$. By Lemma 9.71, there exists $t \in T \subset M_{12}$ with $tx_i = y_i$ for $i = 1, \cdots, 6$. The blocks $t\beta_0$ and $tg\beta_0$ still have five common points. But $t\beta_0 = \beta_0$ since β_0 is a T-orbit, and $tg\beta_0 = \beta_0$ since tg fixes five points, hence is the identity. Therefore $t\beta_0 = tg\beta_0$ and $\beta_0 = g\beta_0$. ∎

Corollary 9.73 *A Steiner system of type $S(5, 6, 12)$ exists.*

Let us examine the blocks of this Steiner system from a geometric viewpoint. Regard $GF(9)$ as $V(2, 3)$, an affine plane over $GF(3)$.

Lemma 9.74 *Let (X, B) be the Steiner system constructed from M_{12}, and let $Y = \{\infty, \omega, \Omega\}$. Every block β containing Y has the form $\beta = Y \cup L$, where L is a line in $GF(9) = V(2, 3)$; conversely, every subset of this form is a block.*

Proof Observe that $\beta_0 = Y \cup L_0$, where $L_0 = \{1, -1, 0\}$, and L_0 is a line in $V(2, 3)$. By Exercise 9.51, M_{12} contains a subgroup $W \cong \mathrm{Aut}(3, 4)$ and each $g \in W$ permutes Y. Therefore $g\beta_0 = Y \cup gL_0$ for every $g \in W$, and gL_0 is a line. One may check that $V(2, 3)$ has 12 lines, so there are 12 blocks of the form

$Y \cup L$: on the other hand, the remark after Theorem 9.58 shows there are exactly 12 blocks containing the 3-point set Y. ∎

Theorem 9.75 M_{12} *is isomorphic to* $\mathrm{Aut}(X, B)$, *where* (X, B) *is a Steiner system of type* $S(5, 6, 12)$.

REMARK   It is known that a Steiner system with these parameters is unique to isomorphism.

Proof   Every $g \in M_{12}$ is a permutation of X carrying blocks to blocks; therefore $M_{12} \subset \mathrm{Aut}(X, B)$. Assume $\varphi \in \mathrm{Aut}(X, B)$. Composing with an element of M_{12} if necessary, we may assume φ permutes $\{\infty, \omega, \Omega\}$ and $\varphi \colon V(2, 3) \to V(2, 3)$ [we are viewing $GF(9)$ as an affine plane $V(2, 3)$ over $GF(3)$]. By Lemma 9.74, $\varphi|V(2, 3)$ is an affine automorphism. But Exercise 9.51 shows there is $g \in M_{12}$ permuting $\{\infty, \omega, \Omega\}$ and with $g|V(2, 3) = \varphi|V(2, 3)$. Since $M_{12} \subset \mathrm{Aut}(X, B)$, it follows that $\varphi g^{-1} \in \mathrm{Aut}(X, B)$. But φg^{-1} permutes $\{\infty, \omega, \Omega\}$ and fixes the other nine points of X. We claim that φg^{-1} fixes every block β; this is clear when $\beta \cap \{\infty, \omega, \Omega\}$ has cardinal 0, 1, or 3. In the remaining case, say, $\beta = \{\infty, \omega, x_1, x_2, x_3, x_4\}$, then $\varphi g^{-1}(\beta)$ must contain ∞ or ω as well as the other four points, and so $\varphi g^{-1}(\beta) = \beta$ because five points determine a block. Theorem 9.60 forces $\varphi g^{-1} = 1$, and therefore $\varphi = g \in M_{12}$. ∎

Theorem 9.76 M_{11} *is isomorphic to* $\mathrm{Aut}(X', B')$, *where* (X', B') *is a Steiner system of type* $S(4, 5, 11)$.

REMARK   It is known that a Steiner system with these parameters is unique to isomorphism.

Proof   Let $X' = X - \{\Omega\}$, $\beta_0' = \beta_0 - \{\Omega\}$, and $B' = \{g\beta_0' \colon g \in M_{11}\}$. It is easy to see (X', B') is contracted from the system (X, B) of M_{12}, hence is a Steiner system of type $S(4, 5, 11)$.

     Clearly $M_{11} \subset \mathrm{Aut}(X', B')$. For the reverse inclusion, let $\varphi \in \mathrm{Aut}(X', B')$ and regard φ as a function on X with $\varphi(\Omega) = \Omega$. Multiplying by an element of M_{11} if necessary, we may assume φ permutes $\{\infty, \omega\}$. Since (X', B') is a contraction of (X, B), a block β' in B' containing $\{\infty, \omega\}$ has the form $\{\infty, \omega\} \cup L$, where L is a line in $V(2, 3)$. As in the proof of Theorem 9.75, $\varphi|V(2, 3)$ is an affine automorphism. Since M_{12} contains a copy of $\mathrm{Aut}(2, 3)$, there is $g \in M_{12}$ with $\varphi|V(2, 3) = g|V(2, 3)$. As in the proof of Theorem 9.68, one may prove $g(\Omega) = \Omega$, whence $g \in (M_{12})_\Omega = M_{11}$. The argument now ends as that of Theorem 9.75: $\varphi g^{-1} \in \mathrm{Aut}(X', B')$ since $M_{11} \subset \mathrm{Aut}(X', B')$; φg^{-1} fixes B'; $\varphi = g \in M_{11}$. ∎

     The subgroup structures of the Mathieu groups are interesting: for example, other simple groups are imbedded in them (a forerunner of this phenomenon in other sporadic groups). Thus, M_{12} contains copies of A_6, $PSL(2, 9)$, and $PSL(2, 11)$, while M_{24} contains copies of A_8, M_{12}, and $PSL(2, 23)$.

     There is an isomorphic copy T of S_6 inside M_{12}. We sketch another proof of the existence of the outer automorphism of S_6. Recall from the construction of T in Lemma 9.71 that there are two T-orbits of size 6, say, β_0 and β_1. An element $\sigma \in T$ of order 5 is a product of two disjoint 5-cycles and it fixes two points; each β_i, $i = 0, 1$, consists of one σ-orbit of size 5 and one σ-orbit of size 1. Regarding M_{12} as 2-transitive, note that $\langle \sigma \rangle \subset M_{10}$, the stabilizer of 2 points; as $\langle \sigma \rangle$ is a Sylow 5-subgroup of M_{10}, Theorem 9.61 shows $N(\sigma)$, the normalizer of $\langle \sigma \rangle$ in M_{12}, acts 2-transitively on the 2-point set $F(\langle \sigma \rangle)$. There is thus an element $\tau \in N(\sigma)$ of order 2, and one may show τ interchanges the orbits β_0 and β_1. It follows that τ normalizes T and conjugation by τ is an outer automorphism of T.

One may show that there is a subgroup G of M_{24} with $G \cong M_{12}$ and there are two G-orbits of size 12. Starting with $g \in G$ of order 11, an argument similar to the one sketched above shows there is an element γ of order 2 in $N(g)$ that interchanges the two G-orbits and which normalizes G. Conjugation by γ is an outer automorphism of M_{12}.

Finally, let us comment on the Steiner systems of types $S(5, 8, 24)$ and $S(5, 6, 12)$. Each of these arises in algebraic coding theory, being the key ingredients in the constructions of the binary and ternary **Golay codes**, respectively. The Steiner system of type $S(5, 8, 24)$ is also used to define the **Leech lattice** in $\mathbf{R}^{24}$, a configuration arising in the study of certain sphere-packing problems as well as in the construction of other sporadic simple groups as subgroups of the (simple) sporadic Conway group of order 4, 157, 776, 806, 543, 360, 000.

chapter ten

Infinite Abelian Groups

THE FIRST REDUCTION: TORSION AND TORSION-FREE

A valuable viewpoint in studying an abelian group G is to consider G as an extension of simpler groups. Of course this reduces the study of G to the study of an extension problem and the study of the simpler groups. In our first reduction, the simpler groups are torsion and torsion-free.

Before we begin, it is convenient to agree upon notation and to make some quite formal definitions. First, all groups are abelian and are written additively; second, the trivial group having one element is denoted by 0 (instead of by {0}).

Definition In the following diagram, capital letters denote groups and arrows denote homomorphisms.

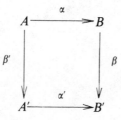

This diagram **commutes** if $\beta\alpha = \alpha'\beta'$.

A special case of such a diagram is a triangular diagram, i.e., one of the homomorphisms is an identity. A common example is

where A is a subgroup of B and i is the inclusion. This diagram commutes if $gi = f$, i.e., $g|A = f$. One also says g **extends** f. A larger diagram composed of squares and triangles **commutes** if any two paths (= composites) from any group to another are equal as functions.

Here is a second formal definition.

Definition A sequence of groups and homomorphisms

$$(*) \qquad \cdots \to A_{k+1} \xrightarrow{f_{k+1}} A_k \xrightarrow{f_k} A_{k-1} \xrightarrow{f_{k-1}} A_{k-2} \to \cdots$$

is **exact** if the image of each map is equal to the kernel of the next map:

$$\operatorname{im} f_k = \ker f_{k-1} \qquad \text{for all } k.$$

EXERCISES

10.1. If $0 \to A \xrightarrow{f} B$ is an exact sequence, then f is one-one. (There is no need to label arrows $0 \to A$ or $C \to 0$, for any such map must be the zero map.)

10.2. If $B \xrightarrow{g} C \to 0$ is an exact sequence, then g is onto.

10.3. If $0 \to A \to B \to C \to 0$ is an exact sequence, then B is an extension of A by C.

10.4. In the exact sequence (*), f_{k+1} is onto if and only if f_{k-1} is one-one.

Definition If G is abelian, the **torsion† subgroup** of G, denoted by tG, is the set of all elements in G of finite order.

Since G is abelian, tG is a subgroup of G.

Definition A group G is **torsion** if $tG = G$; it is **torsion-free** if $tG = 0$.

Theorem 10.1 *Every abelian group G is an extension of a torsion group by a torsion-free group.*

† This terminology comes from algebraic topology, where a space is "twisted" if it has homology groups containing elements of finite order.

Proof It suffices to prove that G/tG is torsion-free. Suppose $n\bar{x} = 0$ for some $\bar{x} \in G/tG$ and some integer $n \neq 0$. If we lift $\bar{x}$ to $x \in G$, then $nx \in tG$; hence, there is an integer $m \neq 0$ with $mnx = 0$. Since $mn \neq 0$, $x \in tG$, and so $\bar{x} = 0$. ∎

Theorem 10.1 reduces the study of arbitrary abelian groups to the study of torsion groups, torsion-free groups, and an extension problem. Our first question is to determine whether this particular extension problem is only virtual or if there is a group whose torsion subgroup is not a direct summand†. Let us first generalize one of our methods of manufacturing groups.

Definition Let $\{A_k: k \in K\}$ be a family of groups indexed by a nonempty set K (K may be infinite). The **direct product**†† of the A_k, denoted by $\Pi_{k \in K} A_k$, is the group consisting of all elements (a_k) in the cartesian product of the A_k under the binary operation:

$$(a_k) + (a_k') = (a_k + a_k').$$

Definition The **direct sum** of the A_k, denoted by $\Sigma_{k \in K} A_k$, is the subgroup of $\Pi_{k \in K} A_k$ consisting of all elements (a_k) almost all of whose coordinates are 0, i.e., only finitely many a_k are nonzero.

If the index set K is finite, then $\Pi_{k \in K} A_k = \Sigma_{k \in K} A_k$; if the index set K is infinite and infinitely many $A_k \neq 0$, the product and the sum are distinct (and are isomorphic only in rare cases).

EXERCISES

10.5. Let $\{A_k: k \in K\}$ be a family of abelian groups. Prove that $t(\Pi A_k) \subset \Pi tA_k$ and $t(\Sigma A_k) = \Sigma tA_k$. Show that the first inclusion is proper when the index set K is the positive integers and $A_k = Z(p^k)$ for some prime p.

10.6. Define functions $p_i: \Pi_{k \in K} A_k \to A_i$ by $(a_k) \mapsto a_i$. Prove that each p_i is a homomorphism of $\Pi_{k \in K} A_k$ onto A_i (p_i is called the ith **projection**). The restriction of p_i (also called a projection) maps $\Sigma_{k \in K} A_k$ onto A_i as well.

As with finite direct sums, there is an internal version of infinite direct sums.

Definition Let $\{A_k: k \in K\}$ be a family of subgroups of G. Then $G = \Sigma A_k$ (**internal direct sum**) if $G = \langle \bigcup_{k \in K} A_k \rangle$ and, for each i, $A_i \cap \langle \bigcup_{k \neq i} A_k \rangle = 0$.

If G is the (external) direct sum ΣA_k, then G is the internal direct sum of its subgroups $\{A_k': k \in K\}$, where A_k' consists of all those "vectors" having elements of A_k in the kth coordinate and 0 everywhere else.

† If an abelian group is a semidirect product, then it is a direct product (or, in additive terminology, a direct sum).
†† Also called *strong direct sum* or *complete direct sum*.

EXERCISES

**10.7. Let $\{A_k: k \in K\}$ be a family of subgroups of G. Prove that $G = \Sigma A_k$ (internal) if and only if every nonzero element $g \in G$ has a unique expression of the form $g = a_{k_1} + \cdots + a_{k_n}$, where $a_{k_i} \in A_{k_i}$, the k_i are distinct, and each $a_{k_i} \neq 0$ (compare Exercise 2.82).

**10.8. Let A be a subgroup of G. Prove that A is a direct summand of G (i.e., there is a subgroup B of G so that G is the internal direct sum of A and B) if and only if there is a (projection) homomorphism $p: G \to A$ with $p(a) = a$ for all $a \in A$ (compare Exercise 7.33). (HINT: Let $B = \ker p$.)

**10.9. Let $\{A_k: k \in K\}$ be a family of subgroups of G. Prove that $G \cong \Sigma A_k$ if and only if, given any abelian group H and any set of homomorphisms $f_k: A_k \to H$, there exists a unique homomorphism $f: G \to H$ that extends each f_k (if $i_k: A_k \to G$ is the inclusion, there is a unique f such that $f i_k = f_k$ for each k), i.e., the following diagrams commute:

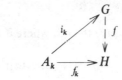

(HINT: First consider the case $H = \Sigma A_k$ to get maps $f: G \to \Sigma A_k$ and $g: \Sigma A_k \to G$. To show that f is an isomorphism with inverse g, use the uniqueness hypothesis in a second diagram in which $H = G$.)

**10.10. Let $\{A_k: k \in K\}$ be a family of subgroups of G. Prove that $G \cong \Pi A_k$ if and only if there exist homomorphisms $p_k: G \to A_k$ with $p_k | A_k = $ identity and, given any abelian group H and any set of homomorphisms $f_k: H \to A_k$, there is a unique homomorphism $f: H \to G$ such that $p_k f = f_k$ for all k, i.e., there is a unique f such that all the following diagrams commute:

Definition   Let $x \in G$ and let n be an integer; x is **divisible by** n if there is an element $y \in G$ with $ny = x$.

EXERCISES

10.11.   Let $x \in G$. Any two solutions in G to the equation $ny = x$ differ by an element z with $nz = 0$.

**10.12.   Let $x \in G$ have order n; if $(m, n) = 1$, then x is divisible by m.

Theorem 10.2   *There exists an abelian group G whose torsion subgroup is not a direct summand.*

Proof   Let P be the set of all primes and let $G = \prod_{p \in P} \mathbf{Z}(p)$.

We claim that there is no nonzero element $x = (x_p)$ in G that is divisible by every prime p. If $qy = x$, then $(qy_p) = (x_p)$, i.e., $qy_p = x_p$ for every p. In particular, if $p = q$, then $x_p = 0$. Therefore, if x is divisible by every prime p, then each coordinate of x is 0, and so $x = 0$.

We shall now show that G/tG contains a nonzero element that is divisible by every prime. Were tG a direct summand of G, then $G \cong tG \oplus (G/tG)$ would also contain such a nonzero element, contradicting the first part of the proof. Consider the element $(a_p) + tG$ in G/tG, where a_p is a generator of $\mathbf{Z}(p)$. If q is a prime, then, by Exercise 10.12, for each $p \neq q$ there is an element $x_p \in \mathbf{Z}(p)$ with $qx_p = a_p$. Hence, if we define $x_q = 0$, then

$$q(x_p) = (a_p) - y,$$

where y has 0 in each coordinate save the qth where it has a_q. Therefore, $y \in tG$ and

$$q((x_p) + tG) = (a_p) - y + tG = (a_p) + tG.$$

We have shown that $(a_p) + tG$ is divisible by every prime; since this element is nonzero (because (a_p) has infinite order), tG cannot be a direct summand of G. ∎

Theorem 10.3   *Every torsion group G is a direct sum of p-primary groups.*

Proof   For any prime p, let

$$G_p = \{x \in G: x \text{ has order a power of } p\}$$

(G_p is called the **p-primary component** of G.) The reader may now prove that $G = \Sigma G_p$ using the proof of the primary decomposition theorem (Theorem 6.1) as a model. ∎

Theorem 10.4   *If G and H are torsion, then $G \cong H$ if and only if $G_p \cong H_p$ for every prime p.*

Proof   Let $f: G \to H$ be an isomorphism and let $g: H \to G$ be its inverse. One checks easily that $f(G_p) \subset H_p$ and, by symmetry, that $g(H_p) \subset G_p$. Let $f_p = f|G_p$ and $g_p = g|H_p$. Both $f_p g_p$ and $g_p f_p$ are identities, so that $G_p \cong H_p$.

Conversely, if $f_p: G_p \to H_p$ are isomorphisms, then there is an isomorphism $f: G \to H$ defined by $(x_p) \mapsto (f_p(x_p))$. ∎

Because of these two theorems, the study of torsion groups is reduced to the study of p-primary groups.

THE SECOND REDUCTION: DIVISIBLE AND REDUCED

A reader asked in Chapter 1 to give examples of infinite abelian groups would probably have responded with the integers, the rationals, and the reals. In this section, we study a common generalization of the latter two groups, the divisible groups. We shall see that every group is an extension of a divisible group by a group having no divisible subgroups.

Definition   A group G is **divisible** if each $x \in G$ is divisible by every $n > 0$.

EXERCISES

10.13.   Prove that the following groups are divisible: additive group of rationals Q; additive group of real numbers R; additive group of complex numbers C; multiplicative group of positive reals; multiplicative group of nonzero elements of an algebraically closed field; the circle group T, i.e., the multiplicative group of all complex numbers z with $|z| = 1$.

*10.14.   The group G/tG constructed in Theorem 10.2 is divisible.

**10.15.   A quotient of a divisible group is divisible.

10.16.   A direct sum (direct product) of groups is divisible if and only if each summand (factor) is divisible.

**10.17.   A torsion-free divisible group is a vector space over Q. (HINT: Verify the axioms after noting that a solution to $ny = x$ exists and is unique.)

The reader knows that every finite-dimensional vector space V has a basis and that any two bases of V have the same number of elements. We now prove the infinite analogs of these theorems; because this may be the reader's first contact with Zorn's lemma (see Appendix IV), we proceed leisurely.

Definition   Let V be a vector space over a field F. A subset X of V is **dependent** if there exist a finite number of vectors $x_1, x_2, \cdots, x_n$ in X and nonzero scalars $a_1, a_2, \cdots, a_n \in F$ such that $\Sigma \, a_i x_i = 0$; otherwise, X is **independent**. A subset X **spans** V if every vector in V is a finite linear combination of vectors in X. A **basis** of V is an independent subset that spans V.

Example 1   Let $V = F[x]$, the vector space of all polynomials in x over a field F. A basis of V is the set $\{1, x, x^2, \cdots, x^m, \cdots\}$.

Theorem 10.5   *If V is a vector space over a field F, then V has a basis. In fact, every independent subset I of V is contained in a basis.*

Proof   Let $\mathscr{S}$ be the family of all independent subsets of V containing I; $\mathscr{S} \neq \varnothing$ because $I \in \mathscr{S}$; partially order $\mathscr{S}$ by ordinary inclusion. Let $\{X_\alpha \colon \alpha \in A\}$

be a simply ordered subset of $\mathscr{S}$, i.e., the X_α are independent subsets of V containing I and, given any two of them, one contains the other. From this it follows that, given any finite number of these X_α, one contains all the others. Now let X be the union of these X_α. It is trivial that X contains each X_α, but we must verify that $X \in \mathscr{S}$ in order that it be an upper bound. Suppose $x_1, x_2, \cdots,$ $x_n \in X$ and $\Sigma\, a_i x_i = 0$, where $a_i \in F$. Now each x_i got into X by being in X_{α_i} for some α_i. There being only finitely many X_{α_i}, one contains all the others; hence, $x_1, x_2, \cdots, x_n$ all lie in this one X_α which is, by hypothesis, independent. Therefore, each $a_i = 0$ and X is independent.

By Zorn's lemma, there is a maximal independent subset Y of V that contains I. We claim that Y is a basis of V, for which it now suffices to prove that Y spans V. Suppose $x \in V$ and x is not a linear combination of elements in Y. Consider the set $Y' = Y \cup \{x\}$. But Y' is an independent set containing I, as the reader may easily check, contradicting the maximality of Y. ∎

Corollary 10.6   *Every subspace W of a vector space V is a direct summand of V.*

Proof   Let X be a basis of W. Since X is independent, Theorem 10.5 says there exists a subset X' of V such that $X \cup X'$ is a basis of V. If W' is the subspace spanned by X', then it is easy to see that $V = W \oplus W'$. ∎

Theorem 10.7   *Let V be a vector space over F. As an abelian group, V is a direct sum of copies of F.*

Proof   Let $B = \{x_k : k \in K\}$ be a basis of V and let F_k denote the one-dimensional subspace generated by x_k. Clearly, each F_k is isomorphic, as a group, to the additive group F.

We claim that the additive group V is isomorphic to $\Sigma_{k \in K}\, F_k$. Since B spans V, every nonzero vector $\alpha \in V$ has an expression $\alpha = \Sigma\, r_{k_i} x_{k_i}$, where the r are nonzero elements of F and all the x are distinct; furthermore, each $r_{k_i} x_{k_i} \in F_{k_i}$. Since B is independent, this is the only expression for α of this kind. By Exercise 10.7, $V \cong \Sigma F_k$. ∎

Corollary 10.8

(i) *Every torsion-free divisible group G is a direct sum of copies of $\mathbf{Q}$;*

(ii) *An abelian group G in which every nonzero element has prime order p is a direct sum of copies of $\mathbf{Z}(p)$.*

Proof

(i) By Exercise 10.17, G is a vector space over $\mathbf{Q}$.

(ii) In the proof of Lemma 6.2 it was verified that G admits a scalar multiplication by elements in $\mathbf{Z}/p\mathbf{Z}$. ∎

We see in particular that the additive group of real numbers is a vector space over $\mathbf{Q}$. A basis of this vector space is usually called a **Hamel basis**, and it is useful for constructing certain analytical counterexamples. For example, a Hamel basis may be used to exhibit a discontinuous function f on the reals satisfying the functional equation

$$f(x + y) = f(x) + f(y)$$

for all real numbers x and y. (If c is the cardinal of the continuum, there are 2^c such functions f but only c continuous real-valued functions.)

EXERCISES

10.18. Let G be a (not necessarily abelian) group, and suppose $|\text{Aut}(G)| = 1$. Prove that G has at most two elements. (The reader now has the power to complete Exercises 7.10 and 7.20.)

**10.19. Let V be a vector space over a field F. Prove that any two bases of V have the same number of elements. [HINTS: (1) The reader need consider only the case in which V is infinite-dimensional, for the finite-dimensional case is well known. (2) The following theorem of set theory may be used: If X is an infinite set and S is the collection of all finite subsets of X, then X and S have the same number of elements.]

Definition Two vector spaces V_1 and V_2 **have the same dimension** if there are bases B_1 of V_1 and B_2 of V_2 having the same number of elements.

EXERCISE

10.20. Let V and W be vector spaces over F, where F is isomorphic to either $\mathbf{Q}$ or $\mathbf{Z}/p\mathbf{Z}$. Show that, as abelian groups, $V \cong W$ if and only if V and W have the same dimension. (This exercise is false for arbitrary fields F. For example, a one-dimensional real vector space is isomorphic, as an abelian group, to a two-dimensional real vector space.)

The easiest example to exhibit of a torsion divisible group is $\mathbf{Q}/\mathbf{Z}$; in particular, its p-primary components are p-primary divisible groups.

Notation If p is a prime, $\mathbf{Z}(p^\infty)$ denotes the p-primary component of $\mathbf{Q}/\mathbf{Z}$.

EXERCISES

10.21. Let $A^{(p)}$ denote the set of all rationals between 0 and 1 of the form m/p^n, where $m, n \geq 0$, under the binary operation "addition modulo 1". For example, if $p = 2$, then $\frac{1}{2} + \frac{1}{2} = 0$, $\frac{1}{2} + \frac{3}{4} = \frac{1}{4}$, etc. Prove that $A^{(p)}$ is a p-primary group and that $\mathbf{Q}/\mathbf{Z} \cong \Sigma \, A^{(p)}$; conclude that $A^{(p)} \cong \mathbf{Z}(p^\infty)$.

10.22.   For a fixed prime p, let G be the set of all pth power roots of unity, i.e., all complex numbers of the form $\exp(2\pi i k/p^n)$, where $k \in \mathbf{Z}$ and $n \geq 0$. Prove that G is a multiplicative group isomorphic to $\mathbf{Z}(p^\infty)$.

10.23.   Show that $\mathbf{Z}(p^\infty)$ is generated by elements $a_1, a_2, a_3, \cdots$, where $pa_1 = 0$, $pa_2 = a_1, \cdots, pa_{n+1} = a_n, \cdots$. If $\langle a_n \rangle$ is the cyclic subgroup of $\mathbf{Z}(p^\infty)$ generated by a_n, then $\langle a_n \rangle \cong \mathbf{Z}(p^n)$, $\langle a_n \rangle \subset \langle a_{n+1} \rangle$ for all n, and

$$\mathbf{Z}(p^\infty) = \bigcup_{n=1}^{\infty} \langle a_n \rangle.$$

10.24.   Prove $\mathbf{Z}(p^\infty)$ is a p-group containing a unique subgroup of order p. Conclude that Theorem 5.36 is false for infinite p-groups.

10.25.   For each $n > 0$, $\mathbf{Q}/\mathbf{Z}$ contains a unique subgroup of order n (which must be cyclic).

10.26.   Every proper subgroup of $\mathbf{Z}(p^\infty)$ is finite, and the set of subgroups is well-ordered by inclusion.

10.27.   Prove that $\mathbf{Z}(p^\infty)$ has the DCC but not the ACC.

Our immediate goal is the classification of divisible groups.

Theorem 10.9   (Injective Property)   *Let A be a subgroup of B and let $f\colon A \to D$ be a homomorphism, where D is divisible. Then f can be extended to a homomorphism $F\colon B \to D$, i.e., an F exists making the diagram below commute.*

$$
\begin{array}{c}
D \\
{\scriptstyle f}\nearrow \ \ \uparrow {\scriptstyle F} \\
0 \to A \hookrightarrow B
\end{array}
$$

Proof   We use Zorn's lemma. Consider the set $\mathscr{S}$ of all pairs (S, h), where S is a subgroup of B containing A and $h\colon S \to D$ extends f; $\mathscr{S}$ is nonempty, for $(A, f) \in \mathscr{S}$. We partially order $\mathscr{S}$ by decreeing $(S_1, h_1) \leq (S_2, h_2)$ if $S_1 \subset S_2$ and h_2 extends h_1. If $\{(S_\alpha, h_\alpha)\}$ is a simply ordered subset of $\mathscr{S}$, define (S_0, h_0) as follows: $S_0 = \cup_\alpha S_\alpha$; if $s \in S_0$, then $s \in S_\alpha$ for some α; define $h_0(s) = h_\alpha(s)$. We leave to the reader the proof that $(S_0, h_0) \in \mathscr{S}$ and that it is an upper bound of $\{(S_\alpha, h_\alpha)\}$. By Zorn's lemma, there exists a maximal pair, (M, h). We shall show that $M = B$, which will complete the proof.

Suppose there is an element $b \in B$ that is not in M. Define $M_1 = M + \langle b \rangle$; clearly, $M \subsetneqq M_1$, so that it suffices to extend h to M_1 to reach a contradiction.

CASE (i) $M \cap \langle b \rangle = 0$. Then $M_1 = M \oplus \langle b \rangle$, and the map $\tilde{h} = h\pi\colon M_1 \to D$, where $\pi\colon M_1 \to M$ is the projection, extends h.

CASE (ii) $M \cap \langle b \rangle \neq 0$. Let k be the smallest positive integer for which $kb \in M$; then every element y in M_1 has the unique expression $y = m + tb$, where $0 \leq t < k$. Since D is divisible, there is an element $x \in D$ with $kx = h(kb)$ (which is defined because $kb \in M$). Define $\tilde{h}: M_1 \to D$ by $\tilde{h}(m + tb) = h(m) + tx$. We leave to the reader the straightforward computation that $\tilde{h}$ is a homomorphism extending h. ∎

Corollary 10.10 *If a divisible group D is a subgroup of a group G, then D is a direct summand of G.*

Proof Consider the diagram

$$
\begin{array}{c}
D \\
{\scriptstyle 1} \nearrow \ \ \uparrow \\
0 \to D \to G
\end{array}
$$

where 1 is the identity map. By Theorem 10.9, there is a homomorphism $p: G \to D$ such that $p(d) = d$ for every $d \in D$. By Exercise 10.8, D is a direct summand of G. ∎

EXERCISES

**10.28. An abelian group G is divisible if and only if it has the injective property. (HINT: Extend homomorphisms $n\mathbf{Z} \to G$ to homomorphisms $\mathbf{Z} \to G$.)

**10.29. G is divisible if and only if $pG = G$ for every prime p; if G is p-primary, G is divisible if and only if $G = pG$.

*10.30. Let G be a nonzero group. Prove that there exists a nontrivial homomorphism $G \to \mathbf{Q}/\mathbf{Z}$.

10.31. A group G is divisible if and only if every nonzero quotient of G is infinite.

10.32. A group G is divisible if and only if G has no maximal subgroups.

Definition If G is an abelian group, dG is the subgroup of G generated by all the divisible subgroups of G (a **divisible subgroup** is a divisible group that is a subgroup).

Lemma 10.11 *dG is a divisible subgroup of G containing every divisible subgroup of G.*

Proof It suffices to prove that dG is divisible. Let $n > 0$ and let $x \in dG$; then $x = x_1 + \cdots + x_k$, where each x_i is in a divisible subgroup D_i of G. Since D_i is divisible, there is an element $y_i \in D_i$ with $ny_i = x_i$. Hence, $y_1 + \cdots + y_k \in dG$ and $n(y_1 + \cdots + y_k) = x$. ∎

Definition An abelian group G is **reduced** if $dG = 0$.

Theorem 10.12   *Every abelian group* $G = dG \oplus R$, *where R is reduced.*

Proof   Since dG is divisible, $G = dG \oplus R$ for some subgroup R (by Corollary 10.10). If R contains a divisible group D, then $dG \oplus D$ is a divisible subgroup of G. But dG contains every divisible subgroup of G, so $D = 0$ and R is reduced.   ∎

EXERCISE

10.33.   If G and H are abelian groups, then $G \cong H$ if and only if $dG \cong dH$ and $G/dG \cong H/dH$.

The reader should note the similarity of the roles of the subgroups tG and dG. We have seen that every abelian group is an extension of a torsion group by a torsion-free group. Now we see that every abelian group is also an extension of a divisible group by a reduced group.

Lemma 10.13   *If G and H are divisible p-primary groups, then* $G \cong H$ *if and only if* $G[p] \cong H[p]$.†

Proof   Necessity is simple, so we need prove only sufficiency. Let $f: G[p] \to H[p]$ be an isomorphism. We may consider f as mapping $G[p] \to H$, so that the injective property implies the existence of a map $F: G \to H$ extending f; we claim that F is an isomorphism.

(i) F is one-one. Let x be a nonzero element of G of order p^n; we show by induction on n that if $F(x) = 0$, then $x = 0$. If $n = 1$, then $x \in G[p]$ and $F(x) = f(x)$; since f is one-one, $x = 0$. Suppose now that x has order p^{n+1} and $F(x) = 0$. Now px has order p^n and $F(px) = 0$. By induction, $px = 0$, which contradicts the fact that x has order p^{n+1}.

(ii) F is onto. Let y be a nonzero element of H of order p^n; we show by induction on n that y is in the image of F. If $n = 1$, then $y \in H[p]$ so that $y \in \text{image } f \subset \text{image } F$. Suppose $p^{n+1} y = 0$ and $p^n y \neq 0$. Since $p^n y \in H[p]$, there is $x \in G$ with $F(x) = p^n y$. Since G is divisible, there is $z \in G$ with $p^n z = x$. Thus, $p^n(y - F(z)) = 0$. Using induction, there is an element $z' \in G$ with $F(z') = y - F(z)$, and so $F(z' + z) = y$.   ∎

Theorem 10.14   *Every divisible group D is a direct sum of copies of* **Q** *and of copies of* $\mathbf{Z}(p^\infty)$ *(for various primes p).*

Proof   It is easy to check that tD is divisible, so we may assume $D \cong tD \oplus (D/tD)$. Since D/tD is a torsion-free divisible group, it is a direct sum of copies of **Q**, by Corollary 10.8. The p-primary component G of tD is divisible. If dimension $G[p]$ (as a vector space over the integers modulo p) is r,

† Recall that $G[p] = \{x \in G: px = 0\}$.

let H be the direct sum of r copies of $\mathbf{Z}(p^\infty)$. Now H is a p-primary divisible group with $H[p] \cong G[p]$. By Lemma 10.13, $G \cong H$. ∎

The structure of divisible groups is not very complicated. One question yet remains: When are two divisible groups isomorphic? If D is a divisible group, let $D_\infty = D/tD$ and $D_p = (tD)[p]$; observe that D_∞ is a vector space over $\mathbf{Q}$ and D_p is a vector space over $\mathbf{Z}/p\mathbf{Z}$.

Theorem 10.15 *If D and D' are divisible groups, then $D \cong D'$ if and only if (i) D_∞ and D'_∞ have the same dimension; (ii) for each prime p, D_p and D'_p have the same dimension.*

Proof Left to the reader. ∎

EXERCISES

10.34. Let G and H be divisible groups, each of which is isomorphic to a subgroup of the other. Prove that $G \cong H$. Is this true if we drop the adjective "divisible"?

10.35. If G and H are divisible and $G \oplus G \cong H \oplus H$, prove that $G \cong H$.

10.36. Prove that the following groups are all isomorphic: $\mathbf{R}/\mathbf{Z}$; the circle group $\mathbf{T}$; $\Pi_p \mathbf{Z}(p^\infty)$; $\mathbf{Q}/\mathbf{Z} \oplus \mathbf{R}$.

*10.37. Let $0 \to A \to B \to C \to 0$ be an exact sequence. If A and C are reduced, then B is reduced.

The reader who knows enough field theory is now able to describe the multiplicative group of the algebraic closure of $\mathbf{Z}/p\mathbf{Z}$. [It is $\Sigma_{q \neq p} \mathbf{Z}(q^\infty)$.]

FREE ABELIAN GROUPS

We now consider an important class of groups that is, in a certain sense, dual to the divisible groups. The reader will see properties of these groups that will remind him of several of the theorems we have just proved.

Definition F is a **free abelian group** on $\{x_k : k \in K\}$ if F is a direct sum of infinite cyclic groups $\langle x_k \rangle$; the set $\{x_k : k \in K\}$ is called a **free set of generators** of F.

Theorem 10.16 *If F is free abelian on $\{x_k : k \in K\}$, every nonzero element $x \in F$ has a unique expression.*

$$x = m_{k_1} x_{k_1} + \cdots + m_{k_n} x_{k_n},$$

where the m are nonzero integers and the k_i are distinct.

Proof Exercise 10.7. ∎

Theorem 10.17   *Let*

$$F = \sum_{i \in I} \mathbf{Z}_i \quad and \quad G = \sum_{j \in J} \mathbf{Z}_j$$

be free abelian groups. Then $F \cong G$ if and only if I and J have the same number of elements.

Proof   Suppose F is free abelian on $\{x_i : i \in I\}$. If p is a prime, then F/pF is a vector space over $\mathbf{Z}/p\mathbf{Z}$. For $a \in F$, set $\bar{a} = a + pF$. We claim that $\{\bar{x}_i : i \in I\}$ is a basis of F/pF. It is clear they span; let us prove the set of them is independent.

Let $\Sigma \bar{m}_i \bar{x}_i = 0$, where $\bar{m}_i \in \mathbf{Z}/p\mathbf{Z}$ and not all $\bar{m}_i = \bar{0}$. If m_i is the representative of $\bar{m}_i$ with $0 \leq m_i < p$, then $\Sigma m_i \bar{x}_i = 0$. In F this equation becomes $\Sigma m_i x_i \in pF$, i.e., there are integers n_i with

$$\sum m_i x_i = p \sum n_i x_i.$$

By Theorem 10.16, $m_i = pn_i$ for all i, so that $\bar{m}_i = \bar{0}$ for all i. This contradiction proves independence. Exercise 10.19 gives $|I| =$ dimension F/pF. Therefore, if $F \cong G$, then $|I| = \dim F/pF = |J|$.

Sufficiency is easy and is left to the reader. ∎

Definition   Let F be free abelian on $\{x_i : i \in I\}$ and let G be free abelian on $\{y_j : j \in J\}$; F and G **have the same rank** if I and J have the same number of elements. If I is finite and has n elements, we say that F has **rank** n.

Theorem 10.17 says that two free abelian groups F and G are isomorphic if and only if they have the same rank. In particular, if $F = G$, any two free sets of generators of F have the same number of elements. The reader will not be misled by the analogy: vector space—free abelian group; basis—free set of generators; dimension—rank. To stress this analogy, we make the following definition.

Definition   A **basis** of a free abelian group F is a free set of generators of F.

Theorem 10.18   *Let F be free abelian with basis $\{x_k : k \in K\}$, G an arbitrary abelian group, and $f: \{x_k : k \in K\} \to G$ any function. There is a unique homomorphism $g: F \to G$ such that*

$$g(x_k) = f(x_k) \quad for\ all\ k.$$

Proof   Define $f_k: \langle x_k \rangle \to G$ by $f_k(mx_k) = mf(x_k)$. It is clear that each f_k is a homomorphism, so that the hypotheses of Exercise 10.9 are satisfied. ∎

It is easy to see that $g(\Sigma m_k x_k) = \Sigma m_k f(x_k)$.

Corollary 10.19   *Every abelian group G is a quotient of a free abelian group.*

Proof   We first show that if X is any set, there exists a free abelian group F having X as a basis. If X consists of one element x, an infinite cyclic group $\mathbf{Z}x$

that has x as a generator can be constructed. For the general case, set $F = \Sigma_{x \in X} \mathbf{Z}x$.

To prove the corollary, let F be the free abelian group with basis G. By Theorem 10.18, the identity function on G extends to a homomorphism $g: F \to G$, and g is clearly onto. Therefore G is a quotient of F. ∎

The construction of a free abelian group on an arbitrary set is quite convenient. For example, in algebraic topology, one wishes to add and subtract continuous functions from an n-simplex to a topological space. This is done by forming the free abelian group on all such functions.

The last corollary provides a convenient way of describing abelian groups.

Definition An abelian group G has **generators** $X = \{x_k : k \in K\}$ and **relations** $\{r_j = 0 : j \in J\}$ if $G \cong F/R$, where F is free abelian on X and R is the subgroup generated by $\{r_j : j \in J\}$.

In each of the following examples, we present abelian groups by generators and relations.

Example 2 $G = \mathbf{Z}(6)$ has generator x and relation $6x = 0$.

Example 3 $G = \mathbf{Z}(6)$ has generators $\{x, y\}$ and relations $\{2x = 0, 3y = 0\}$.

Example 4 $G = \mathbf{Z}(p^\infty)$ has generators

$$\{a_1, a_2, \cdots, a_n, \cdots\}$$

and relations

$$\{pa_1 = 0, \ pa_n = a_{n-1} \text{ if } n > 1\}.$$

Example 5 If G is free abelian on $\{x_k : k \in K\}$, then G has generators $\{x_k : k \in K\}$ and no relations (recall that 0 is the subgroup generated by the empty set). The etymology of the term *free* is apparent now.

We have seen that we can describe an existing group by generators and relations. We can also use generators and relations to construct a group having prescribed properties. For example, is there a reduced p-primary abelian group G such that

$$\bigcap_{n=1}^{\infty} p^n G \neq 0?$$

Construct such a group G as the abelian group with generators

$$a, b_1, b_2, \cdots, b_n, \cdots$$

and relations

$$pa = 0 \quad \text{and} \quad p^n b_n = a \quad \text{for all } n.$$

It is easy to see that G is a p-primary abelian group and that $a \in p^n G$ for all n. We shall prove that $a \neq 0$, leaving the proof that G is reduced, using Exercise 10.37, to the reader.

Let F be the free abelian group on generators

$$a, b_1, b_2, \cdots, b_n, \cdots$$

and let R be the subgroup of F generated by

$$\{pa, p^n b_n - a, \quad n \geq 1\}.$$

We must prove that $a \notin R$. If, on the contrary, $a \in R$, then

$$a = mpa + \sum m_n(p^n b_n - a)$$

for integers m and m_n. Collecting terms gives

$$(1 - mp + \sum m_n)a = \sum m_n p^n b_n.$$

Since $\{a, b_1, b_2, \cdots\}$ is a basis, each $m_n p^n = 0$ and $1 - mp + \sum m_n = 0$. From these equations it follows that each $m_n = 0$, and so $1 = mp$, a contradiction.

Theorem 10.20   (Projective Property)   *Let $\beta: B \to C$ be a homomorphism of B onto C. If F is free abelian and $\alpha: F \to C$ is a homomorphism, then there is a homomorphism $\gamma: F \to B$ with $\beta\gamma = \alpha$, i.e., there is γ making the diagram below commute.*

Proof   Let $X = \{x_k : k \in K\}$ be a basis of F. Since β is onto, for each k there is an element $b_k \in B$ with $\beta(b_k) = \alpha(x_k)$. Define a function $f: X \to B$ by $f(x_k) = b_k$. By Theorem 10.18, there is a homomorphism $\gamma: F \to B$ such that $\gamma(x_k) = b_k$. To check that $\beta\gamma = \alpha$, it suffices to evaluate each on a set of generators of F, e.g., on X. But $\beta\gamma(x_k) = \beta(b_k) = \alpha(x_k)$, as desired.   ∎

Corollary 10.21   *Let G be an abelian group and let $\beta: G \to F$ be onto, where F is free abelian. Then*

$$G = \ker \beta \oplus S,$$

where $S \cong F$.

Proof   Consider the diagram

$$
\begin{array}{c}
F \\
\gamma \swarrow \ \ \downarrow 1 \\
G \xrightarrow{\ \beta\ } F \longrightarrow 0,
\end{array}
$$

where 1 is the identity map. Since F has the projective property, there is a homomorphism $\gamma \colon F \to G$ with $\beta \gamma = 1$. By set theory, γ is one-one, so $S = \operatorname{im} \gamma \cong F$. We let the reader prove, using Theorem 2.24, that $G = \ker \beta \oplus S$. ∎

Another way of stating Corollary 10.21 is that G/K free abelian implies K is a direct summand of G.

Theorem 10.22 *Every subgroup H of a free abelian group F is free abelian; moreover, rank $H \leq$ rank F.*

Proof We give two proofs; the first proof works only when F has finite rank, but it allows us to focus on essentials.

Suppose F has finite rank n; we prove the theorem by induction on n. If $n = 1$, then $F \cong \mathbf{Z}$ and the division algorithm shows any subgroup H of F is 0 or isomorphic to $\mathbf{Z}$. Thus H is free abelian and rank $H \leq 1 =$ rank F. For the inductive step, let $\{x_1, \cdots, x_n\}$ be a basis of F, $F_n = \langle x_1, \cdots, x_{n-1} \rangle$, and $H_n = H \cap F_n$. By induction, H_n is free abelian of rank $\leq n-1$. Now

$$H/H_n = H/(H \cap F_n) \cong (H + F_n)/F_n \subset F/F_n \cong \mathbf{Z}.$$

By Corollary 10.21, $H = H_n$ or $H = H_n \oplus \langle h \rangle$, where $\langle h \rangle \cong \mathbf{Z}$. Therefore H is free abelian of rank $\leq n$.

We now give a second proof that does not assume the rank of F is finite. Let $\{x_k \colon k \in K\}$ be a basis of F which we assume is well-ordered. (That every nonempty set can be somehow well-ordered is equivalent to the axiom of choice.)

For each $k \in K$, define $F_k = \Sigma_{j < k} \langle x_j \rangle$ and $\overline{F}_k = \Sigma_{j \leq k} \langle x_j \rangle$; define $H_k = H \cap F_k$ and $\overline{H}_k = H \cap \overline{F}_k$. Now $F = \cup \overline{F}_k$ and $H = \cup \overline{H}_k$; also, $H_k = H \cap F_k = \overline{H}_k \cap F_k$. Hence

$$\overline{H}_k/H_k = \overline{H}_k/(\overline{H}_k \cap F_k) \cong (\overline{H}_k + F_k)/F_k \subseteq \overline{F}_k/F_k \cong \mathbf{Z}.$$

By Corollary 10.21, either $\overline{H}_k = H_k$ or

$$\overline{H}_k = H_k \quad \langle h_k \rangle, \qquad \text{where } \langle h_k \rangle \cong \mathbf{Z}.$$

We claim H is free abelian on the set of h_k's; note that it will then follow that rank $H \leq$ rank F, for the set of h_k clearly has cardinality $\leq |K| =$ rank F.

Let H^0 be the subgroup of H generated by the h_k. Since $F = \cup \overline{F}_k$, each $h \in H$ (as any element of F) lies in some $\overline{F}_k$. Let $\mu(h)$ be the least index k with $h \in \overline{F}_k$. Suppose $H \neq H^0$ and consider $\{\mu(h) \colon h \in H \text{ and } h \in H^0\}$. There is a least such index j, for K is well-ordered. Choose $h' \in H$ with $\mu(h') = j$ and $h' \in H^0$. Now $\mu(h') = j$ says $h' \in H \cap F_{j+1}$, so that

$$h' = a + mh_j, \qquad a \in H_j, \ m \in \mathbf{Z}.$$

Therefore $a = h' - mh_j \in H$, $a \notin H^0$ (lest $h' \in H^0$), and $\mu(a) < j$, a contradiction. Hence $H = H^0$.

Next, we show that linear combinations of the h_k are unique. It suffices to show that if

$$m_1 h_{k_1} + \cdots + m_n h_{k_n} = 0, \qquad k_1 < \cdots < k_n,$$

then each $m_i = 0$. We may assume $m_n \neq 0$. But then $m_n h_{k_n} \in \langle h_{k_n} \rangle \cap H_{k_n} = 0$, a contradiction. This shows that H is free abelian on the h_k. ∎

EXERCISES

10.38. An abelian group is finitely generated if and only if it is a quotient of a free abelian group of finite rank. Conclude that a direct summand of a finitely generated abelian group is also finitely generated.

10.39. Every subgroup H of a finitely generated abelian group G is itself finitely generated; if G can be generated by r elements, so can H. (We shall show later that both statements are false if we delete the adjective "abelian".)

10.40. The multiplicative group of positive rationals is a free abelian group of (countably) infinite rank. (See Exercise 1.28.)

10.41. If F is a free abelian group of finite rank n, then $\text{Aut}(F)$ is isomorphic to the multiplicative group of all $n \times n$ matrices of determinant ± 1 that have entries in $\mathbf{Z}$.

*10.42. Let $A \xrightarrow{f} B \xrightarrow{g} C \xrightarrow{h} D$ be an exact sequence of free abelian groups. Prove that $B \cong \text{im} f \oplus \ker h$.

**10.43. Let F be a free abelian group of rank n and let H be a subgroup of rank $k < n$. Prove that F/H contains an element of infinite order.

*10.44. An abelian group is free abelian if and only if it has the projective property.

10.45. Let $0 \to F_1 \to F_2 \to \cdots \to F_n \to 0$ be an exact sequence of finitely generated free abelian groups. Prove that

$$\sum_{i=1}^{n} (-1)^i \, \text{rank} \, F_i = 0.$$

Theorem 10.23 *Every abelian group G can be imbedded in a divisible group.*

Proof By Corollary 10.19, there is a free abelian group F with $G \cong F/R$ for some subgroup R of F. Now $F = \Sigma \mathbf{Z}$, so that $F \subset \Sigma \mathbf{Q}$ (just imbed each $\mathbf{Z}$ in a copy of the rationals, $\mathbf{Q}$). Therefore,

$$G \cong F/R \subset (\Sigma \mathbf{Q})/R,$$

and this last group is divisible, being a quotient of a divisible group. ∎

Corollary 10.24 *An abelian group G is divisible if and only if it is a direct summand of every group containing it.*

Proof Necessity is Corollary 10.10. To prove sufficiency, first imbed G in a divisible group D, and then recall that every direct summand of a divisible group is divisible. ∎

There is an analogy between theorems about free abelian groups and divisible groups that may be formalized as follows: Given a commutative diagram containing exact sequences, its **dual** is the commutative diagram containing exact sequences with all arrows reversed. For example, the dual of "$0 \to A \to B$" is "$B \to A \to 0$", i.e., subgroup and quotient group are dual. We let the reader prove that "direct summand" is its own dual, sum and product are dual (see Exercises 10.9 and 10.10), and projective and injective are dual.

FINITELY GENERATED ABELIAN GROUPS

In this section we apply our techniques to classify an important class of abelian groups.

Theorem 10.25 *Every finitely generated torsion-free abelian group G is free abelian.*

Proof We prove the theorem by induction on n, where $G = \langle x_1, \cdots, x_n \rangle$. If $n = 1$, then G is infinite cyclic (or 0 if $x_1 = 0$) and we are done.

Define $\langle x_n \rangle_* = \{y \in G : my \in \langle x_n \rangle$ for some $m \neq 0\}$. It is easy to check that $\langle x_n \rangle_*$ is a subgroup of G and that $G/\langle x_n \rangle_*$ is torsion-free, for $\langle x_n \rangle_*$ is just the inverse image of $t(G/\langle x_n \rangle)$. By induction, $G/\langle x_n \rangle_*$ is free abelian, so that $G = \langle x_n \rangle_* \oplus (\text{free abelian})$. We need only show $\langle x_n \rangle_* \cong \mathbf{Z}$ to complete the proof. Note $\langle x_n \rangle_*$ is finitely generated, for it is a quotient of G.

If $y \in \langle x_n \rangle_*$, then $my = kx_n$ for some $m \neq 0$; the reader may verify that the map $\langle x_n \rangle_* \to \mathbf{Q}$ given by $y \mapsto k/m$ is a well defined homomorphism that is one-one. Thus, $\langle x_n \rangle_*$ is isomorphic to some finitely generated subgroup of the rationals, say, H. Let $H = \langle a_1/b_1, \cdots, a_t/b_t \rangle$ and let $b = \Pi b_i$. Then the map $f : H \to \mathbf{Z}$ defined by $h \to bh$ is one-one, so H, hence $\langle x_n \rangle_*$, is isomorphic to a subgroup of $\mathbf{Z}$, and hence is infinite cyclic. ∎

Theorem 10.26 **(Fundamental Theorem)**† *Every finitely generated abelian group G is a direct sum of primary and infinite cyclic groups, and the number of summands of each kind depends only on G.*

Proof Theorem 10.25 shows the finitely generated torsion-free group G/tG is free abelian; by Corollary 10.21,

$$G = tG \oplus F,$$

where $F \cong G/tG$ is free abelian. Now tG is finitely generated, being a quotient of the finitely generated group G, so Exercise 6.29 shows tG is finite. The basis theorem for finite abelian groups shows tG is a direct sum of primary cyclic groups. The uniqueness assertion for tG is precisely the fundamental theorem of finite abelian groups; the uniqueness of the number of infinite cyclic summands is Theorem 10.17. ∎

---

† It is not clear who discovered this theorem. In 1911, E. Steinitz discovered the structure of finitely generated modules over Dedekind rings, a result that implies the fundamental theorem above.

EXERCISES

**10.46. Prove there is a countable abelian group G that contains an isomorphic copy of every countable abelian group as a subgroup. (HINT: Let G be the direct sum of countably many copies of $\mathbf{Q} \oplus \mathbf{Q}/\mathbf{Z}$.)

10.47. Let F be a free abelian group of finite rank n. Prove that a subgroup H of F has finite index if and only if H is free abelian of rank n.

10.48. Let $\{x_1, \cdots, x_n\}$ be a basis of a free abelian group F. If $k_1, \cdots, k_n$ are integers with $\gcd(k_1, \cdots, k_n) = 1$, then there are elements $y_2, \cdots, y_n$ such that $\{k_1 x_1 + \cdots + k_n x_n, y_2, \cdots, y_n\}$ is a basis of F.

Here is a useful theorem closely related to these ideas. The reader will note that the proof extends to modules over euclidean rings (e.g., $F[x]$ for F a field).†

Theorem 10.27   (Simultaneous Bases)   *Let F be free abelian of finite rank n and let H be a subgroup of finite index. Then there exist bases $\{y_1, \cdots, y_n\}$ of F and $\{h_1, \cdots, h_n\}$ of H such that $h_i \in \langle y_i \rangle$ for all i, i.e., there are integers k_i with $h_i = k_i y_i$.*

Proof   Each $h \in H$ has coordinates with respect to any ordered basis $\{x_1, \cdots, x_n\}$ of F. Choose an ordered basis $\{x_1, \cdots, x_n\}$ of F and an element $h_1 \in H$ so that, among all such choices, the first coordinate is positive and minimal such. If $h_1 = k_1 x_1 + \cdots + k_n x_n$, we claim k_1 divides k_i for each $i \geq 2$. The division algorithm gives $k_i = q_i k_1 + r_i$, where $0 \leq r_i < k_1$. Therefore,

$$h_1 = k_1(x_1 + q_2 x_2 + \cdots + q_n x_n) + r_2 x_2 + \cdots + r_n x_n.$$

Define $y_1 = x_1 + q_2 x_2 + \cdots + q_n x_n$, and note that $\{y_1, x_2, \cdots, x_n\}$ is a basis of F. Since $h_1 = k_1 y_1 + r_2 x_2 + \cdots + r_n x_n$ and we are allowed to reorder bases, our initial "minimal" choice forces $r_2 = \cdots = r_n = 0$; thus, $h_1 = k_1 y_1$.

Now let $h = m_1 y_1 + m_2 x_2 + \cdots + m_n x_n$ be any element of H; we claim k_1 divides m_1. If $m_1 = q k_1 + r$, where $0 \leq r < k_1$, then $h - q h_1 \in H$ has first coordinate $r < k_1$, a contradiction. It follows that $h \mapsto m_1 y_1$ is a projection with image $\langle h_1 \rangle$; therefore, $H = \langle h_1 \rangle \oplus (H \cap \langle x_2, \cdots, x_n \rangle)$ (Exercise 10.8). Since $\langle x_2, \cdots, x_n \rangle$ is free abelian of rank $n-1$ and $H \cap \langle x_2, \cdots, x_n \rangle$ is a subgroup of finite index, the procedure above may be iterated. ∎

It is plain that every finite abelian group G is isomorphic to a quotient F/H, where F is a finitely generated free abelian group and H is a subgroup of finite index; thus Theorem 10.27 gives a new proof of the basis theorem for finite abelian groups.

---

† Every euclidean ring is a principal ideal domain, but not conversely. Nevertheless, Theorem 10.27 does extend to modules over principal ideal domains.

EXERCISE

10.49. Let F be free abelian of rank n, and H a subgroup of the same rank; let $\{x_1, \cdots, x_n\}$ be a basis of F, and $\{y_1, \cdots, y_n\}$ a basis of H, so that $y_i = \Sigma m_{ij}x_j$. Prove that $[F:H] = |\det[m_{ij}]|$. (HINT: First show $|\det[m_{ij}]|$ is independent of the choice of bases of F and of H.)

TORSION GROUPS

Torsion groups can be quite complicated, but there are two special classes of torsion groups that are quite manageable: divisible torsion groups and direct sums of cyclic groups. We shall prove that every torsion group is an extension of a direct sum of cyclics by a divisible group. The proof requires an investigation of a distinguished kind of subgroup.

Definition   A subgroup S of G is **pure** in G if

$$nG \cap S = nS \qquad \text{for every integer } n.$$

It is always true that $nG \cap S \supset nS$, so it is only the reverse inclusion that is significant. Therefore, purity says that whenever $s = ng$ (where $s \in S$ and $g \in G$), there is an element $s' \in S$ with $s = ns'$. In other words, if an element of S is divisible by n in the big group, it is also divisible by n in the subgroup S.

If $G = \langle x \rangle$ is infinite cyclic and if $S = \langle 2x \rangle$, then S is not pure, for $s = 2x$ is divisible by 2 in G, but there is no element y in S with $2y = 2x$.

EXERCISES

10.50. Any direct summand of G is pure in G.

10.51. If G/S is torsion-free, then S is pure; conclude that tG is a pure subgroup of G. Conclude further that a pure subgroup need not be a direct summand.

10.52. If G is torsion-free, a subgroup S of G is pure if and only if G/S is torsion-free.

10.53. Purity is transitive: if K is pure in H and H is pure in G, then K is pure in G.

*10.54. If G is torsion-free, any intersection of pure subgroups is pure. Conclude that if X is a subset of G (where G is torsion-free), then there is a smallest pure subgroup of G containing X; we denote this subgroup by $\langle X \rangle_*$ and call it the **pure subgroup** of G **generated** by X. [See Exercise 10.57 (a).]

10.55. Let $x \in G$, where G is torsion-free. Show that the pure subgroup generated by x is

$$\{y \in G: my \in \langle x \rangle \text{ for some } m \neq 0\}.$$

(We have rediscovered the subgroup $\langle \overset{\prime}{x} \rangle_*$ in the proof of Theorem 10.25.)

10.56.   A pure subgroup of a divisible group is divisible, hence is a direct summand.

*10.57.   (a) Give an example of a group in which the intersection of two pure subgroups is not pure. Conclude Exercise 10.54 does not apply when G has elements of finite order. [HINT: Look in $\mathbf{Z}(p) \oplus \mathbf{Z}(p^3)$.]

(b) Give an example of a group G in which the subgroup generated by two pure subgroups is not pure. (HINT: Take $G = \mathbf{Z} \oplus \mathbf{Z}$.)

**10.58.   Every ascending union of pure subgroups is pure.

**10.59.   Show that an ascending union of direct summands need not be a direct summand. [HINT: Consider the group $G = \Pi \mathbf{Z}(p)$, and let $S_n = \mathbf{Z}(p_1) \oplus \cdots \oplus \mathbf{Z}(p_n)$.]

**10.60.   Let S be pure in G and let $y \in G/S$. Show that y can be lifted to $x \in G$ where x and y have the same order.

Lemma 10.28   *Let T be pure in G. If $T \subset S \subset G$, then S/T is pure in G/T if and only if S is pure in G.*

Proof   If $x \in G$, we shall denote its coset in G/T by $\bar{x}$. Suppose that $ng = s$, where $s \in S$. Then $n\bar{g} = \bar{s}$, so that the purity of S/T guarantees an element $\bar{s}' \in S/T$ with $n\bar{s}' = \bar{s}$. Lifting this equation to G gives

$$ns' - s = t, \qquad \text{for some } t \in T.$$

Hence $ns' - ng = t$, so the purity of T yields $t' \in T$ with $ns' - ng = nt'$. Juggling, we obtain $s = n(s' - t')$; since $T \subset S$, $s' - t' \in S$, and so S is pure in G.

Conversely, assume S is pure in G. If $n\bar{x} = \bar{s}$, then $nx - s = t \in T$, where x and s are liftings of $\bar{x}$ and $\bar{s}$, respectively. There is an element $s_1 \in S$ with $ns_1 = s + t$ (because $T \subset S$), whence $n\bar{s}_1 = \bar{s}$, as desired.   ∎

Lemma 10.29   *A p-primary group G that is not divisible contains a pure cyclic subgroup.*

Proof   Suppose there is an $x \in G[p]$ that is divisible by p^k but not by p^{k+1}; let $p^k y = x$. We let the reader prove that $\langle y \rangle$ is pure in G (Exercise 10.12 says that one need check only powers of p).

We may, therefore, assume that each $x \in G[p]$ is divisible by every power of p. We shall prove, by induction on k, that if $p^k x = 0$ then x is divisible by p. If $k = 1$, $x \in G[p]$ and our claim is certainly true. Suppose $p^{k+1}x = 0$. If $y = p^k x$, then $y \in G[p]$; hence, there is an element $z \in G$ with

$$p^{k+1}z = y = p^k x.$$

Thus $p^k(pz - x) = 0$; by induction, there is $w \in G$ with $pw = pz - x$. Therefore

$$x = p(z - w),$$

as desired. We have shown that $G = pG$, so Exercise 10.29 shows G is divisible. This contradiction completes the proof.   ∎

Definition A subset X of nonzero elements of a group G is **independent** if $\Sigma\, m_\alpha x_\alpha = 0$ implies each $m_\alpha x_\alpha = 0$, where $x_\alpha \in X$ and $m_\alpha \in \mathbf{Z}$.

Our earlier definition of independence (in a vector space) had the m_α scalars in a field, but more important, the conclusion then was that each $m_\alpha = 0$; now we may have elements of finite order and can conclude only that each $m_\alpha x_\alpha = 0$.

Lemma 10.30 *A set X of nonzero elements of G is independent if and only if*

$$\langle X \rangle = \sum_{x \in X} \langle x \rangle.$$

Proof Let $x_0 \in X$ and let $y \in \langle x_0 \rangle \cap \langle X - \{x_0\}\rangle$. Then $y = mx_0$ and $y = \Sigma\, m_\alpha x_\alpha$, where each $x_\alpha \neq x_0$. Therefore,

$$-mx_0 + \sum m_\alpha x_\alpha = 0;$$

thus each term is 0, by independence. Hence, $0 = mx_0 = y$.

The proof of the converse is left to the reader. ∎

Definition A subset X of G is **pure-independent** if X is independent and $\langle X \rangle$ is a pure subgroup of G.

Lemma 10.31 *Let G be a p-primary group. If X is a maximal pure-independent subset of G (i.e., if X is contained in no larger such), then $G/\langle X \rangle$ is divisible.*

Proof By Lemma 10.29, if $G/\langle X \rangle$ is not divisible, it contains a pure cyclic subgroup $\langle \bar{y} \rangle$. Since $\langle X \rangle$ is pure in G, $\bar{y}$ may be lifted to an element $y \in G$, where y and $\bar{y}$ have the same order (Exercise 10.60). We claim that $X^* = \{X, y\}$ is pure-independent, which will contradict the maximality of X. First of all,

$$\langle X \rangle \subset \langle X^* \rangle \subset G$$

and $\langle X^* \rangle / \langle X \rangle = \langle \bar{y} \rangle$, which is pure in $G/\langle X \rangle$. By Lemma 10.28, $\langle X^* \rangle$ is pure in G. Second, suppose

$$my + \sum m_\alpha x_\alpha = 0, \quad x_\alpha \in X, \quad m_\alpha, m \in \mathbf{Z}.$$

In $G/\langle X \rangle$, this equation becomes $m\bar{y} = 0$. Since y and $\bar{y}$ have the same order, $my = 0$; since X is independent, each $m_\alpha x_\alpha = 0$. Hence, X^* is independent and thus pure-independent. ∎

Definition Let G be a torsion group. A subgroup B of G is a **basic subgroup** of G if:

(i) B is a direct sum of cyclic groups;
(ii) B is pure in G;
(iii) G/B is divisible.

Theorem 10.32 (**Kulikov, 1945**) *Every torsion group G contains a basic subgroup.*†

† See Exercise 10.63.

Proof If we show that every p-primary group has a basic subgroup, then it follows from the primary decomposition that every torsion group has a basic subgroup. Assume, therefore, that G is p-primary.

If G is divisible, then $B = 0$ is a basic subgroup. If G is not divisible, then G does contain pure-independent subsets (Lemma 10.29). Since both purity and independence are preserved by ascending unions, so is pure-independence. Therefore, Zorn's lemma may be applied to provide a maximal pure-independent subset X of G. The previous two lemmas show that $B = \langle X \rangle$ is a basic subgroup. ∎

Corollary 10.33 *Every torsion group is an extension of a direct sum of cyclic groups by a divisible group.*

Corollary 10.34 (Prüfer-Baer)† *Let G be a group of bounded order, i.e., $nG = 0$ for some integer $n > 0$. Then G is a direct sum of cyclic groups.*

Proof Let B be a basic subgroup of G. Then G/B is divisible and $n(G/B) = 0$. It follows that $G/B = 0$, i.e., $B = G$. ∎

We have already classified divisible groups. If we can classify direct sums of cyclic groups, then we have classified all torsion groups modulo the extension problem. The question we ask is: If G and H are each direct sums of cyclic groups, when is $G \cong H$? The answer is essentially the same as that for finite groups given in Chapter 6. Recall that if H is a group with $pH = 0$, then it may be viewed as a vector space over $\mathbf{Z}/p\mathbf{Z}$ with dimension $d(H)$, say. We have proved that when G is a finite p-primary group, the number of cyclic summands of order p^n occurring in a decomposition of G is given by $U(n, G) = d(p^{n-1}G/p^nG) - d(p^nG/p^{n+1}G)$. This formula does not generalize to infinite groups because one cannot subtract infinite cardinals.

Assume $G = \Sigma\, C_i$ is a direct sum of p-primary cyclic groups. One can still prove that $d(p^nG/p^{n+1}G)$ is the number of cyclic summands of order $\geq p^{n+1}$. How can we distinguish between those elements in p^nG coming from C_i of order p^{n+1} and those elements in p^nG coming from C_i of larger order? Let a_i be a generator of C_i. If $|C_i| = p^{n+1}$, then p^na_i has order p; if $|C_i| > p^{n+1}$, then p^na_i does not have order p. This elementary observation suggests that we amend our original idea by replacing p^nG by $p^nG \cap G[p]$, i.e., we shall be interested only in elements of order p.

Definition $U\{n, G\} = d(p^nG \cap G[p])/(p^{n+1}G \cap G[p])$.

Of course, the cardinal number $U\{n, G\}$ may be infinite. We let the reader prove that $U\{n, G\} = U(n, G)$ for all $n \geq 0$ when G is finite; indeed, this follows from the next theorem.

Theorem 10.35 *If G is a direct sum of p-primary cyclic groups, then the number of cyclic summands of order p^{n+1} is $U\{n, G\}$.*

---

† Prüfer proved this theorem in 1923 for countable groups G; in 1934, Baer proved the theorem with no cardinality restriction.

Proof Let B_k denote the direct sum of all those cyclic summands of order p^k occurring in a given decomposition of G, and let b_k be the number of summands in B_k (b_k may be 0, finite, or infinite). Thus,

$$G = B_1 \oplus B_2 \oplus \cdots \oplus B_k \oplus \cdots.$$

Now it is easy to see that

$$G[p] = B_1 \oplus pB_2 \oplus \cdots \oplus p^{k-1}B_k \oplus \cdots$$

and

$$p^n G = p^n B_{n+1} \oplus \cdots \oplus p^n B_k \oplus \cdots.$$

Therefore, for all n,

$$p^n G \cap G[p] = p^n B_{n+1} \oplus p^{n+1} B_{n+2} \oplus \cdots,$$

and so

$$(p^n G \cap G[p])/(p^{n+1}G \cap G[p]) \cong p^n B_{n+1}.$$

Hence $U\{n, G\} = d(p^n B_{n+1}) = b_{n+1}$, as desired. ∎

EXERCISES

10.61. Let G and H be direct sums of p-primary cyclic groups. Prove that $G \cong H$ if and only if $U\{n, G\} = U\{n, H\}$ for all $n \geq 0$.

10.62. Prove that $t(\Pi_{n=1}^{\infty} \mathbf{Z}(p^n))$ is not a direct sum of cyclic groups.

*10.63. If G is a p-primary group, then any two basic subgroups of G are isomorphic.

10.64. Give an example of a p-primary group G containing basic subgroups B_1 and B_2 for which $G/B_1 \not\cong G/B_2$. (HINT: Let G be a direct sum of cyclic groups.)

Here is an instance when one can guarantee a pure subgroup is a direct summand.

Theorem 10.36 *Let S be a pure subgroup of G with $nS = 0$ for some $n > 0$; then S is a direct summand of G.*

Proof Let $\pi: G \to G/(S + nG)$ be the natural map. Clearly, this quotient is of bounded order (n times it is 0), so it is a direct sum of cyclic groups, by the Prüfer-Baer theorem. Let $G/(S + nG) = \Sigma \mathbf{Z}(r_\alpha)$, and let $\bar{x}_\alpha$ be a generator of $\mathbf{Z}(r_\alpha)$. For each α, lift $\bar{x}_\alpha$ to $x_\alpha \in G$. Then $r_\alpha x_\alpha \in S + nG$, and so

$$r_\alpha x_\alpha = s_\alpha + nh_\alpha,$$

where $s_\alpha \in S$ and $h_\alpha \in G$. Now r_α divides n, so

$$s_\alpha = r_\alpha \left(x_\alpha - \frac{n}{r_\alpha} h_\alpha \right).$$

Since S is pure, there is an element $s'_\alpha \in S$ with $s_\alpha = r_\alpha s'_\alpha$. Set

$$y_\alpha = x_\alpha - s'_\alpha.$$

We have lifted and adjusted so that $r_\alpha y_\alpha = nh_\alpha$ and $\pi(y_\alpha) = \bar{x}_\alpha$. Let $K = \langle nG, \text{the } y_\alpha \rangle$; we claim $G = S \oplus K$.

(i) $S \cap K = 0$. Let $x \in S \cap K$. Since $x \in K$, $x = \Sigma m_\alpha y_\alpha + nh$; since $x \in S$, $\pi(x) = 0$. Hence, $0 = \Sigma m_\alpha \bar{x}_\alpha$, so that r_α divides m_α for each α (because the set of x_α is independent). But we know that $r_\alpha y_\alpha \in nG$, so surely $m_\alpha y_\alpha \in nG$. Therefore, $x = \Sigma m_\alpha y_\alpha + nh \in nG$; since S is pure, there is an element $s' \in S$ with $x = ns'$. But $nS = 0$; consequently $0 = ns' = x$.

(ii) $S + K = G$. If $g \in G$, then $\pi(g) = \Sigma m_\alpha \bar{x}_\alpha$. Since $\pi(\Sigma m_\alpha y_\alpha) = \Sigma m_\alpha \bar{x}_\alpha$, $g - \Sigma m_\alpha y_\alpha = s + nh \in S + nG$. Therefore, $g = s + (nh + \Sigma m_\alpha y_\alpha) \in S + K$. ∎

Corollary 10.37 *If tG is of bounded order, then tG is a direct summand of G. In particular, tG is a direct summand of G if it is finite.*

Corollary 10.38 *A torsion group G that is not divisible has a p-primary cyclic direct summand for some prime p.*

Proof G is the direct sum of its p-primary components: $G = \Sigma G_p$; since G is not divisible, at least one G_p is not divisible. By Lemma 10.29, G_p has a pure cyclic subgroup S, and S is a direct summand of G_p, hence of G, by Theorem 10.36. ∎

Corollary 10.39 *An indecomposable abelian group G is either torsion or torsion-free.*

Proof Let us suppose that G is an indecomposable group that is neither torsion nor torsion-free, i.e., tG is a proper subgroup of G. Now tG is not divisible, lest it be a summand, and so tG contains a pure cyclic subgroup C, by Lemma 10.29. It follows from Theorem 10.36 that C is a summand of G, a contradiction. ∎

EXERCISES

10.65. A torsion group is indecomposable if and only if it is isomorphic to a subgroup of $\mathbf{Z}(p^\infty)$ for some prime p.

10.66. Show that $\Pi_p \mathbf{Z}(p)$ is not a direct sum of (possibly infinitely many) indecomposable groups.

10.67. (**Kaplansky**) In the following, G is an infinite abelian group.

(a) If every proper subgroup of G is finite, then $G \cong \mathbf{Z}(p^\infty)$ for some p.

(b) If G is isomorphic to every proper subgroup, then $G \cong \mathbf{Z}$.

(c) If G is isomorphic to every proper quotient, then $G \cong \mathbf{Z}(p^\infty)$ for some p.

(d) If every proper quotient of G is finite, then $G \cong \mathbf{Z}$.

A complete classification of all countable torsion groups exists; it is due to Ulm (1933) and Zippin (1935). The proof presupposes a knowledge of ordinal and cardinal numbers. Here are two other interesting results: a *countable* torsion group G is a direct sum of cyclic groups if and only if $\bigcap_n nG = 0$ (Prüfer, 1921); every subgroup of a direct sum of cyclic groups is itself a direct sum of cyclic groups (Kulikov, 1945). The reader is referred to the books of Fuchs, Griffith, and Kaplansky (see bibliography) for details.

TORSION-FREE GROUPS

When we restrict our attention to torsion-free groups, our two previous definitions of independence are equivalent, for $mx = 0$ if and only if $m = 0$ or $x = 0$.

EXERCISES

*10.68. Every torsion-free group G can be imbedded in a vector space V over **Q**. (HINT: First imbed G in a divisible group D and then consider the natural map $D \to D/tD$.)

*10.69. A torsion-free group G has a maximal independent subset with at most r elements if and only if G can be imbedded in an r-dimensional vector space over **Q**. Conclude that any two maximal independent subsets of G have the same number of elements.

Definition The **rank** of a torsion-free abelian group G is the number of elements in a maximal independent subset of G.

Exercise 10.69 shows rank G does not depend on the choice of maximal independent subset of G.

EXERCISE

*10.70. If $0 \to A \to B \to C \to 0$ is an exact sequence of torsion-free groups, then

$$\text{rank } A + \text{rank } C = \text{rank } B.$$

Conclude that any torsion-free group of rank 1 is indecomposable.

Our starting point in examining groups of rank 1 is that each of them is isomorphic to a subgroup of **Q** (Exercise 10.69). Let us first present three nonisomorphic subgroups of **Q**.

G_1: All rationals whose denominator is square-free.
G_2: All **dyadic rationals**, i.e., all rationals of the form $m/2^k$.
G_3: All rationals whose decimal expansion is finite.

EXERCISE

*10.71.   Prove that no two of the groups $G_1, G_2, G_3, \mathbf{Z}, \mathbf{Q}$ are isomorphic.

A perceptive observation is that each of the five groups of Exercise 10.71 can be described by the numbers that are allowed to be denominators (G_3 may be alternatively described as those rationals whose denominators are restricted to be powers of 10).
Let $p_1, p_2, \cdots, p_n, \cdots$ be the sequence of primes.

Definition   A **characteristic** is a sequence

$$(k_1, k_2, \cdots, k_n, \cdots)$$

where each k_n is a nonnegative integer or the symbol ∞.

If G is a subgroup of $\mathbf{Q}$ and if $x \in G$ is nonzero, then x determines a characteristic in the following way: What is the highest power of p_n that divides x in G? That is, for which nonnegative integers k is there an element $y \in G$ satisfying

$$p_n^k y = x?$$

If there is a largest such exponent k, we call it k_n; if there is no largest such exponent, we set $k_n = \infty$.†
It is convenient to write each nonzero integer as a formal infinite product $\Pi p_i^{\alpha_i}$, where the p_i range over all the primes and $\alpha_i \geq 0$ (of course, almost all the $\alpha_i = 0$). Let $m = \Pi p_i^{\alpha_i}$ and $n = \Pi p_i^{\beta_i}$ be given integers. If $a \in G$ has characteristic $(k_1, k_2, \cdots)$, then the definition of characteristic says that there is an $x \in G$ satisfying $mx = na$ if and only if $\alpha_i \leq k_i + \beta_i$ for all i (by convention, $\infty + \beta_i = \infty$).
Each of the five groups in Exercise 10.71 contains $x = 1$. Its characteristic in each group is

$$\mathbf{Z}: \quad (0, 0, 0, \cdots);$$
$$\mathbf{Q}: \quad (\infty, \infty, \infty, \cdots);$$
$$G_1: \quad (1, 1, 1, \cdots);$$
$$G_2: \quad (\infty, 0, 0, \cdots);$$
$$G_3: \quad (\infty, 0, \infty, 0, 0, 0, \cdots).$$

Unfortunately, distinct nonzero elements of the same group G may give rise to distinct characteristics. For example, if $G = G_2 = $ dyadic rationals, the characteristic of 1 is

$$(\infty, 0, 0, \cdots)$$

† k_n is called the p_n-**height** of x.

while the characteristic of $126 = 2 \cdot 3^2 \cdot 7$ is

$$(\infty, 2, 0, 1, 0, 0, \cdots).$$

We are led to the following definition.

Definition Two characteristics are **equivalent** if

(i) they have ∞ in the same coordinates;
(ii) they differ in at most a finite number of coordinates.

It is easy to check that this is an equivalence relation; an equivalence class of characteristics is called a **type**.

Lemma 10.40 *Let G be a subgroup of* $\mathbf{Q}$ *and let x and x' be nonzero elements of G. Then the characteristics of x and of x' are equivalent.*

Proof First of all, if $x' = mx$ for some integer m, then the characteristics of x and x' are equivalent: x' is divisible by every power of p_i that divides x (plus only a few more); x' is divisible by every power of p_i if and only if x is.

Let us now pass to the general case. Since G is a subgroup of $\mathbf{Q}$, there are integers m and n such that

$$mx = nx'.$$

The characteristic of x is equivalent to that of $mx = nx'$ which is equivalent to that of x'. ∎

As a result of this lemma, we may define the **type** $\tau(G)$ of a torsion-free abelian group G of rank 1 (i.e., a subgroup of $\mathbf{Q}$) as the type of the characteristic of any nonzero element of G.

Theorem 10.41 *If G and G' are torsion-free abelian groups of rank* 1, *then* $G \cong G'$ *if and only if* $\tau(G) = \tau(G')$.

Proof Suppose $f: G \to G'$ is an isomorphism. If $x \in G$ is nonzero, then one verifies easily that x and $f(x)$ have the same characteristic. Therefore, $\tau(G) = \tau(G')$.

Assume that $\tau(G) = \tau(G')$ and (without loss of generality) that G and G' are subgroups of $\mathbf{Q}$. If a and a' are nonzero elements of G and G', respectively, then their characteristics $(k_1, k_2, \cdots)$ and $(k'_1, k'_2, \cdots)$ differ in only a finite number of places. If we agree that the notation $\infty - \infty$ means 0, then we may define a nonzero rational number λ by

$$\lambda = \prod p_i^{k_i - k'_i},$$

for it follows from the definition of equivalence and our convention concerning ∞ that almost all the $k_i - k'_i = 0$.

Define $f: G \to \mathbf{Q}$ by $f(x) = \mu x$, where $\mu = \lambda a'/a$. Note that distributivity implies that f is a homomorphism. Now a rational x is in G if and only if there are integers $m = \Pi p_i^{\alpha_i}$ and $n = \Pi p_i^{\beta_i}$ with $mx = na$ and $\alpha_i \le \beta_i + k_i$ for all i; a rational y is in G' if and only if there are integers m and n with $my = na'$ and $\alpha_i \le \beta_i + k_i'$ for all i. We claim that image $f \subset G'$. If $x \in G$, then $mx = na$ and $\alpha_i \le \beta_i + k_i$; hence, $m(\mu x) = n\mu a = (n\lambda)a'$. Since $\alpha_i \le (\beta_i + k_i - k_i') + k_i'$, it follows that $\mu x = f(x) \in G'$. In a similar manner, the reader may see that if $g: G' \to \mathbf{Q}$ is defined by $g(x') = \mu^{-1}x'$, then image $g \subset G$. Therefore, f and g are inverse, and $G \cong G'$. ∎

Theorem 10.42 *If τ is a type, then there exists a group G of rank 1 with $\tau(G) = \tau$.*

Proof Let $(k_1, k_2, \cdots, k_n, \cdots)$ be a characteristic in τ. We define a group G as the subgroup of $\mathbf{Q}$ generated by all rationals of the form $1/m$, where, for all n, p_n^t divides m if and only if $t \le k_n$. It is easy to check that G is a group of rank 1 and that the element 1 in G has the given characteristic. ∎

EXERCISES

10.72. If G is a subring of $\mathbf{Q}$, then G is also a torsion-free (additive) group of rank 1. Prove that the characteristic of 1 in G has only 0 and ∞ as entries.

10.73. Prove that two distinct subrings of $\mathbf{Q}$ are not isomorphic as additive groups, hence they are not isomorphic as rings. Conclude there are uncountably many nonisomorphic subgroups and subrings of $\mathbf{Q}$.

*10.74. If A and B are subgroups of $\mathbf{Q}$, prove there is an exact sequence

$$0 \to A \cap B \to A \oplus B \to A + B \to 0.$$

Definition The **endomorphism ring** of an abelian group G, denoted by End (G), is the set of all endomorphisms of G with multiplication defined as composition and addition defined by

$$(f + g)(x) = f(x) + g(x).$$

It is an easy verification that End(G) is a ring.

EXERCISES

10.75. If G is torsion-free of rank 1, prove that the endomorphism ring of G is isomorphic to a subring of $\mathbf{Q}$. (HINT: Use the injective property of $\mathbf{Q}$.) Given $\tau(G)$, what is the type of the additive group of End(G)?

10.76. Give an example of two nonisomorphic groups having isomorphic endomorphism rings.

10.77. Let G and H be torsion-free of rank 1. If there are nonzero elements $x \in G$ and $y \in H$ with $G/\langle x \rangle \cong H/\langle y \rangle$, prove that $G \cong H$. What is the relation between $G/\langle x \rangle$ and the characteristic of x?

10.78. Let A be the dyadic rationals and let B be the "triadic rationals", i.e., $B = \{m/3^n \in \mathbf{Q}: m, n \in \mathbf{Z}\}$. If G is the subgroup of $\mathbf{Q} \oplus \mathbf{Q}$ generated by $\{(a, 0): a \in A\}$, $\{(0, b): b \in B\}$ and $(1/5, 1/5)$, prove that G is an indecomposable group of rank 2.

It is a theorem of Baer (1937) that there is a Remak-Krull-Schmidt theorem for groups that are direct sums of torsion-free groups of rank 1. Jónsson (1945) proved this is no longer true if one merely assumes the direct summands are indecomposable.

GROTHENDIECK GROUPS

In this last section, we give a construction that will allow us to state the Jordan-Hölder and Remak-Krull-Schmidt theorems in a very elegant way. The basic idea is that the isomorphism classes of a set of groups form a semigroup under the binary operation of direct product; we force this semigroup to be a group.

Definition A set $\mathscr{S}$ of (not necessarily abelian) groups is **closed under direct products** if $\{1\} \in \mathscr{S}$ and $A, B \in \mathscr{S}$ implies $A \times B \in \mathscr{S}$.

Definition If $\mathscr{S}$ is a set of groups closed under direct products, then the **Grothendieck group**† of $\mathscr{S}$, denoted by $K(\mathscr{S})$, is the abelian group with generators the elements of $\mathscr{S}$ and relations

$$A + C = B \quad \text{if} \quad B \cong A \times C.$$

In more detail, let $\mathscr{F}$ be the free abelian group with basis $\mathscr{S}$, and let $\mathscr{R}$ be the subgroup of $\mathscr{F}$ generated by all elements of the form $A + C - (A \times C)$. Then $K(\mathscr{S}) = \mathscr{F}/\mathscr{R}$. If $A \in \mathscr{S}$, we denote $A + \mathscr{R}$ by $[A]$. Note that $[\{1\}] = 0$ (because $\{1\} \cong \{1\} \times \{1\}$), and that if A, $B \in \mathscr{S}$ and $A \cong B$, then $[A] = [B]$ (because $B \cong A \times \{1\}$).

Every element in $K(\mathscr{S})$ has an inverse; if $A \in \mathscr{S}$, what is $-[A]$? That element which, when added to $[A]$, gives 0! The point is that there may be elements of $K(\mathscr{S})$ that are not of the form $[A]$ for some A in $\mathscr{S}$.

Lemma 10.43 *Every element of $K(\mathscr{S})$ has the form $[A] - [B]$, where $A, B \in \mathscr{S}$.*

Proof If $x \in K(\mathscr{S})$, then $x = \sum_{i=1}^{n} A_i - \sum_{j=1}^{m} B_j + \mathscr{R}$, where $A_i, B_j \in \mathscr{S}$ (we do not need integer coefficients for we are willing to repeat an A_i or a B_j several times if necessary). In bracket notation,

$$x = \sum[A_i] - \sum[B_j]$$
$$= [A_1 \times \cdots \times A_n] - [B_1 \times \cdots \times B_m]. \quad \blacksquare$$

---

† After A. Grothendieck, one of the leading algebraic geometers of the twentieth century.

Lemma 10.44 *If A, $B \in \mathscr{S}$, then $[A] = [B] \in K(\mathscr{S})$ if and only if there is a group $C \in \mathscr{S}$ with $A \times C \cong B \times C$.*

Proof If $A \times C \cong B \times C$, then

$$[A] + [C] = [A \times C] = [B \times C] = [B] + [C].$$

Since $K(\mathscr{S})$ is a group, we may cancel $[C]$ to obtain $[A] = [B]$.

Suppose, conversely, that $[A] = [B]$. Then $A - B \in \mathscr{R}$, so there is an equation in $\mathscr{F}$:

$$A - B = \sum (X'_i + X''_i - X_i) - \sum (Y'_j + Y''_j - Y_j),$$

where $X_i = X'_i \times X''_i$ and $Y_j = Y'_j \times Y''_j$. Now transpose to eliminate negative coefficients:

$$A + \sum X_i + \sum (Y'_j + Y''_j) = B + \sum (X'_i + X''_i) + \sum Y_j.$$

Since this equation is a relation among the members of a basis of $\mathscr{F}$, each term appearing on one side occurs on the other side and with equal frequency. It follows that the direct product of all the groups on the left is isomorphic to the direct product of those on the right. If $X = \Pi X_i$, $X' = \Pi X'_i$, etc., then

$$A \times X \times Y' \times Y'' \cong B \times X' \times X'' \times Y.$$

But, by definition, $X \cong X' \times X''$ and $Y \cong Y' \times Y''$, so defining $C = X \times Y' \times Y''$ gives $A \times C \cong B \times C$. ∎

Definition Let $\mathscr{S}$ be a set of groups closed under direct products. A group $G \in \mathscr{S}$ is $\mathscr{S}$-**indecomposable** if $G \neq \{1\}$ and if $G \cong A \times B$ with A, $B \in \mathscr{S}$ implies $A = \{1\}$ or $B = \{1\}$. The **Remak-Krull-Schmidt theorem** holds for $\mathscr{S}$ if each $G \in \mathscr{S}$ is a direct product of $\mathscr{S}$-indecomposables and this factorization is unique: if

$$G \cong A_1 \times \cdots \times A_m \cong B_1 \times \cdots \times B_n$$

with the A's and B's $\mathscr{S}$-indecomposable, then $m = n$ and there is a permutation σ of $\{1, \cdots, n\}$ with $A_i \cong B_{\sigma i}$ for all i.

When we say a group G is a direct product of $\mathscr{S}$-indecomposables, we allow only one factor (G is $\mathscr{S}$-indecomposable) or no factors ($G = \{1\}$).

Theorem 10.45 *Assume $\mathscr{S}$ is a set of groups closed under direct products and such that each $G \in \mathscr{S}$ is a direct product of $\mathscr{S}$-indecomposables. The Remak-Krull-Schmidt theorem holds for $\mathscr{S}$ if and only if $K(\mathscr{S})$ is free abelian with basis X, where X consists of all $[A]$ as A ranges over a complete set of nonisomorphic $\mathscr{S}$-indecomposable groups in $\mathscr{S}$.*

Proof Assume $K(\mathscr{S})$ is free abelian with basis X. Since nonisomorphic $\mathscr{S}$-indecomposable groups determine distinct elements of X, it follows for A and B $\mathscr{S}$-indecomposable that $A \cong B$ if and only if $[A] = [B]$. The hypothesis says each $G \in \mathscr{S}$ is a direct product of $\mathscr{S}$-indecomposables. Suppose

$$G \cong A_1 \times \cdots \times A_n \cong B_1 \times \cdots \times B_m,$$

where the A_i and B_j are $\mathscr{S}$-indecomposable. In $K(\mathscr{S})$,

$$\sum_{i=1}^{n} [A_i] = [A_1 \times \cdots \times A_n] = [B_1 \times \cdots \times B_m] = \sum_{j=1}^{m} [B_j].$$

The independence of X implies $n = m$ and the existence of a permutation σ of $\{1, \cdots, n\}$ with $[A_i] = [B_{\sigma i}]$ for all i. As A_i and $B_{\sigma i}$ are $\mathscr{S}$-indecomposable, our initial observation shows $A_i \cong B_{\sigma i}$ for all i, and thus the Remak-Krull-Schmidt theorem holds for $\mathscr{S}$.

If the Remak-Krull-Schmidt theorem holds for $\mathscr{S}$, it is easy to prove the cancellation law: If $A, B, C \in \mathscr{S}$ and $A \times C \cong B \times C$, then $A \cong B$ (perform an induction on n, where $C = C_1 \times \cdots \times C_n$ and each C_i is $\mathscr{S}$-indecomposable). If $G \in \mathscr{S}$, then $G \cong A_1 \times \cdots \times A_n$, where each A_i is $\mathscr{S}$-indecomposable. In $K(\mathscr{S})$, $[G] = \Sigma[A_i]$; since $K(\mathscr{S})$ is generated by all $[G]$, it follows that X generates $K(\mathscr{S})$. Finally, assume $\sum_{i=1}^{r} m_i[A_i] - \sum_{j=1}^{s} n_j[B_j] = 0$, where $m_i, n_j \geq 0$ and the set $\{A_1, \cdots, A_r, B_1, \cdots, B_s\}$ consists of nonisomorphic $\mathscr{S}$-indecomposable groups. Now $\Sigma m_i[A_i] = \Sigma n_j[B_j]$ implies $[\prod A_i^{m_i}] = [\prod B_j^{n_j}]$ (where A^m denotes the direct product of m copies of A). By Lemma 10.44, there is $C \in \mathscr{S}$ with $C \times \prod A_i^{m_i} \cong C \times \prod B_j^{n_j}$. The cancellation law gives $\prod A_i^{m_i} \cong \prod B_j^{n_j}$ and this plainly violates unique factorization. We have shown $K(\mathscr{S})$ is free abelian with basis X. ∎

EXERCISES

10.79. If $\mathscr{S}$ is a set of groups closed under infinite direct sums (or infinite direct products), then $K(\mathscr{S}) = 0$.

10.80. Compute $K(\mathscr{S})$ when $\mathscr{S}$ is the class† of all finitely generated abelian groups.

10.81. Compute $K(\mathscr{S})$ when $\mathscr{S}$ is the class of all finite abelian groups.

**10.82. If $1 \to X_i' \to X_i \to X_i'' \to 1$ is exact for $i = 1, \cdots, n$, then there is an exact sequence $1 \to \prod X_i' \to \prod X_i \to \prod X_i'' \to 1$.

**10.83. Let A be a group and let $1 \to X' \to X \to X'' \to 1$ be an exact sequence of groups. Then there are exact sequences of the form

$$1 \to X' \to A \times X \to A \times X'' \to 1$$

and

$$1 \to A \times X' \to A \times X \to X'' \to 1.$$

---

† There is a fussy set theoretic problem implicit in this exercise: here $\mathscr{S}$ is a proper class and not a set (see Appendix V). For example, every one-point set can be taken as the underlying set of a group of order 1. The problem is solved by observing that there are only countably many nonisomorphic finitely generated abelian groups, and isomorphic groups determine the same element of $K(\mathscr{S})$. One may thus compute $K(\mathscr{S}')$, where $\mathscr{S}' \subset \mathscr{S}$ is a *set* containing at least one group from every isomorphism class of groups in $\mathscr{S}$. [A more sophisticated solution is to redefine $K(\mathscr{S})$ so its generators are isomorphism classes.]

Definition   A set $\mathscr{S}$ of (not necessarily abelian) groups is **closed under extensions** if $\{1\} \in \mathscr{S}$ and, whenever there is an exact sequence $1 \to A \to E \to B \to 1$ with $A, B \in \mathscr{S}$, then $E \in \mathscr{S}$.

Note that a set closed under extensions is closed under direct products. We now modify the definition of $K(\mathscr{S})$ by imposing different relations on the generators.

Definition   Let $\mathscr{S}$ be a set of groups closed under extensions. The **Grothendieck group** $K^*(\mathscr{S})$ is the abelian group with generators the elements of $\mathscr{S}$ and relations $A + C = B$ when there is an exact sequence $1 \to A \to B \to C \to 1$, i.e., when B is an extension of A by C.

We maintain the bracket notation: if $A \in \mathscr{S}$, then $[A] = A + \mathscr{R}^* \in K^*(\mathscr{S})$, where $\mathscr{R}^*$ is the subgroup of $\mathscr{F}$ generated by the relations in the definition of K^*.

Addition in $K^*(\mathscr{S})$ is interesting.

Lemma 10.46   Let $G = G_0 \supset G_1 \supset \cdots \supset G_n = \{1\}$ be a normal series with factor groups $Q_i = G_{i-1}/G_i$. In $K^*(\mathscr{S})$,

$$[G] = [Q_1] + [Q_2] + \cdots + [Q_n].$$

Proof   The equation $Q_i = G_{i-1}/G_i$ becomes $[Q_i] = [G_{i-1}] - [G_i]$ in $K^*(\mathscr{S})$. Therefore

$$\sum_{i=1}^n [Q_i] = \sum_{i=1}^n ([G_{i-1}] - [G_i]) = [G_0] - [G_n] = [G]. \quad \blacksquare$$

It follows from this lemma that $[A_1 \times \cdots \times A_n] = [A_1] + \cdots + [A_n]$ in $K^*(\mathscr{S})$ since there is a normal series

$$A_1 \times \cdots \times A_n \supset A_2 \times \cdots \times A_n \supset \cdots \supset A_n \supset \{1\}.$$

Lemma 10.47   If $A, B \in \mathscr{S}$, then $[A] = [B] \in K^*(\mathscr{S})$ if and only if there are groups C, U, and V in $\mathscr{S}$ and exact sequences $1 \to U \to A \times C \to V \to 1$ and $1 \to U \to B \times C \to V \to 1$.

Proof   Assume there is a group C in $\mathscr{S}$ such that $A \times C$ and $B \times C$ are extensions of U by V, where U, V are in $\mathscr{S}$. Lemma 10.46 gives the equation in $K^*(\mathscr{S})$:

$$[A \times C] = [U] + [V] = [B \times C].$$

But $A \times C$ is an extension of A by C; hence

$$[A \times C] = [A] + [C]$$

and, similarly,

$$[B \times C] = [B] + [C].$$

Therefore $[A] = [B]$.

Assume $[A] = [B]$. As in the proof of Lemma 10.44, $A - B \in \mathscr{R}^*$ gives an equation in $\mathscr{F}$:

$$A + \sum X_i + \sum (Y'_j + Y''_j) = B + \sum (X'_i + X''_i) + \sum Y_j,$$

where now we know there are exact sequences $1 \to X'_i \to X_i \to X''_i \to 1$ and $1 \to Y'_j \to Y_j \to Y''_j \to 1$. Setting $X = \Pi X_i$, $X' = \Pi X'_i$, etc., the proof of Lemma 10.44 gives

$$A \times X \times Y' \times Y'' \cong B \times X' \times X'' \times Y.$$

Let C be a group isomorphic to both of these. Observe that by Exercise 10.82 there is an exact sequence $1 \to X' \to X \to X'' \to 1$ arising from the definition of X'_i, X_i, X''_i. This gives rise to exact sequences (by Exercise 10.83)

$$1 \to X' \times Y' \to X \times Y' \to X'' \to 1$$

and

$$1 \to X' \times Y' \to A \times [(X \times Y') \times Y''] \to A \times [X'' \times Y''] \to 1.$$

Since the middle term is C, this gives rise to a third exact sequence (Exercise 10.83)

$$1 \to X' \times Y' \to B \times C \to B \times (A \times X'' \times Y'') \to 1.$$

If we define $U = X' \times Y'$ and $V = B \times A \times X'' \times Y''$, then we have an exact sequence $1 \to U \to B \times C \to V \to 1$. We let the reader perform similar manipulations to get an exact sequence $1 \to U \to A \times C \to V \to 1$. ∎

Definition Let $\mathscr{S}$ be a set of groups closed under extensions. A group $G \in \mathscr{S}$ is $\mathscr{S}$-**simple** if $G \neq \{1\}$ and there is no extension $1 \to A \to G \to B \to 1$ with $A, B \in \mathscr{S}$, where $A \neq \{1\}$ and $B \neq \{1\}$. The **Jordan-Hölder theorem** holds for $\mathscr{S}$ if: (i) every $G \in \mathscr{S}$ has a normal series each of whose factor groups is $\mathscr{S}$-simple; (ii) if $\{A_1, \cdots, A_n\}$ and $\{B_1, \cdots, B_m\}$ are the factor groups arising from two such normal series, then $n = m$ and there is a permutation σ of $\{1, \cdots, n\}$ with $A_i \cong B_{\sigma i}$ for all i.

Theorem 10.48 *Let $\mathscr{S}$ be a set of groups closed under extensions and such that each $G \in \mathscr{S}$ has a normal series with $\mathscr{S}$-simple factor groups. Then the Jordan-Hölder theorem holds for $\mathscr{S}$ if and only if $K^*(\mathscr{S})$ is free abelian with basis X, where X consists of all $[A]$ as A ranges over a complete set of nonisomorphic $\mathscr{S}$-simple groups in $\mathscr{S}$.*

Proof Assume $K^*(\mathscr{S})$ is free abelian with basis X. This implies that if A and B are $\mathscr{S}$-simple groups in $\mathscr{S}$, then $[A] = [B]$ if and only if $A \cong B$. Suppose $G \in \mathscr{S}$ has two normal series whose $\mathscr{S}$-simple factor groups are, respectively, $\{A_1, \cdots, A_n\}$ and $\{B_1, \cdots, B_m\}$ (by hypothesis, at least one such normal series for G exists). Lemma 10.46 gives an equation in $K^*(\mathscr{S})$:

$$[A_1] + \cdots + [A_n] = [B_1] + \cdots + [B_m].$$

The independence of X implies $n = m$ and the existence of a permutation σ of

$\{1, \cdots, n\}$ with $[A_i] = [B_{\sigma i}]$ for all i. As A_i and $B_{\sigma i}$ are $\mathscr{S}$-simple, our initial observation shows $A_i \cong B_{\sigma i}$ for all i, and so the Jordan-Hölder theorem holds for $\mathscr{S}$.

If the Jordan-Hölder theorem holds for $\mathscr{S}$, we may speak of *the* $\mathscr{S}$-simple factor groups of G. Moreover, it is easy to see that if A and C lie in $\mathscr{S}$, then the $\mathscr{S}$-simple factor groups of $A \times C$ are the $\mathscr{S}$-simple factor groups of A together with those of C. The existence for each $G \in \mathscr{S}$ of a normal series having $\mathscr{S}$-simple factor groups $A_1, \cdots, A_n$, say, shows (by Lemma 10.46) that in $K^*(\mathscr{S})$ we have $[G] = [A_1] + \cdots + [A_n]$; therefore X generates $K^*(\mathscr{S})$. To show independence of X, suppose

$$\sum_{i=1}^{n} [A_i] - \sum_{j=1}^{m} [B_j] = 0,$$

where the A_i and B_j are $\mathscr{S}$-simple and no A_i is isomorphic to a B_j (we do allow repetitions of A_i or B_j, however). Now $\Sigma\,[A_i] = \Sigma\,[B_j]$ implies $[\Pi A_i] = [\Pi B_j]$, by the remark after Lemma 10.46; by Lemma 10.47, there exists $C \in \mathscr{S}$ so that $C \times \Pi A_i$ and $C \times \Pi B_j$ have the same $\mathscr{S}$-simple factor groups. But our earlier remarks show ΠA_i and ΠB_j have the same $\mathscr{S}$-simple factor groups. Using the remark after Lemma 10.46 once again, we see the $\mathscr{S}$-simple factor groups of ΠA_i are $A_1, \cdots, A_n$ and those of ΠB_j are $B_1, \cdots, B_m$. By the Jordan-Hölder theorem for $\mathscr{S}$, $n = m$ and there is a permutation σ of $\{1, \cdots, n\}$ with $A_i \cong B_{\sigma i}$ for all i. Therefore X is independent and $K^*(\mathscr{S})$ is free abelian with basis X. ∎

The idea that an extension of A by C is a product has now been formalized so that it is, in fact, a product (rather, a sum) of two elements in a group.

EXERCISES

10.84. Prove that the function $K(\mathscr{S}) \to K^*(\mathscr{S})$ that sends $G + \mathscr{R}$ into $G + \mathscr{R}^*$ is a homomorphism onto.

10.85. Compute $K^*(\mathscr{S})$ when $\mathscr{S}$ is the class of all finite abelian groups.

10.86. If $\mathscr{S}$ is the class of all finitely generated abelian groups, prove that $K^*(\mathscr{S}) \cong \mathbf{Z}$.

10.87. Compute $K^*(\mathscr{S})$ when $\mathscr{S}$ is the class of all finite solvable groups.

10.88. If $\mathscr{S}$ is the class of all torsion-free abelian groups of finite rank, prove $G \in \mathscr{S}$ is $\mathscr{S}$-simple if and only if G has rank 1. Prove that $K^*(\mathscr{S}) \neq 0$ by showing that rank induces a homomorphism of $K^*(\mathscr{S})$ onto $\mathbf{Z}$. (HINT: Exercise 10.70.)

10.89. If $\mathscr{S}$ is the class of all torsion-free abelian groups of finite rank, show that the Jordan-Hölder theorem does not hold for $\mathscr{S}$. (HINT: Use Exercise 10.74.) It is known that $K^*(\mathscr{S})$ is nevertheless free abelian in this case.

chapter eleven

Homological
Algebra

THE HOM FUNCTORS

In Chapter 1 we raised the twin questions of describing groups and describing homomorphisms, but our emphasis to this point has been upon groups. In this chapter we focus upon homomorphisms with the ultimate goal of computing some groups of extensions. We restrict ourselves here to abelian groups, but the reader should regard this study as an introduction to the more general theory that includes $H^2(Q, {}_\theta K)$ and the Schur-Zassenhaus lemma (Theorem 7.24).

In this chapter, *group* means abelian group unless we say otherwise.

The fundamental abstraction we need is that of a functor, which we define at once. Let $\mathscr{A}$ be the class of all abelian groups; if $A \in \mathscr{A}$, we write 1_A for the identity map on A.

Definition A function $T: \mathscr{A} \to \mathscr{A}$ is a **covariant** (additive) **functor** if whenever $\alpha: A \to B$ is a homomorphism, there is a homomorphism $T(\alpha): T(A) \to T(B)$ such that:

(i) $T(1_A) = 1_{T(A)}$;

(ii) $T(\beta\alpha) = T(\beta)T(\alpha)$ (where $\beta: B \to C$);

(iii) $T(\alpha + \gamma) = T(\alpha) + T(\gamma)$ (where $\gamma: A \to B$).

The following construction gives an important example of a covariant functor.

275

Definition Let G be a fixed group. Define a function $T: \mathscr{A} \to \mathscr{A}$ by $T(A)$ $= \text{Hom}(G, A)$, the group of all homomorphisms of G into A under the binary operation

$$(f + g)(x) = f(x) + g(x).$$

If $\alpha: A \to B$, define $T(\alpha): T(A) \to T(B)$ by $f \mapsto \alpha f$, where $f \in T(A) = \text{Hom}(G, A)$.

Theorem 11.1 *If G is a group, then $T = \text{Hom}(G, \quad)$ is a covariant functor.*

Proof The verifications of the axioms are immediate. ∎

Definition A function $T: \mathscr{A} \to \mathscr{A}$ is a **contravariant** (additive) **functor** if whenever $\alpha: A \to B$ is a homomorphism, there is a homomorphism $T(\alpha): T(B) \to T(A)$ such that:

   (i) $T(1_A) = 1_{T(A)}$;
  (ii) $T(\beta\alpha) = T(\alpha)T(\beta)$ (where $\beta: B \to C$);
 (iii) $T(\alpha + \gamma) = T(\alpha) + T(\gamma)$ (where $\gamma: A \to B$).

   Observe that the main difference between covariant and contravariant functors is that the latter changes the direction of arrows.

Theorem 11.2 *If K is a group, then $S = \text{Hom}(\quad, K)$ is a contravariant functor.*

Proof If $A \in \mathscr{A}$, define $S(A) = \text{Hom}(A, K)$. If $\alpha: A \to B$ is a homomorphism, define $S(\alpha): \text{Hom}(B, K) \to \text{Hom}(A, K)$ on $g: B \to K$ by $g \mapsto g\alpha$. The remaining details are left to the reader. ∎

   We now have examples of functors of either variance; if we use the term "functor" without a modifier, then what we say is to be true for every functor, contra or co.

EXERCISES

11.1.   The identity functor $J: \mathscr{A} \to \mathscr{A}$, defined by $J(A) = A$ and $J(\alpha) = \alpha$, is a covariant functor.

11.2.   The **zero map** $\alpha: A \to B$ is the homomorphism defined by $\alpha(x) = 0$ for all $x \in A$. If T is a functor, then $T(\alpha) = 0$ for every zero map α, and $T(0) = 0$, where 0 is the zero group.

*11.3.   If $\alpha: A \to B$ is an isomorphism and T is a functor, then $T(\alpha)$ is an isomorphism.

11.4.   Which of the following are functors?

   (a) $T(G) = tG$ and $T(f) = f | tG$.
   (b) $T(G) = dG$ and $T(f) = f | dG$.

(c) $T(G) = G \oplus A$ (where A is fixed) and $T(f) = f \oplus 1_A$.

(d) $T(G) = G$ and $T(f) = -f$.

11.5. Let $A_1, A_2, \cdots, A_n$ be a finite set of groups and let T be a functor; prove that $T(\Sigma A_k) \cong \Sigma T(A_k)$. (HINT: Use Exercise 2.79.) (We shall soon see that this is false if we delete the adjective "finite".)

Definition A **short exact sequence** is an exact sequence of the form

(*) $$0 \to A \overset{\alpha}{\to} B \overset{\beta}{\to} C \to 0;$$

(*) **splits** if there is a map $\delta: C \to B$ with $\beta\delta = 1_C$.

EXERCISES

11.6. The short exact sequence (*) splits if and only if there is a map $\gamma: B \to A$ with $\gamma\alpha = 1_A$.

11.7. If the short exact sequence (*) splits, then $B \cong A \oplus C$.

11.8. If T is a covariant functor and (*) splits, then

$$0 \to T(A) \to T(B) \to T(C) \to 0$$

is a split short exact sequence (similarly for contravariant functors).

To see how functors behave, one must recast definitions in a form recognizable by them. For example, instead of saying an isomorphism $f: A \to B$ is a homomorphism that is one-one and onto, one says instead that there exists a homomorphism $g: B \to A$ with $fg = 1_B$ and $gf = 1_A$ (this is how one solves Exercise 11.3).

Definition A covariant functor T is **left exact** if exactness of

$$0 \to A \overset{\alpha}{\to} B \overset{\beta}{\to} C$$

implies exactness of

$$0 \longrightarrow T(A) \overset{T(\alpha)}{\longrightarrow} T(B) \overset{T(\beta)}{\longrightarrow} T(C);$$

a contravariant functor S is **left exact** if exactness of

$$A \overset{\alpha}{\to} B \overset{\beta}{\to} C \to 0$$

implies exactness of

$$0 \longrightarrow S(C) \overset{S(\beta)}{\longrightarrow} S(B) \overset{S(\alpha)}{\longrightarrow} S(A).$$

Theorem 11.3 $S = \mathrm{Hom}(\ , K)$ *is a left exact functor.*

Proof Let $A \xrightarrow{\alpha} B \xrightarrow{\beta} C \to 0$ be exact. Since S is contravariant, there is a sequence

$$0 \xrightarrow{\quad} S(C) \xrightarrow{S(\beta)} S(B) \xrightarrow{S(\alpha)} S(A),$$

i.e.,

$$0 \xrightarrow{\quad} \mathrm{Hom}(C, K) \xrightarrow{S(\beta)} \mathrm{Hom}(B, K) \xrightarrow{S(\alpha)} \mathrm{Hom}(A, K),$$

which we claim is exact. It must be shown that $S(\beta)$ is one-one and that $\mathrm{im}\, S(\beta) = \ker S(\alpha)$.

(i) $\mathrm{Ker}\, S(\beta) = 0$. Suppose $f: C \to K$ and $S(\beta)f = 0$, i.e., $f\beta = 0$. Then f annihilates image $\beta = C$, since β is onto, and so $f = 0$.

(ii) $\mathrm{Im}\, S(\beta) \subset \ker S(\alpha)$. If $f: C \to K$, then $S(\alpha)S(\beta)(f) = S(\alpha)(f\beta) = f(\beta\alpha) = 0$, since $\beta\alpha = 0$.

(iii) $\mathrm{Ker}\, S(\alpha) \subset \mathrm{im}\, S(\beta)$. Suppose $S(\alpha)(g) = 0$ where $g: B \to K$; thus $g\alpha = 0$. Define $g_{\#}: C \to K$ by $g_{\#}(c) = g(b)$ where $\beta(b) = c$. Now $g_{\#}$ is well defined, for if $\beta(b') = c$, then $b - b' \in \ker \beta = \mathrm{im}\,\alpha$, so that $b - b' = \alpha(a)$ for some $a \in A$. Therefore $g(b - b') = g\alpha(a) = 0$, so that $g(b) = g(b')$. But $S(\beta)(g_{\#}) = g_{\#}\beta = g$, for if $b \in B$ and $c = \beta(b)$, then $g_{\#}\beta(b) = g_{\#}(c) = g(b)$. ∎

Theorem 11.4 $T = \mathrm{Hom}(G, \ )$ *is a left exact functor.*

Proof Similar to the proof given above. ∎

The answer to the question of when "$\to 0$" can be tagged on the end of the functored sequence is given below.

Theorem 11.5 *A group G is free abelian if and only if, whenever* $0 \to A \xrightarrow{\alpha} B \xrightarrow{\beta} C \to 0$ *is exact, the sequence*

$$0 \to \mathrm{Hom}(G, A) \xrightarrow{T(\alpha)} \mathrm{Hom}(G, B) \xrightarrow{T(\beta)} \mathrm{Hom}(G, C) \to 0$$

is exact.

Proof Note that the critical assumption is that $T(\beta)$ is onto. Suppose G is free abelian; consider the diagram above, where $f \in \mathrm{Hom}(G, C)$. By Theorem 10.20, G has the projective property; there is thus a map $g: G \to B$ with $\beta g = f$. But $\beta g = T(\beta)(g)$, so $f \in \mathrm{image}\, T(\beta)$, as desired.

If $T(\beta)$ is onto, then every $f: G \to C$ is of the form $T(\beta)(g)$ for some $g \in \operatorname{Hom}(G, B)$, i.e., $f = g\beta$. Since we are assuming $T(\beta)$ is onto for every exact sequence $A \to B \to C \to 0$, the abelian group G has the projective property. Therefore, G is free abelian, by Exercise 10.44. ∎

Theorem 11.6 *A group K is divisible if and only if, whenever* $0 \to A \overset{\alpha}{\to} B \overset{\beta}{\to} C \to 0$ *is exact, the sequence*

$$0 \longrightarrow \operatorname{Hom}(C, K) \overset{S(\beta)}{\longrightarrow} \operatorname{Hom}(B, K) \overset{S(\alpha)}{\longrightarrow} \operatorname{Hom}(A, K) \longrightarrow 0$$

is exact.

Proof Use the injective property and Exercise 10.28. ∎

Theorem 11.7 *Let $\{A_j : j \in J\}$ be a family of groups; for any group K,*

$$\operatorname{Hom}(\Sigma A_j, K) \cong \Pi \operatorname{Hom}(A_j, K).$$

Proof For each j, let $i_j: A_j \to \Sigma A_j$ be the inclusion. Define a map $\theta: \operatorname{Hom}(\Sigma A_j, K) \to \Pi \operatorname{Hom}(A_j, K)$ by $f \mapsto (fi_j)$. Define a map ψ in the reverse direction by $(f_j) \mapsto f$, where f is the unique map of ΣA_j into K (Exercise 10.9) such that $fi_j = f_j$ for all j. A straightforward check shows that θ and ψ are inverse. ∎

We now have an example of a functor that does not preserve infinite direct sums.

Theorem 11.8 *Let $\{A_j : j \in J\}$ be a family of groups; for any group G,*

$$\operatorname{Hom}(G, \Pi A_j) \cong \Pi \operatorname{Hom}(G, A_j).$$

Proof For each i, let $p_i: \Pi A_j \to A_i$ be the ith projection. An argument similar to the argument given above, using Exercise 10.10, shows that $\theta: \operatorname{Hom}(G, \Pi A_j) \to \Pi \operatorname{Hom}(G, A_j)$ defined by $f \mapsto (p_j f)$ is an isomorphism. ∎

Theorem 11.9 *Let $\mu: A \to A$ be multiplication by m, i.e., $\mu(a) = ma$ for all $a \in A$. If T is a functor, then $T(\mu): T(A) \to T(A)$ is also multiplication by m.*

Proof First note that

$$0 = T(0) = T(-1_A + 1_A) = T(-1_A) + T(1_A) = T(-1_A) + 1_{T(A)};$$

it follows that $T(-1_A) = -1_{T(A)}$. If $m > 0$, then $\mu = \alpha + \cdots + \alpha$ (m times) where $\alpha = 1_A$; if $m < 0$, there is a similar expression with $\alpha = -1_A$. In either case, $T(\mu) = T(\alpha) + \cdots + T(\alpha)$ ($|m|$ times), and $T(\mu)$ is multiplication by m. ∎

We now present some sample computations to illustrate how these theorems may be used.

Example 1 For any group G, $\text{Hom}(G, \mathbf{Q})$ is a vector space over $\mathbf{Q}$.

A group H is torsion-free if and only if every nonzero multiplication $\mu\colon h \mapsto mh$ is one-one; a group H is divisible if and only if every nonzero multiplication $\mu\colon H \to H$ is onto. Thus, a group is torsion-free and divisible, i.e., is a vector space over $\mathbf{Q}$, if and only if every nonzero multiplication is an automorphism.

If $\mu\colon \mathbf{Q} \to \mathbf{Q}$ is multiplication by $m \neq 0$, then μ is an isomorphism. If T is the functor $\text{Hom}(G, \quad)$, then $T(\mu)$ is an isomorphism (Exercise 11.3) that is also multiplication by m. Therefore, $T(G) = \text{Hom}(G, \mathbf{Q})$ is torsion-free and divisible, hence is a vector space over $\mathbf{Q}$.

Example 2 For any group G, $\text{Hom}(\mathbf{Z}, G) \cong G$.

The map $\theta\colon \text{Hom}(\mathbf{Z}, G) \to G$ defined by $f \mapsto f(1)$ is an isomorphism.

Example 3 For any group G, $\text{Hom}(\mathbf{Z}(n), G) \cong G[n] = \{x \in G\colon nx = 0\}$.

Consider the exact sequence

$$0 \to \mathbf{Z} \overset{\mu}{\to} \mathbf{Z} \to \mathbf{Z}(n) \to 0$$

where μ is multiplication by n. If we apply the functor $S = \text{Hom}(\quad, G)$ to this sequence, we obtain the exact sequence

$$0 \to \text{Hom}(\mathbf{Z}(n), G) \to \text{Hom}(\mathbf{Z}, G) \overset{S(\mu)}{\longrightarrow} \text{Hom}(\mathbf{Z}, G);$$

thus $\text{Hom}(\mathbf{Z}(n), G) = \ker S(\mu)$. Since $\text{Hom}(\mathbf{Z}, G) \cong G$ and $S(\mu)$ is multiplication by n, $\text{Hom}(\mathbf{Z}(n), G) \cong G[n]$.

Example 4 For any group G, let $G^* = \text{Hom}(G, \mathbf{T})$, where $\mathbf{T}$ is the circle group; G^* is called the **character group** of G. If

$$0 \to A \to B \to C \to 0$$

is exact, then so is

$$0 \to C^* \to B^* \to A^* \to 0.$$

We are applying the functor $\text{Hom}(\quad, \mathbf{T})$, and $\mathbf{T}$ is a divisible group; exactness of the sequence of character groups thus follows from Theorem 11.6.

EXERCISES

11.9. Show that $\text{Hom}(A, B) = 0$ in each of the following cases:
(a) A is torsion, B is torsion-free.
(b) A is divisible, B is reduced.
(c) A is p-primary, B is q-primary, $p \neq q$.

11.10. Prove that $\operatorname{Hom}(G, \mathbf{T}) = 0$ if and only if $G = 0$.

11.11. If G is finite, prove that $G \cong G^*$ where G^* is the character group of G.

11.12. If S is a subgroup of the finite group G, set

$$S^0 = \{f \in G^*: f(s) = 0 \text{ for all } s \in S\}.$$

Prove that S^0 is a subgroup of G^* and that $S^0 \cong (G/S)^*$.

11.13. Let G be finite of order n and let k divide n. Prove that G has the same number of subgroups of order k as of subgroups of index k. (HINT: Prove that $S \mapsto S^0$ is a one-one correspondence.)

*11.14. Use character groups to give an alternative proof of Exercise 6.16.

11.15. We now present counterexamples showing that Theorems 11.7 and 11.8 cannot be extended. Let p_i denote the ith prime.

(a) If $A_i = \mathbf{Z}(p_i)$, prove that $\operatorname{Hom}(\Pi A_i, \mathbf{Q}) \neq 0$. Conclude that $\operatorname{Hom}(\Pi A_i, \mathbf{Q})$ is isomorphic to neither $\Pi \operatorname{Hom}(A_i, \mathbf{Q})$ nor $\Sigma \operatorname{Hom}(A_i, \mathbf{Q})$.

(b) If $A_i = \mathbf{Z}(p_i)$ and $G = \Sigma A_i$, then $\operatorname{Hom}(G, \Sigma A_i)$ and $\Sigma \operatorname{Hom}(G, A_i)$ are not isomorphic.

(c) If $G = \mathbf{Z}$ and $A_i = \mathbf{Z}(p_i)$, show that $\operatorname{Hom}(G, \Sigma A_i)$ and $\Pi \operatorname{Hom}(G, A_i)$ are not isomorphic.

DEFINITION OF EXT

We wish to consider all abelian extensions B of a group A by a group C, and so we recall the following definition from Chapter 7.

Definition An (**abelian**) **factor set** is a function $f: C \times C \to A$ such that, for all $x, y, z \in C$:

(i) $f(y, z) - f(x + y, z) + f(x, y + z) - f(x, y) = 0$;
(ii) $f(0, y) = 0 = f(x, 0)$;
(iii) $f(x, y) = f(y, x)$.

The main reason we used both additive and multiplicative notations in our previous discussion of extensions was that the quotient could act nontrivially on the kernel: if $c \in C$, ca was a convenient notation for the element $l(c) + a - l(c)$. Since we are now considering only abelian extensions, as is ensured by condition (iii), $ca = a$ always; consequently, there is no need for separate notations. Furthermore, there is no longer a need to keep track of $\theta: C \to \operatorname{Aut}(A)$, for θ must be the trivial map here.

Definition $Z(C, A)$ is the additive group of all abelian factor sets.

Definition A function $g: C \times C \to A$ is a **coboundary** if there is a function

$\alpha\colon C \to A$ with $\alpha(0) = 0$ for which

$$g(x, y) = \alpha(y) - \alpha(x + y) + \alpha(x).$$

Observe that $g(x, y) = g(y, x)$ because C is abelian.

Definition $B(C, A)$ is the set of all coboundaries.

As in Chapter 7, the reader may prove that $B(C, A)$ is a subgroup of $Z(C, A)$.

Definition $\mathrm{Ext}(C, A) = Z(C, A)/B(C, A)$.

Recall Exercise 7.39, which says that two extensions B and B' of A by C are equivalent in case there is a commutative diagram with exact rows

$$
\begin{array}{ccccccccc}
0 & \to & A & \to & B & \to & C & \to & 0 \\
& & \downarrow & & \downarrow & & \downarrow & & \\
0 & \to & A & \to & B' & \to & C & \to & 0
\end{array}
$$

where the outer downward maps are identities.

Notation   The set of equivalence classes of abelian extensions of A by C is denoted by $e(C, A)$.

Theorem 11.10   *$e(C, A)$ is a group isomorphic to* $\mathrm{Ext}(C, A)$ *whose zero element is the class of the split short exact sequence.*

Proof   Let us write $H^2(C, A)$ for the group $Z^2(C, A)/B^2(C, A)$ of Chapter 7 [the map $\theta\colon C \to \mathrm{Aut}(A)$ is trivial here]. Since $B(C, A) = B^2(C, A)$ and $Z(C, A)$ is a subgroup of $Z^2(C, A)$, it follows that $\mathrm{Ext}(C, A)$ is a subgroup of $H^2(C, A)$. A glance at the proof of Theorem 7.20 shows that the one-one correspondence φ there restricts to a one-one correspondence $\mathrm{Ext}(C, A) \xrightarrow{\sim} e(C, A)$. We now apply Exercise 1.20.   ∎

We shall examine the binary operation on $e(C, A)$ in the next section.

Corollary 11.11 $\mathrm{Ext}(C, A) = 0$ *if and only if every short exact sequence* $0 \to A \to B \to C \to 0$ *splits.*

Corollary 11.12   *A group D is divisible if and only if* $\mathrm{Ext}(A, D) = 0$ *for every group A; a group F is free abelian if and only if* $\mathrm{Ext}(F, A) = 0$ *for every group A.*

Proof   If D is divisible and $0 \to D \to E \to A \to 0$ is exact, then D is a direct summand of E, by Corollary 10.10, and so this sequence splits. Therefore $\mathrm{Ext}(A, D) = 0$.

Conversely, if every exact sequence $0 \to D \to E \to A \to 0$ splits, then D is divisible, by Corollary 10.24.

If F is free abelian and $0 \to A \to B \to F \to 0$ is exact, then Corollary 10.21 says this sequence must split; hence $\mathrm{Ext}(F, A) = 0$.

Conversely, choose an exact sequence $0 \to A \to B \to F \to 0$ in which B is free abelian (Corollary 10.19 says this is always possible). Since this sequence splits, F is a direct summand of B. Therefore F is free abelian, by Theorem 10.22. ∎

EXERCISES

11.16. If C is torsion-free and A is of bounded order, then $\mathrm{Ext}(C, A) = 0$.

11.17. If $\mathbf{R}$ denotes the additive group of real numbers, then $\mathrm{Ext}(\mathbf{R}, \Sigma \mathbf{Z}(p)) \neq 0$. (HINT: Use Exercise 10.14.)

11.18. Prove that $\mathrm{Ext}(\mathbf{Z}(p), \mathbf{Z}(p)) \neq 0$.

11.19. If C and A are groups with C p-primary and A q-primary, where $p \neq q$, then $\mathrm{Ext}(C, A) = 0$. (HINT: Use Corollary 10.33.)

PULLBACKS AND PUSHOUTS

Let us give an explicit construction of the sum of two (classes of) extensions in $e(C, A)$. Rather than say that B is an extension of A by C, we redefine an **extension** as a short exact sequence

$$0 \to A \to B \to C \to 0.$$

Suppose now that we have two extensions ($i = 1, 2$):

$$0 \to A \xrightarrow{\lambda_i} B_i \xrightarrow{\pi_i} C \to 0.$$

Suppose further that l_i are transversals and $f_i \colon C \times C \to A$ are the corresponding factor sets. What is the extension of A by C determined by the factor set $f_1 + f_2$?

Every $b_i \in B_i$ has a unique expression

$$b_i = a + l_i(c), \quad a \in A, \ c \in C,$$

and addition obeys the rule

$$[a + l_i(c)] + [a' + l_i(c')] = a + a' + l_i(c + c') + f_i(c, c').$$

Let H be the subset of $B_1 \oplus B_2$ consisting of all elements (b_1, b_2) such that $\pi_1(b_1) = \pi_2(b_2)$. It will be verified that H is a subgroup and that the following sequence is exact:

$$0 \to A \oplus A \xrightarrow{\lambda} H \xrightarrow{\mu} C \to 0,$$

where $\lambda(a, a') = (\lambda_1 a, \lambda_2 a')$ and $\mu(b_1, b_2) = \pi_1(b_1) = \pi_2(b_2)$. Define a transversal l by $l(c) = (l_1(c), l_2(c))$; the usual computation defines a factor set $F \colon C \times C \to A \oplus A$ by $F(c, c') = (f_1(c, c'), f_2(c, c'))$. To produce an extension with a factor set that is the sum of the two coordinates, first define $\nabla \colon A \oplus A \to A$ by $\nabla(a, a') = a + a'$ and let S be the subgroup of $A \oplus H$ consisting of all elements of the form

$$(\nabla (a, a'), -\lambda (a, a')).$$

We shall prove that $(A \oplus H)/S$ is an extension of A by C; moreover, with respect to the transversal $\{l(c) + S\}$, the factor set of this extension is $f_1 + f_2$. The complete construction thus consists of taking a subgroup of a direct sum and then taking a quotient of another direct sum. In the language of the preceding chapter, we perform dual constructions. We proceed to a systematic account of these constructions.

Who would expect that a lemma about 10 groups and 13 homomorphisms could be of any use? Here is the "Five Lemma"; the method of proof is called "diagram-chasing".

Lemma 11.13 **(Five Lemma)** *Consider the commutative diagram with exact rows:*

$$
\begin{array}{ccccccccc}
A_1 & \xrightarrow{f_1} & A_2 & \xrightarrow{f_2} & A_3 & \xrightarrow{f_3} & A_4 & \xrightarrow{f_4} & A_5 \\
\downarrow{\alpha_1} & & \downarrow{\alpha_2} & & \downarrow{\alpha_3} & & \downarrow{\alpha_4} & & \downarrow{\alpha_5} \\
B_1 & \xrightarrow{g_1} & B_2 & \xrightarrow{g_2} & B_3 & \xrightarrow{g_3} & B_4 & \xrightarrow{g_4} & B_5
\end{array}
$$

(i) *If α_2 and α_4 are onto and α_5 is one-one, then α_3 is onto.*
(ii) *If α_2 and α_4 are one-one and α_1 is onto, then α_3 is one-one.*

In particular, if α_1, α_2, α_4, and α_5 are isomorphisms, so is α_3.

REMARK   The way to understand the proof is to chase elements around the diagram without pausing to give names to everything; we approximate this by using transparent notation, e.g., b_2 denotes an element of B_2. The reader should observe that each step in the proof is automatic in the sense that there is only one reasonable way to proceed.

Proof   We shall prove only (i), leaving the proof of the dual statement (ii) to the reader.

If $b_3 \in B_3$, we must find a_3 with $\alpha_3 (a_3) = b_3$. Now $b_3 \mapsto b_4 = g_3 (b_3)$; since α_4 is onto, we may lift b_4 to a_4. Commutativity gives $\alpha_5 f_4 (a_4) = g_4 \alpha_4 (a_4)$ $= g_4 (b_4) = g_4 g_3 (b_3)$, and this last is 0 because the bottom row is exact. Since $\alpha_5 f_4 (a_4) = 0$ and α_5 is one-one, $a_4 \in \ker f_4 = \operatorname{im} f_3$. Hence, there is $a_3 \in A_3$ with $f_3 (a_3) = a_4$. Using commutativity in the next to last square, we see $g_3 \alpha_3 (a_3)$ $= \alpha_4 f_3 (a_3) = \alpha_4 (a_4) = b_4 = g_3 (b_3)$. Conclusion: $b_3 - \alpha_3 (a_3)$ is in $\ker g_3$ $= \operatorname{im} g_2$. Therefore, $b_3 - \alpha_3 (a_3) = g_2 (b_2)$ for some b_2. Since α_2 is onto, we may lift b_2 to a_2. Using commutativity once more, we see $\alpha_3 f_2 (a_2) = g_2 \alpha_2 (a_2)$ $= g_2 (b_2) = b_3 - \alpha_3 (a_3)$. Thus, $b_3 = \alpha_3 (f_2 (a_2) + a_3)$ and α_3 is onto.   ∎

Corollary 11.14   *If B and B′ are extensions of A by C, then B is equivalent to B′ if and only if there is a commutative diagram with exact rows:*

$$0 \rightarrow A \rightarrow B \rightarrow C \rightarrow 0$$

$$1_A \downarrow \qquad \gamma \downarrow \qquad 1_C \downarrow$$

$$0 \rightarrow A \rightarrow B' \rightarrow C \rightarrow 0$$

Proof   The five lemma tells us γ is necessarily an isomorphism.   ∎

   A weaker form of the five lemma says that if the four outer maps are isomorphisms and if there is a middle map α_3 such that all commutes, then α_3 is an isomorphism. We remark that a map α_3 need not exist, for consider the diagram with exact rows

$$0 \rightarrow \mathbf{Z}(p) \rightarrow \mathbf{Z}(p) \oplus \mathbf{Z}(p) \rightarrow \mathbf{Z}(p) \rightarrow 0$$

$$\downarrow \qquad \qquad \qquad \qquad \downarrow$$

$$0 \rightarrow \mathbf{Z}(p) \longrightarrow \mathbf{Z}(p^2) \longrightarrow \mathbf{Z}(p) \rightarrow 0,$$

where the outer maps are identities. One cannot insert a homomorphism in the middle making the resulting diagram commute, lest it be an isomorphism between the two nonisomorphic middle groups.

EXERCISES

**11.20.   Consider the commutative diagram with exact rows:

$$A_1 \rightarrow A_2 \rightarrow A_3 \rightarrow 0$$

$$\alpha_1 \downarrow \qquad \alpha_2 \downarrow$$

$$B_1 \rightarrow B_2 \rightarrow B_3 \rightarrow 0$$

Prove there is a homomorphism from A_3 to B_3 making the augmented diagram commute; this map is an isomorphism if α_1 and α_2 are.

**11.21.   Consider the diagram

$$0 \rightarrow R \rightarrow F \rightarrow C \rightarrow 0$$

$$1_C \downarrow$$

$$E \rightarrow B \rightarrow C \rightarrow 0$$

where the rows are exact and F is free abelian. There exist maps $\alpha\colon F \rightarrow B$ and $\beta\colon R \rightarrow E$ such that the resulting diagram commutes.

**11.22.   Consider the diagram

$$0 \;\to\; A \;\to\; B \;\to\; E$$

$$1_A \Big\downarrow$$

$$0 \;\to\; A \;\to\; D \;\to\; D' \;\to\; 0$$

where the rows are exact and D is divisible. There exist maps $\alpha\colon B \to D$ and $\beta\colon E \to D'$ such that the resulting diagram commutes.

Definition   Consider the extensions

$$E\colon 0 \to A \overset{\lambda}{\to} B \overset{\mu}{\to} C \to 0$$

and

$$E'\colon 0 \to A' \overset{\lambda'}{\to} B' \overset{\mu'}{\to} C' \to 0;$$

a **map** from E to E' is an ordered triple (α, β, γ) of homomorphisms such that the following diagram commutes:

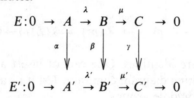

$$E\colon 0 \;\to\; A \overset{\lambda}{\to} B \overset{\mu}{\to} C \;\to\; 0$$

$$\alpha \Big\downarrow \quad \beta \Big\downarrow \quad \gamma \Big\downarrow$$

$$E'\colon 0 \;\to\; A' \overset{\lambda'}{\to} B' \overset{\mu'}{\to} C' \;\to\; 0$$

Our notation is $(\alpha, \beta, \gamma)\colon E \to E'$.

One composes maps of extensions "coordinatewise". The definition of equivalence of extensions E and E' states that there is a map $(1_A, \beta, 1_C)\colon E \to E'$; Corollary 11.14 implies that equivalence is, indeed, an equivalence relation. The equivalence class of an extension E is denoted by $[E]$.

There are two dual operations on extensions. We now prepare the first one.

A **solution** of the diagram

(*)

$$\begin{array}{c} C' \\ \downarrow \gamma \\ B \overset{\mu}{\longrightarrow} C \end{array}$$

is a group B' and maps μ' and γ' such that

$$\begin{array}{ccc} B' & \overset{\mu'}{\longrightarrow} & C' \\ \gamma' \downarrow & & \downarrow \gamma \\ B & \overset{\mu}{\longrightarrow} & C \end{array}$$

commutes.

A solution of a diagram need not be unique: if X is any group and $\beta: X \to B'$ any homomorphism, then

is also a solution.

Definition A solution of the diagram (*) is a **pullback** if, given any solution (B'', μ'', γ'') of (*), there is a unique map $\theta: B'' \to B'$ making the following diagram commute:

(**)

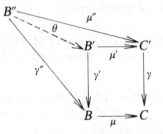

Lemma 11.15 *A pullback exists for the diagram* (*).

Proof Let B' be the subgroup of $B \oplus C'$ consisting of all (b, c') such that $\mu b = \gamma c'$, and define μ' and γ' as projections. It is trivial to verify that we have defined a solution of (*).

Suppose now that there is a second solution (B'', μ'', γ''). Define $\theta: B'' \to B'$ as follows: if $x \in B''$, then $\theta x = (\gamma'' x, \mu'' x)$. One checks quickly that (**) commutes and this is the only θ that works. ∎

REMARK If nonabelian groups are allowed in diagrams, then the construction just given proves the existence of pullbacks in this larger setting as well.

Theorem 11.16 *Let E be an extension of A by C and let $\gamma: C' \to C$ be a homomorphism. There is an extension E' of A by C' and a map $(1_A, \gamma', \gamma): E' \to E$.*

Proof We are given the following diagram to complete:

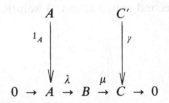

Let (B', μ', γ') be the pullback obtained in Lemma 11.15 and define $\lambda': A \to B'$ by $\lambda' a = (\lambda a, 0)$. It is easy to check that

$$E': 0 \to A \xrightarrow{\lambda'} B' \xrightarrow{\mu'} C' \to 0$$

is an extension and that $(1_A, \gamma', \gamma): E' \to E$. ∎

Notation   The extension E' just constructed is denoted by $E\gamma$.

Notation   If $\gamma: C' \to C$ is a function, define

$$\gamma \times \gamma: C' \times C' \to C \times C \text{ by } (\gamma \times \gamma)(c_1', c_2') = (\gamma c_1', \gamma c_2').$$

If $\lambda: A \to B$ and $\lambda': A' \to B'$ are homomorphisms, define

$$\lambda \oplus \lambda': A \oplus A' \to B \oplus B' \text{ by } (\lambda \oplus \lambda')(a, a') = (\lambda a, \lambda' a').$$

It is useful to do a bit of bookkeeping.

Lemma 11.17   *Let E be an extension of A by C and let $\gamma: C' \to C$ be a homomorphism. If $f: C \times C \to A$ is a factor set of E, then $f(\gamma \times \gamma)$ is a factor set of $E\gamma$.*

Proof   We use the notation of the preceding theorem. Let l be a transversal determining f. Define a transversal L by $L(c') = (l(\gamma c'), c')$. The reader may now compute that the corresponding factor set† is $F = f(\gamma \times \gamma)$, i.e., $F(c_1', c_2') = f(\gamma c_1', \gamma c_2')$. ∎

Corollary 11.18   *Let E and E'' be extensions of A by C, and let $\gamma: C' \to C$. If $[E] = [E'']$, then $[E\gamma] = [E''\gamma]$.*

Proof   Let f and f'' be factor sets of E and E'', respectively; then $f - f'' \in B(C, A)$. By Lemma 11.17, $f(\gamma \times \gamma)$ and $f''(\gamma \times \gamma)$ are factor sets of $E\gamma$ and $E''\gamma$, respectively, and the reader may check that $f(\gamma \times \gamma) - f''(\gamma \times \gamma) \in B(C', A)$. Hence, $[E\gamma] = [E''\gamma]$. ∎

Corollary 11.19   *Let E be an extension of A by C and let $\gamma: C' \to C$ and $\gamma_1: C'' \to C'$ be homomorphisms. Then $[(E\gamma)\gamma_1] = [E(\gamma\gamma_1)]$.*

Proof   If f is a factor set of E, then $(f \circ (\gamma \times \gamma)) \circ (\gamma_1 \times \gamma_1)$ is a factor set of $(E\gamma)\gamma_1$ and $f \circ ((\gamma \times \gamma) \circ (\gamma_1 \times \gamma_1))$ is a factor set of $E(\gamma\gamma_1)$. Since composition of functions is associative, $[(E\gamma)\gamma_1] = [E(\gamma\gamma_1)]$. ∎

We now prepare the second construction. A **solution** of the diagram

$$(\#) \qquad \begin{array}{ccc} & \lambda & \\ A & \to & B \\ \alpha \downarrow & & \\ A' & & \end{array}$$

---

† Recall $F(c_1', c_2') = L(c_1') + L(c_2') - L(c_1' + c_2')$.

is a group B' and maps λ' and α' making the following diagram commute:

Definition A solution of the diagram ($\#$) is a **pushout** if, given any other solution $(B'', \lambda'', \alpha'')$ of ($\#$), there is a unique map $\theta\colon B' \to B''$ making the following diagram commute:

($\#\#$)

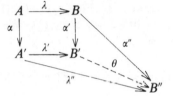

Lemma 11.20 *A pushout exists for the diagram* ($\#$).

Proof Since the dual of a subgroup is a quotient, one should expect the pushout to be a quotient of $A' \oplus B$. If S is the subgroup of $A' \oplus B$ consisting of all $(\alpha a, -\lambda a)$ where $a \in A$, define $B' = (A' \oplus B)/S$, define $\lambda'\colon A' \to B'$ by $\lambda'a' = (a', 0) + S$, and define $\alpha'\colon B \to B'$ by $\alpha'b = (0, b) + S$. It is an easy matter to verify that (B', λ', α') is a solution of ($\#$).

Suppose now that there is a second solution $(B'', \lambda'', \alpha'')$. Define $\theta\colon B' \to B''$ by $\theta((a', b) + S) = \lambda''a' + \alpha''b$. One checks easily that θ is well defined, that ($\#\#$) commutes, and that this is the only θ that works. $\blacksquare$

REMARK Pushouts exist in the larger setting of all (not necessarily abelian) groups, but the construction just given is inadequate (the subgroup S may not be a normal subgroup). The appropriate construction is given in the next chapter.

Theorem 11.21 *Let E be an extension of A by C and let $\alpha\colon A \to A'$ be a homomorphism. There is an extension E' of A' by C and a map $(\alpha, \alpha', 1_C)\colon E \to E'$.*

Proof We are given the following diagram to complete:

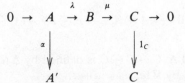

Let (B', λ', α') be a pushout, and define $\mu': B' \to C$ by $\mu'((a', b) + S) = \mu b$. It is a simple matter to verify that μ' is well defined,

$$E': 0 \to A' \xrightarrow{\lambda'} B' \xrightarrow{\mu'} C' \to 0$$

is an extension, and $(\alpha, \alpha', 1_C): E \to E'$. ∎

Notation   Denote the extension E' just constructed by αE.

Lemma 11.22   *Let E be an extension of A by C and let $\alpha: A \to A'$ be a homomorphism. If $f: C \times C \to A$ is a factor set of E, then αf is a factor set of αE.*

Proof   We use the notation of Theorem 11.21. If l is a transversal determining f, define a transversal L by $L(c) = (0, l(c)) + S$. Now the factor set F determined by this choice is $F(c_1, c_2) = L(c_1) + L(c_2) - L(c_1 + c_2) = (0, f(c_1, c_2)) + S = (\alpha f(c_1, c_2), 0) + S = \lambda' \alpha f(c_1, c_2)$. We are done, for λ' is only the inclusion map. ∎

Corollary 11.23   *If $[E] = [E'']$, then $[\alpha E] = [\alpha E'']$.*

Corollary 11.24   *If E is an extension of A by C and if $\alpha: A \to A'$ and $\alpha_1: A' \to A''$ are homomorphisms, then $[\alpha_1(\alpha E)] = [(\alpha_1 \alpha)E]$.*

Corollary 11.25   *Let E be an extension of A by C and let $\alpha: A \to A'$ and $\gamma: C' \to C$ be homomorphisms. Then $[(\alpha E)\gamma] = [\alpha(E\gamma)]$.*

Proof   If f is a factor set of E, then $(\alpha f)(\gamma \times \gamma)$ is a factor set of $(\alpha E)\gamma$ and $\alpha(f(\gamma \times \gamma))$ is a factor set of $\alpha(E\gamma)$. ∎

We have demonstrated that homomorphisms act on extensions and that all possible associativity laws hold.

Definition   Consider the extensions

$$E: 0 \to A \xrightarrow{\lambda} B \xrightarrow{\mu} C \to 0$$

and

$$E': 0 \to A' \xrightarrow{\lambda'} B' \xrightarrow{\mu'} C' \to 0.$$

Their **direct sum** is the extension

$$E \oplus E': 0 \longrightarrow A \oplus A' \xrightarrow{\lambda \oplus \lambda'} B \oplus B' \xrightarrow{\mu \oplus \mu'} C \oplus C' \to 0.$$

Notation   The **diagonal** $\Delta: C \to C \oplus C$ is defined by $\Delta(c) = (c, c)$; the **codiagonal** $\nabla: A \oplus A \to A$ is defined by $\nabla(a, a') = a + a'$.

Definition If E and E' are extensions of A by C, their **sum** $E + E'$ is the extension $(\nabla(E \oplus E'))\Delta$.

The reader should observe, using Corollary 11.25, that both $E + E'$ and $\nabla((E \oplus E')\Delta)$ are equivalent extensions of A by C.

EXERCISES

11.23. If E and E' are extensions with factor sets f and f', respectively, then $f \oplus f'$ is a factor set of $E \oplus E'$.

11.24. If $[E] = [E']$ and $[E_0] = [E_0']$, then $[E \oplus E_0] = [E' \oplus E_0']$.

11.25. Let E be an extension of A by C, E' an extension of A' by C', and $\gamma : C_1 \to C$, $\gamma' : C_2 \to C'$ homomorphisms. Then

$$[(E \oplus E')(\gamma \oplus \gamma')] = [E\gamma \oplus E'\gamma'].$$

A similar equation holds on the left.

*11.26. If $[E] = [E']$ and $[E_0] = [E_0']$, then $[E + E_0] = [E' + E_0']$.

11.27. $[E(\gamma + \gamma_1)] = [E\gamma + E\gamma_1]$; similarly on the left.

*11.28. $[(E + E')\gamma] = [E\gamma + E'\gamma]$; similarly on the left.

The following result shows that these constructions do give an extension corresponding to the sum of two factor sets.

Lemma 11.26 *If f is a factor set, let E_f be an extension it determines (so that $\varphi(f + B(C, A)) = [E_f]$ defines the one-one correspondence of Theorem 11.10). Then $[E_{f + f'}] = [E_f + E_{f'}]$.*

Proof By definition, $E_{f + f'}$ has $f + f'$ as a factor set; $E_f + E_f$ has $\nabla(f \oplus f')\Delta$ as a factor set. But these are the same:

$$\begin{aligned}
\nabla(f \oplus f')\Delta(x, y) &= \nabla(f \oplus f')(x, y, x, y) \\
&= \nabla(f(x, y), f'(x, y)) \\
&= f(x, y) + f'(x, y) \\
&= (f + f')(x, y). \quad \blacksquare
\end{aligned}$$

Definition If $[E]$ and $[E'] \in e(C, A)$, their **Baer sum** is $[E] + [E'] = [E + E']$.

By Exercise 11.26, the Baer sum is a well defined binary operation on $e(C, A)$.

Theorem 11.27 *$e(C, A)$ is an abelian group under Baer sum, and $\varphi : \operatorname{Ext}(C, A) \to e(C, A)$ defined by $\varphi(f + B(C, A)) = [E_f]$ is an isomorphism.*

Proof Lemma 11.26 shows that Baer sum on $e(C, A)$ is the binary operation induced by the one-one correspondence φ. $\quad \blacksquare$

THE EXT FUNCTORS

We had an ulterior motive in examining the addition of extensions, for we are now in a position to prove that Ext defines functors that are closely related to the Hom functors.

Theorem 11.28 *If G is a fixed group, then* $\text{Ext}(G, \ )$ *is a covariant functor.*

Proof If $\alpha: A \to A'$, define a homomorphism

$$\alpha_*: \text{Ext}(G, A) \to \text{Ext}(G, A')$$

by

$$\alpha_*(f + B(G, A)) = \alpha f + B(G, A').$$

It is straightforward to prove that α_* is a well defined homomorphism and that the axioms of the definition of functor are satisfied. ∎

Theorem 11.29 *If K is a fixed group, then* $\text{Ext}(\ , K)$ *is a contravariant functor.*

Proof If $\alpha: A \to A'$, define $\alpha^*: \text{Ext}(A', K) \to \text{Ext}(A, K)$ by $\alpha^*(f + B(A', K)) = f(\alpha \times \alpha) + B(A, K)$. ∎

We have another way of looking at extensions, $e(C, A)$, and this also gives rise to functors. If $\alpha: A \to A'$, define a function $\alpha_\#: e(C, A) \to e(C, A')$ by

$$\alpha_\#([E]) = [\alpha E];$$

if $\gamma: C \to C'$, define a function $\gamma^\#: e(C', A) \to e(C, A)$ by

$$\gamma^\#([E]) = [E\gamma].$$

The following lemma will imply that these functions are homomorphisms (this also follows from Exercise 11.28).

Lemma 11.30† *If* $\alpha: A \to A'$, *then the following diagram commutes for every G:*

$$
\begin{array}{ccc}
\text{Ext}(G, A) & \xrightarrow{\alpha_*} & \text{Ext}(G, A') \\
\varphi \downarrow & & \downarrow \varphi \\
e(G, A) & \xrightarrow{\alpha_\#} & e(G, A')
\end{array}
$$

where φ *is the isomorphism of Theorem 11.10. A dual theorem holds if the second variable is fixed.*

Proof We must show that $\varphi\alpha_* = \alpha_\#\varphi$. If $f \in Z(G, A)$, then $\varphi\alpha_*(f + B) = \varphi(\alpha f + B') = [E_{\alpha f}]$, i.e., the class of an extension with factor set αf. On the

† This lemma says the functors $\text{Ext}(G, \ )$ and $e(G, \ )$ are **naturally equivalent**.

other hand, $\alpha_{\#}\,\varphi(f+B) = \alpha_{\#}[E_f] = [\alpha E_f]$. By Lemma 11.22, αf is a factor set of αE_f, and so $[\alpha E_f] = [E_{\alpha f}]$. ∎

Corollary 11.31 *The functions $\alpha_{\#}$ and $\gamma^{\#}$ are homomorphisms.*

Proof By the lemma, $\alpha_{\#} = \varphi\alpha_{*}\varphi^{-1}$ is the composite of homomorphisms. The dual of Lemma 11.30 shows $\gamma^{\#}$ is a homomorphism. ∎

We now connect Hom to Ext.

Definition Let f be a factor set of the extension

$$E:0 \to A \overset{\lambda}{\to} B \overset{\mu}{\to} C \to 0.$$

If G is a group, define the **connecting homomorphisms**:

$\partial\colon \operatorname{Hom}(G, C) \to \operatorname{Ext}(G, A)$ by $\partial(h) = f(h \times h) + B(G, A)$;
$\delta\colon \operatorname{Hom}(A, G) \to \operatorname{Ext}(C, G)$ by $\delta(h') = h'f + B(C, G)$.

EXERCISE

11.29. Define $\partial_{\#}\colon \operatorname{Hom}(G, C) \to e(G, A)$ by $\partial_{\#}(h) = [Eh]$, and define $\delta^{\#}\colon \operatorname{Hom}(A, G) \to e(C, G)$ by $\delta^{\#}(h') = [h'E]$. Prove the following diagrams commute:

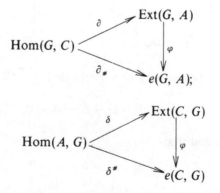

Two lemmas are needed before the main tool for calculating Ext can be presented. The proofs are dual, so the second proof is left as an exercise.

Lemma 11.32 *If $0 \to A \overset{\lambda}{\to} B \overset{\mu}{\to} C \to 0$ is exact, then, for every group G, the map $\lambda^{\#}\colon e(B, G) \to e(A, G)$ is onto.*

Proof Let $0 \to G \overset{\alpha}{\to} E \overset{\beta}{\to} A \to 0$ represent an element of $e(A, G)$. Now hook this sequence to the given sequence to get the following exact sequence:

$$0 \to G \overset{\alpha}{\to} E \overset{\lambda\beta}{\to} B \overset{\mu}{\to} C \to 0.$$

There exists an exact sequence

$$0 \to R \to F \to C \to 0,$$

where F is free abelian. By Exercise 11.21, there are maps such that the following diagram commutes:

$$
\begin{array}{ccccccccc}
0 & \to & R & \to & F & \to & C & \to & 0 \\
& & \downarrow \gamma & & \downarrow \varepsilon & & \downarrow 1_C & & \\
& & & \lambda\beta & & & & & \\
0 & \to & G & \to & E & \to & B & \to & C & \to & 0.
\end{array}
$$

Focusing on the first square in this diagram, we see that $(B, \lambda\beta, \varepsilon)$ is a solution of the diagram

$$
\begin{array}{ccc}
R & \to & F \\
\downarrow & & \\
E & &
\end{array}
$$

If (E', τ, σ) is the pushout for this diagram, then there is a map $\theta: E' \to B$ with

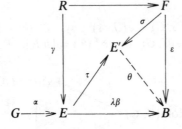

commuting. The reader may now verify that

$$0 \to G \xrightarrow{\tau\alpha} E' \xrightarrow{\theta} B \to 0$$

is an extension and that

$$
\begin{array}{ccccccc}
0 & \to & G & \to & E & \to & A & \to & 0 \\
& & \downarrow 1_G & & \downarrow \tau & & \downarrow \lambda & & \\
0 & \to & G & \to & E' & \to & B & \to & 0
\end{array}
$$

commutes. In other words, $\lambda^{\#}$ sends the class of the bottom extension into the class of the top extension. ∎

Lemma 11.33 If $0 \to A \xrightarrow{\lambda} B \xrightarrow{\mu} C \to 0$ is exact, then for every group G the map $\mu_{\#}: e(G, B) \to e(G, C)$ is onto.

Proof Dual to the preceding proof, using pullbacks instead of pushouts and Exercise 11.22 instead of Exercise 11.21. ∎

Theorem 11.34   Let $0 \to A \overset{\lambda}{\to} B \overset{\mu}{\to} C \to 0$ be exact. For any group G, the following two sequences are exact:

(i) $0 \longrightarrow \mathrm{Hom}(G, A) \overset{T(\lambda)}{\longrightarrow} \mathrm{Hom}(G, B) \overset{T(\mu)}{\longrightarrow} \mathrm{Hom}(G, C) \overset{\partial}{\longrightarrow}$

$\mathrm{Ext}(G, A) \overset{\lambda_*}{\longrightarrow} \mathrm{Ext}(G, B) \overset{\mu_*}{\longrightarrow} \mathrm{Ext}(G, C) \longrightarrow 0;$

(ii) $0 \longrightarrow \mathrm{Hom}(C, G) \overset{S(\mu)}{\longrightarrow} \mathrm{Hom}(B, G) \overset{S(\lambda)}{\longrightarrow} \mathrm{Hom}(A, G) \overset{\delta}{\longrightarrow}$

$\mathrm{Ext}(C, G) \overset{\mu^*}{\longrightarrow} \mathrm{Ext}(B, G) \overset{\lambda^*}{\longrightarrow} \mathrm{Ext}(A, G) \longrightarrow 0.$

(*T and S are the* Hom *functors of Theorems* 11.3 *and* 11.4.)

Proof   The difficult portions are the proofs that μ_* and λ^* are onto, and they follow easily from Lemmas 11.32 and 11.33 and Lemma 11.30. The other steps of the proof are utterly uninspiring, proceeding inexorably in the manner of the proof of Theorem 11.3.   ∎

The groups Ext thus repair the exactness we may have lost by applying Hom. Consider the question: If A is a subgroup of B, when can a homomorphism $h: A \to G$ be extended to a homomorphism of B into G? The exact sequence answers this question. If $0 \to A \overset{i}{\to} B \to B/A \to 0$ is exact (where i is the inclusion), then we have exactness of

$$\mathrm{Hom}(B, G) \overset{S(i)}{\longrightarrow} \mathrm{Hom}(A, G) \overset{\delta}{\longrightarrow} \mathrm{Ext}(B/A, G).$$

Definition   The element $\delta(h)$ is called the **obstruction** of h.

Corollary 11.35   Let A be a subgroup of B and $h: A \to G$ a homomorphism. Then h can be extended to B if and only if its obstruction is 0.

Proof   If h can be extended to B, there is a map $h': B \to G$ such that $h'i = h$. Hence, $h \in \mathrm{im}\, S(i) = \ker \delta$, so that $\delta(h) = 0$.

   If $\delta(h) = 0$, then $h \in \ker \delta = \mathrm{im}\, S(i)$, so that there is an $h' \in \mathrm{Hom}(B, G)$ with $S(i)(h') = h$, i.e., $h'i = h$, as desired.   ∎

Corollary 11.36   If $\mathrm{Ext}(B/A, G) = 0$, every map $h: A \to G$ can be extended to B.

We now prove that Ext behaves very much like Hom.

Theorem 11.37   Let $\{A_i : i \in I\}$ be a family of groups. For any K,

$$\mathrm{Ext}(\Sigma A_i, K) \cong \Pi\, \mathrm{Ext}(A_i, K).$$

Proof For each i, let $0 \to R_i \to F_i \to A_i \to 0$ be exact, where F_i is free abelian. Consider the commutative diagram with exact rows:

$$\text{Hom}(\Sigma F_i, K) \to \text{Hom}(\Sigma R_i, K) \to \text{Ext}(\Sigma A_i, K) \to \text{Ext}(\Sigma F_i, K)$$
$$\downarrow \qquad\qquad \downarrow$$
$$\Pi\,\text{Hom}(F_i, K) \to \Pi\,\text{Hom}(R_i, K) \to \Pi\,\text{Ext}(A_i, K) \to \Pi\,\text{Ext}(F_i, K)$$

where the downward maps are the isomorphisms of Theorem 11.7. By Corollary 11.12, the terms on the far right are 0, so that Exercise 11.20 yields the desired result. ∎

Theorem 11.38 *Let $\{A_i : i \in I\}$ be a family of groups. For any G,*

$$\text{Ext}(G, \Pi A_i) \cong \Pi\,\text{Ext}(G, A_i).$$

Proof Left to the reader. ∎

Theorem 11.39 *For any group A, $\text{Ext}(\mathbf{Z}(n), A) \cong A/nA$.*

Proof Consider the exact sequence

$$0 \to \mathbf{Z} \xrightarrow{\mu} \mathbf{Z} \to \mathbf{Z}(n) \to 0,$$

where μ is multiplication by n. There is exactness of

$$\text{Hom}(\mathbf{Z}, A) \xrightarrow{\;S(\mu)\;} \text{Hom}(\mathbf{Z}, A) \to \text{Ext}(\mathbf{Z}(n), A) \to \text{Ext}(\mathbf{Z}, A) = 0.$$

Now $\text{Hom}(\mathbf{Z}, A) \cong A$ and $S(\mu)$ is multiplication by n, so the result follows. ∎

It follows from Theorem 11.39 that $\text{Ext}(\mathbf{Z}(p), \mathbf{Z}(p)) \cong \mathbf{Z}(p)$, where p is a prime, so that this Ext has p elements. On the other hand, if $0 \to \mathbf{Z}(p) \to B \to \mathbf{Z}(p) \to 0$ is exact, then B has order p^2 so that $B \cong \mathbf{Z}(p^2)$ or $B \cong \mathbf{Z}(p) \oplus \mathbf{Z}(p)$. Therefore, if p is an odd prime, we have an example of inequivalent extensions of A by C with isomorphic middle terms (see Exercise 7.41). The number of nonisomorphic middle groups B (where $0 \to A \to B \to C \to 0$ is exact) is at most $|\text{Ext}(C, A)|$. The example above shows that this inequality can be strict.

Example 5 If A is torsion-free, then $\text{Ext}(A, G)$ is divisible for any group G.

Since A is torsion-free, for each $n > 0$ there is an exact sequence

$$0 \to A \xrightarrow{n} A \to A/nA \to 0,$$

where the first nontrivial map is multiplication by n. This gives rise to an exact sequence

$$\cdots \to \text{Ext}(A, G) \xrightarrow{n} \text{Ext}(A, G) \to 0,$$

so that multiplication by n maps $\text{Ext}(A, G)$ onto itself. This says that $\text{Ext}(A, G)$ is divisible.

Example 6   If $mG = 0$, then $\text{Ext}(A, G) = 0$ for every torsion-free group A.

If $\mu: G \to G$ is multiplication by m, then $\mu_*: \text{Ext}(A, G) \to \text{Ext}(A, G)$ is also multiplication by m. Since $mG = 0$, μ is the zero map; thus μ_* is also the zero map. Therefore

$$\mu_* \text{Ext}(A, G) = m \text{Ext}(A, G) = 0.$$

On the other hand, we have just seen that $\text{Ext}(A, G)$ is divisible, for A is torsion-free. It follows that $\text{Ext}(A, G) = 0$.

We have given a proof by homological algebra of Corollary 10.37: If B is a group with tB of bounded order, then tB is a direct summand of B. The group B/tB is torsion-free, and by Corollary 11.11, $\text{Ext}(B/tB, tB) = 0$ implies that the sequence

$$0 \to tB \to B \to B/tB \to 0$$

splits.

EXERCISES

11.30.   Prove that $\text{Ext}(V, G)$ is torsion-free and divisible whenever V is a vector space over $\mathbf{Q}$.

11.31.   Prove that $\text{Ext}(\mathbf{Q}, \mathbf{Z}) \neq 0$.

11.32.   Prove that

$$\text{Ext}\left(\mathbf{Q}, \sum_{n=1}^{\infty} \mathbf{Z}(p^n) \right) \neq 0.$$

Conclude that there exists a (countable) group G whose torsion subgroup $tG \cong \Sigma \mathbf{Z}(p^n)$ and such that tG is not a summand of G.

11.33.   Let $0 \to R \to F \to A \to 0$ be exact, where F is free abelian. Prove that $\text{Ext}(A, G) \cong \text{Hom}(R, G)/\text{image Hom}(F, G)$. (This is often taken as the definition of $\text{Ext}(A, G)$.)

11.34.   Prove that $\text{Ext}(A, G) = 0$ for every torsion-free group A if and only if $\text{Ext}(\mathbf{Q}, G) = 0$.

11.35.   (**Nunke**) If $\text{Hom}(A, \mathbf{Z}) = 0$ and $\text{Ext}(A, \mathbf{Z}) = 0$, then $A = 0$.

11.36.   If A is a torsion-free group of rank 1 such that $\text{Ext}(A, \mathbf{Z}) = 0$, then $A \cong \mathbf{Z}$.

11.37.   A group G is divisible if and only if $\text{Ext}(\mathbf{Q}/\mathbf{Z}, G) = 0$.

11.38.   Prove that an abelian group A is free abelian if and only if $\text{Ext}(A, F) = 0$ for every free abelian group F.

11.39.   If G is a torsion group with character group G^*, then $G^* \cong \text{Ext}(G, \mathbf{Z})$.

11.40.   Let W be a group with $\text{Ext}(W, \mathbf{Z}) = 0$. Prove that W is torsion-free. If W has finite rank, prove that W is free abelian.

If W is a countable group with $\text{Ext}(W, \mathbf{Z}) = 0$, it is not difficult to prove W is free abelian. However, Shelah (1974) proved the uncountable version of this theorem is undecidable: For W of cardinality $\aleph_1$ (the first uncountable cardinal), the statement "$\text{Ext}(W, \mathbf{Z}) = 0$ implies W is free abelian" and its negation are each consistent with ZFC, the usual axioms of set theory.

When K is an abelian group and Q is a (not necessarily abelian) group, we remarked in Chapter 7 that every extension of K by Q gives K the added structure of a module over the group ring $\mathbf{Z}Q$. In this chapter we constructed a functor Ext so that $\text{Ext}(C, A)$ classifies all (abelian) extensions of A by C (where A and C are abelian). In the general situation, one constructs two *sequences* of functors H^i and H_i, $i \geq 0$. If Q is a group and K is a $\mathbf{Z}Q$-module, $H^i(Q, K)$ is called the ith **cohomology group** of Q with "coefficients" K. When $i = 2$, $H^2(Q, K)$ is the group $H^2(Q, {}_\theta K)$ constructed in Chapter 7 that classifies nonabelian extensions (the homomorphism θ actually defines the module structure of K). When $i = 1$, $H^1(Q, K)$ is a quotient of $\text{Der}(Q, K)$, the group of all crossed homomorphisms from Q to K, which arises in certain conjugacy problems. When $i = 0$, $H^0(Q, K)$ is the subgroup of all fixed points: $H^0(Q, K) = \{k \in K : xk = k \text{ for all } x \in Q\}$. The group $H_i(Q, K)$ is called the ith **homology group** of Q with "coefficients" K. When $i = 2$, $H_2(Q, \mathbf{Z})$ is the **Schur multiplier**, a group arising in studying (matrix) representations of Q. For a prime p, $H_1(Q, \mathbf{Z}/p\mathbf{Z}) \cong Q/Q'Q^p$ (where Q^p is the set of pth powers in Q); when Q is a finite p-group, we thus have $H_1(Q, \mathbf{Z}/p\mathbf{Z}) \cong Q/\Phi(Q)$ (see Exercise 5.61). Also, $H_1(Q, \mathbf{Z}) \cong Q/Q'$; when Q is a subgroup of finite index in a group G, one can see the transfer arising as a map $H_1(G, \mathbf{Z}) \to H_1(Q, \mathbf{Z})$. For a discussion of cohomology and homology groups with applications to group theory, the reader is referred to the books of Cartan-Eilenberg, Cassels-Fröhlich, Gruenberg, Mac Lane, Robinson (1982), Rotman, and Stammbach.

chapter twelve

Free Groups
and Free Products

In this chapter, we shall consider several special classes of infinite groups.

GENERATORS AND RELATIONS

We wish to extend the idea of generators and relations from abelian groups to arbitrary groups. To do this, we need the nonabelian analog of free abelian groups, and so we focus on the basic property of a free set of generators.

Definition A group F is **free** on a subset X if, for every group G and every function $f: X \to G$, there is a unique homomorphism $\tilde{f}: F \to G$ extending f. One calls X a **basis** of F.

We shall prove later that the uniqueness of the extension f is equivalent to saying the subset X generates F.

The first question is whether there are any free groups.

Theorem 12.1 *If X is a set, then there exists a group F that is free on X.*

Proof Let X' be a set disjoint from X and in one-one correspondence with it; denote this correspondence

$$x \leftrightarrow x^{-1}.$$

Let X'' be a set disjoint from $X \cup X'$ that contains only one element we denote by "1". Call $X \cup X' \cup X''$ the *alphabet*, and call its elements *letters*. Let S be the set of all sequences of letters $(a_1, a_2, \cdots)$, i.e., each $a_k = 1$ or $x^{\pm 1}$ for some $x \in X$ (we agree that x^1 may denote x). A **word** on X is a sequence $(a_1, a_2, \cdots) \in S$ such that all coordinates are 1 from some point on, i.e., there is an integer n such that $a_k = 1$ for all $k \geq n$. In particular, the constant sequence

$$(1, 1, 1, \cdots)$$

is a word; it is called the **empty word** and is also denoted by 1. A **reduced word** on X is a word on X that satisfies the extra conditions:

(i) x and x^{-1} are never adjacent;
(ii) If $a_m = 1$ for some m, then $a_k = 1$ for all $k > m$.

In particular, the empty word is a reduced word. Since words contain only a finite number of letters before they become constant, we use the more economical (and suggestive) notation:

$$w = x_1^{\varepsilon_1} x_2^{\varepsilon_2} \cdots x_n^{\varepsilon_n},$$

where $\varepsilon_i = \pm 1$. Observe that this spelling of a reduced word is unique, for this is just the definition of equality of sequences.†

The idea of the construction of the free group F is just this: the elements of F are the reduced words and the binary operation is juxtaposition. Unfortunately, the juxtaposition of two reduced words need not be reduced. This tiny fact incurs tedious case analyses in verifying associativity, so we use a device†† to bypass the tedium.

Let W be the set of reduced words on X. For each $x \in X$, consider the two functions mapping W into itself, $|x|$ and $|x^{-1}|$, defined as follows:

$$|x^{\varepsilon}|(x_1^{\varepsilon_1} \cdots x_n^{\varepsilon_n}) = \begin{cases} x^{\varepsilon} x_1^{\varepsilon_1} \cdots x_n^{\varepsilon_n} & \text{if } x^{\varepsilon} \neq x_1^{-\varepsilon_1}; \\ x_2^{\varepsilon_2} \cdots x_n^{\varepsilon_n} & \text{if } x^{\varepsilon} = x_1^{-\varepsilon_1}, \end{cases}$$

where $\varepsilon = \pm 1$.

Since $|x| \, |x^{-1}|$ and $|x^{-1}| \, |x|$ are each equal to the identity on W, each $|x|$ is a permutation of W (with inverse $|x^{-1}|$). Let S_W be the group of all permutations of W, and let F_0 be the subgroup of S_W generated by $X_0 = \{|x| : x \in X\}$. We claim that F_0 is free on X_0.

An arbitrary element (other than the identity) of F_0 has a factorization $|x_1^{\varepsilon_1}| \cdots |x_n^{\varepsilon_n}|$, where $\varepsilon_i = \pm 1$ and $|x^{\varepsilon}|$ and $|x^{-\varepsilon}|$ are never adjacent (since otherwise we may cancel). Such a factorization is unique, for applying this

---

† See Appendix III.
†† This construction is due to van der Waerden.

function to the reduced word 1 yields the reduced word $x_1^{\varepsilon_1} \cdots x_n^{\varepsilon_n}$, and we have already noted that a reduced word has a unique spelling. Hence, two different factorizations are distinct permutations.

Suppose G is a group and $f: X_0 \to G$ is a function. Define $\tilde{f}: F_0 \to G$ by

$$\tilde{f}(|x_1^{\varepsilon_1}| \cdots |x_n^{\varepsilon_n}|) = f(|x_1|)^{\varepsilon_1} \cdots f(|x_n|)^{\varepsilon_n};$$

$\tilde{f}$ is well defined, since the factorization of an element of F_0 into powers of $|x|$ is unique. Now if α and β are in F_0, then

$$\alpha = |x_1^{\varepsilon_1}| \cdots |x_n^{\varepsilon_n}| \quad \text{and} \quad \beta = |y_1^{\eta_1}| \cdots |y_m^{\eta_m}|;$$

thus, after possible cancellations,

$$\alpha\beta = |x_1^{\varepsilon_1}| \cdots |x_s^{\varepsilon_s}| \, |y_t^{\eta_t}| \cdots |y_m^{\eta_m}| \qquad s \le n, \qquad 1 \le t$$

(our notation is careless, for we do not wish to overlook the cases in which all the $|x|$ are canceled by $|y|$ or all the $|y|$ are canceled by $|x|$). Since cancellation in F_0 implies cancellation in the group G, we have $\tilde{f}(\alpha\beta) = \tilde{f}(\alpha)\tilde{f}(\beta)$. Thus, $\tilde{f}$ is a homomorphism; it is uniquely determined by f because X_0 generates F_0. Therefore, F_0 is free on X_0.

By Exercise 1.20, the set of reduced words in X under juxtaposition is a group isomorphic to F_0, and thus it is free on X. ∎

Corollary 12.2 *Every group G is a quotient of a free group.*

Proof Consider G as a set, and let F be free on G.

$$\begin{array}{ccc} F & \xrightarrow{\tilde{f}} & \\ \cup & \searrow & \\ G & \xrightarrow{f} & G \end{array}$$

If $f: G \to G$ is the identity map, then there is a homomorphism $\tilde{f}: F \to G$ extending f; $\tilde{f}$ is onto because f is onto. ∎

Definition A group G is defined by **generators** $X = \{x_k: k \in K\}$ and **relations** $\Delta = \{r_j = 1: j \in J\}$ if $G \cong F/R$, where F is free on X and R is the *normal* subgroup of F generated by $\{r_j: j \in J\}$. The ordered pair $(X|\Delta)$ is called a **presentation** of G.

There are two reasons forcing us to use the normal subgroup R generated by $\{r_j: j \in J\}$: if $r_j = 1$ and $w \in F$, then $wr_jw^{-1} = 1$; we wish to form the quotient group F/R.

Example 1 $G = \mathbf{Z}(6)$ has generators x and y and relations $x^2 = 1$, $y^3 = 1$, and $xyx^{-1}y^{-1} = 1$.

Example 2 The dihedral group D_n has generators s and t and relations $s^n = 1$, $t^2 = 1$, and $tst = s^{-1}$ (compare D_3 with $\mathbf{Z}(6)$).

Example 3 The group of quaternions has generators a and b and relations $a^4 = 1$, $b^2 = a^2$, and $bab^{-1} = a^{-1}$.

Example 4 The group of quaternions has a presentation

$$(x, y \mid xyx = y, \quad x^2 = y^2).$$

Example 5 If G has the presentation $(X \mid \Delta)$ and H has the presentation $(X \mid \Delta \cup \Delta')$, then there is a well defined homomorphism $G \to H$ taking each generator $x \in X$ to itself [this consequence of the third isomorphism theorem is known as **von Dyck's theorem** (1882)].

Example 6 Consider the group $P(l, m, n)$ having the presentation

$$(s, t \mid s^l = t^m = (st)^n = 1).$$

Using Exercise 3.58, one may show that $P(2, 3, 3) \cong A_4$, $P(2, 3, 4) \cong S_4$, and $P(2, 3, 5) \cong A_5$. Note also that $P(n, 2, 2) \cong D_n$.

Example 7 A free abelian group on $X = \{x_k : k \in K\}$ has generators X and relations $\{x_i x_j x_i^{-1} x_j^{-1} : i, j \in K\}$.

Example 8 A free group on $\{x_k : k \in K\}$ has generators $\{x_k : k \in K\}$ and no relations.

The existence of a group described by specifying its generators and relations has now been established. The description, however, is incomplete in that the order of the constructed group may be difficult to determine. This is not a minor difficulty, for it is even an unsolvable problem to determine from an arbitrary finite presentation whether or not the presented group has only one element (see Corollary 13.32).

EXERCISES

12.1. Using presentations, prove the existence of the following groups: (a) the generalized quaternion groups Q_n of order 2^n; (b) the groups of order p^3 described in Exercise 4.41.

12.2. A free group on no generators has one element; a free group on one generator is infinite cyclic; a free group on more than one generator is a centerless group in which every element (save the identity) has infinite order.

12.3. The group G with the presentation

$$(x, y \mid x^2 = 1, \quad y^3 = 1)$$

is infinite.

Definition Let X be a set and F a semigroup containing X; F is a **free semigroup** on X if, for every semigroup S, every function $f: X \to S$ has a unique extension to a homomorphism of F into S.

Define a (**positive**) **word** on X to be an element of X^n for some $n \geq 1$, where X^n denotes the cartesian product of X with itself n times. If F is the set of all positive words on X, then juxtaposition is an associative binary operation on F (this is easy to see because there is no cancellation here) and F is a semigroup.

EXERCISE

*12.4. (i) The semigroup F of all positive words on X is free on X.
         (ii) Every semigroup is a homomorphic image of a free semigroup.

REMARK   There is a notion of **quotient semigroup**. One first defines a **congruence** on a semigroup S to be an equivalence relation R on S such that

$$xRx' \quad \text{and} \quad yRy' \quad \text{imply} \quad (xy)R(x'y').$$

Given a congruence R on a semigroup S, one may check that the set of all equivalence classes, denoted by S/R, becomes a semigroup with binary operation

$$[x][y] = [xy];$$

this operation is well defined because R is a congruence. There are two general constructions of congruences. The first arises from a homomorphism $\varphi: S \to S'$ between semigroups; define xRy to mean $\varphi(x) = \varphi(y)$. This congruence is called the **kernel** of φ, and it is straightforward to prove the isomorphism theorem:

$$S/\ker \varphi \cong \operatorname{im} \varphi.$$

Here is the second construction. If one views congruences on S as subsets of $S \times S$, then it is easy to see that any intersection of congruences is again a congruence. One may thus speak of the **congruence generated by a subset** of $S \times S$. In particular, let F be the free semigroup on a set X and let $\{a_i = b_i, i \in I\}$ be a family of equations, where $a_i, b_i \in F$. Define R to be the congruence generated by $\{(a_i, b_i): i \in I\} \subset F \times F$. The quotient semigroup $S = F/R$ is said to have the **presentation**

$$S = (X \mid a_i = b_i, i \in I).$$

Theorem 12.3   (Projective Property)   *Let β be a homomorphism of B onto C. If F is a free group and $\alpha: F \to C$ is a homomorphism, then there exists a homomorphism $\gamma: F \to B$ with $\beta\gamma = \alpha$.*

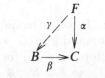

Proof   Let $X = \{x_k : k \in K\}$ be a basis of F. Since β is onto, for each k there is an element $b_k \in B$ with $\beta(b_k) = \alpha(x_k)$. Define a function $f: X \to B$ by $f(x_k) = b_k$.

There is a homomorphism $\gamma: F \to B$ with $\gamma(x_k) = b_k$ for all k, since F is free on X. One checks easily that $\beta\gamma | X = \alpha | X$, so the uniqueness part of the definition of free group gives $\beta\gamma = \alpha$. ∎

EXERCISES

*12.5.   (**Baer**) Prove that a group with the injective property must have order 1. [HINT (**D. L. Johnson**):   Consider $A \subset B$, where A is free on $\{a, b\}$ and B is the semidirect product $A \rtimes \langle c \rangle$ in which c is an element of order 2 acting by $c^{-1}ac = b$ and $c^{-1}bc = a$.] (See Exercise 12.43.)

**12.6.   Let F be free on $\{x_k : k \in K\}$ and let F' denote the commutator subgroup of F; show F/F' is free abelian on $\{F'x_k : k \in K\}$, a set having the same number of elements as $\{x_k : k \in K\}$.

**12.7.   Let $X = \{x_k : k \in K\}$ and let $Y \subset X$. If F is free on X and H is the normal subgroup generated by Y, then F/H is free.

*12.8.   Show that a free group F on $\{x, y\}$ has an automorphism f with $f(f(a)) = a$ for all $a \in F$ and with the further property that $f(a) = 1$ if and only if $a = 1$. (Compare Exercise 1.38.)

Now that we are assured of the existence of free groups, let us examine their uniqueness.

Theorem 12.4   *Let X and Y be nonempty sets, F free on X, and G free on Y. Then $F \cong G$ if and only if X and Y have the same number of elements.*

Proof   If $F \cong G$, then $F/F' \cong G/G'$. By Exercise 12.6, F/F' is free abelian on the cosets of X and G/G' is free abelian on the cosets of Y. By Theorem 10.17, X and Y have the same number of elements.

Suppose $f: X \to Y$ is a one-one correspondence. Consider the diagram

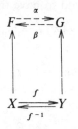

There is a homomorphism $\alpha: F \to G$ extending f and a homomorphism $\beta: G \to F$ extending f^{-1}. The composite $\alpha\beta: G \to G$ is thus an endomorphism of G fixing Y pointwise. Therefore $\alpha\beta = 1_G$, for both are homomorphisms agreeing on Y. Similarly $\beta\alpha = 1_F$, so $F \cong G$. ∎

Definition   If F is free on X, then **rank** F is the number of elements in X.

Theorem 12.4 says that the definition of rank is independent of the choice of basis X.

Corollary 12.5 *If F is free on X, then F is generated by X.*

Proof Let Y be a set having the same number of elements as X and let $f: X \to Y$ be a one-one correspondence. Let G be the free group on Y as constructed in Theorem 12.1. In the proof of Theorem 12.4, we exhibited an isomorphism $\beta: G \to F$ with $\beta(Y) = X$. Since Y generates G, it follows that X generates F. ∎

EDGEPATH GROUPS

We are going to study an analog of Galois theory that arises in algebraic topology: there is a one-one correspondence between the subgroups of the fundamental group of a topological space X and certain spaces $\tilde{X}$ mapping onto X. One consequence of this study, first noticed by Baer and Levi, is the Nielsen-Schreier theorem: Every subgroup of a free group is itself free. We mimic the topological theorems here in a completely algebraic setting.

Definition A **complex** K is a set $V(K)$, whose points are called **vertices**, and a family of nonempty subsets of $V(K)$, called **simplexes**, such that

(i) if $v \in V(K)$, then $\{v\}$ is a simplex;
(ii) if s is a simplex, so is every nonempty subset of s.

A complex is called an *abstract simplicial complex* in the literature.

If a simplex s has $q + 1$ distinct elements, say, $s = \{v_0, v_1, \cdots, v_q\}$, we call s a **q-simplex** and say s has **dimension** q. We write $\dim K = n$ and call K an **n-complex** if n is the largest dimension of a simplex in K. (If there is no such largest n, we say $\dim K = \infty$.)

The reader should know the geometric background behind the definition of a complex, even though no geometry enters the forthcoming discussion. Think of a 0-simplex $\{v_0\}$ as a point, a 1-simplex $\{v_0, v_1\}$ as a line segment with endpoints v_0, v_1, a 2-simplex $\{v_0, v_1, v_2\}$ as a triangle with vertices v_0, v_1, v_2, a 3-simplex as a tetrahedron, and so forth. A complex is then a space that is built of simplexes that are assembled by gluing some of their faces together.

A complex L is a **subcomplex** of a complex K if $V(L) \subset V(K)$ and every simplex of L is also a simplex of K. For example, if s is a simplex in K, then s together with all its nonempty subsets is a subcomplex of K. Another example is provided by the **q-skeleton** of K, denoted by $K^{(q)}$, defined as follows:

$$K^{(q)} = \{\text{simplexes } s \text{ in } K: \dim s \le q\}.$$

Thus, $K^{(0)} = V(K)$, $K^{(1)} = V(K) \cup \{\text{all 1-simplexes in } K\}$, and so forth. Visibly dim $K^{(q)} \le q$. A subcomplex L of K is **full** if each simplex in K having all its vertices in $V(L)$ also belongs to L. Thus, a full subcomplex L of K is determined by its vertices $V(L)$: it is the largest subcomplex of K having vertices $V(L)$.

If $\{L_i: i \in I\}$ is a family of subcomplexes of K, then $\cup L_i$ and $\cap L_i$ are defined in the obvious way: $V(\cup L_i) = \cup V(L_i)$ and a simplex s lies in $\cup L_i$ if s lies in some L_i; $V(\cap L_i) = \cap V(L_i)$ and a simplex s lies in $\cap L_i$ if s lies in every L_i. It is easy to see $\cup L_i$ and $\cap L_i$ are subcomplexes. In order that $\cap L_i$ always be defined, we allow the empty set $\varnothing$ to be a subcomplex (its dimension is -1). In particular, two subcomplexes are **disjoint** if and only if they have no vertices in common.

Definition   An **edge** $e = (u, v)$ in K is an ordered pair of (not necessarily distinct) vertices lying in a simplex of K; u is called the **origin** of e; v is called the **end** of e.

Definition   A **path** α of **length** n is a sequence of edges

$$\alpha = e_1 \cdots e_n$$

where end $e_i = $ origin e_{i+1}, $i = 1, \cdots, n-1$. Define **origin** $\alpha = $ origin e_1 and **end** $\alpha = $ end e_n. We say α is a path from origin α to end α. A path is **closed** at v if it is a path from v to itself.

Definition   A complex K is **connected** if for every pair of vertices u, v in K there is a path in K from u to v.

EXERCISES

12.9.   A subcomplex L of K is connected if and only if $L \cap K^{(1)}$ is connected ($K^{(1)}$ is the 1-skeleton of K).

*12.10.   Call two vertices u, v in K "connectable" if there is a path in K from u to v. Prove that this defines an equivalence relation on $V(K)$.

Definition   The equivalence classes defined in Exercise 12.10 are called the **components** of K.

We may consider a component C as a (full) subcomplex of K by regarding it as all those simplexes of K all of whose vertices lie in C. Thus, every complex K is the disjoint union of connected subcomplexes: its components.

Let $\alpha = e_1 \cdots e_n$ and $\alpha' = e_1' \cdots e_m'$ be paths in K. If end $\alpha = $ origin α', define their **product**

$$\alpha\alpha' = e_1 \cdots e_n e_1' \cdots e_m'.$$

Clearly $\alpha\alpha'$, when defined, is a path from origin α to end α'. Moreover, this product is associative, when defined. We want to construct a group with the above product as multiplication. There are several obstacles: multiplication is not always defined; there is no identity; there are no inverses. The last two obstacles are overcome by imposing an equivalence relation.

Definition   Two paths α and α' are **homotopic**, denoted by $\alpha \sim \alpha'$, if one can be obtained from the other by a finite number of elementary moves consisting of replacing one side of an equation

$$(u, v)(v, w) = (u, w)$$

by the other whenever $\{u, v, w\}$ is a simplex of K.

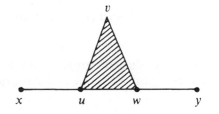

For example, let K be the 2-complex pictured above, let $\alpha = (x, u)(u, w)(w, y)$ and $\alpha' = (x, u)(u, v)(v, w)(w, y)$. If K contains the 2-simplex $\{u, v, w\}$, i.e., the shaded interior of the triangle $\{u, v, w\}$, then $\alpha \sim \alpha'$; if K does not contain this 2-simplex, then $\alpha \not\sim \alpha'$.

It is easy to check that homotopy defines an equivalence relation on the set of all paths in K; denote the equivalence class of α by $[\alpha]$.

EXERCISES

12.11.   If $\alpha \sim \alpha'$, then origin $\alpha = $ origin α' and end $\alpha = $ end α'. Conclude that origin$[\alpha]$ and end$[\alpha]$ make sense.

12.12.   If $\alpha \sim \alpha'$, $\beta \sim \beta'$, and end $\alpha = $ origin β, then $[\alpha\beta] = [\alpha'\beta']$.

**12.13.   Let $\alpha = \alpha_1 \beta \alpha_2$, where β is a closed path lying wholly within a simplex s of K. Prove that $\alpha \sim \alpha_1\alpha_2$.

If $v \in V(K)$, set $i_v = (v, v)$; if $e = (u, v)$, set $e^{-1} = (v, u)$; if $\alpha = e_1 \cdots e_n$, set $\alpha^{-1} = e_n^{-1} \cdots e_1^{-1}$.

Lemma 12.6 *If K is a complex, let $\pi(K)$ be the set of all $[\alpha]$, where α is a path in K. Then*

(i) *each $[\alpha]$ has an origin u and an end v (where u and v are vertices of K) and*

$$[i_u][\alpha] = [\alpha] = [\alpha][i_v];$$

(ii) *the associative law holds when defined;*

(iii) *if origin $[\alpha] = u$ and $end[\alpha] = v$, then*

$$[\alpha][\alpha^{-1}] = [i_u] \quad and \quad [\alpha^{-1}][\alpha] = [i_v].$$

Proof Completely without difficulty. ∎

We force $\pi(K)$ to be a group by choosing a **basepoint** v_0.

Definition Choose a vertex v_0 in K. Then

$$\pi(K, v_0) = \{[\alpha] \in \pi(K): \alpha \text{ is a closed path at } v_0\}.$$

Theorem 12.7 $\pi(K, v_0)$ *is a group.*

Proof This follows immediately from Lemma 12.6 since multiplication is now always defined. ∎

Definition $\pi(K, v_0)$ is called the **edgepath group** of K. (It is isomorphic to the fundamental group of the topological space one may construct from K.)

Corollary 12.8 *Let K be a connected complex with vertices v_0 and u_0. Then*

$$\pi(K, v_0) \cong \pi(K, u_0).$$

Proof Since K is connected, there is a path β in K from v_0 to u_0. Define $f: \pi(K, v_0) \to \pi(K, u_0)$ by $[\alpha] \mapsto [\beta^{-1}\alpha\beta]$. Note that the multiplication $[\beta^{-1}][\alpha][\beta]$ takes place in $\pi(K)$, but that $[\beta^{-1}\alpha\beta] \in \pi(K, u_0)$. It is a simple matter, using Lemma 12.6, to verify that f is a homomorphism with inverse $[\alpha'] \mapsto [\beta\alpha'\beta^{-1}]$. ∎

A 1-complex K is often called a **graph**. Our next aim is to prove that $\pi(K, v_0)$ is free whenever K is a connected graph.

Definition A path $\alpha = e_1 \cdots e_n$ is **reduced** if, in this expression, no edge is adjacent to its inverse. A **circuit** is a reduced closed path.

Definition A **tree** in K is a connected subcomplex T of K with dim $T \leq 1$ and which contains no circuits. (The only tree of dimension 0 consists of a single vertex.) A **maximal tree** is a tree contained in no larger tree.

It is easy to prove that an ascending union of trees is a tree, so it follows from Zorn's lemma that every complex contains a maximal tree. Note also that if u and v

are distinct vertices in a tree T, then there is a unique reduced path in T from u to v (lest T contain a circuit).

Theorem 12.9 *If K is a connected complex, then a tree T in K is a maximal tree if and only if T contains all the vertices of K.*

Proof Suppose there is a vertex v not in T. Choose a vertex v_0 in T. Since K is connected, there is a path $e_1 \cdots e_n$ in K from v_0 to v. Let $e_i = (v_{i-1}, v_i)$. We know that $v_0 \in T$ and $v_n = v \notin T$, so there must be an i with $v_{i-1} \in T$ and $v_i \notin T$. Consider the subcomplex T' with $V(T') = V(T) \cup \{v_i\}$ and with simplexes those of T together with $\{v_{i-1}, v_i\}$. Then T' is connected and contains no circuits, for any circuit in T' must contain v_i, and any path through v_i must contain $(v_{i-1}, v_i)(v_i, v_{i-1})$ or $(v_i, v_{i-1})(v_{i-1}, v_i)$. Therefore T' is a tree, contradicting the maximality of T. Thus, T contains every vertex of K.

The converse is also easy and is left to the reader. ∎

EXERCISES

**12.14. Let K be a complex and let T_1 and T_2 be trees in K with $T_1 \cap T_2 \neq \emptyset$. If $T_1 \cap T_2$ is connected, prove that $T_1 \cup T_2$ is a tree.

**12.15. Let $[\alpha] \in \pi(K, v_0)$. If there exists a tree T in K with each edge of α in T, then $[\alpha] = 1$.

**12.16. Let T be a finite tree. If $n_0(T)$ is the number of vertices in T and $n_1(T)$ is the number of 1-simplexes in T, then†

$$n_0(T) - n_1(T) = 1.$$

Notation Given a connected complex K and a maximal tree T, define a group $G_{K,T}$ having the following presentation:

> *generators*: all edges (u, v) in K;
> *relations*: (a): $(u, v) = 1$ if (u, v) is an edge in T;
>   (b): $(u, v)(v, w) = (u, w)$ if u, v, w lie in a simplex of K.

Lemma 12.10 (Tietze, 1908) *If K is connected with maximal tree T, then*

$$\pi(K, v_0) \cong G_{K,T}.$$

REMARK Since K is connected, we know that different choices of basepoint v_0 yield isomorphic groups.

Proof Let F be the free group with basis all edges (u, v) of K and let R be the normal subgroup of relations, so that $G_{K,T} = F/R$.

For each vertex $v \neq v_0$ in K, there is a unique reduced path α_v in T

---

† If $\dim K = 1$ and K is finite, then the number $n_0(K) - n_1(K)$ is called the **Euler-Poincaré characteristic** of K.

from v_0 to v (since T is a maximal tree); define $\alpha_{v_0} = (v_0, v_0)$. Define a map $G_{K,T} \to \pi(K, v_0)$ as follows. First, define a function on the basis of F by

$$(u, v) \mapsto [\alpha_u(u, v)\alpha_v^{-1}]$$

(which is the class of a closed path at v_0). This function defines a homomorphism $\varphi: F \to \pi(K, v_0)$ which we claim kills the relations R. *Type* (a): if (u, v) is an edge in T, then $\alpha_u(u, v)\alpha_v^{-1}$ lies wholly in T, and so $\varphi((u, v))$ $= [\alpha_u(u, v)\alpha_v^{-1}] = 1$, by Exercise 12.15. *Type* (b): if $\{u, v, w\}$ is a simplex of K, then

$$[\alpha_u(u, v)\alpha_v^{-1}][\alpha_v(v, w)\alpha_w^{-1}] = [\alpha_u(u, v)\alpha_v^{-1}\alpha_v(v, w)\alpha_w^{-1}]$$
$$= [\alpha_u(u, v)(v, w)\alpha_w^{-1}]$$
$$= [\alpha_u(u, w)\alpha_w^{-1}].$$

Therefore φ induces a homomorphism

$$\bar{\varphi}: G_{K,T} \to \pi(K, v_0)$$

by

$$(u, v)R \mapsto \varphi(u, v) = [\alpha_u(u, v)\alpha_v^{-1}].$$

We prove $\bar{\varphi}$ is an isomorphism by constructing its inverse. If $\alpha = e_1 \cdots e_n$ is a closed path in K at v_0, define

$$\theta(\alpha) = e_1 \cdots e_n R \in G_{K,T}.$$

Observe that if α' is another path in K with $\alpha \sim \alpha'$, then the relations in $G_{K,T}$ of type (b) show $\theta(\alpha) = \theta(\alpha')$. There is thus a homomorphism

$$\bar{\theta}: \pi(K, v_0) \to G_{K,T}$$

defined by

$$[e_1 \cdots e_n] = [\alpha] \to \theta(\alpha) = e_1 \cdots e_n.$$

Let us compute. If $[\alpha] \in \pi(K, v_0)$ and $\alpha = e_1 \cdots e_n$, then

$$\bar{\varphi}\bar{\theta}[\alpha] = \bar{\varphi}(\theta(\alpha))$$
$$= \bar{\varphi}(e_1 \cdots e_n R)$$
$$= [\varphi(e_1) \cdots \varphi(e_n)] \text{ (since } \varphi \text{ is a}$$
$$\text{homomorphism killing } R)$$
$$= [\alpha_{v_0} e_1 \cdots e_n \alpha_{v_0}^{-1}].$$

But $\alpha_{v_0} = (v_0, v_0)$, so that $[\alpha_{v_0}] = 1$ in $\pi(K, v_0)$. Therefore, $\bar{\varphi}\bar{\theta}$ is the identity. Finally, suppose (u, v) is a generator of $G_{K,T}$. Then

$$\bar{\theta}\bar{\varphi}((u, v)R) = \bar{\theta}(\varphi(u, v))$$
$$= \bar{\theta}[\alpha_u(u, v)\alpha_v^{-1}]$$
$$= \alpha_u(u, v)\alpha_v^{-1} R.$$

Now α_v^{-1} and α_u lie in R, since their edges do, so that

$$\alpha_u(u, v)\alpha_v^{-1} R = \alpha_u(u, v)R = (u, v)R,$$

the last equation because R is normal in F (Exercise 2.38). Thus $\bar{\theta}\bar{\varphi}$ fixes a set of generators of $G_{K,T}$, hence is the identity. Therefore $\pi(K, v_0) \cong G_{K,T}$. ∎

Theorem 12.11 *If K is a connected 1-complex, then $\pi(K, v_0)$ is a free group. Moreover, if T is a maximal tree in K, then*

$$\text{rank } \pi(K, v_0) = |\{1\text{-simplexes } s\colon s \in K, \ s \notin T\}|.$$

Proof By Lemma 12.10, it suffices to examine $G_{K,T}$. Because of relations of type (a), $G_{K,T}$ is surely generated by all edges (u, v), where (u, v) is not in T. Furthermore, if $e = (u, v)$, the type (b) relation $ee^{-1} = 1$ allows us to choose just one of the two edges determined by the 1-simplex $\{u, v\}$ as part of a generating set of $G_{K,T}$. Next, if u, v, w all lie in a simplex of K, then two of the vertices must be the same, for dim $K = 1$. Thus, the relations of type (b) have the form:

$$(u, u)(u, v) = (u, v);$$
$$(u, v)(v, v) = (u, v);$$
$$(u, v)(v, u) = (u, u).$$

But all of these are trivial: since $(v, v) = 1$ and $(u, v) = (v, u)^{-1}$, the subgroup R of relations is $\{1\}$. Therefore, $G_{K,T}$ is free on the 1-simplexes not in T. ∎

It follows immediately from Theorem 12.11 that the number of edges in the complement of a maximal tree is independent of the choice of maximal tree. However, this fact is easy to see directly, for a tree with n vertices has exactly $n - 1$ edges.

We have established a new criterion for a group to be free: if it is the edgepath group of a connected graph. Note that, in this case, a basis can be given explicitly: If T is a maximal tree in K and α_v is the unique reduced path in T from v_0 to v, then a basis for $\pi(K, v_0)$ is the set of all $[\alpha_u(u, v)\alpha_v^{-1}]$, where $\{u, v\}$ is a 1-simplex not in T. Of course, for any 1-simplex $\{u, v\}$ not in T, only one edge (u, v) is chosen.

Theorem 12.12 *Given a set I, there exists a connected 1-complex K with $\pi(K, v_0)$ free of rank $|I|$.*

Proof For each $i \in I$, choose distinct points u_i, v_i, and let V be the disjoint union of a point v_0 and all $\{u_i, v_i\}$. Let K be the complex with vertices V and 1-simplexes $\{v_0, u_i\}$, $\{v_0, v_i\}$, and $\{u_i, v_i\}$, all $i \in I$. Visibly K is a connected 1-complex. A maximal tree consists of all simplexes $\{v_0, u_i\}$ and $\{v_0, v_i\}$. Therefore, $\pi(K, v_0)$ is free on all 1-simplexes $\{u_i, v_i\}$, a set in one-one correspondence with I. ∎

The complex K just constructed is called a **bouquet of circles** because of the picture

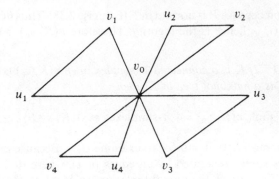

COVERING COMPLEXES AND THE NIELSEN-SCHREIER THEOREM

In the preceding section, we associated a group to every connected complex; in this section we associate a complex to a group.

Definition   Let K and K' be complexes. A **map**† $f: K \longrightarrow K'$ is a function $f: V(K) \longrightarrow V(K')$ such that $\{fv_0, \cdots, fv_q\}$ is a simplex in K' whenever $\{v_0, \cdots, v_q\}$ is a simplex in K. An **isomorphism** is a map that is a one-to-one correspondence and whose inverse is also a map.

Of course, not all the vertices $fv_0, \cdots, fv_q$ need be distinct. When maps may be composed, their composite is again a map.

Definition   If $p: K \to K'$ is a map and L' is a subcomplex of K', then $p^{-1}(L')$ is the subcomplex of K consisting of all those simplexes s in K with $p(s)$ in L'.

Observe that if L' is a full subcomplex of K', then $p^{-1}(L')$ is the full subcomplex of K on $p^{-1}(V(L'))$. This observation applies, in particular, when L' is a simplex in K'.

EXERCISE

12.17.   If $f: K \to K'$ is a map, then the image of f is a subcomplex of K'; if K is connected, then im f is connected.

Let K be a complex and let $\{X_i: i \in I\}$ be a partition of $V(K)$. Define a complex $\overline{K}$ by

$$V(\overline{K}) = \{X_i: i \in I\};$$

$\{X_{i_0}, \cdots, X_{i_q}\}$ is a simplex if there exist vertices $v_j \in X_{i_j}$ with $\{v_0, \cdots, v_q\}$ a simplex in K.

† Maps are called **simplicial maps** in the literature.

Definition The complex $\bar{K}$ constructed from a partition of $V(K)$ is called a **quotient complex** of K.

One may also construct a quotient complex from an equivalence relation on $V(K)$, for the equivalence classes partition $V(K)$.

EXERCISES

**12.18. Prove that $\bar{K}$ is a complex and the function $f: V(K) \to V(\bar{K})$ sending each $v \in V(K)$ into the unique X_i containing it is a map.

**12.19. Let K be connected and let L be a subcomplex that is a disjoint union of trees. Prove that L is contained in a maximal tree of K.

12.20. Let I_n be the 1-complex with vertices $\{t_0, t_1, \cdots, t_n\}$ and 1-simplexes $\{t_0, t_1\}, \{t_1, t_2\}, \cdots, \{t_{n-1}, t_n\}$. Prove that a path in K from u to v is a map $\alpha: I_n \to K$ (some n) with $\alpha(t_0) = u$ and $\alpha(t_n) = v$.

12.21. Let $f: K \to K'$ be a map with $f(v_0) = v_0'$.
(i) Prove that

$$f_\#: \pi(K, v_0) \to \pi(K', v_0')$$

defined by

$$[\alpha] \mapsto [f \circ \alpha]$$

is a homomorphism, where $\alpha: I_n \to K$ is a path with $\alpha(t_0) = v_0 = \alpha(t_n)$.
[Note that if $\alpha = (v_0, v_1)(v_1, v_2) \cdots (v_n, v_0)$, then $f \circ \alpha = (v_0', fv_1)(fv_1, fv_2) \cdots (fv_n, v_0')$.]
(ii) If $f: K \to K$ is the identity, then $f_\#$ is the identity.
(iii) Given maps $f: K \to K'$ and $g: K' \to K''$ with $f(v_0) = v_0'$ and $g(v_0') = v_0''$, then $(g \circ f)_\# = g_\# \circ f_\#: \pi(K, v_0) \to \pi(K'', v_0'')$.
(One can rephrase this exercise by saying π is a functor.)

**12.22. Let K be a connected complex, let L be a full connected subcomplex, and let $v_0 \in V(L)$. If every closed path in K at v_0 is homotopic to a path in L, then the inclusion $L \hookrightarrow K$ induces an isomorphism $\pi(L, v_0) \cong \pi(K, v_0)$.

Definition Let K be a complex. A pair $(\tilde{K}, p)$ is a **covering complex** of K if
(i) $p: \tilde{K} \to K$ is a map;
(ii) $\tilde{K}$ is connected;
(iii) for every simplex s in K, $p^{-1}(s)$ is a union of pairwise disjoint simplexes,

$$p^{-1}(s) = \bigcup \tilde{s}_i,$$

with $p|\tilde{s}_i: \tilde{s}_i \to s$ a one-one correspondence for each i.

The map $p: \tilde{K} \to K$ is called the **projection** and the simplexes $\tilde{s}_i$ are called the **sheets** over s. The picture to keep in mind is

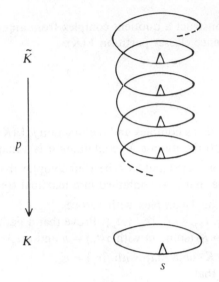

Example 9 Let K be a "triangle": a 1-complex with vertices $\{v_0, v_1, v_2\}$ and 1-simplexes $\{v_0, v_1\}$, $\{v_1, v_2\}$, and $\{v_0, v_2\}$. Let $\tilde{K}$ be the 1-complex with vertices $\{t_i : i \in \mathbf{Z}\}$ and 1-simplexes $\{t_i, t_{i+1}\}$, $i \in \mathbf{Z}$. If we define $p: \tilde{K} \to K$ by $p(t_i) = v_j$ if $i \equiv j$ (mod 3), then $(\tilde{K}, p)$ is a covering complex of K.

EXERCISES

12.23. If $p: \tilde{K} \to K$ is a covering complex, then im $p = K$; conclude that if K has a covering complex, then K is connected.

**12.24. Let $p: \tilde{K} \to K$ be a covering complex and let L be a connected subcomplex of K. If $\tilde{L}$ is a component of $p^{-1}(L)$, then $p|\tilde{L}: \tilde{L} \to L$ is a covering complex.

**12.25. If $p: \tilde{K} \to K$ is a covering complex and T is a tree in K, then $p^{-1}(T)$ is a disjoint union of trees. [HINT: Show each component of $p^{-1}(T)$ is a tree.]

**12.26. If $p: \tilde{K} \to K$ is a covering complex and s is a q-simplex in K, then every sheet $\tilde{s}_i$ over s is a q-simplex. Conclude that

$$\dim \tilde{K} = \dim K.$$

Lemma 12.13 *Let $p: \tilde{K} \to K$ be a covering complex and let $p(\tilde{v}_0) = v_0$. Given a path α in K with origin v_0, there exists a unique path $\tilde{\alpha}$ in $\tilde{K}$ with origin $\tilde{v}_0$ and $p\tilde{\alpha} = \alpha$.*

REMARK   One calls $\tilde{\alpha}$ a **lifting** of α because of the picture

Proof We prove by induction that if α has length n and origin v_0, then there exists a unique $\tilde{\alpha}$ of length n with origin $\tilde{v}_0$ and $p\tilde{\alpha} = \alpha$.

If $n = 1$, then $\alpha = (v_0, v)$, where v_0, v lie in a simplex s of K. We may assume $v_0 \neq v$, so that s is a 1-simplex. If $\tilde{s}_i$ is the sheet over s containing $\tilde{v}_0$, then $\tilde{s}_i$ is a 1-simplex, by Exercise 12.26. Hence $\tilde{s}_i = \{\tilde{v}_0, \tilde{v}\}$ for some vertex $\tilde{v}$ in $\tilde{K}$, so $(\tilde{v}_0, \tilde{v})$ is an edge in $\tilde{K}$ with origin $\tilde{v}_0$ and with $p\tilde{\alpha} = \alpha$. To see that $\tilde{\alpha}$ is unique, suppose $(\tilde{v}_0, \tilde{v})$ and $(\tilde{v}_0, \tilde{u})$ are liftings of α. If $s = \{v_0, v\}$, then the vertices $\tilde{v}_0, \tilde{v}$, and $\tilde{u}$ lie in $p^{-1}(s) = \cup \tilde{s}_i$. Since $p^{-1}(s)$ is the full subcomplex of $\tilde{K}$ on $p^{-1}(\{v_0, v\})$, both edges $(\tilde{v}_0, \tilde{v})$ and $(\tilde{v}_0, \tilde{u})$ lie in $p^{-1}(s)$. Visibly, the sheet containing $\tilde{v}$ and the sheet containing $\tilde{u}$ are not disjoint. It follows that $\tilde{u} = \tilde{v}$.

Assume $n > 1$, so that $\alpha = \alpha_1(v, v')$, where α_1 is a path from v_0 to v of length $n - 1$. By induction, there is a unique lifting $\tilde{\alpha}_1$ from $\tilde{v}_0$ to $\tilde{v}$, where $\tilde{v}$ is some vertex in $\tilde{K}$ with $p(\tilde{v}) = v$. The first step of the induction provides a unique vertex $\tilde{v}'$ in $\tilde{K}$ with $p(\tilde{v}') = v'$ and with $\{\tilde{v}, \tilde{v}'\}$ a simplex in $\tilde{K}$. Therefore $\tilde{\alpha}_1(\tilde{v}, \tilde{v}')$ is the unique lifting of α having origin $\tilde{v}_0$. ∎

Lemma 12.14   *Let $p: \tilde{K} \to K$ be a covering complex and let $p(\tilde{v}_0) = v_0$. If α and β are homotopic paths in K with origin v_0, then their liftings $\tilde{\alpha}$ and $\tilde{\beta}$ having origin $\tilde{v}_0$ are also homotopic.*

Proof Recall that α homotopic to β means that α can be transformed into β by a finite number of elementary moves replacing one side of an equation

$$(u, v)(v, w) = (u, w)$$

by the other whenever u, v, w all lie in a simplex of K. Thus, it suffices to prove that if $(\tilde{u}, \tilde{v})(\tilde{v}, \tilde{w})$ is a lifting of $(u, v)(v, w)$ and if u, v, w lie in a simplex s of K, then $\tilde{u}, \tilde{v}, \tilde{w}$ lie in a simplex of $\tilde{K}$. Let $\tilde{s}$ be the sheet over s containing $\tilde{v}$ and let $\tilde{s}_1$ be the sheet over s containing $\tilde{u}$. Thus, $\tilde{s}_1 = \{\tilde{u}, \tilde{v}_1, \tilde{w}_1\}$, where $p\tilde{v}_1 = v$ and $p\tilde{w}_1 = w$. Now $(\tilde{u}, \tilde{v}_1)$ and $(\tilde{u}, \tilde{v})$ are both liftings of the path (u, v) having origin $\tilde{u}$. By the uniqueness assertion in Lemma 12.13, $\tilde{v} = \tilde{v}_1$. Hence $\tilde{s}$ and $\tilde{s}_1$ are not disjoint, and so $\tilde{u}$ lies in $\tilde{s}$. Similarly, $\tilde{w}$ lies in $\tilde{s}$, so that $\{\tilde{u}, \tilde{v}, \tilde{w}\} = \tilde{s}$ is a simplex in $\tilde{K}$. ∎

Theorem 12.15   *Let $p: \tilde{K} \to K$ be a covering complex with $p(\tilde{v}_0) = v_0$. Then $p_{\#}: \pi(\tilde{K}, \tilde{v}_0) \to \pi(K, v_0)$ is one-one.*

Proof Assume $[\tilde{\alpha}]$ and $[\tilde{\beta}] \in \pi(\tilde{K}, \tilde{v}_0)$ are such that $p_{\#}[\tilde{\alpha}] = p_{\#}[\tilde{\beta}]$. Then $[p\tilde{\alpha}] = [p\tilde{\beta}]$ and $p\tilde{\alpha} \sim p\tilde{\beta}$. Visibly $\tilde{\alpha}$ and $\tilde{\beta}$ are the (unique) liftings of $p\tilde{\alpha}$ and $p\tilde{\beta}$ having origin $\tilde{v}_0$. By Lemma 12.14, $\tilde{\alpha} \sim \tilde{\beta}$, i.e., $[\tilde{\alpha}] = [\tilde{\beta}]$. Therefore $p_{\#}$ is one-one. ∎

What happens to the subgroup $p_\# \pi(\tilde{K}, \tilde{v})$ as the basepoint $\tilde{v}$ is changed?

Theorem 12.16 *Let $p: \tilde{K} \to K$ be a covering complex with $p(\tilde{v}) = v_0$. If $p(\tilde{u})$ $= v_0$, then $p_\# \pi(\tilde{K}, \tilde{v})$ and $p_\# \pi(\tilde{K}, \tilde{u})$ are conjugate subgroups of $\pi(K, v_0)$. Conversely, if H is conjugate to $p_\# \pi(\tilde{K}, \tilde{v})$, then $H = p_\# \pi(\tilde{K}, \tilde{u})$ for some $\tilde{u}$ with $p(\tilde{u}) = v_0$.*

Proof If $\tilde{\beta}$ is a path in $\tilde{K}$ from $\tilde{u}$ to $\tilde{v}$, and if $\beta = p\tilde{\beta}$, then $[\beta] \in \pi(K, v_0)$ and

$$[\beta] p_\# \pi(\tilde{K}, \tilde{v}) [\beta]^{-1} = p_\# \pi(\tilde{K}, \tilde{u}).$$

Conversely, assume $H = [\alpha] p_\# \pi(\tilde{K}, \tilde{v}) [\alpha]^{-1}$. Let $\tilde{\beta}$ be the lifting of α^{-1} with origin $\tilde{v}$. If end $\tilde{\beta} = \tilde{u}$, then $p(\tilde{u}) = v_0$. Moreover,

$$\pi(\tilde{K}, \tilde{u}) = [\tilde{\beta}^{-1}] \pi(\tilde{K}, \tilde{v}) [\tilde{\beta}],$$

so $p_\# \pi(\tilde{K}, \tilde{u}) = [\alpha] p_\# \pi(\tilde{K}, \tilde{v}) [\alpha^{-1}] = H$. ∎

Theorem 12.17 *Let $p: \tilde{K} \to K$ be a covering complex. If v_0 is a vertex in K, then $p^{-1}(v_0)$ is a transitive $\pi(K, v_0)$-set and the stabilizer of a point $\tilde{v}_0 \in p^{-1}(v_0)$ is $p_\# \pi(\tilde{K}, \tilde{v}_0)$.*

Proof For $x \in p^{-1}(v_0)$ and $[\alpha] \in \pi(K, v_0)$, define

$$[\alpha]x = \text{end } \tilde{\alpha},$$

where $\tilde{\alpha}$ is the lifting of α with origin x; Lemma 12.14 shows this definition is independent of the choice of path in $[\alpha]$.

The lifting of $i = (v_0, v_0)$ with origin x is $\tilde{i} = (x, x)$; hence $[i]x = x$ for all $x \in p^{-1}(v_0)$. If $[\beta] \in \pi(K, v_0)$, then $[\beta\alpha]x = \text{end } \widetilde{\beta\alpha}$, where $\widetilde{\beta\alpha}$ has origin x. If $\tilde{\alpha}$ has origin x and end y and if $\tilde{\beta}$ has origin y, then $\tilde{\beta}\tilde{\alpha}$ is a lifting of $\beta\alpha$ with origin x. By Lemma 12.14, end $\widetilde{\beta\alpha} = \text{end } \tilde{\beta}\tilde{\alpha}$, i.e.,

$$[\beta\alpha]x = [\beta] ([\alpha]x).$$

It follows that $p^{-1}(v_0)$ is a $\pi(K, v_0)$-set.

Now $\pi(K, v_0)$ acts transitively: If $x, y \in p^{-1}(v_0)$, then there is a path A in $\tilde{K}$ from x to y (since $\tilde{K}$ is connected). Denoting $p_\# A$ by α, we have $[\alpha] \in \pi(K, v_0)$; moreover $[\alpha]x = y$ because $\tilde{\alpha} = A$ (Lemma 12.13). Finally, the stabilizer of a point x in $p^{-1}(v_0)$ consists of all $[\alpha] \in \pi(\tilde{K}, v_0)$ such that end $\tilde{\alpha} = x$, i.e., $[\tilde{\alpha}] \in \pi(\tilde{K}, x)$, and this subgroup is $p_\# \pi(\tilde{K}, x)$. ∎

Corollary 12.18 *Let $p: \tilde{K} \to K$ be a covering complex.*

(i) *If v_0 is a vertex in K and $\tilde{v}_0 \in p^{-1}(v_0)$, then*

$$[\pi(K, v_0): p_\# \pi(\tilde{K}, \tilde{v}_0)] = |p^{-1}(v_0)|.$$

(ii) *If v_1 and v_2 are vertices in K, then*

$$|p^{-1}(v_1)| = |p^{-1}(v_2)|.$$

Proof

(i) This is immediate from the fact that the cardinal of the orbit of a point x is the index of the stabilizer of x.

(ii) Let $p(\tilde{v}_1) = v_1$ and $p(\tilde{v}_2) = v_2$; let $\tilde{\beta}$ be a path in $\tilde{K}$ from $\tilde{v}_1$ to $\tilde{v}_2$ and let $\beta = p\tilde{\beta}$. It is a simple matter to check that the following diagram commutes:

$$
\begin{array}{ccc}
\pi(\tilde{K}, \tilde{v}_1) & \xrightarrow{G} & \pi(\tilde{K}, \tilde{v}_2) \\
\downarrow{p_*} & & \downarrow{p_*} \\
\pi(K, v_1) & \xrightarrow{g} & \pi(K, v_2),
\end{array}
$$

where $G[\tilde{\alpha}] = [\tilde{\beta}^{-1}\tilde{\alpha}\tilde{\beta}]$ and $g[\alpha] = [\beta^{-1}\alpha\beta]$. Of course, G and g are isomorphisms, which implies that the index on the left equals the index on the right. By part (i), we have $|p^{-1}(v_1)| = |p^{-1}(v_2)|$. ∎

EXERCISE

12.27. Let $p: \tilde{K} \to K$ be a covering complex and suppose there are j points in $p^{-1}(v)$ for each vertex v of K. Prove there are exactly j sheets over every simplex of K.

Let us now consider existence of covering complexes. Suppose first that we have a covering complex $p: \tilde{K} \to K$ with $p(\tilde{v}_0) = v_0$. Each vertex $\tilde{v}$ of $\tilde{K}$ can be described by a path α in K having origin v_0: choose a path $\tilde{\alpha}$ in $\tilde{K}$ from $\tilde{v}_0$ to $\tilde{v}$ and define $\alpha = p\tilde{\alpha}$. Had we chosen a second path, say, $\tilde{\beta}$, from $\tilde{v}_0$ to $\tilde{v}$, then the path $\beta = p\tilde{\beta}$ is also a path in K from v_0 to $p\tilde{v}$; moreover, $[\alpha\beta^{-1}] = p_*[\tilde{\alpha}\tilde{\beta}^{-1}] \in p_*\pi(\tilde{K}, \tilde{v}_0)$. This discussion motivates the next definition.

Definition Let K be a complex with basepoint v_0 and let π be a subgroup of $\pi(K, v_0)$. If α and β are paths in K with origin v_0, then

$$\alpha \equiv_\pi \beta \quad \text{if} \quad \text{end } \alpha = \text{end } \beta \quad \text{and} \quad [\alpha\beta^{-1}] \in \pi.$$

Notation It is easy to see that $\alpha \equiv_\pi \beta$ defines an equivalence relation on the set of all paths in K having origin v_0; denote the equivalence class of such a path α by $\bar{\alpha}$, and denote the family of all $\bar{\alpha}$ by K_π.

We shall make K_π into a complex. Let s be a simplex in K and let α be a path with origin v_0 and end in s. A **continuation** of α in s is a path $\alpha\alpha'$, where α' is a path lying wholly in s. Let

$$[s, \bar{\alpha}] = \{\bar{\beta} \in K_\pi : \beta \text{ is a continuation of } \alpha \text{ in } s\}.$$

Define the simplexes in K_π to be all $[s, \bar\alpha]$, where s is a simplex in K and $\bar\alpha \in K_\pi$ is such that end α is in s.

Theorem 12.19 *Let K be a connected complex, v_0 a vertex of K, and π a subgroup of $\pi(K, v_0)$. Then K_π is a complex and the function $V(K_\pi) \to V(K)$ given by $\bar\alpha \mapsto \text{end } \bar\alpha$ defines a map $p: K_\pi \to K$.*

Proof Straightforward. ∎

There is an obvious choice of a basepoint in K_π: let $\bar v_0 = \bar\alpha$, where $\alpha = (v_0, v_0)$.

Lemma 12.20 *If K is a connected complex, then every path α in K with origin v_0 can be lifted to a path A in K_π from $\bar v_0 = \overline{(v_0, v_0)}$ to $\bar\alpha$.*

Proof Let $\alpha = (v_0, v_1)(v_1, v_2) \cdots (v_{n-1}, v_n)$ be a path in K from v_0 to v_n = end α. Define paths α_i by $\alpha_i = (v_0, v_1)(v_1, v_2) \cdots (v_{i-1}, v_i)$. Observe that if s is the simplex $\{v_i, v_{i+1}\}$, then $\bar\alpha_i$ and $\bar\alpha_{i+1}$ both lie in $[s, \bar\alpha_i]$; hence $(\bar\alpha_i, \bar\alpha_{i+1})$ is an edge in K_π. Therefore $A = (\bar v_0, \bar\alpha_1)(\bar\alpha_1, \bar\alpha_2) \cdots (\bar\alpha_{n-1}, \bar\alpha_n)$ is a path in K_π from $\bar v_0$ to $\bar\alpha_n = \bar\alpha$ which lifts α. ∎

Corollary 12.21 *If K is a connected complex, then K_π is connected.*

Proof There is a path in K_π from $\bar v_0$ to every vertex $\bar\alpha$ of K_π. ∎

Theorem 12.22 *Let K be a connected complex and let π be a subgroup of $\pi(K, v_0)$. Then $p: K_\pi \to K$ is a covering complex and $p_\# \pi(K_\pi, \bar v_0) = \pi$.*

Proof Let us first show that $p: K_\pi \to K$ is a covering complex; only condition (iii) of the definition remains to be checked.

We claim that $p|[s, \bar\alpha]:[s, \bar\alpha] \to s$ is a one-one correspondence. Suppose $\bar\beta$ and $\bar\gamma \in [s, \bar\alpha]$ and $p(\bar\beta) = p(\bar\gamma)$. Then $\beta = \alpha\beta_1$ and $\gamma = \alpha\gamma_1$, where β_1, γ_1 lie wholly in s. Moreover, $\beta\gamma^{-1}$ is defined and $\beta\gamma^{-1} = \alpha\beta_1\gamma_1^{-1}\alpha^{-1} \sim \alpha\alpha^{-1} \sim 1$, by Exercise 12.13, so $[\beta\gamma^{-1}] = 1$ in $\pi(K, v_0)$. Since $1 \in \pi$, $\beta \equiv_\pi \gamma$ and $\bar\beta = \bar\gamma$. Hence $p|[s, \bar\alpha]$ is one-one. To see that $p|[s, \bar\alpha]$ is onto, let v be a vertex in s. If α' is a path in s from end α to v, then $\bar\alpha\bar\alpha' \in [s, \bar\alpha]$ and $p(\bar\alpha\bar\alpha') = v$.

Let s be a simplex in K and let w be a vertex in s. It is easy to check that $p^{-1}(s) = \cup[s, \bar\alpha]$, where the union ranges over all $\bar\alpha \in K_\pi$ with end $\bar\alpha = w$. To prove that $p: K_\pi \to K$ is a covering complex, it suffices to prove the sheets $[s, \bar\alpha]$ are pairwise disjoint. Assume $\bar\gamma \in [s, \bar\alpha] \cap [s, \bar\beta]$. Then $\gamma \equiv_\pi \alpha\alpha_1$ and $\gamma \equiv_\pi \beta\beta_1$, where α_1 and β_1 are paths lying wholly in s. The definition of equivalence gives end $\alpha_1 = $ end β_1, so that $\alpha_1\beta_1^{-1}$ is a path lying wholly in s; moreover, $\alpha_1\beta_1^{-1}$ is a closed path at w. Hence

$$1 = [\gamma\gamma^{-1}] = [\alpha\alpha_1\beta_1^{-1}\beta^{-1}] = [\alpha\beta^{-1}] \in \pi,$$

by Exercise 12.13. It follows that $\alpha \equiv_\pi \beta$, i.e., $\bar\alpha = \bar\beta$, and so $[s, \bar\alpha] = [s, \bar\beta]$.

We claim that $\pi = p_\# \pi(K_\pi, \bar v_0)$. Let $[\alpha] \in \pi(K, v_0)$. Since $p: K_\pi \to K$ is a covering complex, there is a unique lifting $\tilde\alpha$ of α with origin $\bar v_0$. But we constructed such a lifting A in Lemma 12.20; therefore A must be $\tilde\alpha$ and so end $\tilde{\alpha} = $ end $A = \bar{\alpha}$. The following statements are equivalent:

$$[\alpha] \in p_{\#}\, \pi(K_{\pi}, \bar{v}_0);$$
$$[\alpha] = [pA], \text{ where } [A] \in \pi(K_{\pi}, \bar{v}_0);$$
$$\text{end } A = \text{origin } A = \bar{v}_0;$$
$$\bar{\alpha} = \bar{v}_0;$$
$$[\alpha(v_0, v_0)^{-1}] \in \pi;$$
$$[\alpha] \in \pi. \quad \blacksquare$$

Definition   A connected complex K is **simply connected** if $\pi(K, v_0) = \{1\}$.

Corollary 12.23   *Every connected complex K has a simply connected covering complex $\tilde{K}$. In particular, if K is a 1-complex, then $\tilde{K}$ is a tree.*

Proof   Let $\pi = \{1\} \subset \pi(K, v_0)$ and define $\tilde{K} = K_{\pi}$. If $\dim K = 1$, then $\dim \tilde{K} = 1$ (Exercise 12.26); it follows from Theorem 12.11 that $\tilde{K}$ is a tree. $\blacksquare$

We have enough information to prove the Nielsen-Schreier theorem.

Theorem 12.24   **(Nielsen-Schreier)**†   *Every subgroup H of a free group F is itself free.*

Proof   Let F be free of rank $|I|$; let K be the bouquet of $|I|$ circles constructed in Theorem 12.12, so that $\pi(K, v_0)$ may be identified with F. By Theorem 12.22, there is a covering complex $p: K_H \to K$ with $p_{\#}\, \pi(K_H, \bar{v}_0) = H$. Since $p_{\#}$ is one-one, $\pi(K_H, \bar{v}_0) \cong H$. Since $\dim K = 1$, Exercise 12.26 gives $\dim K_H = 1$. Therefore H is free, by Theorem 12.11. $\blacksquare$

Theorem 12.25   *Let F be free of finite rank n and let H be a subgroup of F having finite index j. Then H is free of rank $jn - j + 1$.*

Proof   For a finite connected graph K, let $n_0(K)$ be the number of its vertices and $n_1(K)$ be the number of its 1-simplexes. If T is a maximal tree in K, then Exercise 12.16 gives $n_1(T) = n_0(T) - 1$. Therefore, the number of 1-simplexes in $K - T$ is $n_1(K) - n_1(T) = n_1(K) - n_0(T) + 1$. Since T is a maximal tree, Theorem 12.9 gives $n_0(T) = n_0(K)$. Conclusion: If K is a finite connected graph, then $\pi(K, v_0)$ is free of rank $n_1(K) - n_0(K) + 1$.

Let K be a bouquet of n circles: $n_0(K) = 2n + 1$ and $n_1(K) = 3n$. Let $p: K_H \to K$ be the covering complex corresponding to H. By Corollary 12.18, $[F:H] = j$ is the number of vertices lying over a vertex of K: $n_0(K_H) = jn_0(K)$. Moreover, Exercise 12.27 gives $n_1(K_H) = jn_1(K)$. We compute:

$$n_1(K_H) - n_0(K_H) + 1 = jn_1(K) - jn_0(K) + 1$$
$$= 3jn - j(2n + 1) + 1$$
$$= jn - j + 1.$$

Therefore H is free of rank $jn - j + 1$. $\blacksquare$

† Nielsen (1921) proved this theorem for H finitely generated (he gave an algorithm which decides whether or not a word $w \in F$ lies in H). Schreier (1927) proved the general theorem. This proof is due to Baer and Levi (1936). For other proofs of this theorem, see the book of Lyndon and Schupp.

Suppose a finitely generated group G has a presentation

$$(x_1, \cdots, x_n | r_1 = 1, \cdots, r_m = 1).$$

If F is the free group on $\{x_1, \cdots, x_n\}$ and R is the normal subgroup generated by $\{r_1, \cdots, r_m\}$, then the formula above gives the rank of R. It is possible that rank $R > m$, for more elements may be needed to generate R as a subgroup than to generate R as a normal subgroup (in the latter generation, one is allowed to include conjugates whenever desired).

EXERCISES

12.28. Let G be a finite group that is not cyclic and let $G \cong F/S$, where F is free of finite rank. Prove that rank $S >$ rank F.

12.29. Let F be free of rank 2. Does F have a proper normal subgroup H that is finitely generated?

12.30. A free group of rank > 1 is not solvable.

12.31. Exhibit infinitely many free sets of generators of a free group of rank 2.

12.32. Prove that a group F is free if and only if, for every group G, each extension of G by F is a semidirect product.

12.33. Let G be a finitely generated group containing a subgroup H of finite index. Prove that H is finitely generated.

12.34. Let G be a group having n generators and k relations, where $n > k$; then G contains an element of infinite order. (HINT: Map a free group on n generators onto a free abelian group on n generators, and observe what happens to the relations; use Exercise 10.43.) Conclude that $n \leq k$ when G is finite.

12.35. If F is free and $R \lhd F$, then F/R' is torsion-free, where $R' = [R, R]$. [HINT (**Rosset**): First reduce to the case F/R cyclic of prime order p. Let $x \in F$ satisfy $x^p \in R'$; if $x \in R$, its coset has finite order in R/R'; if $x \notin R$, then $x \notin F'$ (since $F' \subset R$), whence $x^p \notin F'$ and $x^p \notin R'$.]

REMARK   If $p: \tilde{K} \to K$ is a covering space, a map $f: \tilde{K} \to \tilde{K}$ is called a **covering map** (or *deck transformation*) if $pf = p$; one checks easily that

$$\text{Cov}(\tilde{K}/K) = \{\text{all covering maps } f: \tilde{K} \to \tilde{K}\}$$

is a group under composition. It may be proved that if $\tilde{U}$ is a simply connected covering space of K, then $\text{Cov}(\tilde{U}/K) \cong \pi(K, v_0)$ (note that $\text{Cov}(\tilde{U}/K)$ is defined without a choice of basepoint). Indeed, there is an analog of Galois theory here. It may be shown that if $\tilde{U}$ is a simply connected covering complex of K, then $\tilde{U}$ is also a covering complex of every covering complex $\tilde{K}$ of K; moreover, the function $\tilde{K} \mapsto \text{Cov}(\tilde{U}/\tilde{K})$ is a one-one correspondence between the family of all covering complexes of K and the family of all subgroups of $\pi(K, v_0)$. [The interested reader is referred to J. Rotman, *Covering complexes with applications to algebra*, Rocky

Mountain Journal of Mathematics, 1973, for an account in the style of this text.]

The theory of complexes has given a criterion for a group G to be free: Is G the edgepath group of a connected graph? Another way to show that a group is free is to exhibit a basis for it. Let us sketch another proof of the Nielsen-Schreier theorem in order to exhibit a basis for a subgroup H of a free group F on X. Choose a right transversal of H in F, i.e., one representative from each right coset Ha of H in F. Denote the chosen representative of Ha by $\rho(Ha)$. For each $x \in X$, both $\rho(Hax)$ and $\rho(Ha)x$ lie in the coset Hax; therefore

$$t_{a,x} = \rho(Ha)x\rho(Hax)^{-1}$$

is an element of H. It is not difficult to prove that the elements $t_{a,x}$ just defined generate H. The next step is to further restrict the transversal.

Definition   Let F be free on X and let H be a subgroup of F. A **Schreier transversal** of H in F is a right transversal S such that whenever $a = x_1^{\varepsilon_1} \cdots x_n^{\varepsilon_n}$ is a reduced word in S (where $x_i \in X$ and $\varepsilon_i = \pm 1$), then every initial segment $x_1^{\varepsilon_1} \cdots x_k^{\varepsilon_k}, k \le n$, also is in S.

One proves Schreier transversals exist, and then shows that H is free on all $t_{a,x} \ne 1$ arising from a Schreier transversal. Let us prove these assertions using complexes; we shall also see that Schreier transversals arise from maximal trees in a covering complex.

Lemma 12.26   Let $p: \tilde{K} \to K$ be a covering complex and let T be a maximal tree in K. If v_0 is a vertex of K, then each component of $p^{-1}(T)$ contains a vertex in $p^{-1}(v_0)$.

Proof   Suppose, on the contrary, that there exists a component $\tilde{C}$ of $p^{-1}(T)$ that does not meet $p^{-1}(v_0)$. Clearly $p(\tilde{C})$ is not all of $V(K)$: $v_0 \notin p(\tilde{C})$. Since K is connected, it follows that there exists an edge (u, v) in K with $u \in p(\tilde{C})$ and $v \notin p(\tilde{C})$. Since $u \in p(\tilde{C})$, there is a vertex $\tilde{u} \in \tilde{K}$ with $p\tilde{u} = u$. By Lemma 12.13, the path $\alpha = (u, v)$ may be lifted to a path $\tilde{\alpha} = (\tilde{u}, \tilde{v})$ in $\tilde{K}$. Visibly $\tilde{v} \in \tilde{C}$, for we may connect $\tilde{v}$ to $\tilde{u}$. Hence $p\tilde{v} = v \in p(\tilde{C})$, a contradiction. ∎

Theorem 12.27   Let F be free and let H be a subgroup of F. There is a basis X of F and a Schreier transversal $\{\rho(Ha) \in Ha: a \in F\}$ such that a basis for H consists of all $t_{a,x} = \rho(Ha)x\rho(Hax)^{-1}$ that are distinct from 1, where x varies over X.

Proof   Identify F with $\pi(K, v_0)$, where K is a bouquet of circles with $\pi(K, v_0) \cong F$ (Theorem 12.12). Let $p: K_H \to K$ be the covering complex corresponding to H. Choosing a maximal tree T determines a basis of F (Theorem 12.11). Let $\bar{v}_0 = (\overline{v_0, v_0})$ be the "obvious" vertex lying over v_0, so that $p_\# \pi(K_H, \bar{v}_0) = H$. Finally, let $\tilde{T}$ be a maximal tree in K_H containing $p^{-1}(T)$ (that $\tilde{T}$ exists is seen by using Exercises 12.25 and 12.19).

If $\bar{v}$ is any vertex in K_H with $p(\bar{v}) = v_0$, let $\tilde{\gamma}_{\bar{v}}$ be the unique reduced path in $\tilde{T}$

from $\bar{v}_0$ to $\bar{v}$; note that $p_{\#}[\tilde{\gamma}_{\bar{v}}] \in \pi(K, v_0)$. Given a coset $H[\alpha]$, let $\tilde{\alpha}$ be the lifting of α having origin $\bar{v}_0$. If $\bar{v} =$ end $\tilde{\alpha}$, then $\bar{v}$ lies over v_0 and $[\tilde{\alpha}\tilde{\gamma}_{\bar{v}}^{-1}] \in \pi(K_H, \bar{v}_0)$. Applying $p_{\#}$ gives

$$H[\alpha] = Hp_{\#}[\tilde{\gamma}_{\bar{v}}],$$

so the set of $p_{\#}[\tilde{\gamma}_{\bar{v}}]$ is a transversal.

To see the set of all $p_{\#}[\tilde{\gamma}_{\bar{v}}]$ is a Schreier transversal, write $\tilde{\gamma}_{\bar{v}} = \tilde{\alpha}_1 \cdots \tilde{\alpha}_n$, where each $\tilde{\alpha}_i$ contains exactly one edge not in $p^{-1}(T)$. If $\tilde{\gamma}_{\bar{v}}$ has no such factorization, then $p_{\#}[\tilde{\gamma}_{\bar{v}}] = 1$ (Exercise 12.15 and the observation above that $p^{-1}(T) \subset \tilde{T}$). For each $i \geq 1$, let $\tilde{u}_i =$ end $\tilde{\alpha}_i =$ origin $\tilde{\alpha}_{i+1}$. Now $\tilde{u}_i$ lies in some component $\tilde{C}_i$ of $p^{-1}(T)$, and there is a vertex $\dot{v}_i$ in $\tilde{C}_i$ lying over v_0, by Lemma 12.26. Since $\tilde{C}_i$ is a tree, there is a unique reduced path $\tilde{\beta}_i$ in $\tilde{C}_i$ from $\tilde{u}_i$ to $\dot{v}_i$. Consider the new path

$$(\tilde{\alpha}_1\tilde{\beta}_1)(\tilde{\beta}_1^{-1}\tilde{\alpha}_2\tilde{\beta}_2)(\tilde{\beta}_2^{-1}\tilde{\alpha}_3\tilde{\beta}_3) \cdots (\tilde{\beta}_{n-1}^{-1}\tilde{\alpha}_n).$$

Observe that $\tilde{\alpha}_1\tilde{\alpha}_2 \cdots \tilde{\alpha}_i\tilde{\beta}_i \sim \tilde{\gamma}_{\dot{v}_i}$, for both are paths in $\tilde{T}$ from $\bar{v}_0$ to $\dot{v}_i$. Further, for each $\tilde{\delta}_i = \tilde{\beta}_{i-1}^{-1}\tilde{\alpha}_i\tilde{\beta}_i$ we have $p\tilde{\delta}_i$ a closed path in K at v_0 that contains only one edge not in T; thus $\tilde{\delta}_i$ determines a generator of $\pi(K, v_0)$. It follows that each initial segment $\tilde{\delta}_1\tilde{\delta}_2 \cdots \tilde{\delta}_i, i \leq n$, determines a member of the transversal, for $p_{\#}[\tilde{\delta}_1 \cdots \tilde{\delta}_i] = p_{\#}[\tilde{\gamma}_{\dot{v}_i}]$.

Finally, a basis for $\pi(K_H, \bar{v}_0)$ consists of all $[\tilde{\gamma}_{\bar{u}}(\bar{u}, \bar{v})\tilde{\gamma}_{\bar{v}}^{-1}]$, where $(\bar{u}, \bar{v})$ is not in $\tilde{T}$. Since $p_{\#}$ is one-one, the images of these elements comprise a basis for H, and each is of the form $t_{a,x}$. ∎

Theorem 12.28 *Let F be free on $\{x, y\}$. Then F', the commutator subgroup of F, is free of infinite rank.*

Proof By Exercise 12.6, F/F' is a free abelian group with basis $F'x$ and $F'y$. Therefore every right coset $F'a$ has a unique representative of the form $x^m y^n$. The assignment $\rho(F'a) = x^m y^n$ is thus a well defined function whose range is clearly a Schreier transversal of F' in F (write $x^m y^n$ as $x \cdots xy \cdots y$).

If $n > 0$, then $\rho(F'y^n) = y^n$ while $\rho(F'y^n x) \neq y^n x$. Therefore $\rho(F'y^n)x\rho(F'y^n x)^{-1} \neq 1$, so there are infinitely many t distinct from 1. The theorem now follows from Theorem 12.27. ∎

Thus, in contrast to abelian groups, subgroups of finitely generated groups need not be finitely generated.

EXERCISES

**12.36. Let F be free on $\{x, y\}$. Prove that F contains a subgroup that is free on $\{x, y^{-1}xy, \cdots, y^{-n}xy^n, \cdots\}$.

12.37. If F is free of rank > 1, then F' is free of infinite rank.

12.38. Let F be free on $\{x, y\}$. Define $\varphi: F \to S_3$ by $x \mapsto (12)$ and $y \mapsto (123)$. Exhibit a basis for ker φ.

12.39.   Let F be free on $\{a, b, c, d\}$. Prove that $[a, b][c, d]$ is not a commutator. (See Exercise 2.55.)

12.40.   Prove that a subsemigroup of a free semigroup need not be free.

REMARK   The following theorem does not seem to follow easily from the proofs given above: If F is a free group of finite rank n and if $\{x_1, \cdots, x_n\}$ generates F, then $\{x_1, \cdots, x_n\}$ is a basis. We refer the reader to the book of Lyndon and Schupp for a combinatorial proof.

FREE PRODUCTS AND THE KUROŠ THEOREM

We now generalize the notion of a free group to that of a free product. As with the definition of a free group, we shall define a free product as a group having a certain "universal mapping" property. Since it is not obvious that such a group exists, we shall then be obliged to construct one. If the reader does not like the fancy definition, he is urged to look at the normal form theorem (Theorem 12.31) for a more homely, down-to-earth description of the elements in a free product.

Definition   Let $\{A_i : i \in I\}$ be a family of groups. A **free product** of the A_i is a group P such that

(i)   for each i there is an *imbedding*, i.e., a one-one homomorphism, $j_i : A_i \to P$;

(ii)   for every group G and every family of homomorphisms $\{f_i : A_i \to G\}$, there is a unique homomorphism $\psi : P \to G$ that extends every f_i, i.e., $\psi j_i = f_i$.

The definition is easily pictured.

The reader should compare this definition to the analogous property of direct sums of abelian groups (Exercise 10.9). We are obliged to prove existence and uniqueness to show that we really have defined something.

Example 10   A free group F is a free product of infinite cyclic groups.

If F is free on X, then $\langle x \rangle$ is infinite cyclic for each $x \in X$; a family of homomorphisms, $f_x : \langle x \rangle \to G$ determines a function $\varphi : X \to G$ [namely, $\varphi(x) = f_x(x)$] which extends to a unique homomorphism $\psi : F \to G$ with $\psi | \langle x \rangle = f_x$.

Theorem 12.29   *Let $\{A_i : i \in I\}$ be a family of groups. If P and Q are each free products of the A_i, then $P \cong Q$.*

Proof   Let $j_i : A_i \to P$ and $k_i : A_i \to Q$ be the imbeddings

Since P is a free product of the A_i, there is a homomorphism $\psi : P \to Q$ with $\psi j_i = k_i$ for all $i \in I$. Similarly, there is a homomorphism $\theta : Q \to P$ with $\theta k_i = j_i$ for all $i \in I$.

Consider the new diagram

Both $\theta\psi$ and 1_P are homomorphisms $P \to P$ making the diagram commute. Since, by hypothesis, there can be only one such arrow, $\theta\psi = 1_P$. Similarly, $\psi\theta = 1_Q$ and $P \cong Q$.   ∎

Theorem 12.30   *Given a family of groups $\{A_i : i \in I\}$, then a free product of the A_i exists.*

Proof   This proof is so similar to the construction of a free group that we only present its highlights and leave the details to the reader.

Let $A_i^{\#} = A_i - \{1\}$. Assume the subsets $A_i^{\#}$ are pairwise disjoint; call $\{1\} \cup (\cup A_i^{\#})$ the alphabet and its elements letters. Form *words* with these letters; a word w is *reduced* if $w = 1$ or if $w = a_1 \cdots a_n$, where each $a_i \neq 1$ and adjacent letters lie in distinct A_i. Then a free product of the A_i has as elements all reduced words and as multiplication juxtaposition. To verify associativity, it is simplest to use van der Waerden's device of considering all permutations on the set of all reduced words, as in Theorem 12.1.   ∎

Let us describe the homomorphism ψ in the special case of the free product P of two groups A and B. Suppose G is a group and $f : A \to G$ and $g : B \to G$ are homomorphisms. Then $\psi : P \to G$ is defined by

$$\psi(a_1 b_1 a_2 b_2 \cdots a_n b_n) = f(a_1)g(b_1) \cdots f(a_n)g(b_n).$$

Because of this theorem, we may speak of *the* free product of $\{A_i: i \in I\}$; it is denoted by

$$\underset{i \in I}{*} \; A_i;$$

if there are only finitely many groups A_i, one usually writes

$$*A_i = A_1 * \cdots * A_n.$$

Theorem 12.31 (**Normal Form**) *If $w \in *A_i$ and $w \neq 1$, then w has a unique factorization*

$$w = a_1 a_2 \cdots a_n$$

where each $a_j \neq 1$ and adjacent factors lie in distinct A_i.

Proof We saw in the proof of Theorem 12.30 that one may regard the elements of $*A_i$ as reduced words. ∎

EXERCISES

12.41. If G and H have more than one element, then $G * H$ is an infinite centerless group that contains elements of infinite order.

*12.42. Prove that every group G can be imbedded in a centerless group H. (HINT: Take $H = G * \mathbf{Z}$.) Use this result to prove that for every group G there exists a group H with $G \subset \text{Aut}\,(H)$.

*12.43. (= 12.5) Prove there are no groups $G \neq \{1\}$ with the injective property. [HINT (**Humphreys**): Show G is not normal in the semidirect product $H \rtimes G$, where H is as in Exercise 12.42.]

12.44. The operation of free product is commutative and associative: for any groups A, B, C, one has

$$A * B \cong B * A \quad \text{and} \quad (A * B) * C \cong A * (B * C).$$

12.45. If N is the normal subgroup of $A * B$ generated by A, then $(A * B)/N \cong B$ (compare Exercise 12.7).

*12.46. Show there is a (unique) homomorphism of $A_1 * \cdots * A_n$ onto $A_1 \times \cdots \times A_n$ which acts as the identity on each A_i.

12.47. If G' is the commutator subgroup of $G = *A_\alpha$, then $G/G' \cong \Sigma(A_\alpha/A_\alpha')$ (cf. Exercise 12.6).

**12.48. If A_α has a presentation $(X_\alpha | \Delta_\alpha)$, then $*A_\alpha$ has a presentation $(\cup X_\alpha | \cup \Delta_\alpha)$. (Assume the sets X_α, and hence the sets Δ_α, are pairwise disjoint.)

12.49. The **infinite dihedral group** D_∞ is the group with presentation $(s, t | t^2 = 1, tst^{-1} = s^{-1})$. Prove $D_\infty \cong \mathbf{Z}(2) * \mathbf{Z}(2)$.

12.50. If G has a presentation $(x, y | x^2 = 1, y^3 = 1)$, then $G \cong \mathbf{Z}(2) * \mathbf{Z}(3)$.

12.51. Let $M = PSL(2, \mathbf{Z})$ be the **modular group**, i.e., the multiplicative group of all 2×2 unimodular matrices over $\mathbf{Z}$ modulo the subgroup $\{\pm E\}$. Prove that $M \cong \mathbf{Z}(2) * \mathbf{Z}(3)$ (see Exercise 2.20).

12.52. Let $A_1, \cdots, A_n, B_1, \cdots, B_m$ be indecomposable (not a direct product of proper subgroups) groups with both chain conditions. If $A_1 * \cdots * A_n \cong B_1 * \cdots * B_m$, then $n = m$ and there is a permutation σ of $\{1, \cdots, n\}$ with $A_i \cong B_{\sigma(i)}$ for all i. (The same conclusion holds if one assumes each factor is indecomposable in the sense that it is not a free product of proper subgroups.) (HINT: Use Exercise 12.46.)

12.53. **(Baer-Levi)** Prove that no group G can be decomposed as a free product and as a direct product. [HINT (**P. M. Neumann**):   If $G = A * B$ and $a \in A$ and $b \in B$ are nontrivial, then $C_G(ab) \cong \mathbf{Z}$; if $G = C \times D$, then use $ab = cd$, where $c \in C$ and $d \in D$, to show that $C_G(ab)$ is a direct product.]

Kuroš discovered the structure of the subgroups of a free product. We prove this theorem using covering complexes.

Theorem 12.32   *Let K be a connected complex with connected subcomplexes $\{K_i : i \in I\}$. Assume $\cup K_i = K$ and that there is a tree T in K with $K_i \cap K_j = T$, all $i \neq j$. Then*

$$\pi(K, v) \cong \underset{i \in I}{*} \pi(K_i, v_i)$$

for vertices v in K and v_i in K_i.

Proof   For each i there exists a maximal tree T_i in K_i that contains T; we shall show $T' = \bigcup_{i \in I} T_i$ is a maximal tree in K. Let α be a circuit in T', say, with origin u. We may assume $u \in V(T)$: if $u \in V(T_i - T)$ and $t \in V(T)$, choose a reduced path γ in T_i from u to t; the reduced path $\gamma^{-1}\alpha\gamma$ is a circuit with origin t. We claim T' contains no circuit α of the form

(*) $\alpha = \tau_0 \beta \tau_1,$

where τ_0, τ_1 are reduced paths in T and the only vertices of β lying in T are its endpoints. If β lies wholly in some T_i, we contradict T_i being a tree; if β involves vertices u, v with $u \in V(T_i - T)$, $v \in V(T_j - T)$, and $i \neq j$, then $T_k \cap T_l = T$ for all $k \neq l$ implies every path in T' from u to v passes through T, contradicting β having no "interior" vertices in T. Now suppose $\alpha = \tau_0 \beta \tau_1 \gamma$, where τ_0, τ_1 lie in T, β has no interior vertices in T, and the first edge of γ ends at a vertex outside of T. If τ_1 involves t, then an initial segment of α is a circuit of type (*), a contradiction. If τ_1 does not involve t, let σ be a reduced path in T from end β to t; the reduced path $\tau_0 \beta \sigma$ is also a circuit of type (*), another contradiction. We have proved that T' has no circuits, hence is a tree. Finally T' is a maximal tree, for

$$V(T') = V(\bigcup T_i) = V(\bigcup K_i) = V(K).$$

By Lemma 12.10, there is a presentation $(E_i | \Delta_i)$ of $\pi(K_i, v_i)$, where E_i is the set of edges in K_i and Δ_i is the set of relations of the form: (a) $(u, v)(v, w) = (u, w)$ when $\{u, v, w\}$ is a simplex of K_i; (b) $(u, v) = 1$ for every edge in T_i. By Exercise 12.48, a presentation for $*\pi(K_i, v_i)$ is $(\bigcup E_i | \bigcup \Delta_i)$. We claim that $(\bigcup E_i | \bigcup \Delta_i)$ is also a presentation of $\pi(K, v)$: $\bigcup E_i$ consists of all the edges in K; since $T' = \bigcup T_i$, an edge in K lies in T' if and only if it lies in some T_i; $\{u, v, w\}$ lies in a simplex of K if and only if it lies in a simplex of some K_i. Therefore $\pi(K, v) \cong *\pi(K_i, v_i)$. ∎

In the next section we shall prove that every group G is of the form $\pi(K, v_0)$ for some connected complex K (Theorem 12.41).

Theorem 12.33 (Kuroš, 1934) *If H is a subgroup of $\underset{i \in I}{*} G_i$, then $H \cong F * (*II_\alpha)$, where F is free and each H_α is isomorphic to a subgroup of some G_i.*

Proof Let us assume there exist connected complexes K_i with $\pi(K_i, v_i) \cong G_i$, where v_i is a vertex of K_i. Define a new complex K by adding a new vertex v_0 to the disjoint union $\bigcup K_i$ and new 1-simplexes $\{v_0, v_i\}$, all $i \in I$. If T is the tree in K consisting of these new 1-simplexes, then Theorem 12.32 gives

$$\pi(K, v_0) \cong *\pi(K_i \cup T, v_i).$$

But Lemma 12.10 gives $\pi(K_i \cup T, v_i) \cong \pi(K_i, v_i) \cong G_i$. Hence

$$\pi(K, v_0) \cong * G_i.$$

Let $p: K_H \to K$ be the covering complex corresponding to H and let $\tilde{v} \in p^{-1}(v_0)$ be such that $p_\# \pi(K_H, \tilde{v}) = H$ (we have identified $* G_i$ with $\pi(K, v_0)$). For each i, $p^{-1}(K_i)$ is the disjoint union of its components $\tilde{K}_{ij}$; choose a maximal tree $\tilde{T}_{ij}$ in $\tilde{K}_{ij}$. Let $\tilde{L}$ be the 1-subcomplex of K_H:

$$\tilde{L} = \bigcup \tilde{T}_{ij} \cup p^{-1}(T).$$

Finally, let $\tilde{T}$ be a maximal tree in $\tilde{L}$ containing $\bigcup \tilde{T}_{ij}$ (which exists by Exercise 12.19). Observe that $\tilde{T}$ contains no edges in $\tilde{K}_{ij}$ aside from those in $\tilde{T}_{ij}$ lest we violate the maximality of $\tilde{T}_{ij}$ in $\tilde{K}_{ij}$.

Consider the subcomplexes $\tilde{L}$ and $\tilde{K}_{ij} \cup \tilde{T}$, all i, j. Clearly K_H is the union of these, while the intersection of any two of these is the tree $\tilde{T}$. Therefore, Theorem 12.32 gives

$$\pi(K_H, \tilde{v}) \cong \pi(\tilde{L}, \tilde{v}) * (*\pi(\tilde{K}_{ij} \cup \tilde{T}, \tilde{v})).$$

Now $\pi(\tilde{L}, \tilde{v})$ is free because dim $\tilde{L} = 1$. Since $\tilde{T}$ is a maximal tree in $\tilde{K}_{ij} \cup \tilde{T}$, Lemma 12.10 gives $\pi(\tilde{K}_{ij} \cup \tilde{T}, \tilde{v}) \cong \pi(\tilde{K}_{ij}, \tilde{v}_{ij})$ for some vertex $\tilde{v}_{ij} \in \tilde{K}_{ij}$. But $p | \tilde{K}_{ij}: \tilde{K}_{ij} \to K_i$ is a covering complex, by Exercise 12.24. Hence, $\pi(\tilde{K}_{ij}, \tilde{v}_{ij})$ is isomorphic to a subgroup of $\pi(K_i, v_i) \cong G_i$. Therefore $H = p_\# \pi(K_H, \tilde{v})$ is a free product as described since $p_\#$ is one-one. ∎

Let us state the **Gruško-Neumann theorem (1940)** [see Massey (1967) for a topological proof]. Assume $\{G_i : i \in I\}$ is a set of groups and $\varphi: F \to \underset{i \in I}{*} G_i$ is a homomorphism from a free group F onto $\underset{i \in I}{*} G_i$; then there exist subgroups F_i of F, $i \in I$, with $F = \underset{i \in I}{*} F_i$ and with $\varphi(F_i) = G_i$ for each i. Here is one corollary: If G_1 and G_2 are finitely generated groups and if n_1 and n_2 are the minimal numbers of generators of G_1 and G_2, respectively, then the minimal number of generators of $G_1 * G_2$ is $n_1 + n_2$.

EXERCISES

12.54. In Theorem 12.33, we showed that a subgroup H of $* G_i$ is isomorphic to $F * (* H_\alpha)$, where each H_α is a subgroup of some G_i. Prove that H is equal to $F * (* S_\alpha)$, where each S_α is a conjugate of a subgroup of some G_i (assume each G_i is contained in $*G_i$). (HINT: Use Theorem 12.16.)

12.55. If G is a free product of finite groups G_i, then every finite subgroup of G is isomorphic to a subgroup of some G_i.

12.56. If $G = \mathbf{Z}(2) * \mathbf{Z}(4)$, then $\mathbf{Z}(2)$ and $\mathbf{Z}(4)$ are each maximal 2-subgroups of G. Conclude that in an infinite group Sylow subgroups need not be isomorphic, let alone conjugate.

12.57. Prove that the only elements of finite order in the modular group have order 2 or 3.

12.58. Show that the modular group contains a free subgroup of index 6. [HINT: The kernel of $\mathbf{Z}(2) * \mathbf{Z}(3) \to \mathbf{Z}(2) \times \mathbf{Z}(3)$ is torsion-free.]

12.59. Show that the commutator subgroup of the modular group is free.

12.60. Prove that the modular group contains a free subgroup of infinite rank.

**12.61. If $f: G_1 \to G_2$ and $g: H_1 \to H_2$ are homomorphisms, there is a unique homomorphism $h: G_1 * H_1 \to G_2 * H_2$ such that $h|G_1 = f$ and $h|H_1 = g$ (compare Exercise 2.72).

AMALGAMS AND HNN EXTENSIONS

In Chapter 11 we constructed the pushout of a diagram of abelian groups; we now show pushouts of arbitrary groups exist.

Definition   Let B, A_1, A_2 be groups and f_1, f_2 be homomorphisms:

(δ)

$$
\begin{array}{ccc}
B & \xrightarrow{\ f_1\ } & A_1 \\
\downarrow{\scriptstyle f_2} & & \\
A_2 & &
\end{array}
$$

A **solution** of the diagram δ is a group C and homomorphisms g_1, g_2 such that the following diagram commutes (i.e., $g_2 f_2 = g_1 f_1$):

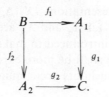

A **pushout** of the diagram δ is a solution (C, g_1, g_2) such that, for any other solution (D, h_1, h_2), there exists a unique homomorphism $\varphi: C \to D$ making the following diagram commute:

(δ^*)

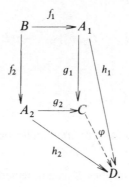

One proves quickly that pushouts are unique to isomorphism if they exist: if (C, g_1, g_2) and (D, h_1, h_2) are both pushouts, then the map $\varphi: C \to D$ is an isomorphism.

Theorem 12.34 *A pushout exists for the diagram δ. Moreover, if for $i = 1, 2$, A_i has presentation $(X_i | \Delta_i)$, then the pushout has presentation*

$$C = (X_1 \cup X_2 | \Delta_1 \cup \Delta_2 \cup \{f_1(b) f_2(b^{-1}) \colon b \in B\}).$$

Proof (Compare Lemma 11.20) Let N be the normal subgroup of $A_1 * A_2$ generated by $\{f_1(b) f_2(b^{-1}) \colon b \in B\}$. Define $C = (A_1 * A_2)/N$ and define $g_i \colon A_i \to C$ by $g_i(a_i) = a_i N$ for $i = 1, 2$. It is easy to verify that (C, g_1, g_2) is a solution of δ.

Suppose (D, h_1, h_2) is a second solution of δ. The definition of free product provides a unique homomorphism $\psi \colon A_1 * A_2 \to D$ with $\psi | A_i = h_i$ for $i = 1, 2$. Since $h_2 f_2 = h_1 f_1$, it follows that $N \subset \ker \psi$ and ψ induces a homomorphism $\varphi \colon C \to D$. One shows easily that the diagram δ^* commutes and that φ is unique. Finally, it is plain from the construction that C has a presentation as described in the statement. ∎

Corollary 12.35 *If $A_2 = \{1\}$ in diagram δ, then the pushout C is A_1/N, where N is the normal subgroup generated by $f_1(B)$.*

An observation is needed. If G is an infinite cyclic group with generator x, we know $G * G$ is a free group of rank 2 and a presentation of $G * G$ is $(x, y|\varnothing)$. It is necessary to write y for the second generator to avoid confusing it with x. More generally, if groups A_i have presentations $(X_i|\Delta_i)$ for $i = 1, 2$, then $A_1 * A_2$ has presentation $(X_1 \cup X_2|\Delta_1 \cup \Delta_2)$ if X_1 and X_2 are disjoint; if X_1 and X_2 are not disjoint, new notation must be introduced to make them disjoint. We have tacitly done this in Theorem 12.34; we shall be more explicit in the next proof.

The next theorem shows pushouts occur quite naturally.

Theorem 12.36 (Seifert–van Kampen, 1931) *Let K be a complex having connected subcomplexes L_1 and L_2 such that $L_1 \cup L_2 = K$ and $L_1 \cap L_2$ is connected. If $v_0 \in V(L_1 \cap L_2)$ (so $L_1 \cap L_2 \neq \varnothing$), then $\pi(K, v_0)$ is the pushout of the diagram*

$$\pi(L_1 \cap L_2, v_0) \longrightarrow \pi(L_1, v_0)$$
$$\downarrow$$
$$\pi(L_2, v_0)$$

where the arrows are induced by the inclusion maps j_i: $L_1 \cap L_2 \hookrightarrow L_i$ for $i = 1, 2$.

REMARK   The hypothesis implies K is connected.

Proof   Denote $L_1 \cap L_2$ by L_0. Choose a maximal tree T_0 in L_0 and for each $i = 1, 2$, choose a maximal tree T_i in L_i containing T_0. By Exercise 12.14, $T_1 \cup T_2$ is a tree in K; moreover, $T_1 \cup T_2$ is a maximal tree because $V(T_1 \cup T_2) = V(T_1) \cup V(T_2) = V(L_1) \cup V(L_2) = V(K)$. Lemma 12.10 says $\pi(K, v_0)$ has a presentation $(E|\Delta' \cup \Delta'')$, where E is the set of edges (u, v) in K, $\Delta' = E \cap (T_1 \cup T_2)$, and

$$\Delta'' = \{ (u, v)(v, w)(u, w)^{-1} : \{u, v, w\} = s \in K \}.$$

There are similar presentations for $\pi(L_i, v_0)$, namely, $(E_i|\Delta_i' \cup \Delta_i'')$, where E_i is the set of edges in L_i.

Denote the set of edges in $L_0 = L_1 \cap L_2$ by E_0. We make E_1 and E_2 disjoint by affixing the symbols j_1 and j_2 (which designate the inclusions). Theorem 12.34 thus gives the presentation for the pushout

$$(j_1 E_1 \cup j_2 E_2 | j_1 \Delta_1' \cup j_1 \Delta_1'' \cup j_2 \Delta_2' \cup j_2 \Delta_2'' \cup \{ (j_1 e)(j_2 e)^{-1} : e \in E_0 \}).$$

The generators may be rewritten as

$$j_1 E_0 \cup j_1 (E_1 - E_0) \cup j_2 E_0 \cup j_2 (E_2 - E_0).$$

The relations include $j_1 E_0 = j_2 E_0$ (so one of these subsets is superfluous). Next, $\Delta_i' = E_i \cap T_i = (E_i \cap T_0) \cup (E_i \cap (T_i - T_0))$, and this gives a decomposition of $j_1 \Delta_1' \cup j_2 \Delta_2'$ into four subsets, one of which is superfluous. Further,

$\Delta'' = \Delta_1'' \cup \Delta_2''$, for if $(u, v)(v, w)(u, w)^{-1} \in \Delta$, then $\{u, v, w\} \in K = L_1 \cup L_2$ and $\{u, v, w\} \in L_i$ for some i. Transform this presentation as follows: (i) isolate those generators and relations involving L_0; (ii) delete superfluous generators and relations involving L_0 (say, delete such having symbol j_2); (iii) erase the now unnecessary symbols j_1 and j_2. It is now apparent that the pushout and $\pi(K, v_0)$ have the same presentation, hence are isomorphic. ∎

Corollary 12.37 *With the hypothesis and notation of the previous theorem, a presentation for $\pi(K, v_0)$ is*

$$(j_1 E_1 \cup j_2 E_2 \,|\, j_1\Delta_1' \cup j_1\Delta_1'' \cup j_2\Delta_2' \cup j_2\Delta_2'' \cup \{(j_1 e)(j_2 e)^{-1} : e \in E_0\}).$$

Corollary 12.38 *If K is a complex having connected subcomplexes L_1 and L_2 such that $L_1 \cup L_2 = K$ and $L_1 \cap L_2$ is simply connected, then for $v_0 \in V(L_1 \cap L_2)$,*

$$\pi(K, v_0) \cong \pi(L_1, v_0) * \pi(L_2, v_0).$$

This corollary should be compared with Theorem 12.32.

Corollary 12.39 *Let K be a complex having connected subcomplexes L_1 and L_2 such that $L_1 \cup L_2 = K$ and $L_1 \cap L_2$ is connected. If $v_0 \in V(L_1 \cap L_2)$ and if L_2 is simply connected, then*

$$\pi(K, v_0) \cong \pi(L_1, v_0)/N,$$

where N is the normal subgroup generated by the image of $\pi(L_1 \cap L_2, v_0)$. Moreover, in the notation of the theorem, $\pi(K, v_0)$ has the presentation

$$(E_1 \,|\, \Delta_1' \cup \Delta_1'' \cup j_1 E_0).$$

Proof Since $\pi(L_2, v_0) = \{1\}$, the first statement is immediate from the Seifert–van Kampen theorem and Corollary 12.35; the second statement is immediate from Corollary 12.37. ∎

We now exploit Corollary 12.39. Let K be a connected 2-complex with basepoint v_0 and let α be a closed path in K at v_0, say

$$\alpha = e_1 \cdots e_n = (v_0, v_1)(v_1, v_2) \cdots (v_{n-1}, v_0).$$

Define a **triangulated polygon** $D(\alpha)$ as the 2-complex with vertices $V(D(\alpha))$ $= \{p_0, \cdots, p_{n-1}, q_0, \cdots, q_{n-1}, r\}$ and 2-simplexes $\{r, q_i, q_{i+1}\}$, $\{q_i, q_{i+1}, p_{i+1}\}$, and $\{q_i, p_i, p_{i+1}\}$, where $0 \le i \le n-1$ and subscripts are read modulo n.

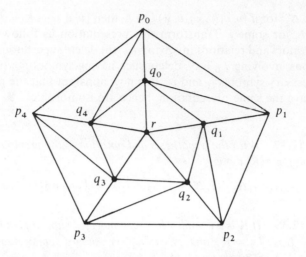

Let $\partial D(\alpha)$ denote the boundary of $D(\alpha)$, i.e., $\partial D(\alpha)$ is the full subcomplex with vertices $\{p_0, \cdots, p_{n-1}\}$. Define the **attaching map** $\varphi_\alpha\colon \partial D(\alpha) \to K$ by $\varphi_\alpha(p_i) = v_i$ for $0 \leq i \leq n-1$. Clearly φ_α carries the boundary path $(p_0, p_1) \cdots (p_{n-1}, p_0)$ onto the path α.

Definition Let α be a closed path in K at v_0, let $D(\alpha)$ be the corresponding triangulated polygon, and let $\varphi_\alpha\colon \partial D(\alpha) \to K$ be the attaching map. The quotient complex $K_\alpha = (K \cup D(\alpha))/\sim$, where $\sim$ identifies each p_i with $\varphi_\alpha(p_i)$, is called the complex obtained from K by **attaching a 2-cell along α**.

Theorem 12.40 *Let α be a closed path in K at v_0 and let K_α be obtained by attaching a 2-cell along α. Then*

$$\pi(K_\alpha, v_0) \cong \pi(K, v_0)/N,$$

where N is the normal subgroup generated by $[\alpha]$.

Proof Define L_1 to be the full subcomplex of K_α with vertices $V(K) \cup \{q_0, \cdots, q_{n-1}\}$ and define L_2 to be the full subcomplex of K_α with vertices $\{r, v_0, q_0, q_1, \cdots, q_{n-1}\}$. Note that $L_1 \cup L_2 = K_\alpha$ and $L_1 \cap L_2$ is the edge (v_0, q_0) and the loop $\{q_0, \cdots, q_{n-1}\}$; it follows that $\pi(L_1 \cap L_2, v_0) \cong \mathbf{Z}$. Now L_2 (isomorphic to the full subcomplex of $D(\alpha)$ with vertices $\{r, q_0, \cdots, q_{n-1}\}$) is simply connected and $\pi(L_1, v_0) \cong \pi(K, v_0)$, by Exercise 12.22. The proof is completed by applying Corollary 12.39, for the image of the infinite cyclic group $\pi(L_1 \cap L_2, v_0)$ is generated by $[\alpha]$. ∎

The following construction is needed to attach a family of 2-cells.

Definition For each $i \in I$, let K_i be a complex with basepoint v_i. The **wedge** of these complexes, denoted by $\bigvee_{i \in I} K_i$, is the disjoint union of the K_i in which the basepoints are all identified to a point b.

A bouquet of circles is an example of a wedge of complexes. Theorem 12.32 shows $\pi(\bigvee K_i, b) \cong * \pi(K_i, v_i)$. The next theorem was used in proving the Kuroš subgroup theorem.

Theorem 12.41   *Given a group G, there exists a connected 2-complex K with $G \cong \pi(K, v_0)$.*

Proof   Let $(X|\Delta)$ be a presentation of G and let B be a bouquet of $|X|$ circles: $V(B) = \{v_0, u_1^x, v_1^x : x \in X\}$. If we identify the closed path $(v_0, u_1^x)(u_1^x, v_1^x)(v_1^x, v_0)$ with x, then each word $w \in \Delta$ may be regarded as a closed path in B at v_0. Let $D(w)$ be the triangulated polygon of w and let

$\varphi_w : \partial D(w) \to B$ be the attaching map; let D be the wedge $\bigvee\limits_{w \in \Delta} D(w)$ and let

$\varphi : \vee\, \partial D(w) \to B$ satisfy $\varphi | \partial D(w) = \varphi_w$. Finally, define K as the quotient complex of $B \cup D$ in which we identify each p^w [in $D(w)$] with $\varphi(p^w) = \varphi_w(p^w)$ [the vertices of $D(w)$ are $r^w, p_0^w, p_1^w, \cdots, q_0^w, q_1^w, \cdots$].

Let T be the tree in K with vertices $\{v_0, u_1^x : x \in X\}$. Define L_1 to be the full subcomplex of K with vertices $V(B) \cup (\bigcup\limits_{w \in \Delta} \{q_0^w, q_1^w, \cdots\})$, and define L_2 to be the full subcomplex of K with vertices $V(T) \cup (\bigcup\limits_{w \in \Delta} \{r^w, q_0^w, q_1^w, \cdots\})$. Note that $L_1 \cup L_2 = K$ and $L_1 \cap L_2$ is the union of T with loops $\{q_0^w, q_1^w, \cdots\}$; it follows that $\pi(L_1 \cap L_2, v_0)$ is free on these loops. Now L_2, being a wedge of simply connected complexes, is simply connected and $\pi(L_1, v_0) \cong \pi(B, v_0)$, by Exercise 12.22. The proof is completed by applying Corollary 12.39, for the image of the free group $\pi(L_1 \cap L_2, v_0)$ is generated by Δ.   ∎

EXERCISE

12.62.   Construct a 2-complex K with $\pi(K, v_0) \cong \mathbf{Z}/2\mathbf{Z}$ (the construction of attaching a 2-cell according to the presentation $(x|x^2)$ yields the "real projective plane").

Amalgams arise from the special case of the Seifert–van Kampen theorem when both induced maps $\pi(L_1 \cap L_2, v_0) \to \pi(L_i, v_0)$, $i = 1, 2$, are one-one (Corollary 12.39 is an instance when this is not so). The advantage of the added hypothesis, as we shall see, is that a normal form is available to describe the elements of $\pi(K, v_0)$. It is convenient to split one corner of the usual diagram into isomorphic parts.

Definition   Let A_1 and A_2 be groups having subgroups B_1 and B_2, respectively; assume $\theta : B_1 \to B_2$ is an isomorphism. The **amalgam**† of A_1 and A_2 over θ is the

---

† Most authors call an amalgam a "free product with amalgamated subgroup".

pushout of the diagram

where $f_i: B_i \to A_i$ are inclusions.

Had we not split the corner, the diagram would be

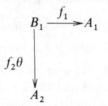

It follows from Theorem 12.34 that the amalgam exists and is unique to isomorphism.

Notation   The amalgam of A_1 and A_2 over θ is denoted by

$$A_1 *_\theta A_2.$$

A less precise notation is $A_1 *_{B_1} A_2$.

We have given an incomplete description of $A_1 *_\theta A_2$, for pushouts are equipped with homomorphisms $g_i: A_i \to A_1 *_\theta A_2$ for $i = 1, 2$, namely, $g_i(a_i) = a_i N$ [where $A_1 *_\theta A_2 = (A_1 * A_2)/N$].

It is clear that, in $A_1 *_\theta A_2$, each $b \in B_1$ is identified with $\theta(b)$; it is not clear whether other identifications are consequences of the amalgamation. For example, is it obvious whether $g_i: A_i \to A_1 *_\theta A_2$ is one-one? Is it even obvious that $A_1 *_\theta A_2 \neq \{1\}$? Our next aim is to give a more concrete description of amalgams in terms of certain types of words. For each $i = 1, 2$, choose a left transversal of B_i in A_i subject only to the condition that the representative of the coset B_i is 1. Let $\bar{a}_i$ denote the chosen representative of $a_i B_i$, so that

$$a_i = \bar{a}_i b_i \qquad \text{for some } b_i \in B_i.$$

Observe that b_i is uniquely determined by a_i (once the elements $\bar{a}_i$ comprising the transversals have been chosen).

Definition A **normal form** is an element of $A_1 * A_2$ of the form

$$\bar{a}_1 \bar{a}_2 \cdots \bar{a}_n b,$$

where $b \in B_1$, $n \geq 0$, the $\bar{a}$ lie in the chosen left transversals of B_i in A_i, and adjacent $\bar{a}$ lie in distinct A_i.

In the special case that B_1 (and hence B_2) is trivial, every reduced word in the free product is a normal form.

Theorem 12.42 **(Normal Form)** *Let B_i be a subgroup of A_i, $i = 1, 2$, and let $\theta: B_1 \to B_2$ be an isomorphism. Every element of $U = A_1 *_\theta A_2$ has a unique expression as a normal form.*

Proof Let N be the normal subgroup of $A_1 * A_2$ generated by $\{b\theta(b^{-1}): b \in B_1\}$. Let $x_1 y_1 \cdots x_n y_n \in A_1 * A_2$, where $x_j \in A_1$, $y_j \in A_2$, and only x_1 and y_n are allowed to be 1. We perform an induction on n that there is a normal form determining the same element as $x_1 y_1 \cdots x_n y_n$ in $A_1 * A_2/N = U$.

Assume $n = 1$. Now $x_1 = \bar{a}_1 b_1$, where $b_1 \in B_1$, so that in U:

$$x_1 y_1 = \bar{a}_1(b_1 y_1) = \bar{a}_1[\theta(b_1)y_1].$$

Since $\theta(b_1)y_1 \in A_2$, we have

$$\theta(b_1)y_1 = \bar{a}_2 b_2, \qquad \text{where } b_2 \in B_2.$$

Conclusion: $x_1 y_1 = \bar{a}_1 \bar{a}_2 b_2 = \bar{a}_1 \bar{a}_2 \theta^{-1}(b_2)$ in U and the last element is a normal form. The inductive step is proved in the same way. (Observe that after the process ends, the last factor b lies in B_1 or B_2; if $b \in B_1$, we have a normal form; if $b \in B_2$, replace b by $\theta^{-1}(b)$ and we have a normal form.) We have thus shown that every $u \in U$ is represented by a normal form. But we have done more: given any $x_1 y_1 \cdots x_n y_n$ in $A_1 * A_2$, the inductive process above assigns a specific normal form to it (denote this normal form by $F(x_1 y_1 \cdots x_n y_n)$).

To prove the uniqueness of this normal form, we construct a homomorphism on U having distinct values on elements described by different normal forms. We use the device of Theorem 12.1. Let M be the set of all normal forms; observe that two normal forms are equal if and only if they have the same spelling (Theorem 12.31). If $a \in A_i$, define a function $|a|: M \to M$ by

$$|a|(\bar{a}_1 \cdots \bar{a}_n b) = F(a\bar{a}_1 \cdots \bar{a}_n b).$$

Clearly, $|1|$ is the identity function on M. Furthermore, consideration of the several cases (depending on possible cancellations) shows that for $a, a' \in A_i$

$$|a| \circ |a'| = |aa'|.$$

Therefore $|a^{-1}| = |a|^{-1}$, so each $|a|$ is a permutation of M. If S_M is the group of all permutations of M, then $a \mapsto |a|$ is a homomorphism $A_i \to S_M$. In particular, if $b \in B_1 \subset A_1$, then $|b|: M \to M$ is defined.

The defining property of free product allows us to assemble these homomorphisms into a homomorphism

$$\Gamma : A_1 * A_2 \to S_M$$

taking a normal form $\bar{a}_1 \cdots \bar{a}_n b$ into $|\bar{a}_1| \cdots |\bar{a}_n| |b|$. Now $b\theta(b^{-1}) \in \ker \Gamma$ for all $b \in B_1$, so Γ induces a homomorphism

$$\Gamma' : U \to S_M$$

by

$$\Gamma'(\bar{a}_1 \cdots \bar{a}_n bN) = \Gamma(\bar{a}_1 \cdots \bar{a}_n b) = |\bar{a}_1| \cdots |\bar{a}_n| |b|.$$

Thus Γ' assigns distinct values in S_M to distinct normal forms, for

$$|\bar{a}_1| \cdots |\bar{a}_n| |b|(1) = \bar{a}_1 \cdots \bar{a}_n b,$$

and the spelling of normal forms is unique. Therefore, distinct normal forms describe distinct elements of U. ∎

Theorem 12.43 *Let $U = A_1 *_\theta A_2$ be an amalgam, where $\theta: B_1 \to B_2$ is an isomorphism.*

(i) *The homomorphisms $g_i: A_i \to U$ are one-one.*
(ii) *If $A'_i = g_i(A_i)$, then $U = \langle A'_1, A'_2 \rangle$ and $A'_1 \cap A'_2 = g_1(B_1) \cong B_1$.*

Proof

(i) If $a_1 \in A_1$, its normal form is $\bar{a}_1 b$ for some $b \in B_1$; moreover, $a_1 \neq 1$ implies $\bar{a}_1 b \neq 1$, by Theorem 12.42. Let $\Gamma': U \to S_M$ be the homomorphism defined in Theorem 12.42. Then $\Gamma' g_1(a_1) = \Gamma'(a_1 N) = \Gamma'(\bar{a}_1 bN) = |\bar{a}_1| |b|$. Since $\Gamma' g_1$ is one-one, it follows that g_1 is one-one. A similar argument shows that g_2 is also one-one.
(ii) Since $U = A_1 * A_2 / N$, it is clear that $U = \langle A'_1, A'_2 \rangle$, where $A'_i = g_i(A_i)$. If $u \in A'_1 \cap A'_2$, then $g_1(a_1) = u = g_2(a_2)$, where $a_i \in A_i$. Let a_1 have normal form $\bar{a}_1 b$ and let a_2 have normal form $\bar{a}_2 b'$. Then applying Γ' gives $|\bar{a}_1| |b| = |\bar{a}_2| |b'|$. It follows from the uniqueness of the normal form that $\bar{a}_1 = 1 = \bar{a}_2$ and that $b = b'$. Conversely, it is easy to see that $bN \in A'_1 \cap A'_2$ for every $b \in B_1$. Therefore, $A'_1 \cap A'_2 = \{bN : b \in B_1\} = g_1(B_1) \cong B_1$. ∎

In view of the last two theorems, it is customary to regard the elements of an amalgam $U = A_1 *_\theta A_2$, where $\theta: B_1 \overset{\sim}{\to} B_2$, as normal forms. The imbeddings g_i are thus regarded as inclusions, and the statement of Theorem 12.43 is simplified to read $U = \langle A_1, A_2 \rangle$ and $A_1 \cap A_2 = B_1 = \theta(B_2)$.

Corollary 12.44 *Let $U = A_1 *_\theta A_2$ be an amalgam, where $\theta: B_1 \to B_2$ is an isomorphism. If $y_1, \cdots, y_r$ are elements of U not in $B_1 = \theta(B_2)$ with $y_j \in A_{i_j}$ and $i_j \neq i_{j+1}$, then $z = y_1 y_2 \cdots y_r \neq 1$.*

Proof It is immediate from the normal form theorem that $z \notin B_1$. ∎

We now apply this last construction to obtain some imbedding theorems.

Theorem 12.45 (Higman, Neumann, and Neumann, 1949)[†] *Let G be a group that contains isomorphic subgroups A and B; let $\varphi: A \to B$ be an isomorphism. There exists a group H containing G and an element t with*

$$\varphi(a) = t^{-1}at \qquad \text{for all } a \in A.$$

Proof Let $\langle u \rangle$ and $\langle v \rangle$ be disjoint infinite cyclic groups; let

$$K_1 = G * \langle u \rangle,$$
$$K_2 = G * \langle v \rangle,$$

and let L_1 be the subgroup of K_1 generated by G and $u^{-1}Au$. Now

$$L_1 = G * u^{-1}Au,$$

for there can be no equation

$$g_1 u^{-1} a_1 u g_2 u^{-1} a_2 u \cdots g_n u^{-1} a_n u = 1$$

in K_1; *a fortiori*, there can be no such equation in L_1. Similarly, the subgroup L_2 of K_2 generated by G and $v^{-1}Bv$ is a free product:

$$L_2 = G * v^{-1} Bv.$$

By Exercise 12.61, there is an isomorphism $\theta: L_1 \to L_2$ such that $\theta | G$ is the identity and $\theta(u^{-1}au) = v^{-1}\varphi(a)v$.

Let $H = K_1 *_\theta K_2$ be the free product of K_1 and K_2 in which we amalgamate L_1 and L_2 via the isomorphism θ. By Theorem 12.43, H contains a subgroup isomorphic to L_1 (and L_1 contains G).

Furthermore, for each $a \in A$,

$$u^{-1}au = v^{-1}\varphi(a)v.$$

If $t = uv^{-1}$, then $t \in H$ and, for all $a \in A$,

$$t^{-1}at = \varphi(a). \quad \blacksquare$$

If G is a countable group, then G is a homomorphic image of a countable free group F of infinite rank:

$$F/R \cong G,$$

where R is a normal subgroup of F. Now we know that F can be imbedded in a free group F_1 on two generators (Theorem 12.28); were R normal in the larger group F_1, then G would be imbedded in F_1/R, a group on two generators. This proof is fictitious, but the theorem is true.

Theorem 12.46 (Higman, Neumann, and Neumann, 1949) *Every countable group G can be imbedded in a group H that can be generated by two elements.*

---

† See Corollary 12.50 for a simpler construction of such a group H.

REMARK   In Exercise 3.44, we have already noted this result in the very special case when G is finite.

Proof† Let $g_0, g_1, g_2, \cdots, g_n, \cdots$ be a list of all the elements of G with $g_0 = 1$. Let $H = G * F$ where F is free on $\{x, y\}$. Consider the subgroups of H

$$A = \langle x, \ g_1 y^{-1}xy, \cdots, g_n y^{-n}xy^n, \cdots \rangle$$

and

$$B = \langle y, \ x^{-1}yx, \cdots, x^{-n}yx^n, \cdots \rangle.$$

Now A and B are isomorphic: indeed, by Exercise 12.36, each is free on the displayed set of generators. Define an isomorphism $\varphi: A \to B$ by

$$\varphi(g_n y^{-n}xy^n) = x^{-n}yx^n \qquad \text{for all } n \geq 0.$$

By Theorem 12.45, there exists a group K containing H and an element t such that

$$\varphi(a) = t^{-1}at \qquad \text{for all } a \in A.$$

We claim that $L = \langle x, t \rangle \subset K$ contains G, which will complete the proof. First of all,

$$y = \varphi(x) = t^{-1}xt \in \langle x, t \rangle.$$

Next, if $n \geq 1$, then

$$t^{-1}(g_n y^{-n}xy^n)t = \varphi(g_n y^{-n}xy^n) = x^{-n}yx^n \in \langle x, t \rangle.$$

It follows that $g_n \in \langle x, t \rangle$, all $n \geq 1$. ∎

EXERCISES

12.63. Let G be a group in which every nonidentity element has infinite order. Show that G can be imbedded in a group H having exactly two conjugacy classes (compare Exercise 3.30). (Of course, H is simple.)

12.64. Prove that there exists a group G having two generators containing an isomorphic copy of every countable abelian group. (HINT: Use Exercise 10.46.)

12.65. Prove that there exists a group G having two generators containing an isomorphic copy of every finite group.

12.66. Let G be a finitely generated group having a finite number of defining relations. Prove that G can be imbedded in a group H having two generators and only finitely many defining relations.

---

† This proof is due to Schupp.

We adopt the following notational convention. Assume $(X|\Delta)$ is a presentation of a group G. Then

$$(G; Y|\Delta')$$

denotes the presentation $(X \cup Y|\Delta \cup \Delta')$, where it is understood that X and Y are disjoint. In particular, when $Y = \varnothing$, we are merely adjoining additional relations, whence $(G|\Delta')$ is a presentation of a quotient group of G.

Definition   Let G be a group having isomorphic subgroups A and B and suppose $\varphi: A \to B$ is an isomorphism. The group G^* with presentation

$$G^* = (G; t|t^{-1}at = \varphi(a) \qquad \text{for all } a \in A)$$

is called an **HNN extension;** G is called the **base** and t is called the **stable letter.** We denote G^* by $G\Omega \quad A$ or, less precisely, by $G\Omega A$.

REMARK   HNN extensions were introduced by G. Higman, B. H. Neumann, and H. Neumann (1949).

The next theorem shows that HNN extensions arise naturally. First of all, consider a connected complex K having disjoint isomorphic subcomplexes A and B; let $\varphi: A \to B$ be an isomorphism. We wish to "add a handle" to K; it is clearer if we assume for the moment that the complexes are topological spaces. Define a space K^* as the quotient space of $K \cup (A \times I)$ (where I is the unit interval) by identifying $a \in A$ with $(a, 0)$ and $\varphi(a)$ with $(a, 1)$. The picture is

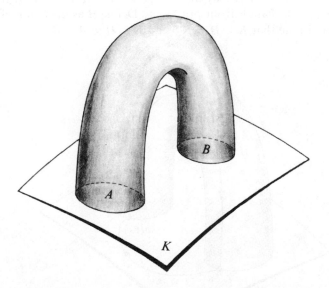

This construction can be carried out with complexes: the role of the unit interval is played by the 1-simplex (also denoted by I) having vertices 0, 1, and the cartesian

product $A \times I$ is made into a complex by "triangulating" it (see Exercise 12.67). One says the quotient K^* is obtained by **adding a handle** to K.

EXERCISES

*12.67. If A is a connected complex, then $A \times I$ is a connected complex. [HINT: Define the $(n + 1)$-simplexes of $A \times I$ to be $\{(a_0, 0), \cdots, (a_i, 0), (a_i, 1), \cdots, (a_n, 1)\}$, where $\{a_0, \cdots, a_n\}$ is an n-simplex in A and $0 \leq i \leq n$.]

12.68. Let A be a connected complex, let $v_0 \in V(A)$, and let $u_0 = (v_0, 0) \in V(A \times I)$. Prove that $\pi(A \times I, u_0) \cong \pi(A, v_0)$.

**12.69. If A is a complex and $v \in V(A)$, denote $(v, 0)$ in $A \times I$ by v^0 and $(v, 1)$ by v^1; for $v_0 \in V(A)$, let β be the edge (v_0^1, v_0^0). If $\alpha = (v_0, v_1) \cdots (v_n, v_0)$ is a path in A, prove $\alpha^0 = (v_0^0, v_1^0) \cdots (v_n^0, v_0^0)$ and $\alpha^1 = (v_0^1, v_1^1) \cdots (v_n^1, v_0^1)$ are paths in $A \times I$ and $\alpha^0 \sim \beta^{-1} \alpha^1 \beta$.

Theorem 12.47 *Let K be a connected complex with disjoint isomorphic subcomplexes A and B; let $v_0 \in V(A)$ be a basepoint and let $\varphi: A \to B$ be an isomorphism. If K^* is obtained from K by adding a handle according to this data, then $\pi(K^*, v_0)$ is an HNN extension with base $\pi(K, v_0)$.*

REMARK   This result suggests the notation $G\Omega A$ for HNN extensions.

Proof As K is connected, we may choose a path γ in K from v_0 to $\varphi(v_0)$; there is also a path β in the handle from $\varphi(v_0)$ to v_0. Define H as the union of γ and the handle, and note that $K \cup H = K^*$ and $K \cap H = A \cup B \cup \gamma$.

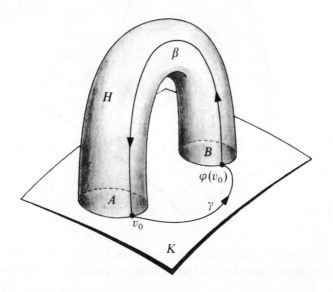

Since $K \cap H$ is connected, the Seifert–van Kampen theorem applies to show $\pi(K^*, v_0)$ is the pushout of the diagram

$$\pi(A \cup B \cup \gamma, v_0) \longrightarrow \pi(H, v_0)$$
$$\downarrow$$
$$\pi(K, v_0)$$

Now $\pi(A \cup B \cup \gamma, v_0) \cong \pi(A, v_0) * \pi(B \cup \gamma, v_0)$, by Corollary 12.38. Indeed, that $\varphi \colon A \to B$ is an isomorphism implies every closed path in $B \cup \gamma$ at v_0 has the form $\gamma(\varphi_{\#}\alpha)\gamma^{-1}$ for some closed path α in A at v_0. Since A and B are disjoint, $H - \gamma \cong A \times I$, and Exercise 12.69 shows that every closed path α in A at v_0 is homotopic to $\beta^{-1}(\varphi_{\#}\alpha)\beta$. In H, therefore,

$$\alpha \sim \beta^{-1}(\varphi_{\#}\alpha)\beta \sim \beta^{-1}\gamma^{-1}(\gamma(\varphi_{\#}\alpha)\gamma^{-1})\gamma\beta.$$

But Corollary 12.38 shows $\pi(H, v_0) \cong \pi(A, v_0) * \langle t \rangle$, where $t = [\gamma\beta]$ and $\langle t \rangle \cong \mathbf{Z}$. Hence $[\alpha] = t^{-1}\varphi_*[\alpha]t$, where $\varphi_* \colon \pi(A, v_0) \to \pi(B \cup \gamma, v_0)$ is the isomorphism induced by φ. It follows easily from the pushout construction that $\pi(K^*, v_0)$ is an HNN extension with base $\pi(K, v_0)$. ■

In contrast to the geometric situation when the isomorphic subcomplexes must be disjoint, the group-theoretic hypothesis for constructing an HNN extension allows the isomorphic subgroups to intersect.

Note a similarity between amalgams and HNN extensions: beginning with two isomorphic subgroups, the first construction gives a group in which the two subgroups are made equal; the second construction gives a group in which the two subgroups are made conjugate. This observation is important in further studies (see the books of Lyndon and Schupp and of Serre in the bibliography).

We need a more general version of HNN extensions in which a set of stable letters may appear.

Definition Let $E = (\Sigma | \Delta)$ be a presentation and let $\{t_i : i \in I\}$ be a nonempty set disjoint from Σ. Assume there is an index set J and, for each $i \in I$, a set of words on Σ, say, $\{a_{ij}, b_{ij} : j \in J\}$. The presentation

$$E^* = (E; t_i, i \in I | t_i^{-1}a_{ij}t_i = b_{ij}, i \in I, j \in J)$$

is said to have **base** E and **stable letters** $\{t_i : i \in I\}$.

We allow a_{ij} and b_{ij} to be 1 so that the number of "honest" relations involving t_i (both a_{ij} and b_{ij} distinct from 1) may be distinct from the number of such relations involving a_{kj}, b_{kj}, and t_k for $k \neq i$.

Example 11 Consider the presentation

$$E^* = (w, x, y, z | y^{-1}xy = w, y^{-1}w^{-1}xwy = xw^{-1}, z^{-1}wxz = w).$$

We may regard E^* as having base $E = (w, x | \varnothing)$, the free group with basis $\{w, x\}$, and stable letters y and z. There are two relations involving the stable letter y and one involving the stable letter z.

We shall henceforth use the following notation.

Notation   If E^* has base E and stable letters $\{t_i : i \in I\}$, define two subgroups of E for each $i \in I$:

$$A_i = \langle a_{ij} : j \in J \rangle \quad \text{and} \quad B_i = \langle b_{ij} : j \in J \rangle;$$

define

$$A = \langle A_i : i \in I \rangle \quad \text{and} \quad B = \langle B_i : i \in I \rangle.$$

Definition   A group with presentation E^* having base E and stable letters $\{t_i : i \in I\}$ is an **HNN extension** if there are isomorphisms $\varphi_i : A_i \to B_i$ for each $i \in I$ with $\varphi_i(a_{ij}) = b_{ij}$ for every j.

If there is only one stable letter, our earlier definition is a special case of this definition. In Example 11 above, $A_y = \langle x, w^{-1}xw \rangle \cong B_y = \langle w, xw^{-1} \rangle$ (both are free of rank 2 with bases the displayed elements) and $A_z = \langle wx \rangle \cong B_z = \langle w \rangle$ (both are infinite cyclic); hence E^* is an HNN extension with base E and stable letters y, z.

EXERCISES

12.70.   If E is a group and F is free with basis $\{t_i : i \in I\}$, then $E^* = E * F$ is an HNN extension with base E and stable letters $\{t_i : i \in I\}$.

**12.71.   Let F be free on $\{x_1, \cdots, x_m\}$ and let $h : G \to F$ be a homomorphism. If there are elements $p_1, \cdots, p_m \in G$ with $h(p_i) = x_i$, $i = 1, \cdots, m$, then $\langle p_1, \cdots, p_m \rangle$ is a free subgroup of G on the displayed generating set.

**12.72.   Let E^* have base E and stable letters $\{p_i : i \in I\}$. Prove that $\langle p_i : i \in I \rangle$ is a free subgroup of E^* with basis $\{p_i : i \in I\}$. (HINT: Use Exercise 12.71.)

Often we use the same letter to denote a presentation and the group it presents. Thus, if $E = (\Sigma | \Delta)$, we may also write $E = F/N$, where F is free on Σ and N is the normal subgroup of F generated by Δ. A word w on Σ may also denote the corresponding element wN of E. For words w_1 and w_2 on Σ,

$$w_1 = w_2 \quad \text{in} \quad E$$

means w_1 and w_2 determine the same element of E. If $E^* = (\Sigma^* | \Delta^*)$ is a presentation with $\Sigma \subset \Sigma^*$ and $\Delta \subset \Delta^*$, there is a (well defined) homomorphism $\lambda : E \to E^*$ with $\lambda(w) = w$ (more precisely, $\lambda(wN) = wN^*$, where we write $E^* = F^*/N^*$ as above). We write

$$E \leq E^*$$

if λ is one-one; this obtains if, for every word w on Σ, $w = 1$ in E if and only if $w = 1$ in E^*. In short, $E \leq E^*$ means that E is imbedded in E^* by the obvious map.

Theorem 12.48 *If E^* is an HNN extension with base $E = (\Sigma|\Delta)$ and stable letters $\{t_i: i \in I\}$, then $E \leq E^*$.*

Proof Since the reader may not be familiar with infinite methods, we prove the theorem only in the special case when I is either finite or countable. Assume $I = \{1, \cdots, n\}$; we perform an induction on n. If $n = 1$, the imbedding theorem of Higman, Neumann, and Neumann (Theorem 12.46) provides a group E_0 containing E and an element τ_1 such that $\tau_1^{-1} a_{1j}\tau_1 = b_{1j}$ for all $j \in J$. Dropping to a subgroup of E_0 if necessary, we may assume E_0 is generated by E and τ_1. A presentation of E_0 can thus be obtained from that of E^* by adding (a possibly empty set of) further relations. Therefore E_0 is a homomorphic image of E^* (in which $t_1 \mapsto \tau_1$). It follows that if w is a word on Σ and $w = 1$ in E^*, then $w = 1$ in E_0 and hence $w = 1$ in E (for E is a subgroup of E_0). Conversely, it is always true that $w = 1$ in E implies $w = 1$ in E^*. We have shown $E \leq E^*$ when $n = 1$.

If $n > 1$, note that E^* is an HNN extension with base

$$E^*_{n-1} = (E; t_1, \cdots, t_{n-1}|t_i^{-1} a_{ij}t_i = b_{ij}, 1 \leq i \leq n-1, j \in J)$$

and stable letter t_n. The inductive hypothesis gives $E \leq E^*_{n-1}$, and the case $n = 1$ gives $E^*_{n-1} \leq E^*$. This completes the finite case.

If $I = \{1, 2, \cdots\}$, define E^*_{n-1} as above for all $n > 1$, and observe that $E \leq E^*_1 \leq E^*_2 \leq \cdots$. Now $E^* = \cup_{n \geq 1} E^*_n$, and this gives the result when I is countable.

(To prove the general case, well-order the index set I and use transfinite induction, taking unions at limit ordinals.) ∎

Corollary 12.49 *If K is a connected complex and K^* is obtained from K by adding a handle, then for every basepoint $v_0 \in V(K)$,*

$$\pi(K, v_0) \leq \pi(K^*, v_0).$$

Proof Immediate from Theorems 12.47 and 12.48. ∎

Here is a sharper form of Theorem 12.45.

Corollary 12.50 *If A and B are subgroups of a group G and $\varphi: A \to B$ is an isomorphism, then $G \leq (G; t|t^{-1}at = \varphi(a), a \in A)$.*

Proof Immediate from Theorems 12.45 and 12.48. ∎

The next result sharpens Theorem 12.46.

Corollary 12.51 *Let $G = (\Sigma|\Delta)$ be a countable group, where $\Sigma = \{g_1, g_2, \cdots\}$. If F is free on $\{x, y\}$, then*

$$H = (G * F; t|t^{-1} xt = y, t^{-1} g_i y^{-i}xy^i t = x^{-i}yx^i, i \geq 1)$$

is a group on two generators t, x that contains G; moreover, t and x each have infinite order.

Proof Plainly H has base $G * F$ and stable letter t. We saw in the proof of Theorem 12.46 that if $A = \langle x, g_i y^{-i} x y^i, i \geq 1 \rangle$ and $B = \langle y, x^{-i} y x^i, i \geq 1 \rangle$, then A and B are each free on the displayed generators; there is thus an isomorphism $\varphi: A \to B$ with $\varphi(x) = y$ and $\varphi(g_i y^{-i} x y^i) = x^{-i} y x^i$ for all $i \geq 1$. Therefore H is an HNN extension of $G * F$, whence $G \leq H$. But it is easy to see $H = \langle t, x \rangle$. We let the reader check that t and x have infinite order. ∎

EXERCISE

12.73. Let $a \in G$ have infinite order. Show there is a group H containing G in which $\langle a \rangle$ and $\langle a^2 \rangle$ are conjugate. (HINT: Theorem 12.48.) Conclude that a conjugate of a subgroup S may be a proper subgroup of S.

We seek a normal form for elements of an HNN extension.

Definition Let W_1 and W_2 be (not necessarily reduced) words on $\{a_1, \cdots, a_n\}$. Then

$$W_1 \equiv W_2$$

if W_1 and W_2 have exactly the same spelling.

If $W_1 \equiv a_1 a_3$ and $W_2 \equiv a_1 a_2^{-1} a_2 a_3$, then $W_1 \not\equiv W_2$.

Definition Let W be a (not necessarily reduced) word on $a_1, \cdots, a_n$. A word Y is a **subword** of W if there are possibly empty words X and Z with $W \equiv XYZ$. A word W **involves** a_i if either a_i or a_i^{-1} is a subword of W.

The next lemma deals with the special HNN extension in which the isomorphism φ is an identity map. Note that the lemma says something about the spelling of a (not necessarily reduced) word as well as the location of the group element corresponding to it.

Lemma 12.52 *Let $E = (S|D)$ and*

$$E^* = (S, t|D, t^{-1} X_i t = X_i, i \in I)$$

be presentations, where the X_i are words on S; let W be a word on $\{S, t\}$ that involves t. If $W = 1$ in E^*, then

  (i) *W contains a subword of the form $t^e C t^{-e}$, where $e = \pm 1$;*
  (ii) *C is a word on S;*
  (iii) *the group element of E determined by C lies in the subgroup of E generated by the X_i.*

Proof We begin by showing that E^* is an amalgam of the form $E *_\psi Y$ for some group Y. The subgroup G generated by the X_i has a presentation

$$G = (X_i, i \in I \mid R_j = 1, j \in J)$$

for some words R_j on the X_i. Let $\{X_i': i \in I\}$ be a set in one-one correspondence with $\{X_i: i \in I\}$ and disjoint from it. Define a group G', another copy of G, by

$$G' = (X_i', i \in I \mid R_j' = 1, j \in J),$$

and let $\psi: G \to G'$ be the isomorphism with $\psi(X_i) = X_i'$ for all i and $\psi(R_j) = R_j'$ for all j. Let $\langle t \rangle$ be an infinite cyclic group and let Y be the direct product

$$Y = G' \times \langle t \rangle.$$

Finally, define

$$A = E *_\psi Y.$$

By Theorem 12.34, one presentation of the amalgam A is

$$A = (S, t, X_i', i \in I \mid D, R_j' = 1, j \in J, t^{-1} X_i' t = X_i', i \in I, X_i' = X_i, i \in I).$$

Erasing superfluous generators and relations, we obtain

$$A = (S, t \mid D, R_j = 1, j \in J, t^{-1} X_i t = X_i, i \in I),$$

for R_j is just R_j' rewritten in unprimed letters X_i. But $G' \subset Y \leq A$, so that $1 = R_j' = \psi^{-1}(R_j') = R_j$ in A. The relations $R_j = 1$ in A are thus superfluous and we obtain $A \cong E^*$.

We now prove the lemma. If W contains a subword tt^{-1} or $t^{-1}t$, we are done. Therefore, we may assume

$$W \equiv W_0 t^{e_1} W_1 \cdots t^{e_n} W_n,$$

where $n \geq 1$, each e_j is a nonzero integer, and the W_j are words on S of which only W_0 and W_n are allowed to be empty.

If $n = 1$, then $W \equiv W_0 t^{e_1} W_1$. Since $W = 1$ in E^*,

$$t^{e_1} = W_0^{-1} W_1^{-1} \in G \cap Y = G',$$

by Theorem 12.43. This is a contradiction, for $Y = G' \times \langle t \rangle \leq E^*$ and $G' \cap \langle t \rangle = \{1\}$. The case $n = 1$ is thus impossible, as the lemma predicts.

For the inductive step, apply Corollary 12.44 to the word W (which is 1 in E^*), and observe that W_j lies in G for some j, where $1 \leq j \leq n - 1$. If e_j and e_{j+1} have opposite sign, we are finished. Suppose e_j and e_{j+1} have the same sign. In E^*,

$$W \equiv \cdots t^{e_j} W_j t^{e_{j+1}} W_{j+1} \cdots = \cdots t^{e_j + e_{j+1}} W_j W_{j+1} \cdots.$$

The second word has one fewer occurrence of a power of t, and it satisfies all the inductive hypotheses. Therefore it, and hence W, has a subword of the desired kind. ∎

Theorem 12.53 (Britton's lemma, 1963)   *Let E^* be an HNN extension of
$E = (S | D)$ with stable letters $\{p_v : v \in V\}$. If W is a word involving at least one
stable letter and if $W = 1$ in E^*, then W contains a subword of the form $p_v^e C p_v^{-e}$,
where $e = \pm 1$ and C is a word on S.*

*Moreover, if $e = -1$, then C is equal in E to a word on $\{a_{vj} : j \in J\}$ (i.e., C is
equal in E to an element of $A_v \subset A$); if $e = +1$, then C is equal in E to a word in
$\{b_{vj} : j \in J\}$ (i.e., C is equal in E to an element of $B_v \subset B$).*

REMARK   A word of the form $p_v^e C p_v^{-e}$ is called a **pinch** if $e = \pm 1$, C is a word on
S, and the "Moreover" paragraph above holds for C. The conclusion of
Britton's lemma is that the word W in the hypothesis must contain a pinch as a
subword.

Proof   We first prove the special case of only one stable letter $p_v = p$. Thus, the
presentation we deal with is

$$E^* = (S, p | D, p^{-1} a_j p = b_j, \quad j \in J).$$

We are done if W contains a subword of the form $p^{-1} p$ or $p p^{-1}$, so we may
assume

$$W \equiv W_0 p^{\alpha_1} W_1 \cdots p^{\alpha_n} W_n,$$

where $n \geq 1$, each α_i is a nonzero integer, and the W_i are words on S of which
only W_0 and W_n may be empty.

We wish to use Lemma 12.52, but it does not apply because a_j and b_j may
be distinct. Since E^* is an HNN extension of E, Theorem 12.48 gives $E \leq E^*$. If
H is the group with presentation

$$H = (S, q | D, q^{-1} a_j q = b_j, \quad j \in J),$$

then clearly $H \cong E^*$. Consider now

$$H^* = (\{S, q\}, t \mid \{D, q^{-1} a_j q = b_j, \quad j \in J\}, t^{-1} b_j t = b_j, \quad j \in J)$$

and the word

$$U \equiv W_0 (qt)^{\alpha_1} W_1 \cdots (qt)^{\alpha_n} W_n.$$

First of all, we claim that $U = 1$ in H^*. To see this, add the new generator p and
the defining relation $p = qt$ to H^* to obtain a group K, where

$$K = (S, q, t, p | D, q^{-1} a_j q = b_j, \ p = qt, \ p^{-1} a_j p = b_j, j \in J).$$

Now $K \cong H^*$ and K contains all the generators and relations of E^*. Therefore,
$W = 1$ in E^* implies $U = 1$ in H^*. Applying Lemma 12.52 to the groups H and
H^* and the word U, we see that U contains a subword $t^e C t^{-e} (e = \pm 1)$, where
C belongs to the subgroup B. If $e = 1$, then $C \equiv W_k$ for some k ($1 \leq k \leq n - 1$),
and we are done. If $e = -1$, then $C \equiv q^{-1} W_k q$ for some k; therefore,
$q^{-1} W_k q \in B$ and so $W_k \in q B q^{-1} = A$.

We now consider the general case. Denote the elements of the index set V
by $1, 2, \cdots$. Let $D(v)$ consist of the relations $p_v^{-1} a_{vj} p_v = b_{vj}$ in E^*. Since $W = 1$

in E^*, we have $W = 1$ in E_r for some r, where

$$E_r = (S, p_1, \cdots, p_r \mid D, D(1), \cdots, D(r)).$$

It is straightforward to show that $E \leq E_1 \leq E_2 \leq \cdots$. Choose s maximal such that W involves p_s; the chain of inequalities implies that $W = 1$ in E_s. Since E_s has base E_{s-1} and stable letter p_s, it follows from the first portion of our proof that W contains a subword $p_s^e C p_s^{-e}$, where, for example, $e = 1$ and C belongs to B_s, the subgroup of E_{s-1} (and hence of E) generated by all $b_{sj}, j \in J$. (A similar conclusion holds if $e = -1$.) If the word C is a word on S, we are done. If, on the other hand, the word C involves some of $p_1, \cdots, p_{s-1}$, then there is a word C' on S (namely, a product of $b_{sj}^{\pm 1}$, $j \in J$) such that $C = C'$ in E_{s-1}. Thus, $CC'^{-1} = 1$ in E_{s-1}, and an induction on s implies that the word CC'^{-1}, hence C, hence W, contains a subword of the desired type. ∎

Definition Let E^* be an HNN extension with base $E = (\Sigma \mid \Delta)$ and stable letters $\{p_v : v \in V\}$. A word W on $\Sigma \cup \{p_v : v \in V\}$ is p_v-**reduced** (for some fixed v) if it contains no pinch $p_v^e C p_v^{-e}$ as a subword.

Corollary 12.54 *Let E^* be an HNN extension of E with stable letters $\{p_v : v \in V\}$. Assume $U \equiv R_0 p_v^{e_1} R_1 \cdots p_v^{e_m} R_m$ and $W \equiv L_0 p_v^{f_1} L_1 \cdots p_v^{f_n} L_n$ are p_v-reduced words, where each $e, f = \pm 1$, and none of the (possibly empty) words R or L involve p_v.*

If $U = W$ in E^*, then $m = n$ and $(e_1, \cdots, e_m) = (f_1, \cdots, f_n)$; moreover, the word $p_v^{e_m} R_m L_n^{-1} p_v^{-f_n}$ is a pinch.

Proof Since $UW^{-1} = 1$ in E^*, Britton's lemma asserts that it contains a pinch; since U and W (hence W^{-1}) are p_v-reduced, the pinch must occur at the interface: the subword $p_v^{e_m} R_m L_n^{-1} p_v^{-f_n}$ is a pinch. In particular, e_m and $-f_n$ must have opposite sign, so that $e_m = f_n$.

The remainder of the corollary is proved by induction on $\max \{m, n\}$. In the pinch $p_v^e R_m L_n^{-1} p_v^{-e}$ displayed above, we know from Britton's lemma that

$$R_m L_n^{-1} = \begin{cases} a_{vj_1}^{\alpha_1} \cdots a_{vj_t}^{\alpha_t} & \text{if } e = -1 \\ b_{vj_1}^{\alpha_1} \cdots b_{vj_t}^{\alpha_t} & \text{if } e = +1, \end{cases}$$

where $\alpha_i = \pm 1$. In the first case,

$$p_v^{-1} R_m L_n^{-1} p_v = p_v^{-1} a_{vj_1}^{\alpha_1} \cdots a_{vj_t}^{\alpha_t} p_v$$

$$= (p_v^{-1} a_{vj_1}^{\alpha_1} p_v)(p_v^{-1} \cdots p_v)(p_v^{-1} a_{vj_t}^{\alpha_t} p_v)$$

$$= b_{vj_1}^{\alpha_1} \cdots b_{vj_t}^{\alpha_t}.$$

We have eliminated one p_v from U and one from W, so the remainder follows from the inductive hypothesis. The proof for the case $e = +1$ is similar. ∎

The normal form theorem has its most succinct statement when an HNN extension has only one stable letter; the statement and proof of the generalization to arbitrary HNN extensions are left to the reader.

Theorem 12.55 (**Normal Form**) *Let E^* be an HNN extension with base $E = (\Sigma \mid \Delta)$ and stable letter t. Each $W \in E^*$ is equal to a t-reduced word:*

$$W = R_0 t^{e_1} R_1 t^{e_2} \cdots t^{e_n} R_n \quad in \quad E^*;$$

moreover, the length n and the sequence of exponents $(e_1, \cdots, e_n)$ are uniquely determined by W.

Proof It is easy to see that such an expression for W exists, for the relations in E^* allow one to replace any pinch by a subword involving two fewer occurrences of the stable letter t. The uniqueness is immediate from Corollary 12.54. ∎

Here is an important class of groups.

Definition A group G is **finitely presented** (or *finitely related*) if it has a presentation with only a finite number of generators and a finite number of relations.

B. H. Neumann (1937) has shown there are uncountably many nonisomorphic finitely generated groups. Since there are only countably many finitely presented groups, it follows that there exist finitely generated groups that are not finitely presented; we shall give an explicit example of such a group below.

Let G be a free abelian group with basis $\{x, y\}$. If F is the free group with basis $\{x, y\}$, then $G \cong F/F'$, where F' is the commutator subgroup of F. We saw in Theorem 12.28 that F' is free of infinite rank. On the other hand, G *is* finitely presented, for $G = (x, y \mid xyx^{-1}y^{-1})$. The normal subgroup of F generated by the single element $xyx^{-1}y^{-1}$ is F' (remember that the normal subgroup generated by a subset X is the subgroup generated by all the conjugates of elements in X).

Theorem 12.56 *A group G is finitely presented if and only if there is a finite connected complex K (i.e., $V(K)$ is finite) with $G \cong \pi(K, v_0)$ for some $v_0 \in V(K)$.*

Proof If $G \cong \pi(K, v_0)$ for a finite complex K, then Lemma 12.10 shows G is finitely presented. Conversely, if G is finitely presented, a finite complex K with $G \cong \pi(K, v_0)$ can be constructed by attaching a finite number of 2-cells to a bouquet of finitely many circles, as in Theorem 12.41. ∎

EXERCISES

12.74. Every free group of finite rank is finitely presented.

12.75. Every finite group G is finitely presented.

12.76. Every finitely generated abelian group is finitely presented.

12.77. If G has a presentation with a finite number of relations, then G is a free product of a finitely presented group and a free group.

****12.78.** Consider the diagram

If A_1 and A_2 are finitely presented and B is finitely generated, then the pushout of this diagram is finitely presented.

We shall now give a nice application of Britton's lemma. First, we need an easy remark.

Lemma 12.57 *Suppose G is a group having a presentation*

$$G_1 = (a_1, \cdots, a_m | r_n = 1, n \in \mathbf{N})$$

as well as a finite presentation

$$G_2 = (b_1, \cdots, b_k | s_1 = 1, \cdots, s_t = 1).$$

Then all but a finite number of the relations $r_n = 1$ in the first presentation are superfluous.

Proof Let $\varphi : G_1 \to G_2$ be an isomorphism with inverse $\psi : G_2 \to G_1$. For each i, $\psi(s_i) = 1$ in G_1, and hence is a word on conjugates of various r_n. Since words have finite length and since there are only finitely many s_i, only a finite number of r_n are needed to prove $\psi(s_i) = 1$ in G_1 for all i. For notational convenience, let us denote these $r_1, r_2, \cdots, r_N$.

For each n, $\varphi(r_n) = 1$ in G_2 so that

$$\varphi(r_n) = w_n(s_1, \cdots, s_t),$$

a word on conjugates of $s_1, \cdots, s_t$. Therefore

$$r_n = \psi\varphi(r_n) = \psi(w_n(s_1, \cdots, s_t)) = w_n(\psi(s_1), \cdots, \psi(s_t)).$$

This equation says that r_n lies in the normal subgroup of G_1 generated by $r_1, r_2, \cdots, r_N$, which is what was to be proved. ∎

As we mentioned earlier, the following theorem was first proved by B. H. Neumann; the following explicit example is due to W. W. Boone.

Theorem 12.58 *There exists a finitely generated group that is not finitely presented.*

Proof Let $\langle a, b \rangle$ be a free group of rank 2. By Theorem 12.28, $\langle a, b \rangle$ contains a free subgroup of infinite rank, say, with free generating set

$\{w_1, w_2, \cdots, w_n, \cdots\}$. Define a group G by the presentation

$$G = (a, b, p \,|\, p^{-1} w_n p = w_n, n = 1, 2, \cdots).$$

This presentation exhibits G as an HNN extension of $\langle a, b \rangle$ as in Lemma 12.52.

    Suppose G were finitely presented. By Lemma 12.57, we may delete all but a finite number of the relations. For notational convenience, we may assume

$$G = (a, b, p \,|\, p^{-1} w_n p = w_n, n = 1, 2, \cdots, M-1).$$

Now $p^{-1} w_M p = w_M$ in G, so that Britton's lemma provides a word Y in $\langle w_1, \cdots, w_{M-1} \rangle$ with

$$w_M = Y \quad \text{in} \quad \langle a, b \rangle.$$

This contradicts the fact that $\{w_1, w_2, \cdots, w_M\}$ freely generates its subgroup. ∎

chapter thirteen

The
Word Problem

STATEMENT OF THE PROBLEM

Novikov and Boone have proved that there exists a group G with unsolvable word problem; there is thus a sequence of elementary questions about G that no one machine can answer. We shall prove this remarkable result in this chapter.

Let Q be a set of questions; a **decision process** (or *algorithm*) for Q is a uniform set of directions which, when applied to any of the questions in Q, produces the correct answer after a finite number of steps, never at any stage of the process leaving the user in doubt as to what to do next.

Suppose G is a group having the presentation

$$G = (x_1, \cdots, x_n | r_j, j \geq 0)$$

By means of this presentation, every (not necessarily reduced) word on the x represents an element of G. We say that the **word problem** for G is **solvable** if there exists a decision process for the set of all questions of the form: Does the word w on the x represent the identity element of G?†

If a word w is written $w = x_1^{\varepsilon_1} \cdots x_n^{\varepsilon_n}$, where $\varepsilon_i = \pm 1$, then the **length** of w is defined to be n. This is in contrast to the usual definition of length which requires

---

† As we have posed it, the word problem for G depends on a particular presentation of G. It is a fact that if the word problem is solvable for some one finite presentation of G, i.e., there are only finitely many generators and finitely many relations, then it is solvable for every finite presentation of G.

that w be reduced. Thus, according to this definition, the empty word has length 0, but $x_1 x_1^{-1}$ has length 2.

To illustrate these ideas, we show that a free group

$$G = (x_1, \cdots, x_n | \varnothing)$$

has solvable word problem. Here is a decision process:

1. If the length of w is 0 or 1, proceed to step 3. If the length of $w \geq 2$, underline the first adjacent pair of letters (if there is any) of the form $x_i x_i^{-1}$ or $x_i^{-1} x_i$; if there is no such pair, underline the final two letters. Proceed to step 2.
2. If the underlined pair has the form $x_i x_i^{-1}$ or $x_i^{-1} x_i$, erase it, and proceed to step 1; otherwise, proceed to step 3.
3. If the word is empty, write $w = 1$ and stop. If the word is not empty, write $w \neq 1$ and stop.

The reader should experiment a bit with this program until convinced it is a decision process.

EXERCISES

13.1. Sketch a proof that every finite group has solvable word problem.
13.2. Sketch a proof that every finitely generated abelian group has solvable word problem.
13.3. Sketch proofs that if each of G and H has solvable word problem, then the same is true of their direct product $G \times H$ and their free product $G * H$.
13.4. Sketch a proof that if $G = (x_1, \cdots, x_n | r_j, j \geq 0)$ has solvable word problem and H is a finitely generated subgroup of G, then H has solvable word problem. (HINT: Write $H = \langle h_1, \cdots, h_m \rangle$ and write each h_i as a word in the x.)

The proof of the Novikov-Boone theorem can be split in half. The initial portion is really mathematical logic, and is a theorem of Markov and Post that there exists a finitely presented semigroup having unsolvable word problem. The more difficult portion consists of constructing a finitely presented group G and showing that solvability of the word problem for G implies solvability of the word problem for the Markov-Post semigroup. Nowhere in the reduction of the group problem to the semigroup problem is a technical definition of unsolvability used, so that the reader knowing only our intuitive description given above can follow this part of the proof. We do, however, include a technical definition below. There are several good reasons for doing this: the word problem can be properly stated; a proof of the Markov-Post theorem can be given; the generators and relations of the Markov-Post semigroup can be accounted for; a proof of a beautiful theorem of G. Higman characterizing those finitely generated groups that can be imbedded in a finitely presented group can be given. This last theorem is then used to give an algebraic characterization of groups having solvable word problem (Theorem 13.29). Finally, we shall show there is no decision process to detect, of an arbitrary finite presentation, whether or not the presented group possesses (almost any) reasonable property.

TURING MACHINES AND THE MARKOV-POST THEOREM

Let us call a subset E of a set X "enumerable" if there is a computer that can recognize every element of E and no others. Of course, the nature of such a well-behaved subset E should not depend on accidental physical constraints affecting a real computer, e.g., whether the number of memory cells may exceed the total number of atoms in the universe. We thus define an idealized computer, called a *Turing machine* (after A. Turing, 1912–1954), which abstracts the essential features of a real computer and which enumerates only those subsets E that, on intuitive grounds, "ought" to be enumerable.

Informally, a Turing machine can be pictured as a box with a tape running through it. The tape consists of a serial collection of squares, which is as long to the right and to the left as desired. The box is capable of printing a finite number of symbols $s_0, s_1, \cdots, s_M$ and of being in a finite number of states $q_0, q_1, \cdots, q_N$. At any fixed moment, the box is in a state q_i and is "scanning" a particular square of the tape that bears a single symbol s_j (we may agree that s_0 means blank). The action of the machine is determined by its initial structure and by q_i and s_j; this action consists of either stopping or going into some state q_l after obeying one of the following instructions:

1. Erase the symbol s_j and print some symbol s_k.
2. Move one square to the right and scan this square.
3. Move one square to the left and scan this square.

The machine is now ready for its next move.

The machine is started in the first place by being given a tape, which may have some nonblank symbols printed on it (one to a square), and by being set to scan some one square while in "starting" state q_1. The machine may eventually stop or it may continue working indefinitely.

We proceed to the formal definitions; after each definition is stated, we shall give an informal interpretation. Let us first choose, once for all, two infinite lists of letters:

$$s_0, s_1, s_2, \cdots \quad \text{and} \quad q_0, q_1, q_2, \cdots.$$

Definition A **quadruple** is a 4-tuple of one of the following three types:

$$q_i s_j s_k q_l;$$
$$q_i s_j R q_l;$$
$$q_i s_j L q_l.$$

Definition A **Turing machine** is a finite nonempty set of quadruples no two of which have the same first two letters.

The three types of quadruples correspond to the three types of moves in our informal description given above. For example, $q_i s_j R q_l$ may be interpreted as being the instruction: "When scanning symbol s_j in state q_i, move right one square and enter state q_l".

Definition A word on a given set of letters is **positive** if it has no negative exponents. We agree the empty word is positive.

Definition An **instantaneous description** α is a positive word on letters s_j and exactly one q_i (which is not at the right end).

As an example, the instantaneous description $s_2 s_0 q_1 s_5 s_2$ is to be interpreted, "The symbols on the tape are $s_2 s_0 s_5 s_2$ (*with blanks everywhere else*) and the machine is in state q_1 scanning s_5".

Definition Let T be a Turing machine and let α, β be instantaneous descriptions. Write $\alpha \to \beta$ if there are (possibly empty) positive words P and Q on s-letters such that one of the following conditions holds:

$$\text{(i)} \quad \left. \begin{array}{l} \alpha = P q_i s_j Q \\ \beta = P q_l s_k Q \end{array} \right\} \quad \text{where } q_i s_j s_k q_l \in T;$$

$$\text{(ii)} \quad \left. \begin{array}{l} \alpha = P q_i s_j s_k Q \\ \beta = P s_j q_l s_k Q \end{array} \right\} \quad \text{where } q_i s_j R q_l \in T;$$

$$\text{(iii)} \quad \left. \begin{array}{l} \alpha = P q_i s_j \\ \beta = P s_j q_l s_0 \end{array} \right\} \quad \text{where } q_i s_j R q_l \in T;$$

$$\text{(iv)} \quad \left. \begin{array}{l} \alpha = P s_k q_i s_j Q \\ \beta = P q_l s_k s_j Q \end{array} \right\} \quad \text{where } q_i s_j L q_l \in T;$$

$$\text{(v)} \quad \left. \begin{array}{l} \alpha = q_i s_j Q \\ \beta = q_l s_0 s_k Q \end{array} \right\} \quad \text{where } q_i s_j L q_l \in T.$$

The reader should interpret $\alpha \to \beta$ as a **basic move** of the machine. Some further explanation is now needed to interpret basic moves described by condition (iii) or (v). The tape is finite, but when the machine comes to an end of the tape, the tape is lengthened by adjoining a blank square. Since s_0 means "blank", the two rules (iii) and (v) thus correspond to the cases when the machine is scanning either the last symbol or the first symbol on the tape.

The proviso in the definition of a Turing machine that no two quadruples have the same first two symbols may be interpreted to mean that there is never ambiguity about a machine's next move. Formally, $\alpha \to \beta$ and $\alpha \to \gamma$ implies $\beta = \gamma$.

Definition A **computation** of a Turing machine is a finite sequence of instantaneous descriptions $\alpha_1, \alpha_2, \cdots, \alpha_t$, where

$$\alpha_i \to \alpha_{i+1}, \qquad i = 1, \cdots, t-1,$$

and α_t is **terminal**, i.e., there is no α with $\alpha_t \to \alpha$.

Definition If T is a Turing machine, its **alphabet** is the set $\{s_0, \cdots, s_M\}$ of all s-letters occurring in its quadruples. If w is a positive word on the alphabet of T, we say that $T(w)$ **exists** if there is a computation of T beginning with $q_1 w$.

Informally, we regard the machine T as being in starting state q_1 and as having w printed on its tape. The running of T should be visualized as a possibly infinite sequence of instantaneous descriptions $q_1 w \to \alpha_2 \to \alpha_3 \to \cdots$. This sequence stops if $T(w)$ exists; otherwise, T runs forever.

Given a Turing machine T, let Ω be the set of all positive words on its alphabet. To the machine T, we associate the following subset of Ω:

$$e(T) = \{w \in \Omega; T(w) \text{ exists}\};$$

we say that T **enumerates** $e(T)$.

Definition Let Ω be the set of positive words on letters $\{s_0, \cdots, s_M\}$. A subset E of Ω is **r.e. (recursively enumerable)** if there is some Turing machine T that enumerates E, i.e., the alphabet of T contains $\{s_0, \cdots, s_M\}$ and $E = e(T)$.

This notion of r.e. subsets of words can be specialized to subsets of the natural numbers $\mathbf{N}$ by considering just the words of the form s_1^{n+1}, where $n \geq 0$. Thus, a subset E of $\mathbf{N}$ is an r.e. subset if there is a Turing machine T with $E = \{n \in \mathbf{N}: T(s_1^{n+1})$ exists$\}$.

EXERCISES

13.5. Prove that there exist subsets of $\mathbf{N}$ that are not r.e. (HINT: There are only countably many Turing machines.)

13.6. Prove that the set of even natural numbers is r.e.

13.7. Give an example of a Turing machine T with s_1 in its alphabet and such that $T(s_1)$ does not exist.

*13.8. If Ω is the set of all positive words on $\{s_0, \cdots, s_M\}$ and if E_1 and E_2 are r.e. subsets of Ω, then $E_1 \cup E_2$ and $E_1 \cap E_2$ are r.e. subsets of Ω.

Every Turing machine T defines an r.e. subset $E = e(T) \subset \Omega$, where Ω is the set of all positive words on its alphabet. How do we tell whether or not a word w lies in E? Feed $q_1 w$ into T and wait (i.e., perform the basic moves specified by the quadruples of T). If $w \in E$, then T will stop eventually. However, for a given w, there is no way of telling, *a priori*, whether T will stop. Certainly this is unsatisfactory for an impatient person. But more: it leads to a surprising fact (Theorem 13.1).

Definition   Let Ω be the set of positive words on $\{s_0, \cdots, s_M\}$. A subset E of Ω is **recursive** if both E and its complement $\bar{E}$ are r.e. subsets.

When E is recursive, there is never an "infinite wait" to decide whether a word w is in E. If T is a Turing machine enumerating E and if $\bar{T}$ is a Turing machine enumerating $\bar{E}$, then, for each $w \in \Omega$, either $T(w)$ or $\bar{T}(w)$ exists. Thus, it can be decided in a finite length of time whether or not a given word w lies in E.

There is a decision process determining whether a word w lies in E if there is a uniform set of directions producing the correct answer (yes or no) after a finite number of steps, and never at any stage leaving the user in doubt as to what to do next. We propose that the recursive sets are precisely those subsets admitting a decision process. Of course, this proposition (called **Church's thesis**) cannot be proved, for it is a question of translating an intuitive notion into precise terms. There have been other attempts to formalize this notion, avoiding Turing machines altogether. Every alternative definition that has been proposed which gives a decision process for recursive sets has been proved to give a decision process for only these sets.

If E is a subset of $\mathbf{N}$, then the usual definition of E being recursive is that both E and $\mathbf{N} - E$ are r.e. subsets of $\mathbf{N}$. According to our definition, both E and $\Omega - E$ are r.e. subsets of Ω, where Ω is the set of positive words on some alphabet. Using Exercise 13.8, the reader may show that these definitions are equivalent.

EXERCISES

**13.9.   Let E_1 and E_2 be recursive subsets of a set of positive words Ω. Prove that $E_1 \cup E_2$ and $E_1 \cap E_2$ are recursive.

**13.10.   If E_1 and E_2 are recursive subsets of $\mathbf{N}$, then $E_1 \times E_2$ is a recursive subset of $\mathbf{N} \times \mathbf{N}$ [first imbed $\mathbf{N} \times \mathbf{N}$ as a recursive subset of $\mathbf{N}$ by "coding" the ordered pair (m, n) into $2^m 3^n$].

Theorem 13.1   *There exists an r.e. subset of the natural numbers $\mathbf{N}$ that is not recursive.*

Proof   Since Turing machines are finite sets of quadruples based on the letters $R, L, q_0, q_1, \cdots; s_0, s_1, \cdots$, there are only countably many Turing machines. We now enumerate them. Assign integer values to the letters in the following way:

$$R \leftrightarrow 0; \; L \leftrightarrow 1; \; q_0 \leftrightarrow 2; \; q_1 \leftrightarrow 4; \; q_2 \leftrightarrow 6; \cdots; \; s_0 \leftrightarrow 3; \; s_1 \leftrightarrow 5; \; s_2 \leftrightarrow 7; \cdots.$$

Let T be a Turing machine consisting of m quadruples. Juxtapose the quadruples in some order to form a word of length $4m$. If $p_1 < p_2 < \cdots$ is the list of all primes, define

$$G(T) = \prod_{i=1}^{4m} p_i^{e_i},$$

where e_i is the integer associated above to the ith letter in our word of length $4m$ ($G(T)$ is called the **Gödel number**† of T). The fundamental theorem of arithmetic implies that distinct Turing machines have distinct Gödel numbers. If we now list these integers $G(T)$ in order of magnitude, we have effectively listed all Turing machines: $T_0, T_1, \cdots, T_n, \cdots$. Define

$$E = \{n \in \mathbf{N} : T_n(s_1^{n+1}) \text{ exists}\}.$$

We claim that E is an r.e. set. Consider the following diagram reminiscent of the proof that the set of rational numbers is countable:

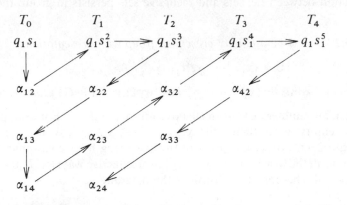

The nth column consists of the sequence of basic moves of the Turing machine T_n beginning with $q_1 s_1^{n+1}$. It is intuitively clear that there is an enumeration of the integers n lying in E: follow the arrows in the diagram; if an instantaneous description α_{ni} in column n is the end of a computation, then $n \in E$. A Turing machine can, in fact, be constructed to carry out these instructions. (Moreover, for later use, there is a specific Turing machine T^* which enumerates E and which satisfies the additional condition that terminal instantaneous descriptions, and only these, involve the letter q_0.) Thus, the set E is an r.e. subset of $\mathbf{N}$.

To show that E is not recursive, it suffices to prove that

$$\bar{E} = \{n \in \mathbf{N} : n \notin E\} = \{n \in \mathbf{N} : T_n(s_1^{n+1}) \text{ does not exist}\}$$

is not an r.e. subset of $\mathbf{N}$. Suppose there were a Turing machine T enumerating $\bar{E}$. Since all Turing machines occur in the list at the beginning of the proof, we have $T = T_{n_0}$ for some $n_0 \in \mathbf{N}$. If $n_0 \in \bar{E}$, then $n_0 \in e(T) = e(T_{n_0})$; hence, $T_{n_0}(s_1^{n_0+1})$ exists and $n_0 \in E$, a contradiction. If $n_0 \notin \bar{E}$, then $n_0 \in E$ and $T_{n_0}(s_1^{n_0+1})$ exists (definition of E); hence $n_0 \in e(T_{n_0}) = \bar{E}$, a contradiction. Therefore $\bar{E}$ is not an r.e. subset and E is not recursive. ∎

† After K. Gödel (1906–1978), the great logician.

Let us link these ideas to algebra. If G is a group with generators $x_1, \cdots, x_m$, then we may regard the set Ω of all (not necessarily positive) words on $x_1, \cdots, x_m$ as the set of all positive words on

$$x_1, x_1^{-1}, \cdots, x_m, x_m^{-1}.$$

We say that the **word problem** for the *group* G is **solvable** if $\{w \in \Omega : w = 1 \text{ in } G\}$ is recursive.

There is a similar definition for semigroups. If H is a semigroup with generators $x_1, \cdots, x_m$ and if Ω is the set of positive words on $x_1, \cdots, x_m$, then the **word problem** for the *semigroup* H is **solvable** if there is a decision process to determine, for an arbitrary pair of words w_1, w_2, in Ω whether $w_1 = w_2$ in H. This gives a criterion for unsolvability suitable for our purposes: The word problem for the semigroup H is unsolvable if there is a word w_0 such that $\{w \in \Omega : w = w_0 \text{ in } H\}$ is not recursive.

The distinction between r.e. sets and recursive sets persists in group theory.

Theorem 13.2   *Let G be a finitely presented group with presentation*

$$G = (x_1, \cdots, x_n | r_1, \cdots, r_m).$$

If Ω is the set of all words on $\{x_1, \cdots, x_n\}$, then $\{w \in \Omega : w = 1 \text{ in } G\}$ is an r.e. set.

Proof   First of all, enumerate Ω in an effective way: $w_1, w_2, \cdots$. For example, first write the empty word, then write $x_1, x_1^{-1}, \cdots, x_n, x_n^{-1}$, then write all words of length 2, etc.; order all words of the same length lexicographically. Next, enumerate all the words on $\{r_1, \cdots, r_m\}$ in an effective way: $R_1, R_2, \cdots$. As in the proof of Theorem 13.1, consider the diagram:

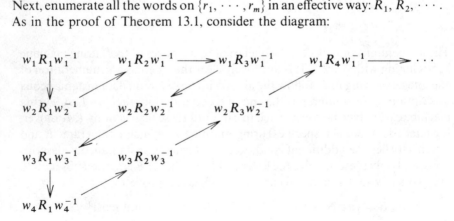

Following the arrows enumerates $\{w \in \Omega : w = 1 \text{ in } G\}$, i.e., the set is r.e.   ∎

The word problem for G is solvable if $\{w \in \Omega : w = 1 \text{ in } G\}$ is a recursive set; therefore the word problem really asks whether $\{w \in \Omega : w \neq 1 \text{ in } G\}$ is also an r.e. set.

Suppose a semigroup G has a presentation

$$G = (X | A_j = B_j, j \in J).$$

The elements of G corresponding to words U and V on X are equal if and only if there is a finite sequence of **elementary operations**

$$U \equiv W_1 \to W_2 \to \cdots \to W_t \equiv V$$

transforming U into V, where $W_i \to W_{i+1}$ consists of replacing one of the words PA_jQ and PB_jQ by the other (where P, Q are positive words on X).

We now associate a semigroup $\gamma(T)$ to a Turing machine T. Assume the letters in the quadruples of T are $q_0, \cdots, q_N, s_0, \cdots, s_M$, and let q, h be new letters. Define $\gamma(T)$ as the semigroup having the following presentation:

generators: $q, h, q_0, q_1, \cdots, q_N, s_0, \cdots, s_M$;

relations: $q_i s_j = q_l s_k$ if $q_i s_j s_k q_l \in T$;

for all $b = 0, 1, \cdots, M$:

$$\left. \begin{aligned} q_i s_j s_b &= s_j q_l s_b \\ q_i s_j h &= s_j q_l s_0 h \end{aligned} \right\} \quad \text{if } q_i s_j R q_l \in T,$$

$$\left. \begin{aligned} s_b q_i s_j &= q_l s_b s_j \\ h q_i s_j &= h q_l s_0 s_j \end{aligned} \right\} \quad \text{if } q_i s_j L q_l \in T,$$

$$\begin{aligned} q_0 s_b &= q_0, \\ s_b q_0 h &= q_0 h, \\ h q_0 h &= q. \end{aligned}$$

Let us denote the set of all these relations by $R(T)$.

The first five types of relations are just the obvious ones suggested by the basic moves of the Turing machine T; the new letter h enables one to distinguish basic move (ii) from basic move (iii) and also (iv) from (v). One may thus interpret h as marking ends of the tape.

Call a word h-**special** if it is of the form $h\alpha h$ for some instantaneous description α. Since h marks ends of the tape, it is quite natural to consider h-special words.

Lemma 13.3 *Let T be a Turing machine with associated semigroup $\gamma(T)$.*

(i) *If U, V are words in $\gamma(T)$, if $U \not\equiv q$ and $V \not\equiv q$, and if $U \to V$ is an elementary operation, then U is h-special if and only if V is h-special.*

(ii) *If $U \equiv h\alpha h$, if $V \not\equiv q$, and if $U \to V$ is an elementary operation of one of the first five types, then $V \equiv h\beta h$, where either $\alpha \to \beta$ or $\beta \to \alpha$ is a basic move of T.*

Proof

(i) This is true because the only relation that creates or destroys h is $hq_0h = q$.

(ii) By the first part, we know that V is h-special, say, $V \equiv h\beta h$. Now an elementary move is a substitution using an equation in a defining relation; such a relation here corresponds to a basic move of T arising from a quadruple, i.e., an elementary operation of one of the first five types. Thus, either $\alpha \to \beta$ or $\beta \to \alpha$, since there is never ambiguity about a Turing machine's next move. ∎

Let us say that q_0 is a **stopping state** for a Turing machine T if an instantaneous description α on its alphabet is terminal if and only if α involves q_0. The Turing machine T^* described in Theorem 13.1 has q_0 as a stopping state.

EXERCISE

**13.11. Let T be a Turing machine enumerating a set E. Prove there is a Turing machine T' with the same alphabet and with stopping state q_0 that also enumerates E. (HINT: This is not difficult!)

Lemma 13.4 *Let T be a Turing machine with stopping state q_0, let Ω be the set of positive words on the alphabet of T, and let E be the subset of Ω enumerated by T. For a word $w \in \Omega$, we have*

$$w \in E \quad \text{if and only if} \quad hq_1wh = q \quad \text{in } \gamma(T).$$

Proof If $w \in E$, there are instantaneous descriptions $\alpha_1, \cdots, \alpha_t$ with $q_1w \equiv \alpha_1 \to \alpha_2 \to \cdots \to \alpha_t$ basic moves of T and with α_t involving q_0. Using those relations in $\gamma(T)$ arising from these basic moves, one sees that $hq_1wh = h\alpha_t h$ in $\gamma(T)$. Now $\alpha_t \equiv Fq_0G$, where F and G are positive words on $s_0, s_1, \cdots, s_M$. The relations $q_0s_b = q_0$ give

$$h\alpha_t h \equiv hFq_0Gh = hFq_0h;$$

the relations $s_bq_0h = q_0h$ now give

$$hFq_0h = hq_0h = q,$$

as desired.

Assume that $hq_1wh = q$ in $\gamma(T)$. There are thus words $W_1, \cdots, W_n$ (on $\{h, q_0, q_1, \cdots, q_N, s_0, s_1, \cdots, s_M\}$) and elementary operations

$$hq_1wh \equiv W_1 \to W_2 \to \cdots \to W_n \equiv hq_0h \to q.$$

By part (i) of Lemma 13.3, each word W_i is h-special, so $W_i \equiv h\alpha_i h$ for some instantaneous description α_i. Let m be the first i for which α_i involves q_0. Note that $m \geq 2$ since α_1 involves q_1, not q_0. The proof is an induction on m. (The reason this proof differs from that of the first half is that equality in $\gamma(T)$ is, of course, a symmetric relation, but $\alpha \to \beta$ being a basic move does not imply $\beta \to \alpha$ is a basic move.)

By part (ii) of Lemma 13.3, for each $i = 1, \cdots, m-1$, we have either $\alpha_i \to \alpha_{i+1}$ or $\alpha_{i+1} \to \alpha_i$. Also we must have $\alpha_{m-1} \to \alpha_m$ since α_m involves q_0, and q_0 is a stopping state for T. If $m = 2$, then $q_1w \equiv \alpha_1, \to \alpha_2$ is a computation of T, so that $T(w)$ exists and $w \in E$. Assume $m > 2$. If all the arrows go to the right, i.e., if $\alpha_i \to \alpha_{i+1}$ for all i, then $q_1w \equiv \alpha_1 \to \alpha_2 \to \cdots \to \alpha_m$ is a computation of T, and so $w \in E$. Otherwise there is an i with

$$\alpha_{i-1} \leftarrow \alpha_i \to \alpha_{i+1}.$$

Since T is a Turing machine, there is never ambiguity about its next move, and so $h\alpha_{i-1}h \equiv h\alpha_{i+1}h$. We may thus eliminate W_i, thereby reducing m. The proof is now completed by induction. ∎

Theorem 13.5 (Markov-Post, 1947) *There is a finitely presented semigroup*

$$\gamma(T^*) = (q, h, q_0, \cdots, q_N, s_0, \cdots, s_M | R(T^*))$$

with unsolvable word problem. There is no decision process to determine, for an arbitrary h-special word $h\alpha h$, whether $h\alpha h = q$ in $\gamma(T^*)$.

Proof If T is a Turing machine with stopping state q_0, let A denote the alphabet of T, let Ω denote the set of positive words on A, and let $E \subset \Omega$ be the subset enumerated by T. Let $\tilde{\Omega}$ be the set of all positive words on $A \cup \{q, h, q_0, \cdots, q_N\}$, where $q_0, \cdots, q_N$ are the q-letters occurring in the quadruples of T, and let

$$\tilde{E} = \{\tilde{w} \in \tilde{\Omega} \colon \tilde{w} = q \quad \text{in } \gamma(T)\},$$

Identify Ω with the following subset Ω_1 of $\tilde{\Omega}$:

$$\Omega_1 = \{hq_1wh \colon w \in \Omega\};$$

under this identification, the subset E of Ω corresponds to

$$E_1 = \{hq_1wh \colon w \in E\}.$$

In this notation, Lemma 13.4 reads:

$$E_1 = \tilde{E} \cap \Omega_1.$$

Assume now that $T = T^*$, a Turing machine as described in Theorem 13.1. Were $\tilde{E}$ recursive, then E_1 would also be recursive, being an intersection of recursive sets (Exercise 13.9). This contradiction shows that $\gamma(T^*)$ has unsolvable word problem.

To prove the final statement, let

$$\tilde{F} = \{h\text{-special words } h\alpha h \colon h\alpha h = q \quad \text{in } \gamma(T^*)\}.$$

If $\tilde{F}$ is a recursive set, then so is $E_1 = \tilde{F} \cap \Omega_1$, a contradiction. ∎

Corollary 13.6 *There is a finitely presented semigroup*

$$\Gamma = (q, q_1, \cdots, q_N, s_1, \cdots, s_M | F_i q_{i_1} G_i = H_i q_{i_2} K_i, i \in I)$$

(where F_i, G_i, H_i, K_i are (possibly empty) positive words on $\{s_1, \cdots, s_M\}$ and $q_{i_1}, q_{i_2} \in \{q, q_1, \cdots, q_N\}$) with unsolvable word problem. There is no decision process to determine, for arbitrary positive words X and Y on $\{s_1, \cdots, s_M\}$ and for $q_j, 1 \leq j \leq N$, whether $Xq_jY = q$ in Γ.

Proof Relabel the generators of the semigroup $\gamma(T^*)$ so that they are now $q, q_1, \cdots, q_N, s_1, \cdots, s_M$ (the q's with subscripts are reindexed so they begin with q_1 instead of q_0; h is now called s_M, and the subscripts are relabeled so that the s-letters now begin with s_1 instead of s_0). Were the set of all $Xq_jY = q$ recursive, then, in the old notation, so would the set of all h-special words $h\alpha h$ such that $h\alpha h = q$ (for it would be an intersection of recursive sets). ∎

THE NOVIKOV-BOONE THEOREM

The history of the word problem for groups is rather involved. It was first considered by M. Dehn (1910) and A. Thue (1914). The solution was given by P. Novikov (1955) and, independently, by W. W. Boone (1954–1957). In 1959, Boone exhibited a much simpler group than any of those previously given, and he proved it has unsolvable word problem. In 1963, Britton proved Britton's lemma in the course of simplifying and shortening Boone's proof. Further improvements were made afterward by Boone, Collins, and Miller; it is this shorter proof we present here.

Theorem 13.7 (**Novikov-Boone, 1955**) *There exists a finitely presented group G having unsolvable word problem.*

We assure the reader that all the mathematical logic required in the proof has already appeared; we need only Corollary 13.6, a paraphrase of the Markov-Post theorem, that exhibits a particular finitely presented semigroup Γ with unsolvable word problem. Recall that Γ has a presentation

$$\Gamma = (q, q_1, \cdots, q_N, s_1, \cdots, s_M \,|\, F_i q_{i_1} G_i = H_i q_{i_2} K_i, \quad i \in I),$$

where F_i, G_i, H_i, K_i are positive words on $S = \{s_b : b = 1, \cdots, M\}$, and $q_{i_1}, q_{i_2} \in \{q, q_1, \cdots, q_N\}$.

The following notation is convenient. If $X \equiv s_1^{\varepsilon_1} \cdots s_m^{\varepsilon_m}$ is a (not necessarily positive) word on S, then $\overline{X} \equiv s_1^{-\varepsilon_1} \cdots s_m^{-\varepsilon_m}$. Note that $\overline{XY} = \overline{X}\,\overline{Y}$. If X and Y are words on S, define

$$(Xq_j Y)^* \equiv \overline{X} q_j Y,$$

where $q_j \in \{q, q_1, \cdots, q_N\}$.

Definition A word Σ is **special** if $\Sigma \equiv \overline{X} q_j Y$, where X and Y are positive words on S, and $q_j \in \{q, q_1, \cdots, q_N\}$.

We now present a group G that will be seen to have unsolvable word problem. It has generators

$$q, q_1, \cdots, q_N, s_1, \cdots, s_M, \{r_i, i \in I\}, x, t, k,$$

and relations, for all $i \in I$ and $b = 1, \cdots, M$,

$$xs_b = s_b x^2 \quad \Big]\Big]\, \Delta_1$$

$$r_i s_b = s_b x r_i x$$
$$r_i^{-1} \overline{F}_i q_{i_1} G_i r_i = \overline{H}_i q_{i_2} K_i \quad \Big]\, \Delta_2$$

$$tr_i = r_i t, \quad tx = xt \,\Big]\, \Delta_3$$
$$kr_i = r_i k, \quad kx = xk$$
$$k(q^{-1} tq) = (q^{-1} tq)k.$$

The subsets $\Delta_1 \subset \Delta_2 \subset \Delta_3$ of the relations are labeled for future reference.

The reduction to the Markov-Post theorem is accomplished by the following lemma.

Lemma 13.8 (**Boone, 1959**) *If* Σ *is a special word, then*

(1) $$k(\Sigma^{-1}t\Sigma) = (\Sigma^{-1}t\Sigma)k \quad in \quad G$$

if and only if $\Sigma^* = q$ *in* Γ.

There is an immediate proof of the Novikov-Boone theorem from Boone's lemma: If there were a decision process to determine whether two words are equal in G, then there would be a decision process to determine, for an arbitrary special word Σ, whether $\Sigma^* = q$ in Γ. But Corollary 13.6 asserts that no such decision process for Γ exists.

Proof of Sufficiency in Boone's Lemma Assume that $\Sigma^* \equiv Xq_jY = q$ in Γ, where X and Y are positive words on $S = \{s_b : b = 1, \cdots, M\}$ and $1 \le j \le N$.

First, we claim that if V is a positive word on S, then

$$r_i V = VR \quad in \quad G,$$

where R is a positive word on r_i and x. We prove this by induction on m, where $V = s_{b_1} \cdots s_{b_m}$. The claim is certainly true when $m = 0$. If $m > 0$, then $V = V_1 s_{b_m}$, where $V_1 = s_{b_1} \cdots s_{b_{m-1}}$. By induction, $r_i V = V_1 R_1 s_{b_m}$, where R_1 is a positive word on r_i and x. Using the relations $xs_b = s_b x^2$ and $r_i s_b = s_b x r_i x$, we see that $R_1 s_{b_m} = s_{b_m} R$ for some positive word R on r_i and x. (A similar argument shows that $r_i^{-1} V = VR'$ in G.)

It follows easily that if U is a positive word on S, then

$$\bar{U}r_i^{-1} = L\bar{U} \quad in \quad G,$$

where L is a word on r_i and x: if $U \equiv s_{b_1} \cdots s_{b_m}$, set $V \equiv s_{b_m} \cdots s_{b_1}$ (so that $\bar{U} = V^{-1}$); then $r_i V = VR$ (as above) and

$$\bar{U}r_i^{-1} = V^{-1}r_i^{-1} = (r_i V)^{-1} = (VR)^{-1} = R^{-1}V^{-1} = R^{-1}\bar{U}.$$

A similar argument shows that $\bar{U}r_i = L'\bar{U}$, where L' is a word on r_i, x.

Assume now that Σ is a special word with $\Sigma^* = q$ in Γ. There is a sequence of elementary operations in Γ

$$\Sigma^* \equiv W_1 \rightarrow W_2 \rightarrow \cdots \rightarrow W_n \equiv q$$

such that, for each v, one of the words W_v, W_{v+1} has the form $UF_i q_{i_1} G_i V$, the other $UH_i q_{i_2} K_i V$, where U and V are positive words on S. Now, in G we have equations

$$\bar{U}(\bar{H}_i q_{i_2} K_i)V = \bar{U}(r_i^{-1}\bar{F}_i q_{i_1} G_i r_i)V$$
$$= L'\bar{U}(\bar{F}_i q_{i_1} G_i)VR',$$

where L' and R' are words on r_i and x. (In a similar manner, one sees that $\bar{U}(\bar{F}_i q_{i_1} G_i)V = \bar{U}(r_i\bar{H}_i q_{i_2} K_i r_i^{-1})V = L''\bar{U}(\bar{H}_i q_{i_2} K_i)VR''$ in G.) Since

$W_v = W_{v+1}$ in Γ implies $W_v^* = W_{v+1}^*$ in G (by relations Δ_2), it follows that, for each v,

$$W_v^* = L_v W_{v+1}^* R_v \quad \text{in} \quad G,$$

for some words L_v and R_v on x and r_i. Setting $L = L_1 L_2 \cdots L_{n-1}$ and $R = R_{n-1} \cdots R_2 R_1$, we see that

$$W_1^* = L W_n^* R \quad \text{in} \quad G.$$

But $W_1^* = (\Sigma^*)^* = \Sigma$ and $W_n^* = q^* = q$. Therefore, if Σ is a special word with $\Sigma^* = q$ in Γ, then

$$\Sigma = LqR \quad \text{in} \quad G,$$

where L and R are words in x and various r_i.

Let us now prove equation (1). We have just seen that there are words L and R such that $\Sigma = LqR$ in G, where L and R both commute with t and k (since x and all r_i commute with t and k). Thus,

$$\begin{aligned}
k\Sigma^{-1}t\Sigma k^{-1}\Sigma^{-1}t^{-1}\Sigma &= kR^{-1}q^{-1}L^{-1}tLqRk^{-1}R^{-1}q^{-1}L^{-1}t^{-1}LqR \\
&= kR^{-1}q^{-1}tqk^{-1}q^{-1}t^{-1}qR \\
&= R^{-1}(kq^{-1}tqk^{-1}q^{-1}t^{-1}q)R \\
&= 1,
\end{aligned}$$

because the last word is a conjugate of a relator.† ∎

Proof of Necessity of Boone's Lemma The proof of necessity of Boone's lemma will be in several steps. After the proof has been completed, we shall exhibit a pictorial sketch that should allow the reader to see the proof more clearly. First of all, we show that the group G is the end result of a chain of HNN extensions. Define groups G_0, G_1, G_2, and G_3 as follows:

$G_0 = \langle x \rangle$, the free group on x;
$G_1 = (G_0; s_b, b = 1, \cdots, M | \text{relations } \Delta_1 \text{ on page 362})$

(recall that this notation means we are adjoining the displayed generators and relations to the given presentation of G_0);

$G_2 = (G_1; r_i, i \in I, q, q_1, \cdots, q_N | \text{relations } \Delta_2)$;
$G_3 = (G_2; t | \text{relations } \Delta_3)$. ∎

Lemma 13.9 *G is an HNN extension of G_3 with stable letter k; G_3 is an HNN extension of G_2 with stable letter t.*

Proof First note that G has base G_3 and stable letter k:

$$G = (G_3; k | k^{-1}r_i k = r_i, k^{-1}xk = x, k^{-1}(q^{-1}tq)k = q^{-1}tq);$$

also, G_3 has base G_2 and stable letter t:

$$G_3 = (G_2; t | t^{-1}r_i t = r_i, t^{-1}xt = x).$$

---

† Observe that the relator $kq^{-1}tqk^{-1}q^{-1}t^{-1}q$ appears only in the last step of the proof. (The words "relator" and "relation" are interchangeable.)

Each of these presentations is of the type occurring in Lemma 12.52, so that each is an HNN extension. ∎

Lemma 13.10 G_2 is an HNN extension of $G_1 * \langle q, q_1, \cdots, q_N \rangle$ with stable letters $\{r_i, i \in I\}$; G_1 is an HNN extension of G_0 with stable letters $\{s_b, b = 1, \cdots, M\}$.

Proof Note that G_2 has base $G_1 * \langle q, q_1, \cdots, q_N \rangle$ and stable letters $\{r_i : i \in I\}$:

$$G_2 = (G_1 * \langle q, q_1, \cdots, q_N \rangle; r_i, i \in I \mid r_i^{-1}(\overline{F}_i q_{i_1} G_i) r_i = \overline{H}_i q_{i_2} K_i,$$

$$r_i^{-1}(s_b x) r_i = s_b x^{-1}).$$

To verify this is an HNN extension, first observe that, for each index i, the subgroup $A_i = \langle \overline{F}_i q_{i_1} G_i, s_b x, b = 1, \cdots, M \rangle$ and the subgroup $B_i = \langle \overline{H}_i q_{i_2} K_i, s_b x^{-1}, b = 1, \cdots, M \rangle$. It is easy to check that A_i and B_i are each free on the displayed generating sets (for example, map A_i onto the free group with free basis $\{q_{i_1}, s_1, \cdots, s_M\}$ by setting x equal to 1 and using Exercise 12.71). There is thus an isomorphism $\varphi_i: A_i \to B_i$ with $\varphi_i(\overline{F}_i q_{i_1} G_i) = \overline{H}_i q_{i_2} K_i$ and $\varphi_i(s_b x) = s_b x^{-1}, b = 1, \cdots, M$.

The presentation

$$G_1 = (x, s_1, \cdots, s_M \mid s_b^{-1} x s_b = x^2, \quad b = 1, \cdots, M)$$

shows that G_1 has base $\langle x \rangle$ and stable letters $s_1, \cdots, s_M$. Since x has infinite order, $\langle x \rangle \cong \langle x^2 \rangle$ and G_1 is an HNN extension of G_0. ∎

Corollary 13.11 *The subset* $\{s_1 x, \cdots, s_M x\}$ *of* G_1 *freely generates its subgroup* $\langle s_1 x, \cdots, s_M x \rangle$.

Proof This is contained in the preceding proof. ∎

EXERCISE

**13.12. There is an automorphism ψ of G_1 sending x to x^{-1} and fixing each s_b.

Lemma 13.12 *Let* Σ *be a fixed special word satisfying the hypothesis of Boone's lemma:*

$$k(\Sigma^{-1} t \Sigma) = (\Sigma^{-1} t \Sigma) k \quad in \quad G.$$

Then there are words L_1 *and* L_2 *on* x *and various* r_i *such that*

$$L_1 \Sigma L_2 = q \quad in \quad G_2.$$

Proof By hypothesis,

$$k^{-1} \Sigma^{-1} t \Sigma k \Sigma^{-1} t^{-1} \Sigma = 1 \quad in \quad G.$$

Since G is an HNN extension of G_3 with stable letter k, we have by Britton's lemma (actually, by the simpler Lemma 12.52)

$$\Sigma^{-1} t \Sigma = C' \quad \text{in} \quad G_3,$$

where C' is a word on various r_i, x, and $q^{-1} t q$. Therefore there exist words W of the form

$$W \equiv \Sigma^{-1} t \Sigma R_0 (q^{-1} t^{e_1} q) R_1 (q^{-1} t^{e_2} q) R_2 \cdots (q^{-1} t^{e_n} q) R_n = 1 \quad \text{in} \quad G_3,$$

where the R_j are (possibly empty) words on x and various r_i, and the exponents $e_j = \pm 1$. We assume this equation is chosen so that n is minimal.

By Britton's lemma (really, by Lemma 12.52 again), W contains a pinch $t^e C t^{-e}$, where

$$C = R \quad \text{in} \quad G_2$$

for some word R on the r_i and x. Note that $C = R$ in G_3 since $G_2 \leq G_3$.

If the initial letter t^e is, in fact, the first t occurring in W, then $e = 1$ and $t^e C t^{-e} \equiv t \Sigma R_0 q^{-1} t^{e_1}$, for C does not involve t. Hence

$$\Sigma R_0 q^{-1} \equiv C = R \quad \text{in} \quad G_2$$

or, equivalently,

$$R^{-1} \Sigma R_0 = q \quad \text{in} \quad G_2,$$

which is of the desired form.

If the initial t^e in $t^e C t^{-e}$ is t^{e_j} for some $j \geq 1$, then we have

$$t^e C t^{-e} \equiv t^{e_j} q R_j q^{-1} t^{e_{j+1}},$$

for C does not involve t. Thus, $e_j = e$ and $e_{j+1} = -e$. Since t commutes with x and the r_i, we have in G_3:

$$
\begin{aligned}
q^{-1} t^{e_j} q R_j q^{-1} t^{e_{j+1}} q &\equiv q^{-1} t^e C t^{-e} q \\
&= q^{-1} t^e R t^{-e} q \\
&= q^{-1} R q \\
&= q^{-1} C q \\
&\equiv q^{-1} (q R_j q^{-1}) q = R_j.
\end{aligned}
$$

Therefore, W has a factorization in G_3 of shorter length, contradicting the choice of minimal n. ∎

Recall that $\Sigma \equiv \bar{X} q_j Y$; we have just shown

$$L_1 \bar{X} q_j Y L_2 = q \quad \text{in} \quad G_2$$

for some words L_1 and L_2 on x and various r_i. We rewrite this last equation:

$$L_1 \bar{X} q_j = q L_2^{-1} Y^{-1} \quad \text{in} \quad G_2.$$

Definition A word w on letters A is **freely reduced** if it contains no subwords of the form aa^{-1} or $a^{-1} a$, where $a \in A$.

Clearly we may assume the words L_1 and L_2 on x and various r_i are freely reduced (if there were such a subword, cancel it, and the equation still holds).

Lemma 13.13 *If L_1 and L_2 are freely reduced, then each of the words $L_1 \overline{X} q_j$ and $q L_2^{-1} Y^{-1}$ are r_i-reduced for each i.*

Proof Let us show that $L_1 \overline{X} q_j$ contains no pinch of the form $r_i^e C r_i^{-e}$ (the proof for the other word is similar). In view of the spelling of $L_1 \overline{X} q_j$, any such pinch is a subword of L_1, so that $C \equiv x^m$ for some $m \neq 0$ (since L_1 is freely reduced). We must show that x^m is not equal in $G_1 * \langle q, q_1, \cdots, q_N \rangle$ to a word on

$$\{\overline{F}_i q_{i_1} G_i, s_1 x, \cdots, s_M x\} \quad \text{or on} \quad \{\overline{H}_i q_{i_2} K_i, s_1 x^{-1}, \cdots, s_M x^{-1}\}.$$

Suppose $x^m = V$ in $G_1 * \langle q, q_1, \cdots, q_N \rangle$, where

$$V \equiv w_0 (\overline{F}_i q_{i_1} G_i)^{e_1} w_1 \cdots (\overline{F}_i q_{i_1} G_i)^{e_n} w_n$$

and each $e_j = \pm 1$, w_j is a word on $\{s_1 x, \cdots, s_M x\}$, and V is reduced as an element in a free product. Since $x^m \in G_1$, one of the free factors, the word V cannot involve q_{i_1}, hence has no occurrence of $\overline{F}_i q_{i_1} G_i$. Therefore

$$x^m = w_0 \text{ in } G_1.$$

Write $w_0 = (s_{b_1} x)^{f_1} \cdots (s_{b_k} x)^{f_k}$, where each $f_v = \pm 1$. Since $x^{-m} w_0 = 1$ in G_1, which is an HNN extension of $\langle x \rangle$, the word $x^{-m} w_0$ must contain a pinch $s_b^f x^\varepsilon s_b^{-f}$, where $\varepsilon = \pm 1$. Britton's lemma says further that if $f = +1$, then $x^\varepsilon \in \langle x^2 \rangle$, which is impossible. Thus the pinch is $s_b^{-1} x^\varepsilon s_b$. Inspection of the spelling of w_0, however, shows that it can contain no such subword. Therefore $w_0 = 1$ and $x^m = 1$. Since x has infinite order, $m = 0$, contradicting our earlier hypothesis that $m \neq 0$.

A similar argument shows that x^m is not equal in

$$G_1 * \langle q, q_1, \cdots, q_N \rangle$$

to a word on $\{\overline{H}_i q_{i_2} K_i, s_1 x^{-1}, \cdots, s_M x^{-1}\}$. ∎

We now know that $L_1 \overline{X} q_j = q L_2^{-1} Y^{-1}$ in G_2 and that each word is r_i-reduced for each i. Because no r_i-letters occur outside of L_1 or L_2, Corollary 12.54 says that the number T of various r_i-letters in L_1 is the same as in L_2.

The following lemma completes the proof of Boone's lemma and, with it, the Novikov-Boone theorem. In view of a further application in the next section, we prove slightly more than we need now.

Lemma 13.14 *Let L_1 and L_2 be words on various r_i and x that are r_i-reduced for each r_i. If X and Y are freely reduced words on $S = \{s_1, \cdots, s_M\}$ and if*

$$L_1 \overline{X} q_j Y L_2 = q \quad \text{in} \quad G_2,$$

then both X and Y are positive words and

$$X q_j Y \equiv (\overline{X} q_j Y)^* = q \quad \text{in} \quad \Gamma.$$

REMARK   The hypothesis that X and Y be freely reduced words on S is automatically true in our case, for we have X and Y positive at the outset.

Proof   The proof is by induction on the total number T of r_i-letters occurring in L_1 (which is the same as for L_2, by our remarks above). If $T = 0$, the equation $L_1 \overline{X} q_j Y L_2 = q$ in G_2 is just

$$x^m \overline{X} q_j Y x^n = q \quad \text{in} \quad G_2.$$

Since no r_i appears and since $G_1 * \langle q, q_1, \cdots, q_N \rangle \leq G_2$, we have

$$x^m \overline{X} q_j Y x^n = q \quad \text{in} \quad G_1 * \langle q, q_1, \cdots, q_N \rangle.$$

Such an equation can hold in a free product only if $q_j = q$ and $x^m \overline{X} = 1 = Y x^n$ in G_1. A glance at the presentation of G_1 shows $m = n = 0$ and X and Y are empty. Thus, we surely have X and Y positive and $X q_j Y = q$ in Γ.

Assume now that $T > 0$. Lemma 13.13 allows us to apply Corollary 12.54; we may write

$$L_1 \overline{X} q_j Y L_2 \equiv L_3 [r_i^e x^m \overline{X} q_j Y x^n r_i^{-e}] L_4 = q \quad \text{in} \quad G_2,$$

where the word in brackets is a pinch. By Britton's lemma, the element $x^m \overline{X} q_j Y x^n$ lies in the subgroup

$$A_i = \langle \overline{F}_i q_{i_1} G_i, s_1 x, \cdots, s_M x \rangle \quad \text{if } e = -1,$$

or in the subgroup

$$B_i = \langle \overline{H}_i q_{i_2} K_i, s_1 x^{-1}, \cdots, s_M x^{-1} \rangle \quad \text{if } e = +1.$$

In the first case, $j = i_1$, i.e.,

(2) $$q_j = q_{i_1};$$

in the second case, $j = i_2$, i.e., $q_j = q_{i_2}$ (because, for example, $x^m \overline{X} q_j Y x^n \in A_i$ is valid in the free product $G_1 * \langle q, q_1, \cdots, q_N \rangle$). Let us focus on the case $e = -1$, the case $e = +1$ being similar. There is a word W arising from $x^m \overline{X} q_j Y x^n \in A_i$:

$$W \equiv x^m \overline{X} q_j Y x^n u_0 (\overline{F}_i q_{i_1} G_i)^{\alpha_1} u_1 \cdots (\overline{F}_i q_{i_1} G_i)^{\alpha_t} u_t$$
$$= 1 \quad \text{in} \quad G_1 * \langle q, q_1, \cdots, q_N \rangle,$$

where $\alpha_j = \pm 1$ and the u_j are (possibly empty) words on

$$\{ s_1 x, \cdots, s_M x \}.$$

Assume that t is minimal and each u_j is reduced as a word on the free generators $\{ s_1 x, \cdots, s_M x \}$ (Corollary 13.11). Since $G_1 * \langle q, q_1, \cdots, q_N \rangle$ is an HNN extension of G_1 and W involves a stable letter (one of the q's), Britton's lemma asserts that W contains a pinch. If a pinch involves the first occurrence of q_j, then $\alpha_1 = -1$ and $Y x^n u_0 G_i^{-1} = 1$ in G_1. We claim that a pinch cannot occur elsewhere in W. Suppose a pinch occurs as a subword of $(\overline{F}_i q_{i_1} G_i)^{\alpha_j} u_j (\overline{F}_i q_{i_1} G_i)^{\alpha_{j+1}}$. If $\alpha_j = +1$, then $\alpha_{j+1} = -1$ and $G_i u_j G_i^{-1} = 1$ in G_1; if $\alpha_j = -1$, then $\alpha_{j+1} = +1$ and $\overline{F}_i^{-1} u_j \overline{F}_i = 1$ in G_1. In either case, we have $u_j = 1$ in G_1. But u_j is a reduced word on the free generators $\{s_1 x, \cdots, s_M x\}$; therefore $u_j \equiv 1$, which violates the minimality of t. We conclude that $t = 1$, $\alpha_1 = -1$ and

$$W \equiv x^m \overline{X} q_j Y x^n u_0 G_i^{-1} q_{i_1}^{-1} \overline{F}_i^{-1} u_1 = 1 \quad \text{in} \quad G_1 * \langle q, q_1, \cdots, q_N \rangle.$$

We have already seen that

$$Y x^n u_0 G_i^{-1} = 1 \quad \text{in} \quad G_1.$$

It follows from W being a word in a free product that

$$\overline{F}_i^{-1} u_1 x^m \overline{X} = 1 \quad \text{in} \quad G_1.$$

We recast these equations into more convenient form (by conjugating):

$$x^n u_0 G_i^{-1} Y = 1 \quad \text{in} \quad G_1$$

and

$$\overline{X} \overline{F}_i^{-1} u_1 x^m = 1 \quad \text{in} \quad G_1.$$

Let us show that, after canceling all subwords (if any) of the form $s_b s_b^{-1}$ or $s_b^{-1} s_b$, the first surviving letter in $G_i^{-1} Y$ is positive, i.e., the whole of G_i^{-1} disappears. Otherwise, the word $G_i^{-1} Y$ begins with s_b^{-1} for some b (recall that G_i is a positive word on $S = \{s_1, \cdots, s_M\}$). Since

$$x^n u_0 G_i^{-1} Y = 1 \quad \text{in} \quad G_1$$

and this word involves s_b, Britton's lemma provides a pinch $s_b^e C s_b^{-e} \equiv s_b^e x^h s_b^{-e}$ (for G_1 is an HNN extension of $\langle x \rangle$ with stable letters S). Since u_0 is a reduced word on $\{s_1 x, \cdots, s_M x\}$, the pinch $s_b^e x^h s_b^{-e}$ cannot be a subword of $x^n u_0$. It follows that the last letter of the pinch $s_b^e x^h s_b^{-e}$ is the first surviving letter of $G_i^{-1} Y$ (for G_i and Y are words on S and do not involve x). If this letter is not positive, as we are assuming, then $e = 1$ and $s_b^e x^h s_b^{-e} \equiv s_b x s_b^{-1}$ (h must be 1). But G_1 has the presentation

$$G_1 = (x, s_1, \cdots, s_M | s_b^{-1} x s_b = x^2, b = 1, \cdots, M).$$

The case $e = 1$ of Britton's lemma thus gives $x \in \langle x^2 \rangle$, and this is a contradiction. It follows that there is a subword Y_1 of Y beginning with a positive letter such that $Y \equiv G_i Y_1$.

In a similar manner, one sees that after cancellation of all subwords of the form $s_b s_b^{-1}$ or $s_b^{-1} s_b$ (if any) in $\overline{X} \overline{F}_i^{-1}$, the last surviving letter is "negative" (in particular, the whole of $\overline{F}_i^{-1}$ disappears). The proof is just as above, inverting the original equation $\overline{X} \overline{F}_i^{-1} u_1 x^m = 1$ in G_1. It follows that there is a subword X_1 of X with $\overline{X}_1$ ending with a negative letter and such that $X \equiv X_1 F_i$.

We have proved that

$$u_0^{-1} = Y_1 x^n \quad \text{in} \quad G_1.$$

Define

$$v_0^{-1} = r_i^{-1} u_0^{-1} r_i.$$

Since $r_i^{-1}s_b x r_i = s_b x^{-1}$, all b, the element v_0^{-1} is a word in $s_1 x^{-1}, \cdots, s_M x^{-1}$. But we may also regard u_0^{-1} and v_0^{-1} as elements of $G_1 = \langle x, s_1, \cdots, s_M \rangle$. By Exercise 13.12, there is an automorphism ψ of G_1 taking x to x^{-1} and fixing all s_b; thus, $\psi(u_0^{-1}) = v_0^{-1}$. It follows that

$$v_0^{-1} = Y_1 x^{-n} \quad \text{in} \quad G_1.$$

A similar argument gives

$$v_1^{-1} = x^{-m} \overline{X}_1 \quad \text{in} \quad G_1,$$

where X_1 is the subword of X defined above and v_1 satisfies the equation

$$v_1^{-1} = r_i^{-1} u_1^{-1} r_i.$$

Let us return to the induction. We have in G_2,

$$
\begin{aligned}
q &= L_1 \overline{X} q_j Y L_2 \equiv L_3 r_i^{-1}(x^m \overline{X}) q_j (Y x^n) r_i L_4 \\
&= L_3 r_i^{-1}(u_1^{-1} \overline{F}_i) q_j (G_i u_0^{-1}) r_i L_4 \\
&= L_3 v_1^{-1} r_i^{-1}(\overline{F}_i q_j G_i) r_i v_0^{-1} L_4 \\
&= L_3 v_1^{-1}(r_i^{-1} \overline{F}_i q_{i_1} G_i r_i) v_0^{-1} L_4 \\
&= (L_3 v_1^{-1}) \overline{H}_i q_{i_2} K_i (v_0^{-1} L_4) \\
&= (L_3 x^{-m} \overline{X}_1) \overline{H}_i q_{i_2} K_i (Y_1 x^{-n} L_4).
\end{aligned}
$$

We have verified that

$$L_3 x^{-m}(\overline{X}_1 \overline{H}_i q_{i_2} K_i Y_1) x^{-n} L_4 = q \quad \text{in} \quad G_2.$$

Now $L_3 x^{-m}$ and $x^{-n} L_4$ are words on x and various r_i having at most $T-1$ occurrences of various r_i. To apply the inductive hypothesis, we must check that $\overline{X}_1 \overline{H}_i$ and $K_i Y_1$ are freely reduced, i.e., contain no subwords of the form $s_b s_b^{-1}$ or $s_b^{-1} s_b$. Now K_i is a positive word on S, hence contains no "forbidden" subwords; further, $Y_1 \equiv G_i^{-1} Y$ is just a subword of Y (since G_i^{-1} disappears), and hence has no forbidden subwords, by hypothesis. The only place where a forbidden subword can occur in $K_i Y_1$ is at the interface; this is impossible, for we have seen that Y_1 begins with a positive letter. A similar argument shows that $\overline{X}_1 \overline{H}_i$ is freely reduced.

By induction, both $X_1 H_i$ and $K_i Y_1$ are positive; hence their subwords X_1 and Y_1 are also positive. Therefore $X \equiv X_1 F_i$ and $Y \equiv G_i Y_1$ are also positive. Induction also gives

$$(\overline{X}_1 \overline{H}_i q_{i_2} K_i Y_1)^* = q \quad \text{in} \quad \Gamma.$$

Therefore, since $\overline{\overline{X}_1 \overline{H}_i} = X_1 H_i$, we have

(3) $$X_1 H_i q_{i_2} K_i Y_1 = q \quad \text{in} \quad \Gamma$$

(it is here we see why the bar operation is used; had we tried inversion instead, we would now only have $H_i X_1 q_{i_2} K_i Y_1 = q$ in Γ). Thus, using (2),

$$
\begin{aligned}
X q_j Y &\equiv X_1 F_i q_j G_i Y_1 \\
&\equiv X_1 F_i q_{i_1} G_i Y_1 = X_1 H_i q_{i_2} K_i Y_1 \quad \text{in} \quad \Gamma.
\end{aligned}
$$

Combining this with equation (3) gives

$$Xq_jY = q \quad \text{in} \quad \Gamma,$$

as desired. ∎

There are other proofs of the unsolvability of the word problem for groups, one due to P. Novikov, another due to G. Higman. Higman's proof is a corollary of his imbedding theorem, which we prove in the next section. The proof we shall give uses our development so far, whereas Higman's original proof of his imbedding theorem does not depend on the Novikov-Boone theorem. Before we review the proof of the Novikov-Boone theorem just given, we must mention a theorem of W. Magnus (1932) proving that every finitely generated group having a single defining relation does have solvable word problem.

There are other group-theoretic questions yielding unsolvable problems (see Theorem 13.31); let us consider one more such question now.

Definition A finitely generated group $G = (X|R)$ has **solvable conjugacy problem** if there is a decision process which determines of an arbitrary pair of words W_1 and W_2 on X whether or not W_1 and W_2 are conjugate as elements of G.

When G is finitely presented one may show that its having solvable conjugacy problem does not depend on the choice of finite presentation. A group with solvable conjugacy problem must have solvable word problem: one can decide whether or not an arbitrary word W is conjugate to 1. The converse is false as we illustrate below.

Let G° denote the group having the same presentation as the group G in the Novikov-Boone theorem except that the relator

$$\rho = kq^{-1}tqk^{-1}q^{-1}t^{-1}q$$

is missing. For a special word Σ, let $w(\Sigma)$ denote $k\Sigma^{-1}t\Sigma k^{-1}\Sigma^{-1}t^{-1}\Sigma$.

Corollary 13.15

(i) *The following statements are equivalent*:

    (a) $w(\Sigma) = 1$ *in* G;
    (b) $\Sigma^* = q$ *in* Γ;
    (c) $w(\Sigma)$ *is conjugate to* ρ *in* G°.

(ii) G° *has unsolvable conjugacy problem. Indeed,* G° *has unsolvable conjugacy problem if and only if* G *has unsolvable word problem.*

REMARK   Boone and A. A. Fridman, independently, proved G° has solvable word problem.

Proof

(i) (a) $\Rightarrow$ (b). This is the necessity of Boone's lemma.
    (b) $\Rightarrow$ (c). As we remarked earlier, the proof of Lemma 13.8 (the sufficiency of Boone's lemma) shows $w(\Sigma)$ is conjugate to ρ in G°,

not merely in G, because the relator ρ appears only in the last step of the proof.

(c) $\Rightarrow$ (a). This is obvious because $\rho = 1$ in G.

(ii) The statement follows immediately from the equivalence of (a) and (c) above. ∎

There is a fruitful geometric method that uses diagrams in the plane to study presentations of groups. E. Rips constructed such diagrams to describe the proof of the Novikov-Boone theorem (as well as that of the Higman theorem in the next section), and he kindly permitted me to give his description here. Our aim is to present enough details so the reader, using pictures, may be able to organize, and hence better understand, the proof just given. Consequently, our remarks shall be informal [the reader may consult Chapter V of the book of Lyndon and Schupp for precise definitions and applications of diagrams (our terminology does not coincide with theirs)].

When we speak of a **polygon** in the plane, we mean the usual two-dimensional geometric figure including its interior; of course, its **boundary** (or perimeter) consists of only finitely many edges and vertices. A **directed polygon** is a polygon each of whose edges is given a direction, indicated by an arrow. A **diagram** is either a directed polygon or a directed polygon whose interior is subdivided into finitely many directed polygons, called **regions**; we insist that any pair of edges that intersect do so in a vertex. Finally, given a presentation $(X \mid R)$ of a group G, a **labeled diagram** is a diagram each of whose (directed) edges is labeled by a generator in X. In particular, one may speak of labeled directed polygons.

Given a presentation $(X \mid R)$, we are going to construct a labeled directed polygon for every word $W = x_1^{e_1} \cdots x_n^{e_n}$, where $x_1, \cdots, x_n$ are (not necessarily distinct) generators in X and each $e_i = \pm 1$. For a technical reason mentioned below, we must restrict the words W. A **cyclic permutation** of a word $W = x_1^{e_1} \cdots x_n^{e_n}$ (notation as above) is a word of the form $x_i^{e_i} \cdots x_n^{e_n} x_1^{e_1} \cdots x_{i-1}^{e_{i-1}}$; note that a cyclic permutation is a conjugate of W, namely, $(x_i^{e_i} \cdots x_n^{e_n}) W (x_i^{e_i} \cdots x_n^{e_n})^{-1}$. A word W (as above) is **cyclically reduced** if every cyclic permutation of W is freely reduced. For a cyclically reduced word $W = x_1^{e_1} \cdots x_n^{e_n}$, construct a labeled directed polygon as follows: draw an n-gon in the plane; choose an edge and label it x_1; label successive edges $x_2, x_3, \cdots$ as one proceeds counterclockwise around the boundary; finally, direct each edge with an arrow according to the sign of e_i (we agree counterclockwise is the positive direction in the plane). For example, assume q, t, $k \in X$ and $W \equiv kq^{-1}tqk^{-1}q^{-1}t^{-1}q$. The labeled directed polygon for W is the octagon:

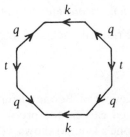

(the first edge is the k-edge on top). Were a word W not cyclically reduced, this construction would yield a polygon having two adjacent edges with the same label and which point in opposite directions; for technical reasons, one avoids such polygons. There is no loss in generality in assuming each relator of a presentation $(X\,|\,R)$ is cyclically reduced: every word has some cyclically reduced conjugate and one may harmlessly replace a relator by any of its conjugates. With this understanding, each relator $r \in R$ yields a labeled directed polygon we call its **relator polygon**. One may now draw a picture of a presentation $(X\,|\,R)$ of a group G by listing the generators X and by displaying the relator polygon of each relator in R. These polygons are easier to grasp (especially when viewing several of them in some diagram) when distinct generators are given distinct colors. The presentation of the group G given in the proof of the Novikov-Boone theorem is pictured as Figure 1 on the inside front cover. There are six types of generators: q; s; r; x; t; k; each has been given a different color. One of the relator polygons is the octagon drawn above. Observe that if G is an HNN extension of a group H with stable letters $p_1, \cdots, p_m$, then a relator involving a p-letter p_i involves exactly two p-letters: p_i and p_i^{-1}. If the corresponding relator polygon is drawn with the edges with label p_i parallel, then both arrows point in the same direction. As a second example, the presentation of the group G_7 defined in the next section is pictured as Figure 3 on the inside back cover.

Assume D is a labeled diagram. Starting at some edge on the boundary of D, we may obtain a word W as we read the edge labels (and see the edge arrows) while making a complete (counterclockwise) tour of D's boundary. Such a word W is called a **boundary word** of D (of course, another choice of starting edge yields a conjugate of W, indeed, a cyclic permutation of W).

Let $(X\,|\,R)$ be a presentation of a group G and let W be a cyclically reduced word on X. The fundamental theorem here is that $W = 1$ in G if and only if there is a labeled diagram whose regions are relator polygons of relators in R and which has W as a boundary word.† An immediate consequence of this theorem is a conjugacy criterion. Assume W and W' are cyclically reduced words on the given generators and consider the annulus with outer boundary word W' and inner boundary word W.

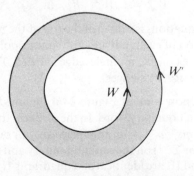

---

† This is not quite accurate; one must assume the set R of relators is **symmetrized**: (i) if $r \in R$, then $r^{-1} \in R$; (ii) each $r \in R$ is cyclically reduced; (iii) every cyclic permutation of $r \in R$ is again in R. It is easy to see every finite set R of relators on X determines a finite symmetrized set R^* of words on X such that $(X\,|\,R)$ and $(X\,|\,R^*)$ present isomorphic groups.

Then W and W' are conjugate in G if and only if the interior of the annulus can be subdivided into relator polygons. The proof consists in finding a path β from W' to W and cutting along β to form a diagram

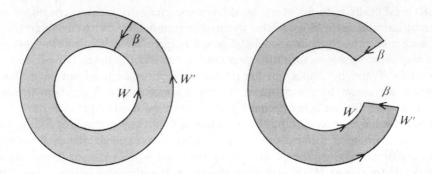

A boundary word of the new diagram is $W'\beta W^{-1}\beta^{-1}$, and the fundamental theorem says this word equals 1 in G. Conversely, given $W'\beta W^{-1}\beta^{-1} = 1$ in G, one may form the desired annulus by identifying the edges labeled β, i.e., start with the diagram on the above right and glue the β's together to obtain the annulus on the left.

An example will reveal how these diagrams illustrate the various steps taken in rewriting a word using the relators of a given presentation. The sufficiency of Boone's Lemma 13.8 requires one to prove, for a special word Σ, that $w(\Sigma) = 1$ in G, where

$$w(\Sigma) \equiv k\Sigma^{-1}t\Sigma k^{-1}\Sigma^{-1}t^{-1}\Sigma.$$

The hypothesis provides a sequence of elementary operations

$$\Sigma^* \equiv W_1 \to W_2 \to \cdots \to W_n \equiv q$$

in the semigroup Γ. The proof begins by showing each W_v^* has the form $W_v^* \equiv \bar{U}_v q_{i_v} V_v$, where $1 \leq v < n$ and U_v, V_v are positive s-words; moreover, there are words L_v and R_v on x and r_i's such that

$$W_v^* = L_v W_{v+1}^* R_v \quad \text{in} \quad G.$$

The diagram for these equations is drawn at the top of the next page, though we have not drawn the subdivision of each polygon into relator polygons (we have taken the liberty of drawing, for example, a single edge labeled Y instead of drawing and labeling each of its constituent s-edges).

The reader should now look at Figure 2 on the inside front cover; one sees a diagram having $w(\Sigma)$ as a boundary word. In the center is the octagon corresponding to the relator $\rho = kq^{-1}tqk^{-1}q^{-1}t^{-1}q$, and there are four quadrants, as drawn on the next page, involving Σ (or Σ^{-1}) on the outer boundary and q (or q^{-1}) on the octagon (actually, quadrants I and III are identical and quadrants II and IV are mirror images of quadrant I). The commutativity of k with x and with each r_i allows one to insert "border" sequences of squares connecting k-edges on the outer boundary to k-edges on the octagon. Similarly, one may connect the t-edges by squares. Since every region in the diagram is a relator polygon, we have shown $w(\Sigma) = 1$ in G.

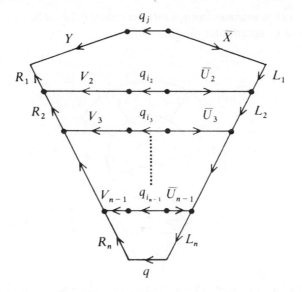

We remark we have really shown a bit more. Recall that G° is the group having the same presentation as G except that the octagonal relator is missing. The proof shows $w(\Sigma)$ is conjugate in G° to ρ (Figure 2 may be regarded as an annulus).

Let us now consider the more difficult necessity of Boone's lemma. We adopt the point of view suggested by our remark above and Corollary 13.15(i): Assuming $w(\Sigma)$ is conjugate in G° to ρ, can one prove $\Sigma^* = q$ in the semigroup Γ? In geometric terms, the hypothesis gives an annulus with $w(\Sigma)$ an outer boundary word, with ρ an inner boundary word, and whose regions are relator polygons of G°, i.e., no such region is the octagon of ρ. The problem is to modify the given annulus so it looks like Figure 2; one can then use one of the quadrants to show $\Sigma^* = q$ in Γ. Consider the given diagram showing $w(\Sigma)$ is conjugate to ρ in G°. Because the only relator polygons of G° involving k are the squares corresponding to $r_i k = k r_i$ and $xk = kx$, one may trace a sequence of adjacent squares connecting different k-edges in the diagram. Now Lemma 13.12 shows there are two such sequences connecting each outer k-edge to each inner k-edge (in particular, the two outer k-edges are not so connected to each other nor are the two inner k-edges so connected to each other). The geometry of the plane prevents any such sequence from intersecting itself:

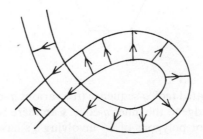

Similarly, two such sequences cannot intersect each other. The only other possible k-edge can occur in an annular "loop":

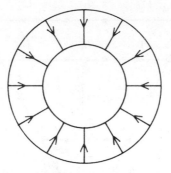

Such an annulus has the same outer and inner boundary words, and hence it may be removed, leaving a diagram as in the hypothesis. A similar argument can be made about sequences connecting t-edges. We conclude one may assume the original annular diagram is divided into four quadrants (two involving Σ and two involving Σ^{-1}) by these k-sequences and t-sequences; moreover, there are no other k-edges or t-edges. The statement of Lemma 13.12 is that the sides L_1 and L_2 of a quadrant involving Σ have only r_i-edges and x-edges:

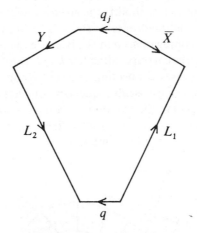

Next, there must be a "dotted red" q-sequence (for various q_i-edges) in this quadrant connecting the outer q_j-edge to the inner q-edge. Can there be any other q-edges in this quadrant? The relator polygons of G° involving q's have the form:

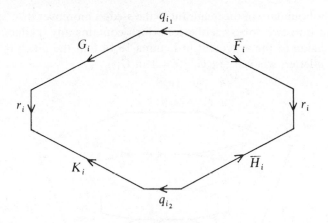

The only other possibility for an occurrence of q-edges is a flower whose eight-sided petals arise from such a relator or its inverse. We have not drawn the relator polygons that subdivide the eye of the flower, but we may assume the eye contains no relator regions having q-edges (otherwise, the eye contains a smaller such flower and we examine it).

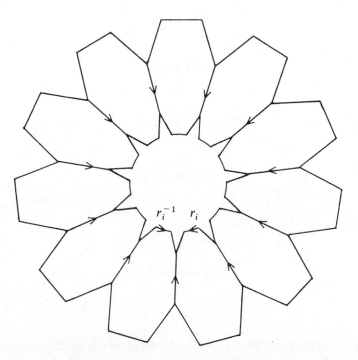

The boundary word of the flower's eye involves r_i's and s_b's and this word is 1 in G_1. By Britton's lemma, this word contains a pinch $r_i^e C r_i^{-e}$, i.e., there are two adjacent petals whose r_i-edges point in opposite directions, and this contradicts the orientations of these petals. We conclude that there is a q-sequence connecting the

q_j-edge on the boundary of the quadrant to the q-edge; moreover this "spine" divides the quadrant into two lobes, neither of which contains any q-edges. (This is the geometric version of the argument in Lemma 13.14 that the pinch in the word W involves no q-letter, whence $x^n u_0 G_i^{-1} Y = 1$ in G_1.)

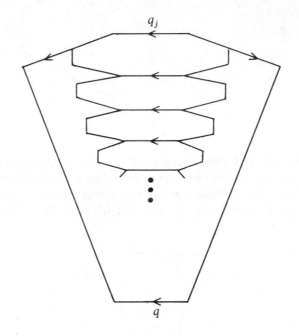

Let us now focus on the top portion of the quadrant:

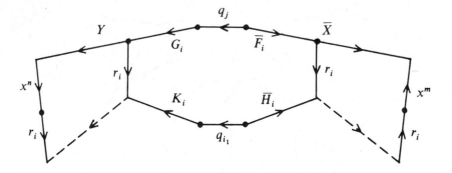

The remainder of the proof shows that the dotted paths can be drawn, each of whose edges is an s-edge ("the first surviving letter . . . is positive"). Actually, the proof shows the rightward path is $\overline{X}_1$ (followed by x^{-m}, which is incorporated into L_3) and the leftward path is Y_1 (followed by x^n, which is incorporated into L_4). One now repeats this procedure down the spine, and the resulting diagram is essentially Figure 2.

The preceding description should aid the reader's understanding of the proof of the Novikov-Boone theorem. If one wishes, one can draw diagrams corresponding to other steps of the proof: indeed, it is probable one could ultimately give a geometric proof of the Novikov-Boone theorem.

AN IMBEDDING THEOREM OF G. HIGMAN

When can a finitely generated group be imbedded in a finitely presented group? The answer to this purely group-theoretic question reveals a harmonic interplay of group theory with mathematical logic. The proof we present here is due to S. Aanderaa (1970).

We need an elementary technical observation before we can state Higman's theorem (compare the first paragraph on page 358).

Lemma 13.16 *Every finitely generated group G has a presentation*

$$G = (a_1, \cdots, a_m | R_i = 1, \quad i \in I)$$

in which each R_i is a positive word on $a_1, \cdots, a_m$.

Proof If $(b_1, \cdots, b_n | d_j = 1, j \in J)$ is a presentation of G, then a new presentation is

$$G = (b_1, \cdots, b_n, c_1, \cdots, c_n | D_j = 1, \quad j \in J, \quad b_k c_k = 1, \quad 1 \leq k \leq n),$$

where D_j is obtained from d_j by replacing each occurrence of b_k^{-1} by c_k. ∎

Definition A group R is **recursively presented** if it has a presentation

$$R = (u_1, \cdots, u_m | w = 1, w \in E),$$

where each w is a positive word on $u_1, \cdots, u_m$ and E is an r.e. set.

We insist that the defining relations be positive words, for, according to our exposition, only sets of positive words are allowed to be r.e. sets. Lemma 13.16 shows that this is really no restriction.

EXERCISES

13.13. If a finitely generated group G has a presentation whose relations form an r.e. set of positive words, then G has a presentation whose relations form a recursive set of positive words. [HINT: Assume the given presentation is

$$G = (u_1, \cdots, u_m | w_i = 1, \quad i = 1, 2, \cdots);$$

define a new presentation

$$G = (u_1, \cdots, u_m, y | y = 1, \quad y^i w_i = 1, \quad i = 1, 2, \cdots).]$$

*13.14. Every finitely generated subgroup of a finitely presented group is recursively presented. (HINT: Consider all words that can be obtained from 1 by a finite number of elementary operations.)

13.15. Show that a recursively presented group G can be imbedded in a group with two generators which is also recursively presented. (HINT: Corollary 12.51.)

Theorem 13.17 (G. Higman, 1961) *Every recursively presented group R can be imbedded in a finitely presented group.*

(A characterization of finitely generated subgroups of finitely presented groups is thus provided by Higman's theorem and Exercise 13.14.)

Proof Assume R has a presentation

$$R = (u_1, \cdots, u_m | w = 1, \quad w \in E),$$

where E is an r.e. set of positive words on $u_1, \cdots, u_m$. There is thus a Turing machine T enumerating E whose alphabet $\{s_0, \cdots, s_M\}$ contains $\{u_1, \cdots, u_m\}$; moreover, Exercise 13.11 allows us to assume T has stopping state q_0. By Lemma 13.4, there is a finitely presented semigroup $\gamma = \gamma(T)$ for which

$$w \in E \quad \text{if and only if} \quad hq_1wh = q \quad \text{in} \quad \gamma.$$

From the semigroup γ, we constructed a semigroup Γ (essentially by relabeling the letter h so that it became s_M). From Γ, we constructed a group G with generators

$$q, q_1, \cdots, q_N, s_1, \cdots, s_M, r_i, x, t, k$$

and the relations Δ_2 of the last section, as well as

$$tr_i = r_it, \qquad tx = xt$$
$$kr_i = r_ik, \qquad kx = xk$$
$$k(q^{-1}\,tq) = (q^{-1}\,tq)k.$$

It is now more convenient to return to the original notation, i.e., from Γ back to γ. Let us first see what we need.

Boone's Lemma 13.8 says that if Σ is a special word, then

$$\Sigma^* = q \quad \text{in } \Gamma \text{ if and only if} \quad k(\Sigma^{-1}t\Sigma) = (\Sigma^{-1}t\Sigma)k \quad \text{in} \quad G.$$

In particular, $\Sigma \equiv h^{-1}q_1wh$ is a special word. Therefore the following are equivalent for a positive word w on $s_0, \cdots, s_M$:

$$w \in E;$$
$$\Sigma^* \equiv hq_1wh = q \quad \text{in} \quad \Gamma;$$
$$k(\Sigma^{-1}t\Sigma) = (\Sigma^{-1}t\Sigma)k \quad \text{in} \quad G;$$

(4) $\qquad k(h^{-1}w^{-1}q_1^{-1}hth^{-1}q_1wh) = (h^{-1}w^{-1}q_1^{-1}hth^{-1}q_1wh)k \quad \text{in} \quad G.$

To simplify equation (4), we introduce a new presentation G' of G using the defining equations

$$k_0 = hkh^{-1} \quad \text{and} \quad t_0 = q_1^{-1}hth^{-1}q_1.$$

Recall that G_2 has generators (in the notation of γ)

$$q, h, q_0, \cdots, q_N, s_0, \cdots, s_M, r_i, x$$

and relations Δ_2:

$$s_b^{-1}xs_b = x^2, \qquad\qquad h^{-1}xh = x^2,$$
$$r_i^{-1}s_bxr_i = xs_b^{-1}, \qquad\quad r_i^{-1}hxr_i = xh^{-1},$$
$$r_i^{-1}\overline{F}_iq_{i_1}G_ir_i = \overline{H}_iq_{i_2}K_i.$$

Define

$$G_3' = (G_2; t_0 | t_0^{-1}(q_1^{-1}hr_ih^{-1}q_1)t_0 = q_1^{-1}hr_ih^{-1}q_1,$$
$$t_0^{-1}(q_1^{-1}hxh^{-1}q_1)t_0 = q_1^{-1}hxh^{-1}q_1).$$

Note that G_3' is another presentation of the group G_3; this is quickly seen by replacing t_0 by its definition. Similarly, we define another presentation of G:

$$G' = (G_3'; k_0 | k_0^{-1}(hr_ih^{-1})k_0 = hr_ih^{-1}, k_0^{-1}(hxh^{-1})k_0 = hxh^{-1},$$
$$k_0^{-1}(hq^{-1}h^{-1}q_1t_0q_1^{-1}hqh^{-1})k_0 = hq^{-1}h^{-1}q_1t_0q_1^{-1}hqh^{-1}).$$

Lemma 13.18 G_3' *is an HNN extension of* G_2 *with stable letter* t_0; G' *is an HNN extension of* G_3' *with stable letter* k_0.

Proof Exactly as the proof of Lemma 13.9. ∎

Lemma 13.19 *Let w be a positive word on* $s_0, \cdots, s_M$. *Then*

$$w \in E \quad \text{if and only if} \quad k_0(w^{-1}t_0w) = (w^{-1}t_0w)k_0 \quad \text{in} \quad G'.$$

Proof Equation (4) above has this simpler form when we change t to t_0 and k to k_0. ∎

Define a new group G_4 as the free product

$$G_4 = G' * R.$$

The generators $u_1, \cdots, u_m$ of R also appear as generators of G', for $\{u_1, \cdots, u_m\} \subset \{s_0, \cdots, s_M\}$. To consider the presentation of G_4 given by the union of the generators and relations of G' and of R, we force the presentations to be disjoint by introducing new letters $\{a_1, \cdots, a_m\}$ for the subset $\{u_1, \cdots, u_m\}$ as it occurs in G'. As a consequence, the set E now consists of certain positive words on $\{a_1, \cdots, a_m\} \subset \{s_1, \cdots, s_M\}$.

Define groups G_5, G_6, and G_7 as follows:

$$G_5 = (G_4; b_1, \cdots, b_m | b_i^{-1}u_jb_i = u_j, b_i^{-1}a_jb_i = a_j,$$
$$b_i^{-1}k_0b_i = k_0u_i^{-1}, \quad \text{all} \quad i, j = 1, \cdots, m)$$

$$G_6 = (G_5; d\,|\,d^{-1}k_0 d = k_0, \quad d^{-1}a_i b_i d = a_i, \quad i = 1, \cdots, m)$$
$$G_7 = (G_6; \sigma\,|\,\sigma^{-1}t_0\sigma = t_0 d, \quad \sigma^{-1}k_0\sigma = k_0,$$
$$\sigma^{-1}a_i\sigma = a_i, \quad i = 1, \cdots, m).$$

We prove Higman's theorem by showing first that each of these groups is an HNN extension of its predecessor. Theorem 12.48 then will give

$$R \subset G_4 \le G_5 \le G_6 \le G_7.$$

The proof is completed by showing G_7 has a finite presentation. After the proof has been given, we shall draw diagrams that show how the presentation of G_7 could have been discovered.

Lemma 13.20 *The subgroups* $\langle a_1, \cdots, a_m, k_0 \rangle_4$ *and* $\langle a_1, \cdots, a_m, t_0 \rangle_4$ *of* G_4 *are free on the displayed generating sets.*

REMARK   The subscript 4 indicates that the subgroups are computed in G_4, i.e., only the relations of G_4 are used.

Proof   Recall our analysis of the group G'; it is the end result of a chain of HNN extensions

$$G_1 * \langle q, q_0, \cdots, q_N \rangle \le G_2 \le G_3' \le G',$$

where G_1 has base $\langle x \rangle$ and stable letters $h, s_0, \cdots, s_M$. Exercise 12.72 gives $\langle h, s_0, \cdots, s_M \rangle_1$ free on these generators, so that Theorem 12.48 gives $\langle h, s_0, \cdots, s_M \rangle_4$ free on these generators. Since $\{a_1, \cdots, a_m\} \subset \{s_0, \cdots, s_M\}$, we have $\langle a_1, \cdots, a_m \rangle_4$ free on $\{a_1, \cdots, a_m\}$.

Let us now prove $\langle a_1, \cdots, a_m, k_0 \rangle_4$ is free on $\{a_1, \cdots, a_m, k_0\}$; a similar argument works for $\langle a_1, \cdots, a_m, t_0 \rangle_4$. Suppose

$$W \equiv C_0 k_0^{e_1} C_1 k_0^{e_2} \cdots C_{n-1} k_0^{e_n} = 1 \quad \text{in} \quad G_4,$$

where $e_v = \pm 1$ and C_v are (possibly empty) reduced words on $a_1, \cdots, a_m$; suppose further that n is minimal (as we range over all words W of this form with $W = 1$ in G_4). Now $G_4 = G' * R$ and W involves only letters in G', so that

$$W = 1 \quad \text{in} \quad G'.$$

Since G' is an HNN extension of G_3' with stable letter k_0, Britton's lemma says that if W involves k_0, i.e., if $W \not\equiv C_0$, then W contains a pinch $k_0^e C_v k_0^{-e}$, where the group element corresponding to C_v is a word on $hr_i h^{-1}$, hxh^{-1}, and $hq^{-1}h^{-1}q_1 t_0 q_1^{-1}hqh^{-1}$. The relations of G' say that k_0 commutes with these words. Therefore, we may replace $k_0^e C_v k_0^{-e}$ in W by C_v, violating the minimality of n. It follows that W does not involve k_0, and therefore is a reduced word C_0 on $a_1, \cdots, a_m$. We have already seen that $\langle a_1, \cdots, a_m \rangle_4$ is free; hence $W \equiv 1$. ∎

Lemma 13.21 G_5 *is an HNN extension of* G_4.

Proof   Clearly G_5 has base G_4 and stable letters $b_1, \cdots, b_m$. To verify it is an

HNN extension, we must show there are isomorphisms $\varphi_i \colon A_i \to B_i$, where

$$A_i = \langle u_1, \cdots, u_m, a_1, \cdots, a_m, k_0 \rangle_4$$

and

$$B_i = \langle u_1, \cdots, u_m, a_1, \cdots, a_m, k_0 u_i^{-1} \rangle_4,$$

and where

$$\varphi_i(u_j) = u_j, \quad \varphi_i(a_j) = a_j, \quad \text{and} \quad \varphi_i(k_0) = k_0 u_i^{-1}.$$

Note that the subgroups A_i and B_i are equal.

Visibly

$$A_i = \langle u_1, \cdots, u_m \rangle * \langle a_1, \cdots, a_m, k_0 \rangle_4$$
$$= R * \langle a_1, \cdots, a_m, k_0 \rangle_4,$$

for $G_4 = R * G'$. Since $\langle a_1, \cdots, a_m, k_0 \rangle_4$ is free on the displayed generators, the map $\varphi_i \colon A_i \to B_i = A_i$ described above is a well defined homomorphism. Similarly, the map $\psi_i \colon A_i \to A_i$ given by

$$\psi_i(u_j) = u_j, \quad \psi_i(a_j) = a_j, \quad \text{and} \quad \psi_i(k_0) = k_0 u_i$$

is well defined. But ψ_i is the inverse of φ_i, and so φ_i is an isomorphism. ∎

Lemma 13.22 G_6 *is an HNN extension of* G_5.

Proof Clearly G_6 has base G_5 and stable letter d. To verify it is an HNN extension, we must show that there is an isomorphism

$$\varphi \colon \langle k_0, a_1 b_1, \cdots, a_m b_m \rangle_5 \to \langle k_0, a_1, \cdots, a_m \rangle_5$$

with $\varphi(k_0) = k_0$ and $\varphi(a_i b_i) = a_i$.

In G_5, $k_0^{-1} b_i k_0 = b_i u_i$. Thus, the function $G_5 \to G_4$, defined by setting each $b_i = 1$ and each $u_i = 1$ and sending the other generators to themselves, is a well defined homomorphism, for all the relations of G_5 are preserved ($b_i = 1$ implies $1 = k_0^{-1} b_i k_0 = b_i u_i = u_i$). This map takes each of $\langle k_0, a_1 b_1, \cdots, a_m b_m \rangle_5$ and $\langle k_0, a_1, \cdots, a_m \rangle_5$ onto the subgroup $\langle k_0, a_1, \cdots, a_m \rangle_4$ of G_4, which is free on the displayed generators (Lemma 13.20). By Exercise 12.71, each of the two subgroups of G_5 is free on the displayed generators. Therefore $k_0 \mapsto k_0$ and $a_i b_i \mapsto a_i$ defines an isomorphism. ∎

The next lemma will be needed in verifying that G_7 is an HNN extension of G_6.

Lemma 13.23 *The subgroup A of G' generated by* $t_0, a_1, \cdots, a_m, k_0$ *has the presentation*

$$A = (t_0, a_1, \cdots, a_m, k_0 \mid k_0^{-1} w^{-1} t_0 w k_0 = w^{-1} t_0 w, \quad w \in E).$$

REMARK Recall our change in notation: Although E originally was a set of positive words on $u_1, \cdots, u_m$, we are now assuming E is composed of positive words on $a_1, \cdots, a_m$.

Proof   The relations $k_0^{-1}w^{-1}t_0wk_0 = w^{-1}t_0w$, for all $w \in E$, do hold in G', by Lemma 13.19, and hence they hold in the subgroup A of G'. To see that no other relations are needed, we shall show if U is a word on $k_0, t_0, a_1, \cdots, a_m$ for which

$$U = 1 \quad \text{in} \quad A,$$

then U can be transformed into 1 via elementary operations using only these relations. The proof consists of several applications of Britton's lemma.

It is easy to see that for every $w \in E$ and $\varepsilon = \pm 1$, $\eta = \pm 1$, the given relations imply

$$t_0^{\varepsilon}wk_0^{\eta} = wk_0^{\eta}w^{-1}t_0^{\varepsilon}w.$$

If U contains a subword of the form $t_0^{\varepsilon}wk_0^{\eta}$, then

$$U \equiv U_1 t_0^{\varepsilon}wk_0^{\eta}U_2 \to U_1 wk_0^{\eta}w^{-1}t_0^{\varepsilon}wU_2$$

is an elementary operation. Now cancel all subwords (if any) of the form yy^{-1} or $y^{-1}y$, where $y \equiv k_0, t_0$, or some a_i. With each such operation, the total number of occurrences of t_0^{ε} that precede k_0^{η}'s goes down. We may therefore assume that U contains no subword of the form $t_0^{\varepsilon}wk_0^{\eta}$, where $w \in E$, and that U is freely reduced.

If U does not involve k_0, then U is a reduced word on $t_0, a_1, \cdots, a_m$. But, by Lemma 13.20, the subgroup $\langle t_0, a_1, \cdots, a_m \rangle_4$ is free on these generators. Hence $U \equiv 1$, and the desired conclusion holds. Similarly, we are done if U does not involve t_0, for $\langle k_0, a_1, \cdots, a_m \rangle_4$ is free on these generators. Therefore, we may assume U involves both k_0 and t_0.

We are in a situation calling for Britton's lemma, for $U = 1$ in G', U involves k_0, and G' is an HNN extension of G_3' with stable letter k_0 (Lemma 13.18). Therefore, U contains a pinch $k_0^{\varepsilon}Vk_0^{-\varepsilon}$, where

$$V = D \quad \text{in} \quad G_3'$$

and D is a word on hr_ih^{-1}, hxh^{-1}, and $hq^{-1}h^{-1}q_1t_0q_1^{-1}hqh^{-1}$. We assume that D is chosen with the minimal number of occurrences of t_0.

Recall that G_3' is an HNN extension of G_2 with stable letter t_0. We claim that both D and V are t_0-reduced. For notation, define $\Delta \equiv hq^{-1}h^{-1}q_1$, so that D is a word on hr_ih^{-1}, hxh^{-1}, and $\Delta t_0\Delta^{-1}$. If D is not t_0-reduced, it contains a pinch, and so

$$D \equiv D_1 \Delta t_0^f \Delta^{-1} D_2 \Delta t_0^{-f} \Delta^{-1} D_3,$$

where D_2 does not involve t_0. By Britton's lemma, we have

$$\Delta^{-1}D_2\Delta = N \quad \text{in} \quad G_2,$$

where N is a word on $q_1^{-1}hr_ih^{-1}q_1$ and $q_1^{-1}hxh^{-1}q_1$ (look at the presentation of G_3'); hence N commutes with t_0 in G_3'. Therefore, in G_3',

$$D = D_1 \Delta t_0^f N t_0^{-f} \Delta^{-1} D_3 = D_1 \Delta N \Delta^{-1} D_3,$$

contradicting our choice of D having the minimum number of occurrences of t_0. It follows that D is t_0-reduced.

We now show that V is t_0-reduced. If not, V contains a pinch $t_0^g C t_0^{-g}$ as a subword, where

$$C = N \quad \text{in} \quad G_2$$

for some word N (as in the paragraph above) that commutes with t_0 in G_3'. Now C is a subword of V not involving t_0, while V is a subword of U not involving k_0. Since U is a word on $k_0, t_0, a_1, \cdots, a_m$, we see that C is a word on $a_1, \cdots, a_m$. Since $G_3' \leq G_4$ and $\langle t_0, a_1, \cdots, a_m \rangle_4$ is free on the displayed generators, by Lemma 13.20, the element C commutes with t_0 in G_3' (hence in G_4) if and only if $C \equiv 1$. The pinch $t_0^g C t_0^{-g}$ in V is thus $t_0^g t_0^{-g}$; since V is a subword of U, we have contradicted the assumption that U is freely reduced.

We claim next that D must involve t_0. As $V = D$ in G_3' and both V and D are t_0-reduced, Corollary 12.54 applies to show that V and D have the same number of occurrences of t_0. If we assume that D does not involve t_0, then V does not involve t_0. Therefore, V is a word on $a_1, \cdots, a_m$, and D is a word on $h r_i h^{-1}$ and $h x h^{-1}$ (we assume D is chosen with the minimal total number of occurrences of all r_i). The equation $V = D$ thus holds in G_2, which is an HNN extension of $G_1 * \langle q, q_0, \cdots, q_N \rangle$ with stable letters r_i. Visibly V is r_i-reduced, all i, for it is a word on $a_1, \cdots, a_m$, and hence does not even involve any r_i. To see that D is r_i-reduced for all i, assume otherwise. As D is a word on $h r_i h^{-1}$ and $h x h^{-1}$ containing a pinch,

$$D \equiv D_1 h r_i^l h^{-1} D_2 h r_i^{-l} h^{-1} D_3,$$

where D_2 involves no r_i's. Thus, $D_2 = h x^m h^{-1}$, and we have

$$D = D_1 h r_i^l h^{-1} x^m r_i^{-l} h D_3 \quad \text{in} \quad G_2.$$

The pinch in D is thus $r_i^l x^m r_i^{-l}$, and Britton's lemma asserts that x^m is equal in $G_1 * \langle q, q_0, \cdots, q_N \rangle$ to a word on $\overline{F}_i q_{i_1} G_i, s_1 x, \cdots, s_M x$ or a word on $\overline{H}_i q_{i_2} K_i, s_1 x^{-1}, \cdots, s_M x^{-1}$. But neither possibility can occur unless $m = 0$, as we have seen in the proof of Lemma 13.13. Therefore, we have $m = 0$ and we may erase $r_i^l r_i^{-l}$ from D, contradicting our assumption that the total number of occurrences of all r_i in D is minimal. We conclude that both V and D are r_i-reduced for all i, so that Corollary 12.54 says that V and D have the same number of occurrences of each r_i. Therefore D involves no r_i and is thus a word on $h x h^{-1}$, i.e., we have $D = h x^n h^{-1}$. The equation $V = D$ thus holds in G_1 and is

$$V = h x^n h^{-1} \quad \text{in} \quad G_1.$$

Suppose V involves a_j. Since G_1 is an HNN extension of $\langle x \rangle$ with stable letters $h, s_0, \cdots, s_M$, and since $\{a_1, \cdots, a_m\} \subset \{s_0, \cdots, s_M\}$, Britton's lemma asserts that $V h x^{-n} h^{-1}$ contains a pinch $a_j^d C a_j^{-d}$, where C is a word on x. Since $h \notin \{s_0, \cdots, s_M\}$, h is not a_j, so that this pinch must be a subword of V. But V does not involve x, so that $C \equiv 1$, contradicting the fact that V is freely reduced. As V is a word on $a_1, \cdots, a_m$, we must have $V \equiv 1$. This is another contradiction, for $k_0^e V k_0^{-e} \equiv k_0^e k_0^{-e}$ is a subword of U, and U is freely reduced.

We have proved that D involves t_0, so that D involves

$hq^{-1}h^{-1}q_1t_0q_1^{-1}hqh^{-1}$. Let us write

$$D \equiv D_1(hq^{-1}h^{-1}q_1t_0^{\alpha}q_1^{-1}hqh^{-1})T,$$

where $\alpha = \pm 1$, the word in parentheses is the final occurrence of the long conjugate of t_0 occurring in D, and T is a word on hr_ih^{-1} and hxh^{-1}. Since D involves t_0, we know that V also involves t_0, so we may write

$$V \equiv V_0t_0^{\varepsilon_1}V_1 \cdots V_{r-1}t_0^{\varepsilon_r}V_r,$$

where each V_j is a word on $a_1, \cdots, a_m$. We noted earlier that V and D are each t_0-reduced, so we may apply Corollary 12.54 to the equation $V = D$ in G_3'. It follows that

$$t_0^{\varepsilon_r}V_rT^{-1}hq^{-1}h^{-1}q_1t_0^{-\alpha}$$

is a pinch, so that

$$V_rT^{-1}hq^{-1}h^{-1}q_1 = K \quad \text{in} \quad G_2,$$

where K is a word on $q_1^{-1}hr_ih^{-1}q_1$ and $q_1^{-1}hxh^{-1}q_1$. Since T is a word on hr_ih^{-1} and hxh^{-1}, we may write

$$T^{-1} = hL_1h^{-1} \quad \text{in} \quad G_2,$$

where L_1 is a word on r_i and x. We may write

$$K = q_1^{-1}hL_2h^{-1}q_1 \quad \text{in} \quad G_2,$$

where L_2 is also a word on r_i and x. Therefore, we have

$$V_rhL_1h^{-1}hq^{-1}h^{-1}q_1 = q_1^{-1}hL_2h^{-1}q_1 \quad \text{in} \quad G_2,$$

which we may rewrite as

$$L_2^{-1}h^{-1}q_1V_rhL_1 = q \quad \text{in} \quad G_2.$$

Note that V_rh is freely reduced, for V_r is a freely reduced word on $\{a_1, \cdots, a_m\}$ and $h \notin \{a_1, \cdots, a_m\}$. Lemma 13.14 applies to give V_rh, hence its subword V_r, positive and

$$hq_1V_rh = q \quad \text{in} \quad \gamma.$$

Since V_r is a positive word on S, Lemma 13.4 says that $V_r \in E$. Let us return to the birthplace of V_r: the word $t_0^{\varepsilon_r}V_r$ is a subword of V, and $k_0^e Vk_0^{-e}$ is a subword of U. Thus $t_0^{\varepsilon_r}V_rk_0^{-e}$ is a subword of U, and, since $V_r \in E$, we have contradicted the condition that U contains no such subword. ∎

Lemma 13.24 G_7 is an HNN extension of G_6.

Proof   Clearly G_7 has base G_6 and stable letter σ. To verify it is an HNN extension, we must show there is an isomorphism

$$\varphi: \langle k_0, t_0, a_1, \cdots, a_m \rangle_6 \to \langle k_0, t_0d, a_1, \cdots, a_m \rangle_6$$

with $\varphi(k_0) = k_0$, $\varphi(t_0) = t_0d$, and $\varphi(a_i) = a_i$. Since $G' \le G_5 \le G_6$, we have

$\langle k_0, t_0, a_1, \cdots, a_m \rangle_6 = \langle k_0, t_0, a_1, \cdots, a_m \rangle_4 = A$, the subgroup whose presentation was determined in Lemma 13.23. Let us denote the subgroup $\langle k_0, t_0 d, a_1, \cdots, a_m \rangle_6$ by B. To show that $\varphi: A \to B$ is a well defined homomorphism, we must check that it preserves the relations of A: if $w \in E$, then is

$$k_0^{-1} w^{-1} t_0 dw k_0 = w^{-1} t_0 dw \quad \text{in} \quad B?$$

We shall show this equation holds in G_6 (which implies that it holds in B).

Let us introduce notation. If w is a word on $a_1, \cdots, a_m$, then w_b is the word obtained from w by replacing each a_i by b_i and w_u is the word obtained from w by replacing each a_i by u_i. If $w \in E$, then $w_u = 1$, for w_u is one of the original defining relations of R. For $w \in E$, each of the following equations holds in G_6:

$$k_0^{-1} w^{-1} t_0 dw k_0 = k_0^{-1} w^{-1} t_0 (dwd^{-1}) dk_0$$
$$= k_0^{-1} w^{-1} t_0 ww_b dk_0$$

(because $da_i d^{-1} = a_i b_i$ in G_6 and a_i, b_j commute in $G_5 \leq G_6$). Since k_0 and d commute in G_6, we have

$$k_0^{-1} w^{-1} t_0 ww_b dk_0 = k_0^{-1} w^{-1} t_0 ww_b k_0 d$$
$$= k_0^{-1} w^{-1} t_0 wk_0 (k_0^{-1} w_b k_0) d$$
$$= k_0^{-1} w^{-1} t_0 wk_0 w_b w_u d$$

(because b_i and u_j commute and $k_0^{-1} b_i k_0 = b_i u_i$)

$$= k_0^{-1} w^{-1} t_0 wk_0 w_b d$$

(since $w_u = 1$). We have shown that

$$k_0^{-1} w^{-1} t_0 dw k_0 = (k_0^{-1} w^{-1} t_0 wk_0) w_b d$$
$$= w^{-1} t_0 ww_b d.$$

On the other hand,

$$w^{-1} t_0 dw = w^{-1} t_0 (dwd^{-1}) d$$
$$= w^{-1} t_0 ww_b d,$$

as we saw above. Therefore

$$k_0^{-1} w^{-1} t_0 dw k_0 = w^{-1} t_0 dw \quad \text{in} \quad G_6$$

and $\varphi: A \to B$ is a well defined homomorphism onto B.

To see that φ is an isomorphism, we construct a homomorphism $\psi: G_6 \to G_6$ whose restriction $\psi | B$ is the inverse of φ. Define ψ by setting

$$\psi(d) = \psi(b_i) = \psi(u_i) = 1$$

and $\psi | G' = 1_{G'}$. Inspection of the various presentations shows that $\psi: G_6 \to G_6$ is a well defined homomorphism. Since $\psi(k_0) = k_0$, $\psi(a_i) = a_i$, and $\psi(t_0 d) = t_0$, we see that $\psi | B$ is the inverse of φ. ∎

We have now verified that $R \subset G_4 \leq G_5 \leq G_6 \leq G_7$, so that R is imbedded in G_7. Note that G_7 is finitely generated, but the presentation we have been working with has infinitely many relations.

Lemma 13.25 G_7 is finitely presented.

Proof   If we delete those relations in G_7 of the form $w_u = 1$, where $w \in E$, only a finite number of relations remain (recall that w_u means replace each a_i in w by u_i); we claim that $w_u = 1$ is a consequence of remaining relations.

If $w \in E$, then $k_0^{-1} w^{-1} t_0 w k_0 = w^{-1} t_0 w$. Therefore.

$$\sigma^{-1}(k_0^{-1} w^{-1} t_0 w k_0)\sigma = \sigma^{-1} w^{-1} t_0 w \sigma.$$

Since σ commutes with k_0 and all a_i, this gives

$$k_0^{-1} w^{-1} \sigma^{-1} t_0 \sigma w k_0 = w^{-1} \sigma^{-1} t_0 \sigma w.$$

Therefore

$$k_0 w^{-1} t_0 d w k_0 = w^{-1} t_0 d w.$$

Inserting $w k_0 k_0^{-1} w^{-1}$ and $w w^{-1}$ gives

$$(k_0 w^{-1} t_0 w k_0) k_0^{-1} w^{-1} d w k_0 = (w^{-1} t_0 w) w^{-1} d w.$$

The terms in parentheses are equal, so that canceling gives

(5) $$k_0^{-1} w^{-1} d w k_0 = w^{-1} d w.$$

Now the relations $d^{-1} a_i b_i d = a_i$ and $a_i b_j = b_j a_i$ give

$$d w d^{-1} = w w_b,$$

and so

(6) $$w^{-1} d w = w^{-1} d d^{-1} w w_b d = w_b d.$$

Substituting equation (6) into equation (5) gives $k_0^{-1} w_b d k_0 = w_b d$. Since k_0 and d commute, cancellation of d gives

(7) $$k_0^{-1} w_b k_0 = w_b.$$

But the relations $k_0^{-1} b_i k_0 = b_i u_i$ and $b_i u_j = u_j b_i$ give

$$k_0^{-1} w_b k_0 = w_b w_u.$$

This last equation, together with (7), gives

$$w_b = w_b w_u,$$

so that $w_u = 1$, as desired. ∎

We have thus imbedded the given recursively presented group R into G_7, a finitely presented group. This completes the proof of G. Higman's imbedding theorem. ∎

Let us review the proof just given. The group G_7 is obtained from $G_4 = G' * R$ by adjoining generators b_i, d, σ and their defining relations. Aanderaa's proof of Higman's theorem amounts to showing two properties of G_7: (i) the given recursively presented group R is imbedded in G_7; (ii) the group G_7 is finitely presented. The first property follows from G_7 being the last group in a chain of HNN extensions starting from $G' * R$. Surely, some of the relations, e.g., $dk_0 = k_0 d$, are present for this reason. Where do the other relations, guaranteeing finite presentation of G_7, come from? E. Rips shows that a diagram (Figure 4 on the inside back cover) of the proof of Lemma 13.25 explains the construction of G_7.

Consider Figure 3 on the inside back cover; it is a picture of the presentation of G_7. Recall that each $w \in E$ is a positive word on $\{a_1, \cdots, a_m\}$ so that each diagram involving a_i begets a similar diagram involving w. Further, each diagram involving b_i or u_i yields similar diagrams involving w_b or w_u (recall w_b denotes the word obtained from w by replacing each a_i by b_i; w_u replaces each a_i by u_i). Now the proof that G_7 is finitely presented amounts to showing $w \in E$ implies $w_u = 1$ in G_7. Clearly a diagram is required.

One may regard a map in the plane as a map lying on the surface of the sphere obtained from the plane by adjoining a point at infinity. This observation allows one to draw a new version of a diagram. For example, assume the diagram below shows $aba^{-1}b^2 = 1$ in some group:

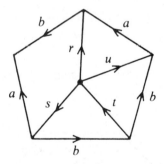

If ∞ denotes the point at infinity, this diagram may be redrawn as follows:

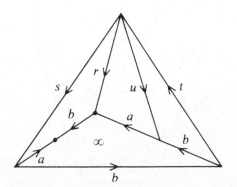

(to redraw, number the vertices and connect them as they are connected in the original diagram). Note that all (bounded) regions are still relator regions (corresponding to the inverses of the original relators) with the exception of the region marked by ∞; note also that the (unbounded) exterior region is also a relator region (the boundary label is *sbt* as in the original diagram). In sum, every not necessarily bounded region is a relator region save the region marked by ∞; moreover, the boundary label w of this region satisfies $w = 1$ in the group.

Let us return to G_7. For a word $w \in E$, let us draw a diagram, new version, showing $w_u = 1$ in G_7.

Certainly one needs the octagonal relator involving w.

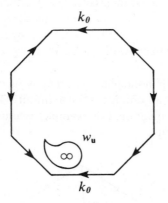

To subdivide this, draw a second such octagon inside and yet a third octagon perturbed by two d-edges.

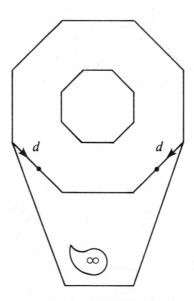

Connect the vertices of the two upper octagons by σ-edges. All that remains is to enclose w_u in a relator region. The lower part of the diagram is subdivided as in Figure 4. Our contention is that this diagram showing $w_u = 1$ in G_7 is reasonably simple, considering that the relations must also be compatible with having HNN extensions.

The corollary of Higman's imbedding theorem we give next may allow us to drop the hypothesis that R be finitely generated. If G is a countable group, let G^{II} denote the group on two generators containing G that was constructed in Corollary 12.51.

Corollary 13.26 *If G is a countable group for which G^{II} is recursively presented, then G can be imbedded in a finitely presented group.*

Proof Higman's theorem asserts that G^{II} can be imbedded in a finitely presented group. ∎

At this point, we omit some details which essentially require accurate bookkeeping in order to give a presentation for G^{II} from a given presentation of G. We assert there is a presentation of the abelian group

$$G = \sum_{\aleph_0} (\mathbf{Q} \oplus \mathbf{Q}/\mathbf{Z}),$$

the direct sum of countably many copies of $\mathbf{Q} \oplus \mathbf{Q}/\mathbf{Z}$, for which G^{II} is recursively presented.

Corollary 13.27 *There exists a finitely presented group that contains an isomorphic copy of every countable abelian group as a subgroup.*

Proof By Exercise 10.46, every countable abelian group may be imbedded in

$$G = \sum_{\aleph_0} (\mathbf{Q} \oplus \mathbf{Q}/\mathbf{Z}).$$

Our assertion above is that G^{II} is recursively presented, so that the result follows from Corollary 13.26. ∎

There are only countably many finitely presented groups, and their free product is a countable group G having a presentation for which G^{II} is recursively presented.

Corollary 13.28 *There exists a universal finitely presented group, i.e., a finitely presented group containing an isomorphic copy of every finitely presented group as a subgroup.*

Proof The result follows from Corollary 13.26 and the assertion above about the free product G. ∎

Let us now indicate how the Boone-Novikov theorem may be derived from the Higman theorem. It is not difficult to construct a recursively presented group G

having unsolvable word problem (there is such a construction that is a variant of the group exhibited in Theorem 12.58). By Higman's theorem, there is a finitely presented group H containing G. Since every finitely generated subgroup of a group having solvable word problem also has solvable word problem (Exercise 13.4), it follows that the word problem for H is unsolvable.

SOME APPLICATIONS

Groups with solvable word problem admit an algebraic characterization. In the course of proving this, we encounter groups which are not finitely generated, yet over whose presentations we still have some control. Let $G = (x_i, i \geq 0 \,|\, r_j, j \geq 0)$ be a presentation, let Ω be the set of all words on $\{x_i, i \geq 0\}$, and let $R = \{w \in \Omega \,|\, w = 1 \text{ in } G\}$. Encode Ω into $\mathbf{N}$ using Gödel numbers: associate to the word $x_{i_1}^{e_{i_1}} \cdots x_{i_n}^{e_{i_n}}$ the positive integer $\Pi_{k=1}^{n} p_{2k}^{i_k} p_{2k+1}^{1+e_{i_k}}$ (the exponent of p_{2k+1} is thus either 0 or 2), where $p_0 < p_1 < \cdots$ is the sequence of primes.

Definition   A presentation $G = (x_i, i \geq 0 \,|\, r_j, j \geq 0)$ is **recursive** if the coded image of the relations R is an r.e. subset of $\mathbf{N}$; this presentation has **solvable word problem** if the coded image of R is a recursive subset of $\mathbf{N}$.

If a presentation has a finite number of generators, then it may be shown that this definition of recursive presentation coincides with our earlier one.

Definition   A group G is **recursively presented** (or has **solvable word problem**) if it has some presentation that is recursive (or has solvable word problem).

We remarked in an earlier footnote that if one finite presentation of a finitely presented group G has solvable word problem, then so does every other finite presentation. The analog of this statement is no longer true when groups are not finitely generated. For example, let S be an r.e. subset of $\mathbf{N}$ that is not recursive, and let G be the free group of (countably) infinite rank. Now the presentation

$$G = (x_i, i \geq 0 \,|\, \varnothing)$$

shows that G has solvable word problem. On the other hand, the presentation

$$G = (x_i, i \geq 0 \,|\, x_i = 1 \text{ if } i \in S)$$

is a recursive presentation whose trivial words are not encoded as a recursive subset of $\mathbf{N}$; the word problem posed by this second presentation is thus unsolvable.

Since we wish to avoid technical logical details, we shall shamelessly declare certain groups arising in the next proof to be recursively presented or to have solvable word problem; the serious reader, of course, cannot be so cavalier.

Theorem 13.29 (Boone-Higman, 1974)† *A finitely generated group G has solvable word problem if and only if G can be imbedded in a simple subgroup of some finitely presented group.*

Proof Assume $G = \langle g_1, \cdots, g_n \rangle \subset S \subset H$, where S is simple and $H = (h_1 \cdots, h_m | r_1, \cdots, r_q)$. Let Ω denote the set of all words on $\{g_1, \cdots, g_n\}$. Since G is finitely generated and H is finitely presented, Theorem 13.2 implies $\{w \in \Omega : w = 1 \text{ in } G\}$ is an r.e. set. We must show $\{w \in \Omega : w \neq 1 \text{ in } G\}$ is an r.e. set. Choose $s_0 \in S$, $s_0 \neq 1$. For $w \in \Omega$, define $N(w)$ to be the normal subgroup of H generated by $\{w, r_1, \cdots, r_q\}$. Because S is simple, the following statements are equivalent for $w \in \Omega$: $w \neq 1$ in G; $w \neq 1$ in H; $N(w) \cap S \neq \{1\}$; $S \subset N(w)$; $s_0 = 1$ in $H/N(w)$. Since $H/N(w)$ is finitely presented, Theorem 13.2 shows the set of all words in Ω equal to 1 in $H/N(w)$ is an r.e. set. A decision process determining whether $w = 1$ in G thus consists in checking whether $s_0 = 1$ in $H/N(w)$.

To prove the converse, assume $G = \langle g_1, \cdots, g_n \rangle$ has solvable word problem. If Ω is the set of all words on $\{g_1, \cdots, g_n\}$, then Exercise 13.10 shows $\{(u, v) \in \Omega \times \Omega | u \neq 1, v \neq 1\}$ is a recursive set; enumerate this set: (u_1, v_1), $(u_2, v_2), \cdots$ (each word u or v has many subscripts in this enumeration). Denote G by G_0; define

$$G_1 = (G; x_1, t_i, i \geq 1 | t_i^{-1} u_i x_1^{-1} u_i x_1 t_i = v_i x_1^{-1} u_i x_1, \quad i \geq 1).$$

It is plain that G_1 has base $G * \langle x_1 \rangle$ and stable letters $\{t_i : i \geq 1\}$; G_1 is an HNN extension, for $A_i = \langle u_i x_1^{-1} u_i x_1 \rangle$ and $B_i = \langle v_i x_1 u_i x_1 \rangle$ are infinite cyclic. Thus $G \leq G_1$. One may show that G_1 is recursively presented and has solvable word problem. Since G_1 has the same properties as G, we may iterate the construction just done. Thus, we construct G_k, an HNN extension with base $G_{k-1} * \langle x_k \rangle$, and we define $S = \bigcup_{k \geq 1} G_k$. Clearly $G \subset S$, and it may be shown S is recursively presented. To see that S is simple, choose u, $v \in S$, $u \neq 1, v \neq 1$. There is an integer k with $u, v \in G_{k-1}$. By construction, there is a stable letter p in G_k with

$$p^{-1} u x_k^{-1} u x_k p = v x_k^{-1} u x_k.$$

Therefore

$$v = (p^{-1} u p)(p^{-1} x_k^{-1} u x_k p)(x_k^{-1} u^{-1} x_k)$$

lies in the normal subgroup generated by u. Since u, v are arbitrary nontrivial elements of S, it follows that S is simple.

Corollary 12.51 shows S can be imbedded in a group K having two generators; moreover, S recursively presented implies K is recursively presented. Finally, the Higman imbedding theorem shows K can be imbedded in a finitely presented group. ∎

---

† Kuznetsov (1958) proved that a recursively presented simple group has solvable word problem.

It is an open question whether a group with solvable word problem can be imbedded in a finitely presented simple group.

Our final result explains why it is usually difficult to extract information about groups from presentations. The following proof is due to C. F. Miller, III.

Lemma 13.30   (Rabin, 1958)   *Let $G = (\Sigma|\Delta)$ be a finitely presented group and let Ω be the set of all words on Σ. There are finite presentations $D(w)$, uniform in w, such that*

(i) *if $w \neq 1$ in G, then $G \leq D(w)$;*
(ii) *if $w = 1$ in G, then $D(w) = \{1\}$.*

Proof Let $\langle x \rangle$ be an infinite cyclic group and imbed $G * \langle x \rangle$ in a group A having two generators a_1, a_2. By Corollary 12.51 and Exercise 12.78, we may assume that A is finitely presented and that a_1 and a_2 have infinite order. Define

$$B = (A; b_1, b_2 | b_1^{-1} a_1 b_1 = a_1^2, \quad b_2^{-1} a_2 b_2 = a_2^2).$$

It is clear B is an HNN extension with base A and stable letters $\{b_1, b_2\}$, whence $G \leq A \leq B$. Define

$$C = (B; c | c^{-1} b_1 c = b_1^2, \quad c^{-1} b_2 c = b_2^2).$$

Plainly C is an HNN extension with base B and stable letter c, whence $G \leq A \leq B \leq C$.

If $w \in \Omega$ and $w \neq 1$ in G, then $[w, x] = wxw^{-1}x^{-1}$ has infinite order in A. We claim $c, [w, x]$ freely generate their subgroup in C. Let V be a nontrivial freely reduced word on $\{c, [w, x]\}$ with $V = 1$ in C. If V does not involve c, then $V \equiv [w, x]^n$ for some $n \neq 0$, and this contradicts $[w, x]$ having infinite order. If V does involve c, then Britton's lemma shows V has a subword $c^\varepsilon W c^{-\varepsilon}$, where $W \in \langle b_1, b_2 \rangle$. Since V does not involve b_1 or b_2, we have $W \equiv 1$ and this contradicts V being freely reduced.

We now construct a second chain of HNN extensions. Begin with an infinite cyclic group $\langle r \rangle$. Define

$$S = (r, s | s^{-1} r s = r^2)$$

and

$$T = (r, s, t | s^{-1} r s = r^2, t^{-1} s t = s^2).$$

It is clear S is an HNN extension with base $\langle r \rangle$ and stable letter s, and that T is an HNN extension with base S and stable letter t. Using Britton's lemma, one may show that r, t freely generate their subgroup of T.

Since $\langle c, [w, x] \rangle$ and $\langle r, t \rangle$ are free groups of rank 2, we may form the amalgam

$$D(w) = (C * T | r = c, \ t = [w, x]).$$

We conclude that if $w \neq 1$ in G, then $G \leq C \leq D(w)$. On the other hand, if $w = 1$ in G, then $D(w)$ is still a presentation (though it is not an amalgam). We watch the dominoes fall in $D(w)$: $w = 1$; $[w, x] = 1$; $t = 1$; $s = 1$; $r = 1$; $c = 1$; $b_1 = 1 = b_2$; $a_1 = 1 = a_2$. Therefore $D(w) = \{1\}$. ∎

REMARK   To see what we mean by the phrase "uniform in w", observe that $D(w)$ has the presentation:

$$\text{generators: } a_1, a_2, b_1, b_2, c, r, s, t;$$
$$\text{relations: the relations of } A, B, C, S, T$$
$$\text{and } r = c, t = [w, x].$$

Definition   A property $\mathscr{M}$ of finitely presented groups is a **Markov property** if

(i) whenever G has property $\mathscr{M}$ and $H \cong G$, then H has property $\mathscr{M}$;
(ii) there exists a finitely presented group G_1 having property $\mathscr{M}$;
(iii) there exists a finitely presented group G_2 that cannot be imbedded in a finitely presented group having property $\mathscr{M}$.

Some examples of Markov properties are: being trivial, finite, abelian, solvable, nilpotent, torsion, torsion-free, free, having solvable word problem. Being simple is also a Markov property, for the Boone-Higman theorem shows finitely presented simple groups have solvable word problem (and hence so do their finitely presented subgroups). Also, having solvable conjugacy problem is a Markov property: a finitely presented group G_2 having unsolvable word problem cannot be imbedded in a finitely presented group H having solvable conjugacy problem, for H, hence G_2, would have solvable word problem.

The following theorem was first proved for semigroups by Markov (1950). Our discussion of the word problem shows that the passage from semigroups to groups may be difficult.

Theorem 13.31   (**Adian-Rabin, 1958**)   *If $\mathscr{M}$ is a Markov property, there does not exist a decision process which will determine of an arbitrarily given finite presentation whether or not the presented group has property $\mathscr{M}$.*

Proof   Let G_1 and G_2 be finitely presented groups as in the definition of Markov property; let H be a finitely presented group having unsolvable word problem. Define $G = G_2 * H$ and construct groups $D(w)$ as in Rabin's lemma; finally, define finitely presented groups $E(w) = D(w) * G_1$. We restrict our attention to words w on the generators of H. If $w \neq 1$ in H, then $G_2 \leq G$ $\leq D(w) \leq E(w)$; the defining property of G_2 shows that $E(w)$ does not have property $\mathscr{M}$; if $w = 1$ in H, then $D(w) = \{1\}$ and $E(w) \cong G_1$ which does have property $\mathscr{M}$. Therefore, any process deciding whether or not $E(w)$ has property $\mathscr{M}$ essentially solves the word problem in H. ∎

Corollary 13.32   *There does not exist a decision process which will determine of an arbitrarily given finite presentation whether or not the presented group is trivial, finite, abelian, solvable, nilpotent, simple, torsion, torsion-free, free, has solvable word problem, or has solvable conjugacy problem.*

Proof   Each of the listed properties is Markov. ∎

While a property of finitely presented groups being Markov is sufficient for the nonexistence of a decision process as in the Adian-Rabin theorem, it is not necessary. For example, the property of being infinite is not Markov, but the existence of a decision process determining whether or not an arbitrary finite presentation defines an infinite group would imply the existence of a similar decision process determining finiteness, contradicting Corollary 13.32. Indeed, this remark shows the Adian-Rabin theorem also holds for the "complement" of a Markov property.

Does every finitely presented group have some Markov property? Recall that a finitely presented group is *universal* if it contains an isomorphic copy of every finitely presented group (we saw in Corollary 13.28 that such groups exist).

Theorem 13.33 *A finitely presented group H satisfies no Markov property if and only if H is a universal group.*

Proof Let H be universal and assume H has some Markov property $\mathcal{M}$. By definition, there is some finitely presented group G_2 that cannot be imbedded in a finitely presented group having property $\mathcal{M}$; that G_2 can be imbedded in H gives a contradiction. Therefore, H can have no Markov property. The converse follows from the observation that the property "not universal" is a Markov property. ∎

appendix one

Some Major
Algebraic Systems

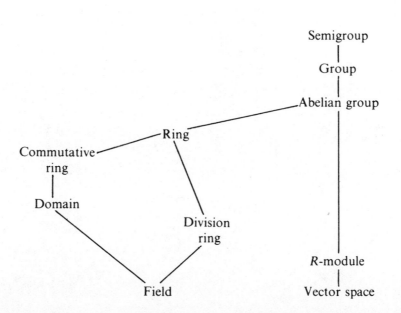

A **ring** (which we always assume to contain $1 \neq 0$) is a set with two binary operations: addition and multiplication. It is an abelian group under addition, a semigroup with 1 under multiplication, and the two operations are linked by the distributive laws.

A **commutative ring** is a ring in which multiplication is commutative.

A **domain** (or *integral domain*) is a commutative ring in which $ab = 0$ implies $a = 0$ or $b = 0$.

A **field** is a commutative ring in which every nonzero element has a multiplicative inverse.

A **division ring** (or *skew field*) is a ring in which every nonzero element has a multiplicative inverse. Thus, a commutative division ring is a field. The nonzero elements of a division ring form a multiplicative group.

If R is a ring, an abelian group M is an **R-module** if there is a scalar multiplication, i.e., a function $R \times M \to M$ [whose value on (r, m) is denoted by rm], which has the properties:

$$(rr')m = r(r'm);$$
$$r(m + m') = rm + rm';$$
$$(r + r')m = rm + r'm;$$
$$1m = m$$

for all r, r', $1 \in R$ and m, $m' \in M$.

A **vector space** is an R-module where R is a field.

appendix two

Equivalence Relations and Equivalence Classes

If X is a nonempty set, a **binary relation** on X is a subset R of $X \times X$. If $(x, y) \in X \times X$, one writes xRy instead of $(x, y) \in R$. For example, the binary relation $<$ on the reals consists of all points in the plane lying above the line $y = x$. Ordinarily, one writes $2 < 3$ instead of $(2, 3) \in <$.

A binary relation $\sim$ on X is an **equivalence relation** if, for all $x, y, z \in X$:

(i) $x \sim x$ (**reflexivity**);
(ii) $x \sim y$ implies $y \sim x$ (**symmetry**);
(iii) $x \sim y$ and $y \sim z$ implies $x \sim z$ (**transitivity**).

If $x \in X$, the subset $\{y \in X : y \sim x\}$, denoted by $[x]$, is called the **equivalence class** containing x.

Proposition *If $\sim$ is an equivalence relation on a set X, then $x \sim a$ (where x, $a \in X$) if and only if $[x] = [a]$.*

Proof Assume $x \sim a$. If $y \in [x]$, then $y \sim x$, so transitivity gives $y \sim a$; we conclude that $[x] \subset [a]$. Now symmetry gives $a \sim x$, and an argument as above shows $[a] \subset [x]$.

Assume $[x] = [a]$. Since $x \in [x]$, by reflexivity, we have $x \in [a]$ and $x \sim a$. ∎

Let X be a nonempty set. A **partition** of X is a family of nonempty subsets $\{S_i : i \in I\}$ of X such that:

$$S_i \cap S_j = \emptyset \qquad \text{if } i \neq j \text{ (\textbf{pairwise disjointness})};$$
$$X = \bigcup S_i.$$

Proposition *If $\sim$ is an equivalence relation on X, then the family of all equivalence classes of $\sim$ is a partition of X.*

Proof Reflexivity gives $x \in [x]$ for every $x \in X$, and this implies each $[x]$ is nonempty and that $X = \bigcup_{x \in X} [x]$. To check disjointness, assume $[x] \cap [y] \neq \emptyset$, i.e., there is an element $z \in [x] \cap [y]$. Then $z \sim x$ and $z \sim y$ imply, by the first proposition, $[z] = [x]$ and $[z] = [y]$, whence $[x] = [y]$. ∎

Proposition *If $\{S_i : i \in I\}$ is a partition of X, then there is an equivalence relation on X whose equivalence classes are the S_i.*

Proof Define $x \sim y$ if there is an S_i containing x and y. It is immediate that the relation is reflexive and symmetric. To prove transitivity, suppose $x, y \in S_i$ and $y, z \in S_j$. Then $y \in S_i \cap S_j$, so that $S_i = S_j$ (pairwise disjointness) and $x \sim z$. Therefore we have defined an equivalence relation. If $x \in S_i$, it follows easily that $[x] = S_i$. ∎

The importance of equivalence relations is just this: Let $\sim$ be an equivalence relation on X and let Y be the set of its equivalence classes. Treating the elements of Y merely as elements (and not as classes), one identifies equivalent elements of X.

For example, we do not wish to distinguish between the fractions $\frac{1}{2}$ and $\frac{2}{4}$, and so we decree that two fractions a/b and c/d are "equal" if $ad = bc$. In reality, we are considering the set X of all ordered pairs of integers (a, b) for which $b \neq 0$, under the equivalence relation given by

$$(a, b) \sim (c, d) \qquad \text{if } ad = bc.$$

The fraction a/b is, by definition, the equivalence class $[(a, b)]$, and so the first proposition above gives $a/b = c/d$ if and only if $ad = bc$. The rational $\frac{1}{2}$ is thus the class of $(1, 2)$ and of $(2, 4)$.

appendix three

Functions

If X and Y are sets, a **relation** from X to Y is a subset of $X \times Y$. A **function** $f: X \to Y$ is a relation from X to Y such that:

(i) for each $x \in X$, there is some $y \in Y$ with $(x, y) \in f$;
(ii) for each $x \in X$, the y above is unique.

If $(x, y) \in f$, the element y is the **image** of x under f and is ordinarily written $f(x)$. With this notation, a function f is the subset of $X \times Y$ consisting of all pairs $(x, f(x))$. In other words, a function is defined by its domain X, its target Y, and its graph.

In practice, one thinks of a function as something dynamic; it assigns elements of Y to elements of X. Indeed, most elementary texts define a function as a "rule of correspondence". Even at a naive level, though, this can be misleading. For example, are $(x+1)^2$ and $x^2 + 2x + 1$ different rules? The reader may prove, using the definition above, that if f and $g: X \to Y$, then $f = g$ if and only if $f(x) = g(x)$ for each $x \in X$. In particular, a **sequence** in a set X is a function $f: J \to X$, where J is the set of positive integers. Writing x_j for $f(j)$, the sequence f is usually denoted by $f = (x_j)$ $= (x_1, x_2, \cdots, x_j, \cdots)$. It follows that sequences (x_j) and (y_j) are equal if and only if $x_j = y_j$ for every j.

Part (ii) of the definition of function deserves a bit more comment; it says that a function is "single-valued", or, as we prefer to say, that a function is **well defined**. When attempting to define a function one must prove that (ii) holds lest only a relation be defined.

The function $f: X \to X$ defined by $f(x) = x$ for all $x \in X$ is called the **identity function** on X and is denoted by 1_X. If Y is a subset of X, the function $i: Y \to X$ defined by $i(y) = y$ for all $y \in Y$ is called the **inclusion**. If Y is a proper subset of X, then 1_Y is distinct from the inclusion i, for $1_Y: Y \to Y$ while $i: Y \to X$. Thus, for two functions to be the same, it is not enough that they have the same graph; they must also have the same domains and the same target sets.

If $f: X \to Y$ and $g: Y \to Z$ are functions, their **composite**, denoted by $g \circ f$, is the function $k: X \to Z$ defined by $k(x) = g(f(x))$ for every $x \in X$. Note that the composite $g \circ f$ is defined only when the target of f (namely, Y) coincides with the domain of g. If $h: Z \to W$ is a function, then the associativity law holds: $h \circ (g \circ f) = (h \circ g) \circ f$.

If $i: Y \to X$ is an inclusion and $f: X \to Z$ is a function, the **restriction** of f to Y, denoted by $f \mid Y$, is the composite $f \circ i$. Thus, $f \mid Y: Y \to Z$, and, for all $y \in Y$, we have $(f \mid Y)(y) = f(y)$.

A function $f: X \to Y$ is **one-one** (or *injective*) if distinct elements of X have distinct images, i.e., if $f(x) = f(x')$, then $x = x'$. This definition may profitably be compared with that of f being well defined, which is the converse: If $x = x'$, then $f(x) = f(x')$. A function $f: X \to Y$ is **onto** (or *surjective*) if each $y \in Y$ is the image of something in X, i.e., if $y \in Y$ there exists an $x \in X$ with $f(x) = y$. A function $f: X \to Y$ is a **one-one correspondence** (or *bijection*) if it is both one-one and onto; such a function always has a unique **inverse**, i.e., a function $g: Y \to X$ such that $f \circ g = 1_Y$ and $g \circ f = 1_X$. One must consider both composites, for $f \circ g = 1_Y$ implies only that f is onto and g is one-one.

appendix four

Zorn's Lemma

Let X be a nonempty set. A binary relation $\leq$ on X is a **partial order** if, for all $x, y, z \in X$:

(i) $x \leq x$ (**reflexivity**);
(ii) $x \leq y$ and $y \leq x$ implies $x = y$ (**antisymmetry**);
(iii) $x \leq y$ and $y \leq z$ implies $x \leq z$ (**transitivity**).

The best example of a partially ordered set is a collection of subsets of a set Y, where $\leq$ means $\subset$. (Some authors write $\subseteq$ in place of $\subset$ so their notation is consistent with $\leq$. However, we use the simpler notation $\subset$ for the relation arising more often, and we write $\subsetneqq$ those few times when strict inclusion must be emphasized.)

A partial order is a **simple order** (or *total order*) if, for each $x, y \in X$, either

$$x \leq y \quad \text{or} \quad y \leq x.$$

If S is a nonempty subset of a partially ordered set X, an **upper bound** of S is an element $x_0 \in X$ (not necessarily in S) such that

$$s \leq x_0 \quad \text{for all } s \in S.$$

Finally, a **maximal element** in X is an element y_0 in X which is smaller than no other element in X, i.e., if $x \in X$ and

$$\text{if } y_0 \leq x, \quad \text{then } y_0 = x.$$

403

There are partially ordered sets having many maximal elements, and there are partially ordered sets having no maximal elements.

Zorn's Lemma   *If X is a partially ordered set in which every simply ordered subset has an upper bound, then there exists a maximal element in X.*

Zorn's lemma is equivalent to a much more intuitive statement, the **axiom of choice**, which says that the cartesian product of nonempty sets is itself nonempty. We regard either of these statements as an axiom of mathematics, and we shall not be ashamed to use either when necessary. There is another statement equivalent to the axiom of choice which is useful. A partially ordered set X is called **well-ordered** if every nonempty subset of X contains a smallest element. (Well-ordered sets must be simply ordered because, in particular, every subset with two elements has a smallest element.) The natural numbers N is well-ordered, but the set Z of all integers is not well-ordered.

Well-ordering Principle   *Given a set X, there is a partial order for which X is well-ordered.*

For example, though Z is not well-ordered under the usual definition of $\leq$, a new ordering can be defined on it so that it is well-ordered: $\{0, 1, -1, 2, -2, \cdots\}$. Here is an example of a well-ordered set in which an element may have infinitely many predecessors: Let X be the following subset of R with the usual definition of $\leq$:

$$X = \left\{1 - \frac{1}{n} : n > 0\right\} \cup \left\{2 - \frac{1}{n} : n > 0\right\}.$$

appendix five

Cardinals
and Classes

It is well known that certain "paradoxes" arise if one is careless about the foundations of set theory. We now sketch some features of the foundational system we accept. Its primitive undefined terms are **class**, **element**, and a **membership relation** denoted by $\in$. The word "set" will receive a special meaning below. Thus, one should replace each occurrence of "set" in our earlier appendix on functions by the word "class". In particular, it is legitimate to speak of functions between classes, to ask whether they are one-one or onto, and to take composites of them.

Define two classes X and Y to be **equipotent**, denoted by $|X| = |Y|$, if there exists a one-one correspondence $f: X \to Y$. It is easily seen that equipotence is an equivalence relation. An ordering relation can also be introduced: define $|X| \leq |Y|$ if there exists a function $f: X \to Y$ that is one-one (but not necessarily onto). This relation can be easily shown to be reflexive and transitive, and the **Cantor-Bernstein theorem** asserts it is antisymmetric: $|X| \leq |Y|$ and $|Y| \leq |X|$ implies $|X| = |Y|$.

The first event is the construction of the **ordinals** (which are certain well-ordered classes), and it begins with the construction of the **natural numbers** N. One defines $0 = \varnothing$, $1 = \{0\}$, and, inductively, $n = \{0, 1, \cdots, n-1\}$. Thus, each natural number n is actually a well-ordered class. Next, one considers N itself; it, too, is a well-ordered class and, as such, it is called the ordinal ω. There is a way to continue this construction in such a way that each ordinal α is a certain class [we denote here by $c(\alpha)$] with a specific well-ordering of its elements.

The next step introduces an ordering $\alpha \leq \beta$ between ordinals. There are two main properties of this ordering. First, the class Λ of all ordinals is itself well-ordered. Second, if $\alpha \leq \beta$, then $|c(\alpha)| \leq |c(\beta)|$. (It follows that this ordering coincides with the usual ordering when α and β are natural numbers.) Given an ordinal α, consider $\{\beta \in \Lambda : |c(\beta)| = |c(\alpha)|\}$. Since Λ is well-ordered, this nonempty subclass of Λ has a smallest element, called the **cardinal** of α. The class of cardinals, being a subclass of Λ, is also well-ordered. In particular, each natural number is its own cardinal, and ω is the next cardinal. Since cardinals have been constructed by stripping away the well-ordering from ordinals α [leaving only the equipotence class of the underlying classes $c(\alpha)$], one renames these "smallest" ordinals in their role as cardinals. Thus, for example, when considering the ordinal ω as a cardinal, one calls it $\aleph_0$ ($\aleph$ is the first letter of the Hebrew alphabet and is pronounced "aleph"). The sequence of cardinals begins with the natural numbers $0, 1, 2, \cdots$ and continues with the infinite cardinals $\aleph_0, \aleph_1, \cdots$ (actually, the sequence of large cardinals is indexed by the ordinals, and if $\alpha < \beta$, then $\aleph_\alpha < \aleph_\beta$).

The cardinals offer us an array of increasing sizes, a system of measures, with which each class can be compared. We say a class X **has a cardinal** if $|X| = |c(\alpha)|$ for some ordinal α (more precisely, if $\aleph_\beta$ is the cardinal of α, we say X has cardinal $\aleph_\beta$). If a class has a cardinal, it is called a **set**; otherwise it is called a **proper class**. The class of all ordinals and the class of all sets are examples of proper classes. Indeed, one thinks of sets as "small" classes, small enough to have a cardinal.

One can avoid paradoxes by obeying the commandment: No proper class shall be an element of a class; for a proper class x, the statement $x \in A$ is always false. However, inclusion of classes is acceptable: for any classes A and B, possibly proper, $A \subset B$ is a statement that may be true.

If X and Y are sets, we say X and Y have the **same cardinal** (or have the **same number of elements**) if $|X| = |Y|$, i.e., X and Y are equipotent. (It is a theorem that $|X| = |Y|$ for any pair of proper classes X and Y.)

Let us come down from the stratosphere. A set X is **finite** if its cardinal is some natural number n, and we say X has cardinal n, or X has n elements, or $|X| = n$; otherwise X is **infinite**. There is a useful result here (sometimes called the *pigeonhole principle*).

Theorem *If X and Y are finite sets of the same cardinal (in particular, if $X \doteq Y$), then a function $f : X \to Y$ is one-one if and only if f is onto.*

A set X is called **countable** if X is finite or $|X| = \aleph_0$ (i.e., there is a one-one correspondence $f : \mathbf{N} \to X$). An infinite set X is countable, therefore, if and only if there is a list (with no repetitions) of all its elements $x_0, x_1, x_2, \cdots$ [define $x_n = f(n)$]. The sets $\mathbf{N}$ of natural numbers, $\mathbf{Z}$ of integers (positive and negative), and $\mathbf{Q}$ of rationals are countable sets. On the other hand, the sets $\mathbf{R}$ of reals and $\mathbf{C}$ of complexes are **uncountable**, that is, they are sets whose cardinal is greater than $\aleph_0$. If X is an infinite countable set, the family of all subsets of X is uncountable; the family of all finite subsets of X, however, is countable.

Of course, set theory is more than this. There is an arithmetic of cardinals: they can be added, multiplied, and exponentiated. There are also some intriguing problems (e.g., the **Continuum Hypothesis** asks whether the cardinal of $\mathbf{R}$ is $\aleph_1$, the first uncountable cardinal; it turns out that this question cannot be answered using the usual axioms of set theory). These matters and more are left to the interested reader to pursue, for such topics are not needed to understand our text.

appendix six

Principal
Ideal Domains

We shall let R denote a commutative ring with 1 throughout this appendix.

If $x_1, \cdots, x_k$ are elements of R, let $(x_1, \cdots, x_k)$ denote the set of all linear combinations of the x with coefficients in R:

$$(x_1, \cdots, x_k) = \{\Sigma r_i x_i : r_i \in R\}.$$

It is easy to check that $(x_1, \cdots, x_k)$ is an ideal in R.

If a and b are in R, one says that a **divides** b in case $ac = b$ for some $c \in R$. If $x_1, \cdots, x_k$ are in R, a **common divisor** of $x_1, \cdots, x_k$ is an element $c \in R$ that divides each x_i; a **greatest common divisor** (**gcd**) is a common divisor that is divisible by every common divisor.

A **principal ideal domain** (**PID**) is a domain R in which every ideal is principal; for each ideal I in R, therefore, there is an element $r_0 \in I$ with

$$I = (r_0) = \{rr_0 : r \in R\}.$$

Theorem A *If R is a PID and if $x_1, \cdots, x_k$ are elements of R, then R contains a gcd of $x_1, \cdots, x_k$ and this gcd is a linear combination of the x.*

Proof Since R is a *PID*, there is an element $d \in R$ with

$$(x_1, \cdots, x_k) = (d);$$

as any element of $(x_1, \cdots, x_k)$, d is a linear combination of the x. It follows that any common divisor of the x divides d. But d is a common divisor of the x, for each $x_i \in (d)$, so that $x_i = r_i d$ for some $r_i \in R$. Therefore, d is a gcd. ∎

A **unit** in R is an element $u \in R$ with a multiplicative inverse in R, i.e., there is an element $v \in R$ with $uv = 1$. Two elements a and b in R are **associates** if there is a unit $u \in R$ with $a = ub$.

Theorem B   *Let R be a PID, and let $x_1, \cdots, x_k \in R$. Any two gcd's of $x_1, \cdots, x_k$ are associates.*

Proof   If a and b are gcd's, then each divides the other. Therefore, $a = ub$ and $b = va$, where $u, v \in R$. Hence, $a = uva$, so that $1 = uv$ since R is a domain. Therefore, u is a unit and a and b are associates. ∎

An element $p \in R$ is **irreducible** if p is not a unit and, in every factorization $p = ab$, either a or b is a unit.

A domain R is a **unique factorization domain (UFD)** if

 (i) every nonzero $a \in R$ that is not a unit is a product of irreducible elements;
(ii) if $p_1 \cdots p_m = q_1 \cdots q_n$, where the p and q are irreducible, then there is a one-one correspondence between the factors ($m = n$) such that corresponding factors are associates.

In short, R is a *UFD* if the fundamental theorem of arithmetic holds in R.

We wish to prove that every *PID R* is a *UFD*; our first task is to show that every nonzero $a \in R$ that is not a unit is a product of irreducibles.

Lemma C   *If R is a PID, there is no infinite sequence of ideals*

$$I_1 \subsetneqq I_2 \subsetneqq \cdots \subsetneqq I_n \subsetneqq I_{n+1} \subsetneqq \cdots.$$

Proof   It is easy to check that

$$I = \bigcup_{n=1}^{\infty} I_n$$

is an ideal. Since R is a *PID*, $I = (d)$ for some $d \in R$. Now d got into I by being in I_n for some n. Hence,

$$I = (d) \subset I_n \subsetneqq I_{n+1} \subset I,$$

a contradiction. ∎

Lemma D   *If R is a PID and $a \in R$ is nonzero and not a unit, then a is a product of irreducibles.*

Proof   If $a = bc$, where neither b nor c is a unit, we say that b is a *proper factor* of a. It is easy to check that if b is a proper factor of a, then $(a) \subsetneqq (b)$.

Call a "good" if it is a product of irreducibles; otherwise, call a "bad". If b and c are good, so is their product bc. Thus, if a is bad, a is not irreducible and it has a proper bad factor. Suppose $a = a_0$ is bad. Assume inductively that there exist $a_0, a_1, \cdots, a_n$ such that each a_{i+1} is a bad, proper factor of a_i. Since a_n is bad, it has a proper, bad factor a_{n+1}. By induction, there is an infinite sequence $a_0, a_1, \cdots$, in which each a_{i+1} is a proper factor of a_i. There is thus an infinite sequence of ideals $(a_0) \subsetneqq (a_1) \subsetneqq \cdots$, and this contradicts Lemma C. Therefore, every nonzero nonunit is good. ∎

Theorem E (**Euclid**) *Let R be a PID and let p be an irreducible element in R. If p divides ab, then p divides a or p divides b.*

Proof If p does not divide a, then the gcd of p and a is 1, for p is irreducible. By Theorem A, there are elements $s, t \in R$ with $1 = sa + tp$. Therefore, $b = sab + tpb$. Since p divides ab, we see that p divides each of the terms on the right; hence p divides b. ∎

Theorem F (**Fundamental Theorem of Arithmetic**) *Every principal ideal domain R is a unique factorization domain.*

Proof By Lemma D, every nonzero $a \in R$ that is not a unit is a product of irreducibles.

If $p_1 \cdots p_m = q_1 \cdots q_n$, where the p and q are irreducibles, then p_1 divides $q_1 \cdots q_n$. By iterated applications of Theorem E, p_1 divides some q_j. Since both p_1 and q_j are irreducible, they must be associates: $p_1 u = q_j$ for some unit u. Since R is a domain,

$$(up_2)p_3 \cdots p_m = \prod_{i \neq j} q_i,$$

and the proof is completed by an induction on $\max\{m, n\}$. ∎

Bibliography

Albert, A. A., *Introduction to Algebraic Theories*, University of Chicago Press, 1941.
Artin, E., *Geometric Algebra*, Interscience, 1957.
———, *Galois Theory*, Notre Dame, 1955.
Babakhanian, A., *Cohomological Methods in Group Theory*, M. Dekker, 1972.
Biggs, N. L., and White, A. T., *Permutation Groups and Combinatorial Structures*, Cambridge University Press, 1979.
Birkhoff, G., and Mac Lane, S., *A Survey of Modern Algebra*, 4th ed., Macmillan, 1977.
Burnside, W., *The Theory of Groups of Finite Order*, 2nd ed., Cambridge University Press, 1911.
Cameron, P. J., *Parallelisms of Complete Designs*, Cambridge University Press, 1976.
Carmichael, R., *An Introduction to the Theory of Groups of Finite Order*, Ginn, 1937.
Carter, R. W., *Simple Groups of Lie Type*, Wiley, 1972.
Cohn, P. M., *Universal Algebra*, Harper & Row, 1965.
Coxeter, H. S. M., and Moser, W. O., *Generators and Relations for Discrete Groups*, Springer-Verlag, 1965.
Curtis, C., and Reiner, I., *Representation Theory of Finite Groups and Associative Algebras*, Wiley, 1962.
Davis, M., *Computability and Unsolvability*, McGraw-Hill, 1958.
Dickson, L. E., *Linear Groups*, Leipzig, 1900.
Dieudonné, J., *Sur les Groupes Classiques*, Hermann, 1958.
Dixon, J. D., *Problems in Group Theory*, Blaisdell, 1967.
Feit, W., *Characters of Finite Groups*, Benjamin, 1969.
Fuchs, L., *Infinite Abelian Groups I*, Academic, 1970.
———, *Infinite Abelian Groups II*, 1973.
Gorenstein, D., *Finite Groups*, Harper & Row, 1968.

————, *Finite Simple Groups*, Plenum, 1982.

Griffith, P. A., *Infinite Abelian Group Theory*, University of Chicago, 1970.

Gruenberg, K. W., *Cohomological Topics in Group Theory*, Springer-Verlag, 1970.

————, and Weir, A. J., *Linear Geometry*, Springer-Verlag, 1977.

Hall, M., Jr., *The Theory of Groups*, Macmillan, 1959.

Hall, P., *Nilpotent Groups*, London University, Queen Mary College Notes, 1969.

Higgins, P. J., *Notes on Categories and Groupoids*, van Nostrand-Reinhold, 1971.

Huppert, B., *Endliche Gruppen* I, Springer-Verlag, 1967.

Huppert, B., and Blackburn, N., *Finite Groups*, II and III, Springer-Verlag, 1982.

Isaacs, I. M., *Character Theory of Finite Groups*, Academic, 1976.

Jacobson, N., *Basic Algebra I*, Freeman, 1974.

————, *Basic Algebra II*, Freeman, 1979.

Johnson, D. L., *Topics in the Theory of Group Presentations*, Cambridge University Press, 1980.

Jordan, C., *Traité des Substitutions et des Équations Algébriques*, Gauthier-Villars, 1870.

Kaplansky, I., *Infinite Abelian Groups*, University of Michigan, revised ed., 1969.

————, *Fields and Rings*, 2nd ed., University of Chicago Press, 1972.

Kurosh, A. G., *The Theory of Groups*, I and II, Chelsea, 1956.

Lang, S., *Algebra*, Addison-Wesley, 1965.

Lyndon, R., and Schupp, P., *Combinatorial Group Theory*, Ergebnisse Series, Springer-Verlag, 1977.

Mac Lane, S., *Homology*, Academic, 1963.

Magnus, W., Karrass, A., and Solitar, D., *Combinatorial Group Theory*, Wiley, 1966.

Massey, W. S., *Algebraic Topology: An Introduction*, Harcourt, Brace and World, 1967.

Miller, C. F., III, *On Group-Theoretic Decision Problems and Their Classification*, Princeton, 1971.

Miller, G. A., Blichfeldt, H. F., and Dickson, L. E., *Theory and Applications of Finite Groups*, Wiley, 1916.

Neumann, H., *Varieties of Groups*, Springer-Verlag, 1967.

Passman, D. S., *Permutation Groups*, Benjamin, 1969.

Pedoe, D., *An Introduction to Projective Geometry*, Pergamon, 1963.

Powell, M. B., and Higman, G. (eds.), *Finite Simple Groups*, Academic, 1971.

Puttaswamaiah, B. M., and Dixon, J. D., *Modular Representations of Finite Groups*, Academic, 1977.

Robinson, D. J. S., *Finiteness Conditions and Generalized Soluble Groups*, Ergebnisse vols. 62 and 63, Springer-Verlag, 1972.

————, *A Course in the Theory of Groups*, Springer-Verlag, 1982.

Rotman, J. J., *An Introduction to Homological Algebra*, Academic, 1979.

Schenkman, E., *Group Theory*, van Nostrand, 1965.

Scott, W. R., *Group Theory*, Prentice-Hall, 1964.

Serre, J.-P., *Trees*, Springer-Verlag, 1980.

Specht, W., *Gruppentheorie*, Springer-Verlag, 1956.

Speiser, A., *Die Theorie der Gruppen von Endlicher Ordnung*, Springer-Verlag, 1927.

Stammbach, U., *Homology in Group Theory*, Springer-Verlag, 1980.

Suzuki, M., *Group Theory* I, Springer-Verlag, 1982.

Van der Waerden, B. L., *Modern Algebra*, Ungar, 1948.

Wehrfritz, B. A. F., *Infinite Linear Groups*, Springer-Verlag, 1973.

Weyl, H., *Symmetry*, Princeton, 1952.

Wielandt, H., *Finite Permutation Groups*, Academic, 1964.

Zassenhaus, H., *The Theory of Groups*, Chelsea, 1956.

Zieschang, H., Vogt, E., and Coldewey, H.-D., *Surfaces and Planar Discontinuous Groups*, Springer-Verlag, 1980.

Notations and Definitions

$X \subset Y$	X is a subset of Y ($X = Y$ is allowed)
$X \subsetneqq Y$	X is a proper subset of Y
1_X	identity function on a set X
$\{x_i : i \in I\}$	set consisting of elements x_i indexed by a set I
$\langle x_i : i \in I \rangle$	subgroup generated by elements x_i
$G^{\#} = G - \{1\}$	(here G is a multiplicative group)
$K^{\#} = K - \{0\}$	(here K is a field)
$V^{\#} = V - \{0\}$	(here V is a vector space)
G/H	quotient group with elements the (left) cosets of the normal subgroup H
$G/\!/H$	set of left cosets of the (not necessarily normal) subgroup H
A_n	alternating group of order $n!/2$ (p. 38)
A_∞	infinite alternating group (p. 45)
$\mathrm{Aff}(V) \cong \mathrm{Aff}(n, K)$ $\cong \mathrm{Aff}(n, q)$	affine group (p. 196)
$\mathrm{Aut}(G)$	automorphism group of a group G (p. 130)
$\mathrm{Aut}(K)$	automorphism group of a field K (p. 162)
$\mathrm{Aut}(V) \cong \mathrm{Aut}(n, K)$ $\cong \mathrm{Aut}(n, q)$	group of affine automorphisms (p. 198)
$\mathrm{Aut}(X, B)$	automorphism group of a Steiner system (X, B) (p. 229)
$B^2(Q, {}_\theta K)$	group of coboundaries (p. 148)
$B_{ij}(\lambda)$	matrix transvection (p. 163)
$\mathbf{C}$	complex numbers

413

$C_G(x)$	centralizer in G of element x (p. 40)
$C_G(H)$	centralizer in G of subgroup H (p. 88)
D_n	dihedral group of order $2n$ (p. 61)
D_∞	infinite dihedral group (p. 325)
dG	maximal divisible subgroup of abelian G (p. 249)
E	identity matrix
End (G)	endomorphism ring of abelian group G (p. 268)
Ext (C, A)	group of abelian factor sets modulo coboundaries (p. 282)
$e(C, A)$	group of abelian extensions of A by C (p. 282)
$F(\alpha)$	field obtained by adjoining α to F (p. 70)
G'	commutator subgroup of G (p. 24)
$G^{(i)}$	ith higher commutator subgroup of G (p. 83)
G_x	stabilizer of x (p. 180)
Gal (K/F)	Galois group of fields $K \supset F$ (p. 72)
$GF(q)$	Galois field having $q = p^n$ elements (p. 161)
$GL(V) \cong GL(n, K)$	
$\cong GL(n, q)$	general linear group (p. 162)
$\Gamma L(V) \cong \Gamma L(n, K)$	
$\cong \Gamma L(n, q)$	group of all nonsingular semilinear transformations (p. 200)
$\Gamma LF(K) \cong \Gamma LF(q)$	group of all semilinear fractional transformations (p. 214)
γ_a	conjugation by a (p. 8)
$\gamma_i(G)$	ith term in descending central series of G (p. 89)
$H^2(Q, {}_\theta K)$	second cohomology group $= Z^2(Q, {}_\theta K)/B^2(Q, {}_\theta K)$ (p. 148)
Hol (G)	holomorph of G (p. 140)
Hom (A, C)	group of all homomorphisms from A to C (p. 276)
Inn (G)	inner automorphism group of G (p. 130)
$\tilde{K}$	covering complex of K (p. 313)
$K(\mathscr{S})$	Grothendieck group of $\mathscr{S}$ (indecomposables) (p. 269)
$K^*(\mathscr{S})$	Grothendieck group of $\mathscr{S}$ (simples) (p. 272)
$LF(K) \cong LF(q)$	group of all linear fractional transformations (p. 214)
M_n	Mathieu group ($n = 10, 11, 12, 22, 23, 24$) (pp. 217–225)
$\mathbf{N}$	natural numbers
$N_G(H)$	normalizer in G of a subgroup H (p. 48)
$P^n(K) \cong P^n(q)$	projective n-space (p. 205)
$PGL(V) \cong PGL(n, K)$	
$\cong PGL(n, q)$	group of all projectivities (p. 212)
$PSL(V) \cong PSL(n, K)$	
$\cong PSL(n, q)$	projective unimodular group (p. 166)
$P\Gamma L(V) \cong P\Gamma L(n, K)$	
$\cong P\Gamma L(n, q)$	collineation group (p. 212)
$\pi(K, v_0)$	edgepath group of complex K with basepoint v_0 (p. 308)
$\mathbf{Q}$	rational numbers
Q	quaternions of order 8 (p. 63)
Q_n	generalized quaternions of order 2^n (p. 66)
$\mathbf{R}$	real numbers
S_n	symmetric group of order $n!$ (p. 32)
S_X	all permutations of a set X (p. 32)
$Sc(V)$	all nonzero scalar transformations (p. 211)
$Sc_1(V)$	all scalar transformations having determinant 1 (p. 213)
$Sh(K) \cong Sh(q)$	sharply 3-transitive group (p. 216)
$SL(V) \cong SL(n, K)$	
$\cong SL(n, q)$	special linear group: all transformations of determinant 1 (p. 163)
$\mathbf{T}$	circle group $= \mathbf{R}/\mathbf{Z}$ (p. 8)
$T_{f, \gamma}$	transvection (transformation) (p. 170)

$\mathrm{Tr}(V)$	group of translations (p. 195)
tG	torsion subgroup of abelian group G (p. 241)
$U(R)$	group of units of ring R (p. 131)
$\mathbf{V}$	4-group (p. 41)
$V(K)$	vertex set of a complex K (p. 305)
$V(n, K) \cong V(n, q)$	vector space of dimension n over K ($GF(q)$) (p. 195)
$(X\,|\,\Delta)$	presentation with generators X and relations Δ (p. 301)
$\mathbf{Z}$	integers
$Z(G)$	center of G (p. 39)
$Z^i(G)$	ith higher center of G (p. 89)
$\mathbf{Z}(n)$	cyclic group of order n (p. 18)
$\mathbf{Z}(p^\infty)$	group of all pth power roots of unity (p. 247)
$Z^2(Q, {}_\theta K)$	group of all factor sets (p. 147)
$Z(n, K) \cong Z(n, q)$	group of all nonzero $n \times n$ scalar matrices (p. 204)
$Z_1(n, K) \cong Z_1(n, q)$	group of all $n \times n$ scalar matrices of determinant 1 (p. 164)
$\Phi(G)$	Frattini subgroup of G (p. 97)
$\varphi(n)$	Euler φ-function (p. 17)
σ_*	collineation induced by field automorphism σ (p. 200)
$A *_\theta A'$ or $A *_S A'$ (where $\theta\colon S \to S'$ is an isomorphism)	amalgam (p. 333)
$A \times B$	dircct product (p. 28)
$\displaystyle\prod_{i=1}^{n} A_i$	direct product (p. 28)
$\displaystyle\prod_{i\in I} A_i$	direct product (p. 242)
$A \oplus B$	direct sum (p. 99)
$\displaystyle\sum_{i=1}^{n} A_i$	direct sum (p. 99)
$\displaystyle\sum_{i\in I} A_i$	direct sum (p. 242)
$A * B$	free product (p. 323)
$\displaystyle *_{i\in I} A_i$	free product (p. 323)
$G\,\Omega_\theta A$ or $G\Omega A$	HNN extension (p. 339)
$A \rtimes_\theta B$ or $A \rtimes B$	semidirect product (p. 137)
$A \wr B$	wreath product (p. 142)

characteristic polynomial of A: $\det(xE - A)$

characteristic root of A = eigenvalue: scalar c with $A\alpha = c\alpha$ for some nonzero vector α

characteristic vector of A = eigenvector: nonzero vector α with $A\alpha = c\alpha$ for some scalar c

division algorithm: if a, b are integers (polynomials) with $a \neq 0$, there are unique integers
  (polynomials) q, r with $b = qa + r$ and $0 \leq r < |a|$ (degree $r <$ degree a)

similar: two matrices A and B are similar if there is a nonsingular matrix P with $PBP^{-1} = A$;
  two matrices are similar if and only if they represent the same linear transformation on a
  vector space relative to (possibly) distinct ordered bases

Index

Figure 3
Presentation of the group G_7 of the Higman imbedding theorem.

Generators:

a_i k_o t_o u_i b_i d σ

Relations:

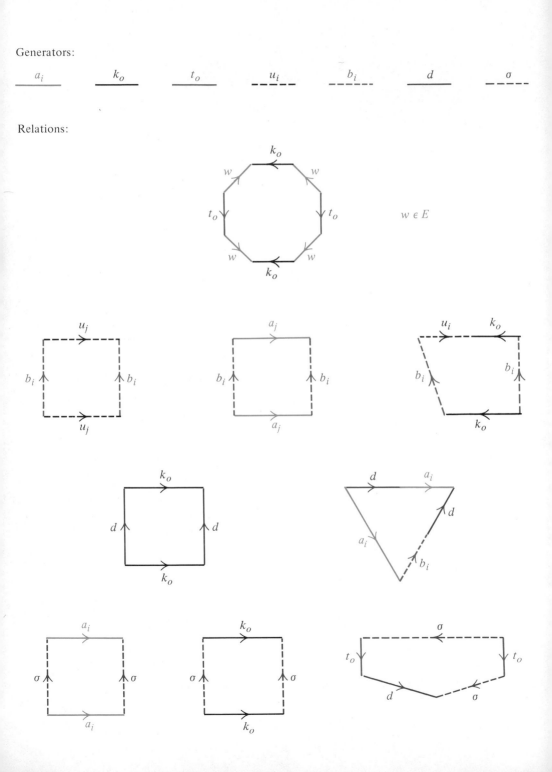

Figure 4
Proof of Lemma 13.25: G_7 is finitely presented.

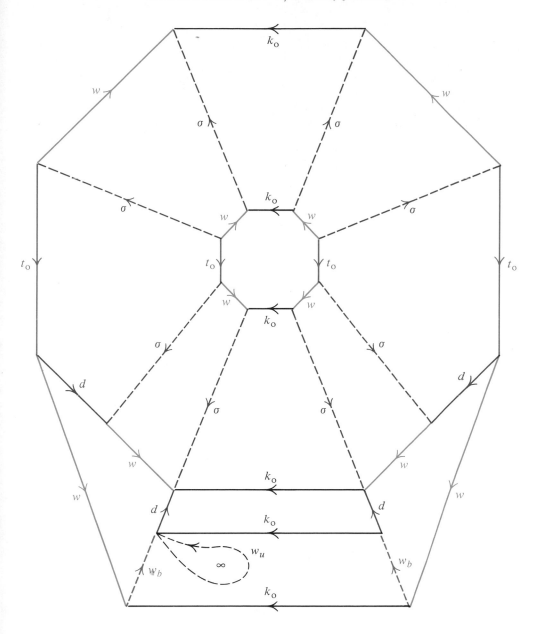